国 家 科 学 技 术 学 术 著 作 出 版 基 金 资 助 出 版
Supported by the National Fund for Academic Publication in Science and Technology

中国东北三省夜蛾总科名录 I：目夜蛾科（部分）、尾夜蛾科、瘤蛾科和夜蛾科

The Catalogue of the Noctuoidea in the Three Provinces of Northeast China I: Families Erebidae(part), Euteliidae, Nolidae and Noctuidae

韩辉林[1, 2] [俄罗斯]科诺申科[3] 李成德[1]
HAN HUILIN[1, 2] VLADIMIR S. KONONENKO[3] Li CHENGDE[1]

1. 东北林业大学林学院，哈尔滨，150040，中国. 邮箱：hanhuilin@aliyun.com

1. *School of Forestry, Northeast Forestry University, Harbin, 150040 China. E-mail: hanhuilin@aliyun.com*

2. 东北林业大学森林生态系统可持续经营教育部重点实验室，哈尔滨，150040，中国. 邮箱：1710312254@qq.com

2. *Key Laboratory of Sustainable Forest Ecosystem Management-Ministry of Education, Northeast Forestry University, Harbin, 150040 China. E-mail: 1710312254@qq.com*

3. 俄罗斯科学院远东分院东亚陆地生物多样性联邦科学中心，海参崴，690022，俄罗斯. 邮箱：vlad_kononenko@mail.ru

3. *Federal Scientific Center of the East Asia Terrestrial Biodiversity, Far Eastern Branch of Russian Academy of Sciences, 159 Prospect stoletiya Vladivostoka, Vladivostok 690022 Russia. E-mail: vlad_kononenko@mail.ru*

黑龙江科学技术出版社
HEILONGJIANG SCIENCE AND TECHNOLOGY PRESS

图书在版编目（CIP）数据

中国东北三省夜蛾总科名录 = The Catalogue of the Noctuoidea in the Northeast China I: Famlies Erebidae (part), Euteliidae, Nolidae and Noctuidae. I, 目夜蛾科(部分)、尾夜蛾科、瘤蛾科和夜蛾科 : 汉、英 / 韩辉林, (俄罗斯) 弗拉迪米尔·S·科诺年科, 李成德编著. -- 哈尔滨 : 黑龙江科学技术出版社, 2020.12（重印 2022.5）

ISBN 978-7-5719-0822-5

Ⅰ. ①中… Ⅱ. ①韩… ②弗… ③李… Ⅲ. ①夜蛾科－东北地区－名录－汉、英 Ⅳ. ①Q969.436.5-62

中国版本图书馆 CIP 数据核字(2020)第 270518 号

中国东北三省夜蛾总科名录 I
目夜蛾科(部分)、尾夜蛾科、瘤蛾科和夜蛾科
ZHONGGUO DONGBEI SAN SHENG YE'E ZONGKE MINGLU I
MUYE'E KE(BUFEN) WEIYE'E KE LIUYE'E KE HE YE'E KE
韩辉林　(俄罗斯)弗拉迪米尔·S·科诺年科　李成德编著

项目总监　侯　擘
责任编辑　焦　琰 回　博 孔　璐 刘松岩
封面设计　孔　璐
出　　版　黑龙江科学技术出版社
地　　址　哈尔滨市南岗区公安街 70-2 号
邮　　编　150007
电　　话　（0451）53642106
传　　真　（0451）53642143
网　　址　www.lkcbs.cn
发　　行　全国新华书店
印　　刷　黑龙江艺德印刷有限责任公司
开　　本　889 mm × 1194mm　1/16
印　　张　33.25
字　　数　1000 千字
版　　次　2020 年 12 月第 1 版
印　　次　2022 年 5 月第 2 次印刷
书　　号　ISBN 978-7-5719-0822-5
定　　价　380.00 元

序

中国东北三省是由黑龙江、吉林、辽宁三省组成，位于北纬 53°至 38°之间，南北延伸可达 1600 多千米；南部与河北的东北部、西部与内蒙古东部相接，北部和东北部与俄罗斯以黑龙江（阿穆尔河）和乌苏里江（乌苏里河）、东部和东南部与朝鲜以鸭绿江和图们江隔江相望。东北三省南北温差大，由北至南为寒温带至温带大陆性季风气候，南部略受海洋性气候影响；植被在北部以针叶林为主、中至南部多以针阔混交林为主，与俄罗斯、西伯利亚东部、南部（阿穆尔盆地和乌苏里盆地）和朝鲜北部相似。东北三省动物区系隶属古北界东北亚区，与俄罗斯西伯利亚、朝鲜、韩国、日本北部同属一亚区，因此有许多俄罗斯西伯利亚北部、东北部，以及一些东洋界热带迁徙物种有规律的或偶然的分布。

中国东北三省夜蛾总科的分类学研究已经持续了 100 多年，但其调查研究还是不够完善。过去 50 年里，虽然国内外学者对中国东北三省的夜蛾总科的分类研究没有间断，但是由于文献资料不全、国际同行间交流欠缺等原因，导致误鉴定、引用陈旧或错误学名等问题时常发生。夜蛾总科系统分类学在近代发生了比较大的重组，特别是传统夜蛾科在亚科、属、种的分类单元上的隶属关系发生重大变化，导致很多亚科、族、属、种至今没有明确的中文种名。当前新的分类系统虽然还存在部分不稳定性，但近年来国际上基本公认其将传统夜蛾科、毒蛾科和灯蛾科重新整合成目夜蛾科、尾夜蛾科、瘤蛾科、夜蛾科等四科系统。

本书在整合已有相关研究成果的基础上，通过 20 多年野外标本采集和借阅相关兄弟单位标本予以补充，共整理出我国东北三省目夜蛾科、尾夜蛾科、瘤蛾科、夜蛾科等 4 科 34 亚科 353 属 883 种，同时列出详细同物异名、国内外分布、相关寄主等信息，并对一些在国内文献资料上沿用错误、归属不清等问题给予了变更和相关说明，能够方便读者清晰地了解新旧分类系统中的相关变动。希望本书能够为专业研究学者及农林业保护工作者进行相关研究工作时提供准确的科学信息而尽绵薄之力。

本书的编著得到德国 G. Behounek、D. Stüning、W. Mey、A. Hausmann，英国 M. Honey、J. Holloway、A. Zilli，立陶宛 A. Saldaitis、匈牙利 L. Ronkay 和 G. Ronkay，俄罗斯 A. Matov 等国际同行专家的学术支持，特别是得到 G. Behounek 先生提供个人数据库对所有数据信息核对的支撑。本书在申报国家科学技术学术著作出版基金时得到河北大学任国栋教授、南开大学李后魂教授、中国农业大学彩万志教授、吉林省林业科学研究院宋丽文研究员的鼎力推荐。在历年采集、借阅标本时得到东北师范大学、延边大学、北华大学、东北农业大学、东北林业大学实践教学基地（帽儿山、凉水林场）、黑龙江省林业有害生物防治检疫站及其基层站（大兴安岭地区、绥化、齐齐哈尔、大庆、佳木斯、尚志、牡丹江）、黑河市林业科学研究所、吉林省林业科学研究院、长白山科学研究院、长白山森工集团、辽宁省林业有害生物防治检疫站及其基层站（沈阳、丹东、抚顺、铁岭、盘锦、锦州、鞍山、建昌、葫芦岛、大连）等同行单位的大力支持；同时，历年来奔波于各地进行野外采集时做了大量工作的东北林业大学林学院森林保护学科昆虫分类研究室的同学们为本书的撰写做出了巨大的贡献，在此一并致以最衷心的感谢。

本研究得到以下基金项目资助：国家自然科学基金（No. 31872261、31572294、31470652、31272355、30700641）、中央高校基本科研业务费专项基金资助（No. 2572019CP11、2572018CJ03、2572014DA02、DL09CA05）、中国留学归国博士后特别资助基金、教育部留学回国基金、黑龙江省博士后科研启动金。

由于作者水平有限，书中难免有错漏和不足之处，万望读者不吝赐教、指正。

作者

2020年12月于哈尔滨、符拉迪沃斯托克（海参崴）

Contents

目录

Introduction（前言）

The present Catalogue is a result of the faunal inventory of four families (Erebidae[part], Euteliidae, Nolidae and Noctuidae) of the Superfamily Noctuoidea in the Three Provinces of Northeast China. Representatives of the Noctuoidea are very diverse over the World and in China particularly and enumerate over 883 described species in the local fauna of the Three Provinces of Northeast China. These groups were formerly treated by researchers as a family Noctuidae, the largest and worldwide distributed group of the order Lepidoptera.

By the reason of their high specific diversity, abundance of numerous species and phytophagous feeding of the larvae, many Noctuoidea have great ecological as well as economic significance as an important component of natural or artificial ecosystems. Some species of the noctuoids have high migratory ability and are able to migrate for a long distance. Due to these reasons they are well known as serious pests of agricultural crops, and some of them often cause considerable injure plants of the economic importance. From the result of human activities, development and changing of the agricultural system during last decades some species of noctuoids became rare or vanishing in the Three Provinces of Northeast China, but some are still serious pests of the agricultural and forestry.

The Three Provinces of Northeast China (东北三省) are located between 53º to 38º of North latitude and extended from North to South for over 1600 km. This territory consists of the three provinces of Heilongjiang, Jilin and Liaoning, and the East part of Nei Mongol Zizhiqu and Northeast part of province Hebei.

The heartland of the region is Northeast China Plain, which is mainly used for the agriculture proposes in the time being. It is bordered with Russia mainly by the Heilong Jiang(=Amur river), Eerguna He(=Argun river) and Wusuli Jiang(=Ussuri river), from North Korea by the Yalu Jiang(=Amnok river) and Tumen Jiang(=Duman river). The north and the south of the Three Provinces of Northeast China are mountainous, belong to different montane systems, with average altitudes of Da Hinggan Ling(=Greater Khingan Range) 1100~1400 m and the highest peak Huanggang Liang reaches 2029 m, and Changbai Shan with the highest peak Baiyunfeng with elevation 2691 m.

The climate of the Three Provinces of Northeast China is temperate, but differs considerably between

Northern and Southern parts. It is strongly continental in the North and marine, partially monsoon in the South.

The vegetation of the Three Provinces of Northeast China is represented by Boreal taiga forests in the North, Northeast deciduous forests in the mid of area and evergreen pine forests mixed with deciduous in the south of the region. Large areas in the Three Provinces of Northeast China Plain primary are covered by steppe grasslands and marches. By the types of the vegetation of the Three Provinces of Northeast China is close partially to the East and Siberia (Amur and Ussuri basins) of Russia. Sixty major forest (including scrub) types were recognized in the Three Provinces of Northeast China, which include twenty-one types of coniferous forests, four types of mixed evergreen coniferous and deciduous broad-leaved forests, and thirty-five types of deciduous broad-leaved forests (Qian, Yuan & Chou 2003).

Zoogeographically the Three Provinces of Northeast China belongs partially to the Boreal and mainly to Northeast China (='Manchurian') subregion of the Palaearctic Region in unit with Russia(southeast of Transbaikalia, Siberia), Korea and northern Japan.

The northern part of the Three Provinces of Northeast China has more faunal similarity with Siberia (Transbaikalia), and some Siberian faunal elements are affected in the high elevation here. The Boreal taiga elements are characteristic for the northernmost of Prov. Heilonjiang along the Heilong Jiang(=Amur river). The Changbai Shan probably is the southernmost boundary of the distribution of Boreal faunal complex where the species of this group occur in higher elevation. On the other hand, the southernmost part of the Three Provinces of Northeast China lies in the northern part of the subtropical climatic zone and contains many Subtropical, Oriental origin genera and species. The brief view for the Three Provinces of Northeast China Noctuoidea fauna shows that it has rather complex composition, but consists of three main faunal complexes as Palaearctic Boreal, Pacific-Palaearctic, and Oriental tropical and subtropical. The Northeast China faunal complex is dominated, although a Palaearctic Boreal complex probably dominated in the North of the region and in high mountain regions and the subtropical complex is represented mainly by the migrating Oriental species.

Due to the transitional position of the Three Provinces of Northeast China between Boreal and Northeast China (='Manchurian') subregions of the Palaearctic and Oriental biogeographic regions, many species with Boreal Siberian, Northeast China pattern of distribution occur here, as well as some the Oriental-tropical migrating species are regularly or occasionally appear here as a migrants, and some of them may have economic importance as a serious quarantine pests.

The first data on the Noctuoidea of the Three Provinces of Northeast China were published in the entomological literature from the end of nineteen century (since 1889) in by H. Leech (1889a, 1900a, 1900b), later they were included to the Catalogue of British Lepidoptera (Hampson 1903-1913) and 'Macrolepidoptera of the World', edited by A. Seitz ([1907]-1914) and in the supplement of this issue. Japanese fauna has been described in the nineteen century by same authors as well as by Butler (1878-1889) and Walker (1854-1866). At same period the intensive description of the Noctuidae fauna started on the neighbouring territories of Korea (Leech 1889a, 1889b, 1900a, 1900b), Alphéraky (1897), Herz (1904) and Russia, mainly along Heilong Jiang(=Amur) and Wusuli Jiang(=Ussuri) basins (Ménétriès, 1859, 1859; Bremer, 1861, 1864; Oberthür, 1879, 1880; Graeser, 1888-1892; Staudinger, 1888, 1892). The intensive study of Chinese Noctuoidea fauna was initiated in 20 century by German entomologists with descriptions of many new noctuids species mainly in Central and Southeast of China (Mell, 1936, 1943; Draudt, 1937, 1939, 1950; Boursin, 1941, 1951, 1954a, 1954b, 1955, 1961, 1963), and many of them were recorded in the Three Provinces of Northeast China by subsequent researchers.

In 1982, two very important books for lepidopterology in East Asia and China particularly have been published in Japan ('Moths of Japan'; most part of Noctuidae subfamilies was written by Sugi [see Sugi, 1982a], the subfamily Herminiinae - by Owada [see Owada, 1982] and family Nolidae by Inoue [see Inoue, 1982]), and in China ('Iconographia Insectorum Sinicorum, Vol. 3, family Noctuidae', written by Chen [1982]). Both issues were illustrated with colour plates and could be used as guides for the identification of Chinese Noctuidae for the time being. Chen (1982) listed 348 species with their detail distribution in the Three Provinces of Northeast China. Later Chen (1999) published the most comprehensive catalogue of Noctuidae (*sensu lato*) of China where he reported 422 species for the Three Provinces of Northeast China.

After publications of these books, the faunal study of noctuoids in the Three Provinces of Northeast China was activated. Since the eighties of the last century and for the time being the faunal study of the Noctuidae *sensu lato* in the Three Provinces of Northeast China were continue a by publication of small local faunal lists for provinces, certain mountain ranges and various parts of the Three Provinces of Northeast China (Zhang & Sun, 1987; Gui, 1987; Li *et al*,. 1989; Ju *et al.*, 1997; Lou *et al.*, 1997; Wu *et al.*, 1999a, 1999b; Pei *et al.*, 2000; Liu *et al*,. 2006; Jin *et al.*, 2009; Jin *et al.*, 2011; Fu *et al.*, 2011; Xia *et al.*, 2012; Jin *et al.*, 2014).

Several tens species were added for the list of the Three Provinces of Northeast China Noctuidae by the review of certain genera or subfamilies (Han *et al.*, 2003, 2005a, 2005b, 2006a, 2008a; Han & Li, 2008; Sui & Han, 2010; Zhou & Han, 2010) or by numerous reports of newly recorded species (Han *et al.*, 2006b, 2007a, 2007b, 2007c, 2008b, 2010; Yang *et al.*, 2007; Guo *et al.*, 2010; Li & Han, 2011; Zhang *et al.*, 2011; Deng & Han, 2012; Zhao *et al.*, 2012; Tang & Han, 2013; Li & Han, 2013; Huang & Han, 2014; Zhang & Han, 2015; Wang & Han, 2017). Four species: *Zanclognatha lui* Han & Park, 2005, *Meganola liaoningensis* Han & Li, 2008, *Lophomilia nekrasovi* Kononenko & Behounek, 2009 and *Paragabara curvicornuta* Kononenko & Matov, 2010 were described as a new for science or recorded from the Three Provinces of Northeast China in taxonomic revision of some genera and species groups (Han *et al.*, 2005; Han & Li, 2008; Kononenko & Behounek, 2009; Kononenko *et al.*, 2010).

Compared with the Three Provinces of Northeast China, the fauna of the Noctuoidea of adjacent countries is studied more complete. There are illustrated monographic publications for Korea (Kononenko *et al.*, 1998; Kononenko & Han, 2007) and a series of the checklist, monographs and colourplate guide books for *the Far Eastern and Siberian fauna of Russia* (Kononenko, 1990, 2003, 2005, 2010, 2016a, 2016b). Because the fauna of the Three Provinces of Northeast China, Korea and Russia(Siberia) belong to the same zoogeographic region and have up to 75% of common species these books could serve for inventory and identification of species as well as the source of information on distribution and biology of common species.

In spite of the fact that the faunal study of the Noctuoidea in the Three Provinces of Northeast China continued over 100 years, the inventory of such large superfamily has not been complete yet. Although numerous faunal articles dealing with the Three Provinces of Northeast China by Chinese and foreign authors have been published in separated articles, which were sometimes published in Chinese in small local journals during last five decades, the new data on the Three Provinces of Northeast China Noctuoidea have not been well known to foreign workers of the Noctuoidea, by this reason as well as by the language barriers. On the other hand, foreign taxonomic and faunal publications often are inaccessible for Chinese researchers. As the result of such situation, until now, the Noctuoidea fauna of the Three Provinces of Northeast China has not been well presented itself to international researcher and Chinese researcher sometimes published old, incorrect taxonomic or faunal data.

The main goal of this book is to summarize our present knowledge about Noctuoidea fauna of the Three Provinces of Northeast China. We would like to present a complete faunal and nomenclature data in the modem taxonomic interpretation and complete a synonymic Catalogue for both domestic and foreign users, as well as present the introduction to rich and characteristic the Three Provinces of Northeast China fauna for students of the Noctuidae.

For the classification adopted in this Catalogue some explanation will be required. The classification system of the Noctuidae (*sensu lato*) adopted in the beginning of twenty century in classic works by Hampson (1903-1913) are based only on external characters and have become outdated.

Superfamily Noctuioidea (in modern treatment) is the largest taxonomic unit in the order Lepidoptera, and enumerated about 4200 genera (Kitching & Rawlins, 1999) and more than 42400 species (Nieukerken *et al.*, 2011) in the World.

In present treatment the superfamily Noctuoidea includes six families: Oenosandridae (4 genera, 8 species presented only in Australia), Notodontidae (704 genera, 3800 species worldwide), Nolidae (186 genera and about 1738 species worldwide), Erebidae (1760 genera, and about 24569 species worldwide), Euteliidae (29 genera, 520 species worldwide) and Noctuidae (1089 genera, and about 11772 species worldwide) (Nieukerken *et al.*, 2011).

The superfamily is determined by a number of morphological characters, the most characteristic of them is the forewing venation with four or three veins (in Notodontidae) extending from the radial cell. The amount of the included families and composition of the superfamily had long time been controversial. Numerous classifications of the family group Noctuoidea, presenting from two to thirteen families have been proposed by different authors. In the most well-established former classification, the superfamily Noctuoidea includes eight families: Oenosandridae, Doidae, Notodontidae, Lymantriidae, Aganaidae, Arctiidae, Nolidae and Noctuidae. In the last two decades classification Noctuoidea and, in particular, the Noctuidae (sensu lato) has undergone dramatic changes on the basis of modern morphological and molecular studies (Speidel, 1998; Kitching & Rawlins, 1999; Fibiger & Lafontaine, 2005; Lafontaine & Fibiger, 2006; Yela *et al.*, 2012, Fibiger *et al.*, 2012; Zahiri *et al.*, 2011, 2012, 2013a, 2013b).

In 2005, Fibiger and Lafontaine (Fibiger & Lafontaine, 2005) proposed a new classification of the Noctuoidea, which includes ten families: Oenosandridae, Doidae, Notodontidae, Strepsimanidae, Nolidae, Lymantriidae, Arctiida, Erebidae, Micronoctuidae and Noctuidae; later the rank of families Nolidae, Strepsimanidae, Arctiidae, Lymantriidae and Erebidae (sensu Fibiger & Lafontaine, 2005) was downgraded to subfamilies of the family Noctuidae (Lafontaine & Fibiger, 2006).

In the next step, on the basis of molecular studies from the former family Noctuidae were allocated the families Erebidae, Euteliidae, Nolidae and Noctuidae. The rank of the Nolidae was defined as a family, which includes subfamilies Nolinae, Sarrothripinae and Chloephorinae, (two later were previously attributed to the Noctuidae). The family Micronoctuidae was downgraded to tribe of the subfamily Hypenodinae (Erebidae), and the status Aganaidae, Arctiidae and Lymantriidae was defined in the rank of subfamilies composed Erebidae (Zahiri *et al.*, 2011, 2012, 2013a, 2013b). At present, this classification accepted by most researchers and used in the Catalogs Noctuoidea North America (Lafontaine & Schmidt, 2010), Europe (Fibiger *et al.*, 2012), Borneo (Holloway, 2011) and Siberia (Kononenko, 2016).

In present Catalogue, the classification of the superfamily Noctuoidea is adopted on the basis of the above directory with account of the latest publications on the molecular phylogeny of the superfamily Noctuoidea (Zahiri *et al.*, 2011, 2012, 2013a, 2013b). The treatment of higher taxa Noctuoidea by different authors since 1982, referred in the section "References" for each family, subfamily and tribe.

The Catalogue is based mainly on material collected by the senior author and students of the College of entomology of the Northeast Forestry University (Harbin) in the years 1990-2018 in the Northeast China, and on that in the collection of the following institutions: Northeast Agricultural University (Harbin, Heilongjiang Prov.), Da Hinggan Ling Forest Disease and Insect Pest Control and Quarantine Station (Jiagdaqi, Heilongjiang Prov.), Yanbian University (Yanji, Jilin Prov.), Beihua University (Jilin, Jilin Prov.), Northeast Normal University(Changchun, Jilin Prov.), Jilin Provincial Academy of Forestry Sciences (Changchun, Jilin Prov.), Liaoning Forest Disease and Insect Pest Control and Quarantine Station (Shenyang, Liaoning Prov.). Some materials from the Three Provinces of Northeast China collected in twenties of the last century by American and Russian entomologists have been examined in the Carnegie Museum (USA, Pittsburgh). Some type materials from the Naturkunde Museum Berlin, Germany Berlin (NKMB) [= Zoologisches Museum der Humboldt Universität, Berlin (ZMHU)]; and Zoologisches Forschunginstitut und Museum Alexander Koenig, Bonn, Germany (ZFMK); Institute of Biology and Pedology, Far Eastern Branch of Academy of Sciences (Russia, Vladivostok) and Zoological Institute of Russian Academy of Sciences, St. Petersburg, Russia (ZISP) have been examined.

The present Catalogue includes data on 883 species of 353 genera of four families of the Noctuoidea (except subfamilies Arctiinae and Lymantriinae in Erebidae) (Han, 2019). The following information is provided for each taxa listed in the Catalogue.

The valid scientific name for each rank of taxa is given by bold font. All scientific names are supplied by Chinese names, for many taxa the Chinese name is created for the first time. The next line presents the origin name of the taxon with year and author, citation of the original description, the type genus for the higher taxa and the type species for the genera; for the species the type locality is presented. The type locality is given in modern spelling and interpretation, if it was changed it is repeated in the original interpretation or spelling in square brackets. The old spelling of the type localities or direct quotation type locality (if its modern name is unclear) from the original description are marked by quotation marks. The nomenclature, spelling and data for type species and type localities for genera and species are presented according catalogues by Nye (1975) and Poole (1989), those for higher taxa—according Speidel & Naumann (2004).

The main sources of the information for the nomenclature of the generic names, authors and type localities were the catalogue 'The generic names of moths of the world'(Nye, 1975), and some addition information since 1975 from the issue 'Moths of Borneo'(Holloway 1985, 1988, 1989, 1998, 2003, 2005, 2008, 2009) and authors archives. The data for information on specific names, type localities and synonymy based on Poole`s Catalogue (Poole, 1989), and site 'The Global Lepidoptera Names Index' of the Nature History Museum (updated version 2018) and authors archives and publications as well as museum and bibliographical work.

The synonyms and homonyms for genera, species or subspecies are listed chronologically in the paragraph 'Synonymy' under the original combination of the valid name. Most important emendations, misspellings, misidentifications or incorrect uses of the generic or specific names are also included to the synonymy.

The geographic distribution of the species is given in separated paragraph 'Distribution'. First the distribution in China (according Chinese provinces) is reported, after the dot and n-dash [;] the distribution outside China is provided.

The known foodplants of the larvae is reported for the species. The main sources for the foodplants were the publications by Matov & Kononenko (2012), Park *et al.* (2011) and the original data.

Because of the Noctuoidea faunas of the Three Provinces of Northeast China and Russian(Siberia), Korea and northern Japan have a large number of species in common, this catalogue can be expected to be useful both for Chinese and Russian lepidopterists, as well as for other colleagues and students working on the faunistic, systematics and taxonomy of the Noctuoidea.

Acknowledgements

We wish to express our cordial thanks to our colleagues Mr. G. Behounek (Germany), Dr. A. Saldaitis (Lithuania), Dr. L. Ronkay and Mr. G. Ronkay (Hungary) for their constant kindness and consultation given us in various ways. Our special thank to Mr. G. Behounek for his kind permitting to use his taxonomic Data Base. We also owe great gratitude to M. Honey, J. Holloway and A. Zilli (Great Britain), Dr. D. Stüning, Dr. W. Mey and Dr. A. Hausmann (Germany), Dr. A. Matov (Russia) for their kind helps in sending and examining type-specimens or formation about them. Our great thanks are due to students of the laboratory of entomology in Northeast Forestry University and Yanbian University for their enthusiastic collecting in the Three Provinces of Northeast China and various help given us in the course of this work.

The present study was supported by the National Nature Science Foundation of China (No. 31872261, 31572294, 31470652, 31272355, 30700641), and the Fundamental Research Funds for the Central Universities(No. 2572019CP11, 2572018CJ03, 2572014DA02, DL09CA05). We thank to the curators of collections in the Northeast Forestry University.

Abbreviation

ab: aberration

Art: Article (of Code)

auct: auctorum – sensu authors – misidentification.

auct., nec: [misidentification by authors]

emend: emendation.

f: form – [in original description].

incorr. spell: incorrect subsequent spelling

missp: misspelling [subsequent misspelling by authors].

nom. nud: nomen nudum.

preocc: preoccupied name (homonym)

r: race

repl. name: replacement name

s. str: sensu stricto

[sic!]: [sci is a Latin word meaning “like this”, “like that”].

ssp: subspecies.

suppr: suppressed

syn: synonymy.

unavail: unavailable name

var: variation – [in original description].

...Hübner, 1822, nec Linnaeus, 1758 – preoccupied, homonym (for species group names).

Family **Erebidae** Leach, [1815] 目夜蛾科

Erebida Leach, [1815], *in* Brewster, *The American Edition of the New Edinburgh Encyclopedia, Conducted by David Brewster — Assisted by upwards of One Hundred Gentlemen in Europe*, 9: 134. Type genus: *Erebus* Latreille, 1810.

Synonymy: Erebidae Guenée, 1852; Erebinae Forbes, 1954; Erebini Forbes, 1954; Thysaniini Grote, 1895.

Erebidae is the biggest family of Lepidoptera with 1760 genera and 24569 described species (Nieukerken *et al.*, 2011). The members of this family are as diverse as can be imagined for a huge group of moth. Due its high diversity and great range of variability the family is difficult to define morphologically. Traditionally the taxa included to modern Erebidae were placed earlier authors in the subfamilies Catocalinae (sensu auctorum), Ophiderinae (sensu auctorum), Acontiinae (sensu auctorum), Herminiinae, Hypeninae, Aganaidae, Arctiidae and Lymantriidae.

Erebidae was separated from Noctuidae (sensu auctorum) by Fibiger & Lafontaine (2005); later Arctiidae and Lymantriidae were included to Erebidae on a subfamily rank. In present treatment it includes 16 subfamilies. The separation of Erebidae as a family is based and confirmed by molecular investigation (Zahiri *et al.*, 2011, 2013).

The family Erebidae is dominated in China, especially in its south part. It is represented here by 11 subfamilies, 90 genera and 230 species in the Three Provinces of Northeast China (without two subfamilies: Arctiinae and Lymantriinae).

References. Hampson, 1913, 1913a (Noctuidae: Stictopterinae, Acontianae, Erastrianae, Catocalinae); Seitz, [1907] –1914; 1931–1938 (Noctuidae: Catocalinae; Acontiinae, part); Chen, 1982, 1999 (Noctuidae); Sugi, 1982 (Noctuidae: Catocalinae, Ophiderinae, Acontiinae, part); Kononenko *et al.*, 1989 (Noctuidae: Ophiderinae, Catocalinae, Acontiinae, part); Kitching, Rawlins, 1999 (Catocalinae, Calpinae, Hypeninae, Herminiinae); Sviridov, 2003; (Noctuidae: Herminiinae, Rivulinae, Hypeninae, Catocalinae); Fibiger, Hacker, 2005 (Erebidae); Fibiger, Lafontaine, 2005 (Erebidae); Kononenko, 2005 (Erebidae), 2010 (Noctuidae: part); Kononenko, Pinratana, 2005 (Noctuidae: Herminiinae, Rivulinae, Hypeninae, Catocalinae); Lafontaine, Fibiger, 2006 (Noctuidae: Erebinae); Kononenko, Han, 2007 (Noctuidae: Rivulinae, Hypenodinae, Araeopteroninae, Eublemminae, Herminiinae, Phytometrinae, Aventiinae, Erebinae, Calpinae, Catocalinae); Matov *et al.*, 2008 (Noctuidae: part); Kononenko, 2016 (Erebidae), 2011 (Erebidae); Fibiger *et al.*, 2011 (Erebidae); Zahiri *et al.*, 2011, 2012 (Erebidae); Kononenko, Pinratana, 2013 (Erebidae).

注：目前国内对本科的中文种名称有“裳夜蛾科”和“裳蛾科”，但 Family Erebidae 源于 Subfamily Erebinae Leach, [1815]，模式属是目夜蛾属 *Erebus* Latreille, 1810，因此将本科中文科名确定为目夜蛾科更合理；同时，考虑到已有中文种名在中文文献记载中长年使用，且当前分类系统还可能有所调整，所以本书中除必要的调整外，一般继续沿用原有中文属、种名。

Subfamily **Arctiinae** 灯蛾亚科 (not treated here)

References. Fang, 1982, 2000 (Arctiidae); Sugi, 1982 (Arctiidae); Fibiger & Lafontaine, 2005 (Arctiidae); Fibiger & Hacker, 2005 (Arctiidae); Lafontaine & Fibiger, 2006 (Erebidae: Arctiinae); Matov *et al.*, 2008 (Arctiidae); Fibiger *et al.*, 2012 (Erebidae: Arctiinae); Speidel, Witt, 2011 (Erebidae: Arctiinae); Zahiri *et al.*, 2011, 2012 (Erebidae: Arctiinae).

Subfamily **Lymantriinae** 毒蛾亚科 (not treated here)

References. Zhao, 1982, 2003 (Lymantriidae); Sugi, 1982 (Lymatriidae); Tschistjakov, 2003 (Lymantriidae); Fibiger, Lafontaine, 2005 (Lymantriidae); Hacker, Fibiger, 2005 (Lymantriidae); Lafontaine, Fibiger, 2006 (Erebidae: Lymantriidae); Matov *et al.*, 2008 (Lymatriidae); Fibiger, *et al.*, 2011 (Erebidae: Lymatriinae); Speidel, Witt, 2011 (Erebidae: Lymantriinae); Zahiri *et al.*, 2011, 2012 (Erebidae: Lymatriinae).

Subfamily **Herminiinae** Leach, [1815] 1830 长须夜蛾亚科

Herminida Leach, [1815] 1830, *in* Brewster, *The American Edition of the New Edinburgh Encyclopedia, Conducted by David Brewster – Assisted by upwards of One Hundred Gentlemen in Europe* 9: 134. Type genus: *Herminia* Latreille, 1802.

Synonymy: Herminites Duponchel, [1845] 1844; Herminidae Herrich-Schäffer, [1851] 1845; Polypogoninae Hampson, 1918; Polypogoninae Lhomme, 1923.

In the treatment of the subfamily and sequence of genera we follow to the revisions of Japanese Herminiinae by Owada (1982, 1987, 1992a, 1992b) and of the Bornean Herminiinae by Holloway (2008) as the most comprehensive publications on Herminiinae of Asian fauna. Following Poole (1989) Holloway (2008) treated the genus *Polypogon* in a broad concept including *Herminia*, *Macrochilo*, *Pechipogo* and *Zanclognatha* as synonyms because their distinction is not well understood and because the Bornean species of this complex show a great range of variability both in external and genitalia characters without clear indications for generic grouping. Holloway (2011) and Zahiri *et al.* (2013) placed Herminiinae as a sister group to Aganainae on the basis of molecular study.

The subfamily includes about 1000 species are in the World fauna. It is most diverse in the subtropical and tropical areas and is associated mainly with forest habitats, in Palaearctic it is most diverse in the Chinese and Sino-Himalayan subregions with about 250 species. 16 genera and 51 species occur in the Three Provinces of Northeast China.

References. Chen, 1982 (Herminiinae); Owada, 1982, 1987, 1988, 1992a, 1992b (Herminiinae); Holloway, 2008 (Herminiinae).

Genus *Edessena* Walker, [1859] 1858 白肾夜蛾属

Edessena Walker, [1859] 1858, *List of the Specimens of Lepidopterous Insects in the Collection of the British Museum*, 16: 162. Type species: *Edessena gentiusalis* Walker, [1859] 1858 [North China].

The genus includes 2 species, *E. hamada* and Oriental *E. gentuisalis*; 1 species is in the Three Provinces of Northeast China.

Edessena hamada (Felder & Rogenhofer, 1874) 钩白肾夜蛾

Renodes hamada Felder & Rogenhofer, 1874, *Reise Österreichischen Fregatte Novara um die Erde in den Jahren 1857, 1859. Zoologischer Theil*, 2 (Abt. 2): pl. 119: 23. Type locality: Japan.

Distribution. China (Heilongjiang, Jilin, Liaoning, Hebei, Fujian, Jiangxi, Hunan, Sichuan, Yunnan); Russia (East), Korea, Japan.

寄主：麻栎（*Quercus acutissima*）（Matov & Kononenko, 2012）。

Genus *Hadennia* Moore, 1887 厚角夜蛾属

Hadennia Moore, 1887, *The Lepidoptera of Ceylon*, 3: 236. Type species: *Bocana hypenalis* Walker, [1859] 1858 [Sri Lanka] (=*Bocana jutalis* Walker, [1859] 1858).

Synonymy: *Wilkara* Swinhoe, 1918; *Walkara*: Swinhoe, 1918, emend.; *Bertula* auct., nec Walker, [1859]1858.

The genus includes 15 species, distributed mainly in the Oriental region; 1 species is in the Three Provinces of Northeast China.

Hadennia incongruens (Butler, 1879) 白点厚角夜蛾

Bocana incongruens Butler, 1879, *Annals and Magazine of Natural History*, (5)4: 448. Type locality: Japan, Yokohama.

Synonymy: *Nodaria amurensis* Staudinger, 1888; *Bertula jutalis* auct..

Distribution. China (Jilin, Hubei); Russia (East), Korea, Japan.

寄主：青冈 (*Quercus glauca*)（Matov & Kononenko, 2012）。

Genus *Mosopia* Walker, [1866] 1865 莫须夜蛾属

Mosopia Walker, [1866] 1865, *List of the Specimens of Lepidopterous Insects in the Collection of the British Museum*, 34: 1187. Type species: *Mosopia megaspila* Walker, [1866] 1865 [Penang].

Synonymy: *Trotosema* Butler, 1879.

The genus includes 6 species, mainly distributed in the Oriental region; 1 species is in the Three Provinces of Northeast China.

***Mosopia sordidum* (Butler, 1879) 暗莫须夜蛾**

Trotosema sordidum Butler, 1879, *Annals and Magazine of Natural History*, (5)4 (24): 449. Type locality: Japan, vic. Yokohama.

Distribution. China (Liaoning); Korea, Japan.

Genus *Cidariplura* Butler, 1879 胸须夜蛾属

Cidariplura Butler, 1879, *Annals and Magazine of Natural History*, (5) 4: 449. Type species: *Cidariplura gladiata* Butler, 1879 [Japan].

The genus includes 22 species, distributed mainly in the East and Southeast Asian; 1 species is in the Three Provinces of Northeast China.

***Cidariplura bilineata* (Wileman & South, 1919) 双带胸须夜蛾**

Mastigophorus bilineata Wileman & South, 1919, *Entomoogist*, 52: 267. Type locality: Taiwan.

Synonymy: *Cidariplura brevivittalis* Moore, 1867, misidentification; *Elyra albifascia* Hampson, 1929.

Distribution. China (Liaoning, Guangxi, Xizang, Taiwan); Korea, Japan, India, Bangladesh.

Genus *Paracolax* Hübner, [1825] 1816 奴夜蛾属

Paracolax Hübner, [1825] 1816, *Verzeichniss bekannter Schmettlinge* [sic！]: 344. Type species: *Pyralis derivalis* Hübner, 1796 [Europe].

Synonymy: *Capnistis* Warren, 1913; *Paraherminia* Richards, 1932; *Crinisinus* Bryk, 1948.

The genus includes 18 species, distributed mainly in the Oriental and the Chinese subregion of the Palaearctic regions; 6 species are in the Three Provinces of Northeast China.

***Paracolax contigua* (Leech, 1900) 邻奴夜蛾**

Bleptina contigua Leech, 1900, *Transactions of the Entomological Society of London*, 1900: 625. Type locality: China.

Synonymy: *Paracolax tokui* Owada, 1982.

Distribution. China (Jilin, Liaoning, Hubei, Sichuan, Taiwan); Korea, Japan.

***Paracolax pryeri* (Butler, 1879) 黄肾奴夜蛾**

Herminia pryeri Butler, 1879, *Illustrations of Typical Specimens of Lepidoptera Heterocera in the Collection of the British Museum*, 3: 63, pl. 56: 11. Type locality: Japan, Yokohama.

Distribution. China (Liaoning, Zhejiang, Fujian, Hunan, Taiwan); Korea, Japan.

***Paracolax tristalis* (Fabricius, 1794) 曲线奴夜蛾**

Phalaena tristalis Fabricius, 1794, *Entomologiae Systematica emendata et Aucta. Secundum Classes, Ordines, Genera, Species Adjectis Synonimis, Locis, Observationibus, Descriptionibus*, 3 (2): 224. Type locality: Italy.

Synonymy: *Crambus glaucinatus* Fabricius, 1794; *Pyralis derivalis* Hübner, 1796; *Crambus tristis* Fabricius, 1798; *Crambus derivatus* Haworth, 1809; *Pyralis glaucinalis* auct..

Distribution. China (Heilongjiang, Jilin, Liaoning); Russia (European part, Siberia), Korea, Japan; Europe.

寄主：栎属、柳属、山楂属、悬钩子属植物（陈一心，1999）；欧洲栗（*Castanea sativa*）、欧洲桤木（*Alnus glutinosa*）、黄花柳（*Salix caprea*）、秦氏悬钩子（*Rubus caesius*）、栎属（*Quercus*）、山楂属（*Crataegus*）、葛藤属（*Pueraria*）、梣属（*Fraxinus*）、蒲公英属（*Taraxacum*）（Matov & Kononenko, 2012）。

注：《中国动物志》（夜蛾科）中引用本种同物异名"*Pyralis derivalis* Hübner, 1796"，在此予以变更。

***Paracolax trilinealis* (Bremer, 1864) 三线奴夜蛾**

Herminia trilinealis Bremer, 1864, *Memoires de l'Académie Imperiale des Sciences de St.-Petersbourg* (7)8, 1: 64, pl. 5: 23. Type locality: Russia, "Ussuri" [Primorye terr.].

Synonymy: *Crinisinus turbo* Bryk, 1948.

Distribution. China (Heilongjiang, Jilin); Russia (East), Korea, Japan.

寄主：双子叶植物新鲜或枯死的叶片（Owada, 1987）。

***Paracolax fascialis* (Leech, 1889) 宽带奴夜蛾**

Herminia fascialis Leech, 1889, *Entomologist*, 22: 64, pl. 2: 3. Type locality: Japan, Hakodate.

Synonymy: *Zanclognatha helialis* Staudinger, 1892.

Distribution. China (Jilin); Russia (East), Korea, Japan.

寄主：针叶树木（罗汉柏、日本柳杉）（陈一心，1999）；日本柳杉（*Cryptomeria japonica*）、罗汉柏（*Thujopsis dolabrata*）、槲树（*Quercus dentata*）（Matov & Kononenko, 2012）。

***Paracolax fentoni* (Butler, 1879) 圆斑奴夜蛾**

Herminia fentoni Butler, 1879, *Illustrations of Typical Specimens of Lepidoptera Heterocera in the Collection of the British Museum*, 3: 63, pl. 56, f. 12. Type locality: Japan, Yokohama.

Distribution. China (Jilin); Russia (East), Korea, Japan.

Genus *Idia* Hübner, [1813] 极夜蛾属

Idia Hübner, [1813], *Zuträge zur Sammlung Exotischer Schmetterlinge*, 1: pl. [212]. Type species: *Idia aemula* Hübner, [1813] [North America].

Synonymy: *Epizeuxis* Hübner, 1818; *Camptylochila* Stephens, 1834; *Helia* Duponchel, 1844; *Campylochila*: Agassiz, 1847, missp.; *Helia* Guenée, 1854, preocc. (Hübner, 1818 [Lepidoptera, Noctuidae]); *Pseudaglossa* Grote, 1874; *Zenomia* Dognin, 1914.

The genus includes 40 species, distributed in the Palaearctic, Oriental, Ethiopian, Nearctic and Neotropical regions, of them 6 species are in the Palaearctic region; 2 species are in the Three Provinces of Northeast China.

Idia quadra **(Graeser, 1888 [1889]) 波纹极夜蛾**

Helia quadra Graeser, 1888 [1889], *Berliner entomologische Zeitschrift*, 32, II: 378. Type locality: Russia: "Raddefka" [Khabarovsk terr., Judish autonomy, Radde], Khabarovsk, Primorye terr., "Suifun" [Razdolnaya river].

Synonymy: *Epizeuxis curvipalps*: Herz, 1904, nec Butler, 1879, misidentification.

Distribution. China (Heilongjiang, Jilin); Russia (East), Korea, Japan.

Idia curvipalpis **(Butler, 1879) 白斑极夜蛾**

Capnodes curvipalpis Butler, 1879, *Annals and Magazine of Natural History*, (5)4: 369. Type locality: Japan, Yokohama.

Synonymy: *Epizeuxis lunulata* Herz, 1904.

Distribution. China (Liaoning, Zhejiang); Russia (East), Korea, Japan.

寄主：幼虫取食双子叶植物的枯叶（Owada, 1987）。

Genus ***Naarda*** Walker, 1866 纳达夜蛾属

Naarda Walker, 1866, *List of specimens of the lepidopterous insects of the collection of the British Museum*, 35: 1694. Type species: *Naarda bisignata* Walker, 1866 [Indonesia].

Synonymy: *Gynaephila* Staudinger, 1892.

The genus includes 39 species, distributed mainly in the Chinese subregion of the Palaearctic and the Oriental regions, 13 species in the Ethiopian region, and 7 species in the Australian region; 1 species is in the Three Provinces of Northeast China.

注：首次给出中文属名。

Naarda maculifera **(Staudinger, 1892) 褐圆纳达夜蛾**

Gynaephila maculifera Staudinger, 1892, *in* N.M. Romanoff (ed.), *Mémoires sur les Lépidoptères*, 6: 629, pl. 14: 13. Type locality: Russia, Vladivostok; Askold I., "Sutschan", [Partizansk].

Distribution. China (Heilongjiang, Jilin, Sichuan); Russia (East), Korea, Japan.

注：李明文、韩辉林 (2013) 将本种归入"*Gynaephila* Staudinger, 1892"，在此予以变更。

Genus *Hydrillodes* Guenée, 1854 亥夜蛾属

Hydrillodes Guenée, 1854, *in* Boisduval & Guenée, *Histoire Naturelle des Insectes. Species Général des Lépidoptères*, 8: 65. Type species: *Hydrillodes lentalis* Guenée, 1854 [Central India].

Synonymy: *Echana* Walker, [1859] 1858; *Olybama* Walker, [1859] 1858; *Bibacta* Moore, 1882; *Ragana* Swinhoe, 1900; *Cellacrinata* Bethune-Baker, 1908.

The genus includes 61 species, distributed mainly in the Oriental, Ethiopian, Australian regions and in the Chinese subregion of the Palaearctic region; 2 species are in the Three Provinces of Northeast China.

Hydrillodes morosa **(Butler, 1879) 阴亥夜蛾**

Bleptina morosa Butler, 1879, *Illustrations of Typical Specimens of Lepidoptera Heterocera in the Collection of the British Museum*, 3: 64, pl. 56: 15. Type locality: Japan.

Synonymy: *Hydrillodes funeralis* Warren, 1913; *Hydrillodes obscurans* Bryk, 1948; *Hydrillodes lentalis* auct..

Distribution. China (Heilongjiang, Jilin, Zhejiang, Xinjiang); Russia (East), Korea, Japan.

注：《中国动物志》（夜蛾科）中引用本种的同物异名“*Hydrillodes funeralis* Warren, 1913”，在此予以变更。

Hydrillodes pacificus **Owada, 1982 拟弓须亥夜蛾**

Hydrillodes pacifica Owada, 1982, *Moths of Japan*, 1: 916, pl. 224: 23-26. [incorrect original spelling]. Type locality: Japan, Amami Isl.

Synonymy: *Hydrillodes pacificus*: Sugi, *in* Hirashima 1989: 7-8; Owada, 1987: 67.

Distribution. China (Jilin, Liaoning); Korea, Japan.

Genus *Bertula* Walker, 1858 拟胸须夜蛾属

Bertula Walker, 1858, *List of the Specimens of Lepidopterous Insects in the Collection of the British Museum*, 16: 162. Type species: *Bertula abjudicalis* Walker, [1859] 1858 [Sri Lanka].

Synonymy: *Elyra* Walker, [1859] 1858; *Neviasca* Walker, 1858, preocc. (Walker, [1859] 1858 [Lepidoptera, Noctuidae]); *Eordaea* Walker, [1859]; *Cardalena* Walker, 1859 unnecessary repl. name; *Gabrisa* Walker, 1865 unnecessary repl. name.

The genus includes 49 species, distributed mainly in the Oriental and the Chinese subregion of the Palaearctic regions; 2 species are in the Three Provinces of Northeast China.

Bertula bistrigata **(Staudinger, 1888) 双拟胸须夜蛾**

Zanclognatha bistrigata Staudinger, 1888, *Entomologische Zeitung. Entomologischen Vereine zu Stettin*, 49:

236. Type locality: Russia, Vladivostok, "Sutschan-Gebiet" [Partizansk].

Synonymy: *Bocana spacoalis chosenana* Bryk, 1948 (ssp.).

Distribution. China (Heilongjiang, Liaoning, Fujian); Russia (East), Korea, Japan.

寄主：光叶碗蕨（*Dennstaedtia scabra*）（Matov & Kononenko, 2012）。

***Bertula spacoalis* (Walker, 1859) 条拟胸须夜蛾**

Bleptina spacoalis Walker, 1859, *List of the Specimens of Lepidopterous Insects in the Collection of the British Museum.*, 19: 872. Type locality: North China.

Distribution. China (Liaoning, Hebei, Hunan, Jiangxi, Fujian, Sichuan); Korea, Japan.

Genus *Nodaria* Guenèe, 1854 疽夜蛾属

Nodaria Guenèe, 1854, *in* Boisduval & Guenèe, *Histoire Naturelle des Insectes. Species Général des Lépidoptères*, 8: 63. Type species: *Nodaria hispanalis* Guenèe, 1854 [Spain].

The genus includes about 50 species, diversified in the Oriental region, some species reach the southeast part of the Palaearctic region. 1 species is in the Three Provinces of Northeast China.

***Nodaria tristis* (Butler, 1879) 悲疽夜蛾**

Bocana tristis Butler, 1879, *Illustration of Typical Specimens of Lepidoptera Heterocera in the Collection of the British Museum*, 3: 65, pl. 57: 2. Type locality: Japan, Yokohama.

Distribution. China (Jilin, Liaoning, Zhejiang, Hunan, Yunnan); Korea, Japan.

Genus *Simplicia* Guenée, 1854 贫夜蛾属

Simplicia Guenée, 1854, *in* Boisduval & Guenée, *Histoire Naturelle des Insectes. Species Général des Lépidoptères*, 8: 15. Type species: *Herminia rectalis* Eversmann, 1842 [Russia: Orenburg, Kasan].

Synonymy: *Libisosa* Walker, [1859] 1858; *Culicula* Walker, 1864; *Aginna* Walker, 1865; *Nabartha* Moore, 1887.

The genus includes 43 species, distributed mainly in the Oriental and the Chinese subregion of the Palaearctic regions; 3 species are in the Three Provinces of Northeast China.

***Simplicia rectalis* (Eversmann, 1842) 黑点贫夜蛾**

Herminia rectalis Eversmann, 1842, *Bulletin de la Société Impériale des Naturalistes de Moscou*, 15 (3): 558. Type locality: Southeast Russia: Orenburg, Kasan.

Synonymy: *Herminia sicca* Butler, 1879; *Simplicia rectalis minoralis* Warren, 1913 (ssp.).

Distribution. China (Heilongjiang, Jilin, Liaoning, Jiangsu); Russia (Ural to East), Korea, Japan, Turkey;

Central and South Europe.

寄主：栎属（*Quercus*）、披碱草属（*Elymus*）植物（Matov & Kononenko, 2012）。

Simplicia niphona (Butler, 1878) 曲线贫夜蛾

Bocana niphona Butler, 1878, *Illustration of Typical Specimens of Lepidoptera Heterocera in the Collection of the British Museum*, 2: 56, pl. 38: 9. Type locality: Japan, Yokohama.

Synonymy: *Agynna similis* Moore, 1882; *Simplicia niphona olivacea* A.E. Prout, 1929 (ssp.); *Simplicia niphona superior* A. E. Prout, 1929 (ssp.).

Distribution. China (Jilin, Nei Mongol Hebei, Zhejiang, Hunan, Fujian, Hainan, Guangxi, Yunnan, Xizang, Taiwan); Korea, Japan, Indonesia (Sumatra), Malaysia (Borneo), Sri Lanka, India, Nepal, Pakistan.

Simplicia caeneusalis (Walker, [1859] 1858) 锯线贫夜蛾

Sophronia caeneusalis Walker, [1859] 1858, *List of the Specimens of Lepidopterous Insects in the Collection of the British Museum*., 16: 94. Type locality: Australia.

Synonymy: *Bocana robustalis* Walker, 1866; *Aginna simulata* Moore, 1882; *Liosa obiana* Swinhoe, 1919; *Simplicia lautokiensis* Prout, 1933; *Simplicia ryukyuensis* Sugi, 1965; *Simplicia caeneusalis buffetti* Holloway, 1977 (ssp.).

Distribution. China (Liaoning, Hainan, Yunnan); Korea, Japan, India, Thailand, Indonesia (Borneo), New Guinea, USA(Hawaii), Australia.

Genus *Zanclognatha* Lederer, 1857 镰须夜蛾属

Zanclognatha Lederer, 1857, *Die Noctuinen Europa's, mit Zuziehung einger bisher mesit dazu gezahlter Arten des Asiatischen Russland's, Kleinasien's, Syrien's, und Labrador's. Wien's und Labrador's*: 45. Type species: *Pyralis tarsiplumalis* Hübner, 1796 [Europe] (=*Phalaena lunalis* Scopoli, 1763).

Synonymy: *Erpyzon* Hübner, 1808, rejected name; *Cleptomita* Grote, 1873; *Megachyta* Grote, 1873; *Pityolita* Grote, 1873; *Mesoplectra* Butler, 1879; *Adrapsoides* Matsumura, 1925; *Treitschkendia* Berio, 1989; *Zellerminia* Beck, 1966.

The genus includes 58-60 species, distributed in the Palaearctic, Oriental, Nearctic and Ethiopian regions; 15 species are in the Three Provinces of Northeast China.

Zanclognatha griselda (Butler, 1879) 杉镰须夜蛾

Herminia griselda Butler, 1879, *Illustrations of Typical Specimens of Lepidoptera Heterocera in the Collection of the British Museum*, 3: 63, pl. 56: 9. Type locality: Japan, Yokohama.

Distribution. China (Liaoning, Fujian); Russia (East), Korea, Japan.

寄主：柳杉属、铁杉属、松属、冷杉属植物（陈一心，1999）；赤松（*Pinus densiflora*）、鱼鳞云杉

（*Picea ajanensis*）、日本柳杉（*Cryptomeria japonica*）、冷杉属（*Abies*）、铁杉属（*Tsuga*）植物（Matov & Kononenko, 2012）。

***Zanclognatha lilacina* (Butler, 1879) 常镰须夜蛾**

Mesoplectra lilacina Butler, 1879, *Illustrations of Typical Specimens of Lepidoptera Heterocera in the Collection of the British Museum*, 3: 65, pl. 57: 3. Type locality: Japan, Yokohama.

Synonymy: *Zanclognatha celatrix* Filipjev, 1927.

Distribution. China (Jilin, Zhejiang, Jiangxi, Fujian); Russia (East), Korea, Japan.

寄主：日本冷杉（Owada, 1987）。

***Zanclognatha lunalis* (Scopoli, 1763) 镰须夜蛾**

Phalaena lunalis Scopoli, 1763, *Entomologia Carniolica Exhibens Insecta Carnioliae Indigena et Distributa in Ordines, Genera, Species, Varietates Methodo Linnaeana*: 241: 612. Type locality: Europe: Slovenia.

Synonymy: *Pyralis tarsiplumalis* Hübner, 1796; *Zanclognatha tarsiplumalis chosensis* Bryk, 1948 (ssp.).

Distribution. China (Heilongjiang, Jilin, Liaoning, Xinjiang, Hubei, Zhejiang); Russia (European part, Siberia), Korea, Japan; Central Asia, Europe.

寄主：萹蓄 (*Polygonum aviculare*)、桦叶鹅耳枥（*Carpinus betulus*）、黑果越橘（*Vaccinium myrtillus*）、黑刺李（*Prunus spinosa*）、山毛榉属（*Fagus*）、栎属（*Quercus*）、莴苣属（*Lactuca*）、蒲公英属（*Taraxacum*）植物（Matov & Kononenko, 2012）。

***Zanclognatha lui* Han & Park, 2005 吕镰须夜蛾**

Zanclognatha lui Han & Park, 2005, *Korean Journal of Systematic Zoology*, 21 (1): 2, figs 1A-D, 3A-B. Type locality: China, Mt. Changbai.

Distribution. China (Jilin); Russia (East), Korea.

***Zanclognatha curvilinea* (Wileman & South, 1917) 曲线镰须夜蛾**

Paracolax curuilinea Wileman & South, 1917, *Entomologist*, 50: 27. Type locality: Japan.

Distribution. China (Heilongjiang, Jilin); Korea, Japan.

***Zanclognatha fumosa* (Butler, 1879) 窄肾镰须夜蛾**

Herminia fumosa Butler, 1879, *Illustrations of Typical Specimens of Lepidoptera Heterocera in the Collection of the British Museum*, 3: 62, pl. 56: 8. Type locality: Japan.

Synonymy: *Zanclognatha assirnilis* Staudinger, 1888.

Distribution. China (Heilongjiang, Jilin); Russia (East), Korea, Japan.

寄主：高羊茅（*Festuca arundinacea*）、长穗刻鳞薹草（*Carex incisa*） (Kogi 1985)；刻叶紫堇（*Carex incisa*）、高羊茅（*Festuca arundinacea*）（Matov & Kononenko, 2012）。

Zanclognatha obliqua **Staudinger, 1892 斜镰须夜蛾**

Zanclognatha obliqua Staudinger, 1892, *in* N.M. Romanoff (ed.), *Mémoires sur les Lépidoptères*, 6: 616, pl. 14: 10. Type locality: Russia, Primorye terr., Askold I..

Distribution. China (Heilongjiang, Xizang); Russia (East), Korea, Japan.

寄主：高羊茅（*Festuca arundinacea*）、长穗刻鳞薹草（*Carex incisa*） (Kogi, 1985)；高羊茅（*Festuca arundinacea*）、薹草属（*Carex*）植物（Matov & Kononenko, 2012）。

Zanclognatha tarsipennalis **(Treitschke, 1835) 扁镰须夜蛾**

Herminia tarsipennalis Treitschke, 1835, *Die Schmetterlinge von Europa*, 10 (3): 5. Type locality: Germany, Brandenburg.

Synonymy: *Herminia denticornalis* Wocke, 1850.

Distribution. China (Jilin, Hubei, Taiwan); Russia (European part, Ural, Siberia), Korea, Japan; Central Asia, Near East, Europe, North Africa.

寄主：圆柏（*Juniperus chinensis*）、欧洲刺柏（*Juniperus communis*）、葡萄叶铁线莲（*Clematis vitalba*）、萹蓄（*Polygonum aviculare*）、黄花柳（*Salix caprea*）、树莓（*Rubus fruticosus*）、覆盆子（*Rubus idaeus*）、火炬树（*Rhus typhina*）、洋常春藤（*Hedera helix*）、山毛榉属（*Fagus*）、栎属（*Quercus*）、桦木属（*Betula*）、越橘属（*Vaccinium*）、车前草属（*Plantago*）、蒲公英属（*Taraxacum*）植物（Matov & Kononenko, 2012）。

Zanclognatha subgriselda **Sugi, 1959 叔灰镰须夜蛾**

Zanclognatha subgriselda Sugi, 1959, *Tinea*, 5: 284, pl. 38: 18. Type locality: Japan, Tokyo, Takao-san.

Distribution. China (Liaoning, Zhejiang, Fujian); Russia (East), Japan.

寄主：取食枯叶（Owada, 1987）。

Zanclognatha triplex **(Leech, 1900) 长阳镰须夜蛾**

Nodaria triplex Leech, 1900, *Transactions of the Entomological Society of London*, 1900: 634. Type locality: China, Changyang.

Synonymy: *Zanclognatha sugii* Owada, 1980.

Distribution. China (Jilin, Hubei); Russia (Transbaikalia, East), Korea, Japan.

注：根据本种模式标本产地为湖北长阳，首次给出中文种名。

Zanclognatha helva (Butler, 1879) 黄镰须夜蛾

Herminia helva Butler, 1879, *Annals and Magazine of Natural History*, (5)4: 477. Type locality: Japan, Yokohama.

Distribution. China (Jilin, Zhejiang, Fujian, Hunan, Taiwan); Russia (East), Korea, Japan.

寄主：青冈（*Quercus glauca*）（Matov & Kononenko, 2012）。

Zanclognatha reticulatis (Leech, 1900) 白斑镰须夜蛾

Adrapsa reticulatis Leech, 1900, *Transactions of the Entomological Society of London*, 1900: 616. Type locality: China, Changyang.

Synonymy: *Zanclognatha reticulatis ruptistigma* Holloway, 1976 (ssp.).

Distribution. China (Liaoning, Zhejiang, Hubei, Fujian, Guangxi, Sichuan, Xizang, Taiwan); Russia (East), Japan, Vietnam, Thailand, Malaysia, Indonesia.

Zanclognatha umbrosalis Staudinger, 1892 暗镰须夜蛾

Zanclognatha umbrosalis Staudinger, 1892, *in* N.M. Romanoff (ed.), *Mémoires sur les Lépidoptères*, 6: 613, pl. 14: 8. Type locality: Russia, “Suifun”, [Primorye terr., Razdolnaya river], “Sutschan-Gebiet” [Partizansk].

Synonymy: *Nodaria leechi* South, 1905; *Zanclignatha triplex koreognatha* Bryk, 1948 (ssp.).

Distribution. China (Jilin); Russia (East), Korea, Japan.

注：首次给出中文种名。

Zanclognatha perfractalis Bryk, 1948 全镰须夜蛾

Zanclognatha perfractalis Bryk, 1948, *Arkiv för Zoologi*, 41 (A) 1: 147, pl. 6: 7. Type locality: North Korea, Hamgyeong bugdo [Shuotsu].

Synonymy: *Zanclognatha sviridovi* Owada, 1992.

Distribution. China (Jilin); Russia (East), Korea.

注：首次给出中文种名。

Zanclognatha violacealis Staudinger, 1892 紫灰镰须夜蛾

Zanclognatha violacealis Staudinger, 1892, *in* N.M. Romanoff (ed.), *Mémoires sur les Lépidoptères*, 6: 612, pl. 14: 6. Type locality: Russia, [Primorye terr.] “Ussuri”, Askold I..

Synonymy: *Zanclognatha stramentacealis*: Leech, 1889: 562, 1900: 562; auct., nec Bremer, 1864.

Distribution. China (Heilongjiang, Jilin); Russia (East), Korea, Japan.

Genus *Pechipogo* Hübner, [1825] 1816 培夜蛾属

Pechipogo Hübner, [1825] 1816, *Verzeichniss bekannter Schmettlinge* [sic!]: 345. Type species: *Pyralis pectitalis* Hübner, 1796 [Europe] =*Phalaena Geometra strigilata* Linnaeus, 1758.

Synonymy: *Erpyzon* Hübner, 1806, rejected name; *Pechipogon*: Stephens, 1834, missp.; *Pechypogon* Agassiz, 1846, emend.; *Herminia* auct..

The genus includes 2 species, distributed in the West and the Chinese subregion of the Palaearctic region; 1 species is in the Three Provinces of Northeast China.

Pechipogo strigilata **(Linnaeus, 1758) 培夜蛾**

Phalaena Geometra strigilata Linnaeus, 1758, *Systema Naturae* (Edn. 10) 1: 528. Type locality: Europe.

Synonymy: *Phalaena barbalis* Clerk, 1759; *Phalaena palpalis* Fabricius, 1775; *Crambus palpatus* Haworth, 1809, emend.; *Crambus barbata* Haworth, 1809, emend..

Distribution. China (Jilin, Liaoning); Russia (European part, Ural to Siberia), Mongolia, Korea, Japan, Kazakhstan; Europe.

寄主：夏栎（*Quercus robur*）、欧洲桤木（*Alnus glutinosa*）、灰桤木（*Alnus incana*）、欧洲山杨（*Populus tremula*）、黄花柳（*Salix caprea*）、覆盆子（*Rubus idaeus*）、金雀儿（*Cytisus scoparius*）、洋常春藤（*Hedera helix*）、欧银花（*Lonicera xylosteum*）、茜草（*Rubia peregrina*）、（*Rubia tinctoria*）、欧洲千里光（*Senecio vulgaris*）、栎属（*Quercus*）、桦木属（*Betula*）、鹅耳枥属（*Carpinus*）、榛属（*Corylus*）、蔷薇属（*Rosa*）、铁线莲属（*Clematis*）植物（Matov & Kononenko, 2012）。

Genus *Polypogon* Schrank, 1802 翅须夜蛾属

Popypogon Schrank, 1802, *Fauna Boica. Durchgedachte Geschichte der in Baiera Einheimischen und Zahmen Thiere*, 1 (2): 161. Type species: *Phalaena tentacularia* Linnaeus, 1758 [Europe].

Synonymy: *Popypogon*: Doubleday, 1850 missp.; *Herminia* auct.; *Macrochilo* auct..

The genus includes 7 species, distributed in the West and the Chinese subregion of the Palaearctic region; 2 species are in the Three Provinces of Northeast China.

Polypogon tentacularia **(Linnaeus, 1758) 淡黄翅须夜蛾**

Phalaena Geometra tentacularia Linnaeus, 1758, *Systema Naturae* (Edn. 10) 1: 522. Type locality: Europe.

Synonymy: *Pyralis tentaculalis* [Denis et Schiffermüller], 1775; *Herminia tentacularia carpathica* Hormuzaki, 1894 (var.).

Distribution. China (Heilongjiang, Jilin, Liaoning); Russia (European part, Ural to Siberia), Mongolia, Korea, Japan, Kazakhstan; Europe.

寄主：蕨菜（*Pteridium aquilinum*）、藜属（*Chenopodium*）、桦木属（*Betula*）、柳属（*Salix*）、悬钩子属（*Rubus*）、一枝黄花属（*Solidago*）、柳兰（*Chamaenerion angustifolium*）、早熟禾属（*Poa*）、

蒲公英属（*Taraxacum*）植物（Matov & Kononenko, 2012）。

注：本种国内无文献记录，根据国外文献资料整理，发现在东北大部有分布。首次给出中文种名。

Polypogon tarsicrinata (Bryk, 1948) 暗翅须夜蛾

Zanclognatha tarsicrinata Bryk, 1948, *Arkiv för Zoologi*, 41 (A)1: 146, pl. 5: 5. Type locality: North Korea, Shuotsu.

Synonymy: ?*Nodaria parallela* Wileman, 1911; *Polypogon tarsicrinata*: Poole, 1989; *Polypogon gryphalis*: Sviridov, 1990, incorrect, misidentification; *Polypogon gryphalis*: Kononenko *et al.*, 1998, 2005, misidentification; *Polypogon tarsicrinata*: Owada, 2006, correction.

Distribution. China (Heilongjiang, Jilin); Russia (European part, East), Korea, Japan; Central Europe.

Genus ***Herminia*** Latreille, 1802 长须夜蛾属

Herminia Latreille, 1802, *in* Sonnini's Buffon, *Histoire Naturelle, Generale et Particuliere des Crustaces et des Insectes. Ouvrage Faisant Suite aux Oeuvres de Leclere de Buffon, et Partie du cours Compet d'Histoire Naturelle Reedige par C.S. Sonnini*, 3: 413. Type species: *Phalaena ventilabris* Fabricius, 1787 [Europe] (=*Phalaena tarsicrinalis* Knoch, 1782).

Synonymy: *Herminea* Sodoffsky, 1837, missp.; *Pogonitis* Sodoffsky, 1837, unnecessary repl. name; *Quaramia* Berio, 1989; *Zanclognatha* auct..

The genus includes 15-18 species, distributed in the Palaearctic, Oriental, and Ethiopian regions; 7 species are in the Three Provinces of Northeast China.

Herminia grisealis ([Denis & Schiffermüller], 1775) 枥长须夜蛾

Pyralis grisealis [Denis & Schiffermüller], 1775, *Ankündung eines Systematisches Werkes von den Schmetterlinge der Wiener Gegend. Wien*: 120. Type locality: Austria, Vienna district.

Synonymy: *Phalaena nemoralis* Fabricius, 1775; *Crambus nemorum* Fabricius, 1794, emend.; *Crambus nemoratus* Haworth, 1809, emend..

Distribution. China (Heilongjiang, Jilin, Nei Mongol, Sichuan, Yunnan, Taiwan); Russia (European part, Ural to East), Korea, Japan; Kazakhstan; Near East, Europe.

寄主：大荨麻（*Urtica dioica*）、锐刺山楂（*Crataegus oxyacantha*）、黑刺李（*Prunus spinosa*）、黑莓（*Rubus fruticosus*）、覆盆子（*Rubus idaeus*）、互叶金腰（*Chrysosplenium alternifolium*）、鸡爪槭（*Acer palmatum*）、欧洲红瑞木（*Cornus sanguinea*）、欧洲绣球荚蒾（*Viburnum opulus*）、西洋接骨木（*Sambucus nigra*）、接骨木（*Sambucus racemosa*）、林生水苏（*Stachys silvatica*）、欧洲栗（*Castanea sativa*）、青冈（*Quercus glauca*）、灰桤木（*Alnus incana*）、欧洲榛（*Corylus avellana*）酸模属（*Rumex*）、越橘属（*Vaccinium*）、椴树属（*Tilia*）、榆属（*Ulmus*）、虎耳草属（*Saxifraga*）、毛核木属（*Symphoricarpos*）、拉拉藤属（*Galium*）、莴苣属（*Lactuca*）、蒲公英属（*Taraxacum*）植物

（Matov & Kononenko, 2012）。

注：《中国动物志》（夜蛾科）中引用其同物异名“*Phalaena nemoralis* Fabricius, 1775”，在此予以变更。

***Herminia robiginosa* (Staudinger, 1888) 四线长须夜蛾**

Zanclognatha robiginosa Staudinger, 1888, *Entomologische Zeitung. Entomologischen Vereine zu Stettin*, 49: 275. Type locality: Russia, Khavbarovsk terr., Judish autonomy, Radde; Primorye terr., Vladivostok, Askold Is., “Sutschan-Gebiet” [Partizansk], “Suifun” [Razdolnaya river].

Synonymy: *Nodaria assimilata* Wileman, 1911.

Distribution. China (Liaoning); Russia (Siberia), Korea, Japan.

***Herminia tarsicrinalis* (Knoch, 1782) 灰长须夜蛾**

Phalaena tarsicrinalis Knoch, 1782, *Beiträge zur Insektengeschichte*, 2: 75, pl. 4: 1-12. Type locality: Germany, Leipzig.

Synonymy: *Phalaena ventilabris* Fabricius, 1787; *Crambus tarsicrinatus* Haworth, 1809; *Zanclognatha biumbralis* Turati et Verity, 1911; *Zanclognatha balatonalis* Osthelder, 1935.

Distribution. China (Heilongjiang, Jilin, Hebei, Jiangsu); Russia (European part, Ural, Siberia), Korea, Japan, Turkey, Kazakhstan; Europe.

寄主：铁线草（*Clematis recta*）、葡萄叶铁线莲（*Clematis vitalba*）、酸模（*Rumex acetosa*）、青冈（*Quercus glauca*）、大荨麻（*Urtica dioica*）、欧洲木莓（*Rubus caesius*）、树莓（*Rubus idaeus*）、欧洲黑莓（*Rubus fruticosus*）、高丽悬钩子（*Rubus coreanus*）、柳兰（*Epilobium angustifolium*）、短柄野芝麻（*Lamium album*）、蒲公英属（*Taraxacum*）、莴苣属（*Lactuca*）植物（Matov & Kononenko, 2012）。

***Herminia arenosa* Butler, 1878 赭黄长须夜蛾**

Herminia arenosa Butler, 1878, *Illustrations of Typical Specimens of Lepidoptera Heterocera in the Collection of the British Museum*, 3: 56, pl. 38: 8. Type locality: Japan, Yokohama.

Synonymy: *Zanclognatha heureka* Bryk, 1942.

Distribution. China (Jilin, Liaoning, Shanxi); Russia (Siberia), Korea, Japan.

***Herminia innocens* Butler, 1878 暗缘长须夜蛾**

Herminia innocens Butler, 1878, *Illustration of Typical Specimens of Lepidoptera Heterocera in the Collection of the British Museum*, 3: 63, pl. 56: 10. Type locality: Japan, Yokohama.

Distribution. China (Jilin, Liaoning, Jiangsu); Korea, Japan.

寄主：蒲公英（*Taraxacum mongolicum*）（Park *et al.*, 2006）。

Herminia stramentacealis Bremer, 1864 窄肾长须夜蛾

Herminia stramentacealis Bremer, 1864, *Memoires de l'Académie Imperiale des Sciences de St.-Petersbourg*, (7)8, 1: 64, pl. 5: 22. Type locality: Russia, "Kengka-See" [Primorye terr., Khanka lake].

Synonymy: *Herminia violacealis*: Remm et Martin, 1979, nec Staudinger, 1892; *Herminia satakei* Owada, 1982.

Distribution. China (Liaoning, Shandong, Jiangxi); Russia (Siberia), Korea, Japan.

寄主：大豆属（*Glycine*）植物（Matov & Kononenko, 2012）。

Herminia dolosa Butler, 1879 滑长须夜蛾

Herminia dolosa Butler, 1879, *Annals and Magazine of Natural History*, (5)4: 447. Type locality: Japan, Yokohama.

Synonymy: *Herminia tomarinia* Bryk, 1942.

Distribution. China (Liaoning, Shandong, Fujian); Russia (Siberia), Korea, Japan.

Genus *Sinarella* Bryk, 1949 辛夜蛾属

Sinarella Bryk, 1949, *Arkiv för Zoologi*, 41 (A) 1: 144, pl. 6: 2. Type species: *Sinarella stigmatophora* Bryk, 1948 [North Korea].

The genus includes 10 species, distributed in the Chinese subregion of the Palaearctic and Oriental regions; 5 species are in the Three Provinces of Northeast China.

Sinarella aegrota (Butler, 1879) 弱辛夜蛾

Bleptina aegrota Butler, 1879, *Illustrations of Typical Specimens of Lepidoptera Heterocera in the Collection of the British Museum*, 3: 65, pl. 57: 1.

Synonymy: *Bleptina incultalis* Leech, 1889; *Sinarella stigmatophora* Bryk, 1948.

Distribution. China (Jilin, Liaoning, Jiangsu); Russia (Siberia), Korea, Japan.

寄主：苔藓植物（Bryophyta）（Matov & Kononenko, 2012）。

Sinarella cristulalis (Staudinger, 1892) 淡色辛夜蛾

Zanclognatha cristulalis Staudinger, 1892, *in* N.M. Romanoff (ed.), *Mémoires sur les Lépidoptères*, 6: 610, pl. 14: 7. Type locality: Russia, Primorye terr., Vladivostok, "Sutschan" [Partizansk].

Distribution. China (Jilin); Russia (Siberia), Korea.

Sinarella nigrisigna (Leech, 1900) 黑斑辛夜蛾

Nodaria nigrisigna Leech, 1900, *Transactions of the Entomological Society of London*, 1900: 640. Type locality: China.

Synonymy: *Nodaria microlepidopteronis* Strand, 1920; *Araeognata sichotensis* Kurentzov, 1950.

Distribution. China (Jilin, Taiwan); Russia (Siberia), Korea, Japan.

Sinarella punctalis **(Herz, 1904) 基暗辛夜蛾**

Zanclognatha punctalis Herz, 1904, *Annuaire du Musée Zoologique de l'Académie Imperiale des Sciences de St.-Petersbourg*, 9: 324, pl. 1: 9. Type locality: North Korea, 38º5′ N, 128º E [Pun-tung village].

Synonymy: *Zanclognatha nigrobasalis* Yamamoto et Sugi, 1955.

Distribution. China (Jilin); Russia (Siberia), Korea, Japan.

寄主：苔藓植物（Bryophyta）（Matov & Kononenko, 2012）。

Sinarella japonica **(Butler, 1881) 重斑辛夜蛾**

Olybama japonica Butler, 1881, *Transactions of the Entomological Society of London*, 1881: 583. Type locality: Japan, Tokyo.

Distribution. China (Jilin); Russia (Siberia), Korea, Japan.

寄主：苔藓植物（Bryophyta）（Matov & Kononenko, 2012）。

Subfamily **Pangraptinae** Grote, 1882 眉夜蛾亚科

Pangraptinae Grote, 1882, *New Checklist of North American Moths*: 42. Type genus: *Pangrapta* Hübner, 1818.

Earlier the genus *Pangrapta* and related genera were placed in Catocalinae (sensu auctorum) (Sugi, 1982, 1992; Kononenko *et al.*, 1998, Kononenko & Pinratana, 2005; Holloway, 2005). The Pangraptinae was considered as a tribe of the Eublemminae by Fibiger and Lafontaine (2005) or as tribe of Catocalinae by Holloway (2005). The rank of the taxon is raised to subfamily level by Lafontaine and Schmidt (2010) and Holloway (2011), later it was supported by molecular study (Zahiri *et al.*, 2013). Except the genus *Pangrapta*, Holloway (2011) placed to Pangraptinae several tropical genera formerly considered in Catocalinae (sensu auctorum). The core of the subfamily is the genus *Pangrapta* with 84 described species. The exact number of genera and species of Pangraptinae is not clear. 1 genus and 9 species are represented in the Three Provinces of Northeast China.

References. Chen, 1982, 1999 (Noctuidae); Sugi, 1982 (Noctuidae: Ophiderinae, part); Kononenko *et al.*, 1998 (Noctuidae: Ophiderinae, part); Sviridov, 2003 (Noctuidae: Catocalinae); Fibiger, Lafontaine, 2005 (Erebidae: Eublemminae, Pangraptini); Kononenko, 2005 (Erebidae: Catocalinae), 2010 (Noctuidae: Pangraptinae); Holloway, 2005 (Noctuidae: Catocalinae, Pangraptini), 2011 (Erebidae: Pangraptinae); Kononenko, Pinratana, 2005 (Noctuidae: Catocalinae, part), 2013 (Erebidae: Pangraptinae); Lafontaine, Fibiger, 2006 (Noctuidae: Pangraptini); Kononenko, Han, 2007 (Noctuidae: Eublemminnae: Pangraptini); Matov *et al.*,

2008 (Noctuidae: Eublemminae, part); Zahiri *et al.*, 2011, 2012 (Erebidae).

注：首次给出中文亚科名。

Genus *Pangrapta* Hübner, 1818 眉夜蛾属

Pangrapta Hübner, 1818, *Zuträge Sammlung Exotischer Schmetterlinge*, 1: 18. Type species: *Pangrapta decoralis* Hübner, 1818 [USA].

Synonymy: *Marmorinia* Guenée, 1852; *Saraca* Walker, [1866] 1865; *Stenozethes* Hampson, 1926; *Zethes* auct..

The genus includes about 84 species, distributed mainly in the tropical and subtropical areas of the Old and New Worlds, most diverse in the Oriental region; 9 species are in the Three Provinces of Northeast China.

Pangrapta perturbans (Walker, 1858) 浓眉夜蛾

Zethes perturbans Walker, 1858, *List of the Specimens of Lepidopterous Insects in the Collection of the British Museum.*, 15: 1525. Type locality: India(North), Assam.

Synonymy: *Egnasia trimantesalis* Walker, [1859] 1858; *Pangrapta duplex* Sugi & Kononenko, 1996; *Pangrapta costaemacula* auct., nec Staudinger, 1888.

Distribution. China (Jilin, Jiangsu, Zhejiang, Fujian, Yunnan); Korea, India.

Pangrapta costaemacula Staudinger, 1888 绿斑眉夜蛾

Pangrapta costaemacula Staudinger, 1888, *Entomologische Zeitung. Entomologischen Vereine zu Stettin*, 49: 282. Type locality: Russia, Primorye terr., Vladivostok, Bikin.

Synonymy: *Pangrapta trimantesalis* auct..

Distribution. China (Heilongjiang, Jilin); Russia (Siberia), Korea.

寄主：东亚女贞（*Ligustrum ibota*）（Matov & Kononenko, 2012）。

注：首次给出中文种名。

Pangrapta suaveola Staudinger, 1888 隐眉夜蛾

Pangrapta suaveola Staudinger, 1888, *Entomologische Zeitung. Entomologischen Vereine zu Stettin*, 49: 281. Type locality: Russia, Primorye terr., Vladivostok, Baranovsky, Sutschan [Partizansk], Bikin.

Distribution. China (Heilongjiang); Russia (Siberia), Korea, Japan.

Pangrapta vasava (Butler, 1881) 点眉夜蛾

Egnasia vasava Butler, 1881, *Transactions of the Entomological Society of London*, 1881: 582. Type locality: Japan, Yokohana.

Synonymy: *Pangrapta incisa* Staudinger, 1888.

Distribution. China (Heilongjiang, Shandong, Henan, Jiangsu, Hubei); Russia (Siberia), Korea, Japan.

寄主：黑榆（*Ulmus davidiana*）、裂叶榆（*Ulmus laciniata*）、榔榆（*Ulmus parvifolia*）、榆（*Ulmus pumila*）（Matov & Kononenko, 2012）。

Pangrapta flavomacula Staudinger, 1888 黄斑眉夜蛾

Pangrapta flavomacula Staudinger, 1888, *Entomologische Zeitung. Entomologischen Vereine zu Stettin*, 49: 279. Type locality: Russia, Primorye terr., Vladivostok, Askold I., Baranovsky, "Sutschan" [Partizansk], "Suifun" [Razdolnaya river].

Synonymy: *Pangrapta flavomacula robiginosa* Kardakoff, 1928 (f.); *Pangrapta flavomacula albata* Kardakoff, 1928 (f.); *Pangrapta flavomacula turbata* Kardakoff, 1928 (f.).

Distribution. China (Heilongjiang, Jiangsu, Fujian, East China); Russia (Siberia), Korea.

Pangrapta lunulata (Sterz, 1915) 白痣眉夜蛾

Pangrapta lunulata Sterz, 1915, *Deutsche Entomologische Zeitschrift Iris*, 29: 131, pl. 3: 6. Type locality: "Amur".

Synonymy: *Pangrapta pseudalbistigma* Yoshimoto, 1993; *Pangrapta albistigma* auct., nec Hampson, 1898.

Distribution. China (Jilin, Hebei, Shaanxi, Zhejiang, Hubei, Sichuan, Taiwan); Russia (Siberia), Korea, Japan.

寄主：梣属（*Fraxinus*）植物（Matov & Kononenko, 2012）。

注：《中国动物志》（夜蛾科）中引用本种同物异名"*Pangrapta albistigma* (Hampson, 1898)"，在此予以变更。

Pangrapta marmorata Staudinger, 1888 纹理眉夜蛾

Pangrapta marmorata Staudinger, 1888, *Entomologische Zeitung. Entomologischen Vereine zu Stettin*, 49: 277. Type locality: Russia, "Raddefka" [Khabarovsk ter., Judish autonomy, Radde]; Vladivostok, Askold I., "Sutschan" [Partizansk], Bikin.

Distribution. China (Jilin); Russia (Siberia), Korea.

注：首次给出中文种名。

Pangrapta obscurata (Butler, 1879) 苹眉夜蛾

Marmorinia obscurata Butler, 1879, *Illustrations of Typical Specimens of Lepidoptera Heterocera in the Collection of the British Museum*, 3: 68, pl. 57: 11. Type locality: Japan, Hakodate.

Distribution. China (Heilongjiang, Jilin, Liaoning, Hebei, Shandong, Hunan, Taiwan); Russia (Siberia), Korea, Japan.

寄主：山楂（*Crataegus pinnatifida*）（Park *et al.*, 2006）；花楸（*Sorbus commixta*）、忍冬（*Lonicera japonica*）、苹果（*Malus pumila*）、三叶海棠（*Malus sieboldii*）、梨属（*Pyrus*）、李属（*Prunus*）植物（Matov & Kononenko, 2012）。

Pangrapta griseola **Staudinger, 1892 暗灰眉夜蛾**

Pangrapta griseola Staudinger, 1892, *in* N.M. Romanoff (ed.), *Mémoires sur les Lépidoptères*, 6: 620, pl. 14: 11. Type locality: Russia, "Ussuri" [Primorye terr.].

Distribution. China (Jilin); Russia (Siberia), Korea.

注：首次给出中文种名。

Subfamily **Hypeninae** Herrich-Schäffer, [1851] 1845 髯须夜蛾亚科

Hypenidae Herrich-Schäffer, [1851] 1845, *Systematische Bearbeitung der Schmetterlinge von Europa*, 2: 424. Type genus: *Hypena* Schrank, 1802.

The Hypeninae is a relatively small subfamily distributed in all parts of the world, except the Arctic and Antarctic regions. Some authors included the genus *Rivula* to Hypeninae (Kitching & Rawlins, 1999 [1998]), but it is currently treated in distinct subfamily Rivulinae. Some Asian genera of Hypeninae were revised by Lödl (1994), Lödl & Mayerl (1998) and Mayerl & Lödl (1999). However, most genera and species of East Asian fauna are still unrevised and the types of many species described from the Oriental region are not illustrated what make difficult their identification. The subfamily subdivided for *Hypena* and *Mecistoptera* generic groups. The composition of the later is not completely resolved.

Worldwide, the subfamily includes about 500 species, distributed mainly in the Oriental and Ethiopian regions; about 150 species are known from the Palaearctic, mainly from Sino-Himalayan subregion. Nine genera and 26 species known from the Three Provinces of Northeast China.

References. Chen, 1982, 1999 (Noctuidae); Sugi, 1982 (Noctuidae: Hypeninae); Lödl, 1994 (Noctuidae: Hypeninae); Kononenko *et al.*, 1998 (Noctuidae: Hypeninae); Sviridov, 2003 (Noctuidae: Hypeninae); Fibiger, Lafontaine, 2005 (Erebidae: Hypeninae); Fibiger, Hacker, 2005 (Erebidae: Hypeninae); Lafontaine, Fibiger, 2006 (Noctuidae: Hypeninae); Kononenko, Pinratana, 2005 (Noctuidae: Hypeninae), 2013 (Erebidae: Hypeninae); Kononenko, Han, 2007 (Noctuidae: Hypeninae); Matov *et al.*, 2008 (Noctuidae: Hypeninae); Holloway, 2008 (Noctuidae: Hypeninae), 2011 (Erebidae: Hypeninae); Kononenko, Behounek, 2009 (Erebidae: Hypeninae); Fibiger *et al.*, 2010 (Noctuidae: Hypeninae), 2011 (Erebidae: Hypeninae); Kononenko, 2010 (Erebidae: Hypeninae); Zahiri *et al.*, 2011, 2012 (Erebidae, Hypeninae).

Genus ***Hypena*** Schrank, 1802 髯须夜蛾属

Hypena Schrank, 1802, *Fauna Boica. Durchgedachte Geschichte der in Baiera Einheimischen und Zahmen*

Thiere, 2 (2): 163. Type species: *Phalaena proboscidalis* Linnaeus, 1758 [Europe].

Synonymy: *Bomolocha* Hübner, [1825] 1816 (subgen.); *Ophiuche* Hübner, [1825] 1816; *Erichila* Billberg, 1820; *Herpyzon* Hübner, 1822; *Dichromia* Grote, 1854; *Badausa* Walker, [1863] 1864; *Peliala* Walker, 1865; *Euhypena* Grote, 1873; *Macrhypena* Grote, 1873; *Meghypena* Grote, 1873; *Plathypena* Grote, 1873; *Apanda* Moore, 1882; *Mathura* Moore, 1882; *Nesamiptis* Meyrick, 1899; *Anepischetos* Smith, 1900; *Placerobela* Turner, 1903; *Rosthypena* Beck, 1996; *Obeshypena* Beck, 1996; *Erchila*: Poole, 1989, missp..

The genus is subdivided into the following subgenera: *Bomolocha* Hübner, [1825] 1816; *Hypena* Schrank, 1802; *Ophiuche* Hübner, 1816; *Tetetrastycpena* Lödl, 1994; *Plathypena* Grote, 1873; *Jussalypena* Lödl, 1994; *Trichypena* Joannis, 1915; *Extremypena* Lödl, 1994; *Biangulypena* Lödl, 1994; *Pseudodichromia* Lödl, 1994. Only two of them, *Bomolocha* and *Hypena*, are known from the temperate zone. The genus includes over 470 species distributed in the Old and New Worlds, mainly in the tropics and subtropics. Two subgenera and 14 species are in the Three Provinces of Northeast China.

Subgenus ***Bomolocha*** Hübner, [1825] 1816 **卜夜蛾亚属**

Bomolocha Hübner, [1825] 1816, *Verzeichniss bekannter Schmettlinge* [sic!]: 343. Type species: *Phalaena crassalis* Fabricius, 1787 [Europe].

Synonymy: *Badausa* Walker, [1863] 1864; *Euhypena* Grote, 1873; *Macrhypena* Grote, 1873; *Meghypena* Grote, 1873.

***Hypena* (*Bomolocha*) *stygiana* Butler, 1878 阴髯须夜蛾**

Hypena stygiana Butler, 1878, *Illustrations of Typical Specimens of Lepidoptera Heterocera in the Collection of the British Museum*, 3: 55, pl. 56: 2. Type locality: Japan, Yokohama.

Distribution. China (Jilin, Liaoning, Zhejiang, Jiangxi, Xizang); Russia (Siberia), Korea, Japan.

寄主：齿叶溲疏（*Deutzia crenata*）、茜草（*Rubia cordifolia*）（Matov & Kononenko, 2012）。

***Hypena* (*Bomolocha*) *zilla* Butler, 1879 齐髯须夜蛾**

Hypena zilla Butler, 1879, *Illustrations of Typical Specimens of Lepidoptera Heterocera in the Collection of the British Museum*, 3: 60, pl. 56: 1. Type locality: Japan, Hakodate.

Synonymy: *Bomolocha nikkensis* Wileman & West, 1930.

Distribution. China (Heilongjiang, Hubei); Russia (Siberia), Korea, Japan.

寄主：黑樱桃（*Prunus maximowiczii*）（Matov & Kononenko, 2012）。

***Hypena* (*Bomolocha*) *squalida* Butler, 1878 污髯须夜蛾**

Hypena squalida Butler, 1878, *Illustrations of Typical Specimens of Lepidoptera Heterocera in the*

Collection of the British Museum, 3: 60, pl. 56: 2. Type locality: Japan, Yokohama.

Distribution. China (Heilongjiang, Jilin Liaoning, Hunan, Jiangxi, Fujian, Sichuan); Russia (Siberia), Korea, Japan.

寄主：三叶海棠（*Malus sieboldii*）、梅（*Prunus mume*）、杜鹃花属（*Rhododendron*）植物（Matov & Kononenko, 2012）。

Hypena (_Bomolocha_) _bicoloralis_ Graeser, [1889] 1888 双色髯须夜蛾

Hypena bicoloralis Graeser, [1889] 1888, *Berliner entomologische Zeitschrift*, 32, II: 381. Type locality: Russia, Khabarovsk.

Synonymy: *Bomolocha rhombalis* auct.; *Bomolocha fontis* auct..

Distribution. China (Heilongjiang, Jilin, Liaoning, Hubei); Russia (Siberia), Korea, Japan.

寄主：毛叶水苎麻（*Boehmeria holosericea*）（Park *et al.*, 2006）；榔榆（*Ulmus parvifolia*）、台湾榉（*Zelkova serrata*）（Matov & Kononenko, 2012）。

Hypena (_Bomolocha_) _nigrobasalis_ (Herz, 1904) 暗基髯须夜蛾

Bomolocha nigrobasalis Herz, 1904, *Annuaire du Musée Zoologique de l'Académie Imperiale des Sciences de St.-Petersbourg*, 9: 329, pl. 1: 13. Type locality: Korea, 38°5′N, 128° E [Pun-tung village].

Distribution. China (Jilin); Russia (Siberia), Korea, Japan.

寄主：杜鹃花属（*Rhododendron*）植物（Matov & Kononenko, 2012）。

注：首次给出中文种名。

Hypena (_Bomolocha_) _bipartita_ Staudinger, 1892 分色髯须夜蛾

Hypena bipartita Staudinger, 1892, *in* N.M. Romanoff (ed.), *Mémoires sur les Lépidoptères*, 6: 625, pl. 14: 12. Type locality: Russia, Khabarovsk terr., Judish autonomy, Radde.

Distribution. China (Heilongjiang, Hubei); Russia (Siberia), Japan.

Subgenus ***Hypena*** Schrank, 1802 髯须夜蛾亚属

Hypena Schrank, 1802, type species: *Phalaena proboscidalis* Linnaeus, 1758: 533 [Europe], by subsequent designation by Curtis, 1829: 288.

Synonymy: *Erichila* Billberg, 1820; *Herpyzon* Hübner, 1822; *Ophiuche* Hübner, [1825]; *Dichromia* Guenée, 1852; *Peliala* Walker, 1865; *Mathura* Moore, 1882; *Apanda* Moore, 1882; *Nesamiptis* Meyrick, 1899; *Anepischetos* Smith, 1900; *Placerobela* Turner, 1903; *Erchila*: Poole, 1989.

Hypena (Hypena) proboscidalis (Linnaeus, 1758) 髯须夜蛾

Phalaena Pyralis proboscidalis Linnaeus, 1758, *Systema Naturae* (Edn. 10) 1: 533. Type locality: Europe.

Synonymy: *Phalaena ensalis* Fabricius, 1794; *Crambus proboscidatus* Haworth, 1809; *Hypena proboscidalis deleta* Staudinger, 1892 (var.); *Hypena proboscidalis flexilinea* Warren, 1913 (ssp.).

Distribution. China (Heilongjiang, Xinjiang, Hubei, Xizang, Taiwan); Russia (European part, Siberia), Korea, Japan, India(North), Central Asia, Near East, Europe.

寄主：蛇麻（*Humulus lupulus*）、大蝎子草（*Girardinia diversifolia*）、异株荨麻（*Urtica dioica*）、欧荨麻（*Urtica urens*）、羊角芹（*Aegopodium podagraria*）、野芝麻属（*Lamium*）、林生水苏（*Stachys silvatica*）、车前草属（*Plantago*）植物（Matov & Kononenko, 2012）。

Hypena (Hypena) tristalis Lederer, 1853 豆髯须夜蛾

Hypena tristalis Lederer, 1853, *Verhandlungen des Zoologisch-Botanischen vereins in Wien*, 3: 376, pl. 5: 4. Type locality: East Kazakhstan, West Altai, Ust-Kamengorsk, Ust-Buchtarminsk.

Synonymy: *Hypena tripunctalis* Bremer, 1864.

Distribution. China (Heilongjiang, Jilin, Liaoning, Nei Mongol, Xinjiang, Hebei, Shanxi, Hubei, Fujian, Yunnan, Xizang); Russia (Siberia), Korea, Japan.

寄主：大叶苎麻（*Boehmeria longispica*）、大荨麻（*Urtica dioica*）、春榆（*Ulmus japonica*）、榔榆（*Ulmus parvifolia*）、（*Desmodium oxyphyllum*）、大豆（*Glycine max*）、野葛（*Pueraria thunbergiana*）（Matov & Kononenko, 2012）。

Hypena (Hypena) kengkalis Bremer, 1864 肯髯须夜蛾

Hypena kengkalis Bremer, 1864, *Memoires de l'Académie Imperiale des Sciences de St.-Petersbourg* (7)8, 1: 63, pl. 5: 21. Type locality: Russia, "Kengka Zee" [Primorye terr., Khanka lake].

Synonymy: *Hypena kengkalis warreni* Bryk, 1948 (ssp.).

Distribution. China (Jilin, liaoning, Hebei, Hubei); Russia (Siberia), Korea, Japan.

寄主：胡枝子属（*Lespedeza*）植物（Matov & Kononenko, 2012）。

Hypena (Hypena) conspersalis Staudinger, 1888 晕髯须夜蛾

Hypena conspersalis Staudinger, 1888, *Entomologische Zeitung. Entomologischen Vereine zu Stettin*, 49: 282. Type locality: Russia, Khabarovsk terr., Judish autonomy, Radde, "Ussuri" [Primorye terr.], Vladivostok, "Suifun" [Razdolnaya river], Bikin.

Synonymy: *Hypena passerinalis* Graeser, 1888; *Hypena chosenula* Bryk, 1948; *Hypena conspersalis koreana* Bryk, 1949 (ssp.).

Distribution. China (Jilin); Russia (Siberia), Korea.

Hypena (Hypena) sagitta (Fabricius, 1775) 马蹄髯须夜蛾

Noctua sagitta Fabricius, 1775, *Entomologiae Systematica emendata et Aucta. Secundum Classes, Ordines, Genera, Species Adjectis Synonimis, Locis, Observationibus, Descriptionibus*: 597. Type locality: India.

Synonymy: *Phalaena orosia* Cramer, 1780.

Distribution. China (Liaoning, Hunan, Fujian, Guangdong, Hainan, Guangxi, Guizhou, Yunnan, Xizang, Taiwan); Korea, Japan, India, Pakistan, Vietnam.

Hypena (Hypena) trigonalis (Guenée,1854) 两色髯须夜蛾

Dichromia trigonalis Guenée, 1854, *in* Boisduval & Guenée, *Histoire Naturelle des Insectes. Species Général des Lépidoptères*, 8: 19. Type locality: Central India.

Synonymy: *Dichromia sextalis* Walker, [1859] 1858.

Distribution. China (Jilin, Liaoning, Shandong, Henan, Zhejiang, Jiangxi, Fujian, Sichuan, Guizhou, Yunnan, Xizang, Taiwan); Korea, Japan, India, Pakistan, Indonesia.

寄主：苎麻（陈一心，1999）。

Hypena (Hypena) claripennis (Butler, 1878) 笋髯须夜蛾

Dichromia claripennis Butler, 1878, *Illustrations of Typical Specimens of Lepidoptera Heterocera in the Collection of the British Museum*, 2: 54, pl. 38: 2. Type locality: Japan, Yokohama.

Distribution. China (Liaoning, Jiangxi); Russia (Siberia), Korea, Japan.

寄主：竹笋（陈一心，1999）；大叶苎麻（*Boehmeria longispica*）、小赤麻（*Boehmeria spicata*）、荨麻属（*Urtica*）植物（Matov & Kononenko, 2012）。

Hypena (Hypena) amica (Butler, 1878) 斜线髯须夜蛾

Dichromia amica Butler, 1878, *Illustrations of Typical Specimens of Lepidoptera Heterocera in the Collection of the British Museum*, 2: 55, pl. 33: 3. Type locality: Japan, Hakodate, Yokohama.

Distribution. China (Liaoning, Zhejiang, Hubei, Taiwan); Russia (Siberia), Korea, Japan, India.

寄主：悬铃叶苎麻（*Boehmeria tricuspis*）、（*Boehmeria platanifolia*）（Park *et al.*, 2006）；密花苎麻（*Boehmeria densiflora*）、大叶苎麻（*Boehmeria longispica*）、伏毛苎麻（*Boehmeria niponoñivea*）、苎麻（*Boehmeria nivea*）、小赤麻（*Boehmeria spicata*）（Matov & Kononenko, 2012）。

Genus *Harita* Moore, 1882 哈利夜蛾属

Harita Moore, 1882, *in* Hewitson & Moore, *Taylor and Francis*, 187. Type species: *Harita rectilinear* Moore, 1882 [India].

Synonymy: *Placerobela* Turner, 1903.

The genus includes 5 species distributed in the Old World and Australia regions; 1 species is in the Three Provinces of Northeast China.

注：首次给出中文属名。

Harita belinda (Butler, 1879) 贝哈利夜蛾

Hypena belinda Butler, 1879, *Illustration of Typical Specimens of Lepidoptera Heterocera in the Collection of the British Museum*, 3: 61, pl. 56: 3. Type locality: Japan, Yokohama.

Distribution. China (Liaoning, Xizang); Korea, Japan.

注：首次给出中文种名。

Genus *Zekelita* Walker, 1863 齿口夜蛾属

Zekelita Walker, 1863, *List of the Specimens of Lepidopterous Insects in the Collection of the British Museum*, 27: 188. Type species: *Zekelita equalisella* Walker, 1863 [South Africa].

Synonymy: *Rhynchodontodes* Warren, 1913.

The genus includes 31 species of 12 subgenus, distributed in the Palaearctic region, mainly in the tropical and subtropical areas; 1 species is in the Three Provinces of Northeast China.

Zekelita plusioides (Butler, 1879) 丑齿口夜蛾

Rhynchina plusioides Butler, 1879, *Illustration of Typical Specimens of Lepidoptera Heterocera in the Collection of the British Museum*, 61, pl. 56: 5. Type locality: Japan, Yokohama.

Distribution. China (Liaoning, Shanxi, Jiangsu, Zhejiang, Jiangxi, Hainan, Sichuan); Korea, Japan, India, Pakistan, Afghanistan.

Genus *Rhynchina* Guenée, 1854 口夜蛾属

Rhynchina Guenée, 1854, *in* Boisduval and Guenée, *Histoire Naturelle des Insectes. Species Général des Lépidoptéres. Tome Huitiéme. Deltoides et Pyralites*, 1854:20. Type species: *Rhynchina pionealis* Guenée, 1854 [India].

Synonymy: *Ceraptila* Guenée, 1854; *Zekelita* Walker, 1863.

The genus includes 55 species, distributed in the Palaearctic region, mainly in the East of the Mt. Himalaya to Indonesia region, 1 species is in the Three Provinces of Northeast China.

Rhynchina cramboides (Butler, 1879) 洁口夜蛾

Hormisa cramboides Butler, 1879, *Illustration of Typical Specimens of Lepidoptera Heterocera in the

Collection of the British Museum, 3: 61, pl. 56: 6. Type locality: Japan, Yokohama.

Synonymy: *Hormisa morosa* Butler, 1879.

Distribution. China (Heilongjiang, Jilin, Liaoning, Shandong, Hunan, Sichuan, Xizang); Korea, Japan, India.

Genus ***Gonepatica*** Sugi 1982 仿角衣夜蛾属

Gonepatica Sugi 1982, *Moths of Japan*, 1: 881. Type species: *Egnasia opalina* Butler, 1879 [Japan, Honshu].

Synonymy: *Ectogonia* auct..

Monotypic genus, distributed in the Sino-Pacific subregion of the Palaearctic region; 1 species is in the Three Provinces of Northeast China.

Gonepatica opalina **(Butler, 1879) 仿角衣夜蛾**

Egnasia opalina Butler, 1879, *Annals and Magazine of Natural History*, (5)4: 451. Type locality: Japan, Yokohama.

Distribution. China (Heilongjiang, Jiangsu, Zhejiang); Russia (Siberia), Korea, Japan.

寄主：麻栎（*Quercus acutissima*）、枹栎（*Quercus serrata*）（Matov & Kononenko, 2012）。

注：《中国动物志》（夜蛾科）将本种归入"*Ectogonia* Hampson, 1896 角衣夜蛾属"，在此予以变更。

Genus ***Hepatica*** Staudinger, 1892 粗胫夜蛾属

Hepatica Staudinger, 1892, *in* N.M. Romanoff (ed.), *Mémoires sur les Lépidoptères*, 6: 634. Type species: *Hepatica anceps* Staudinger, 1892. Type locality: Russia, "Sutschan" [Primorye terr., Partizansk].

The genus includes 13 species, distributed predominantly in the Oriental and the Sino-Pacific subregion of the Palaearctic regions, 1 species is in the Three Provinces of Northeast China.

Hepatica anceps **Staudinger, 1892 双粗胫夜蛾**

Hepatica anceps Staudinger, 1892, *in* N.M. Romanoff (ed.), *Mémoires sur les Lépidoptères*, 6: 634, pl. 12: 11. Type locality: Russia, "Sutschan-Gebiet" [Primorye terr., Partizansk].

Distribution. China (Jilin); Russia (Siberia), Korea, Japan.

Genus ***Lophomilia*** Warren, 1913 微夜蛾属

Lophomilia Warren, 1913, *in* A.Seitz (ed.), *Die Gross-Schmetterlinge der Erde*, 3: 411. Type species: *Egnasia polybapta* Butler, 1879 [Japan].

Synonymy: *Atuntsea* Berio, 1977; *Bryograpta* Sugi, 1977.

The genus includes 13 species, distributed predominantly in the Chinese and Sino-Himalayan subregions of the Palaearctic and Oriental regions; 3 species are in the Three Provinces of Northeast China.

Lophomilia polybapta (Butler, 1879) 小冠微夜蛾

Egnasia polybapta Butler, 1879, *Illustrations of Typical Specimens of Lepidoptera Heterocera in the Collection of the British Museum*, 3: 66, pl. 57: 7. Type locality: Japan, Yokohama.

Distribution. China (Liaoning, Jiangsu, Yunnan); Russia (Siberia), Korea, Japan.

寄主：麻栎（*Quercus acutissima*）、枹栎（*Quercus serrata*）（Matov & Kononenko, 2012）。

Lophomilia flaviplaga (Warren, 1912) 座黄微夜蛾

Micardia flaviplaga Warren, 1912, *Novitates zoologicae*, 19: 38. Type locality: Japan.

Distribution. China (Jilin, Liaoning, Jiangxi); Russia (Siberia), Korea, Japan.

寄主：枹栎（*Quercus serrata*）（Park *et al.*, 2006）；大桴栎（*Quercus grosseserrata*）、蒙古栎（*Quercus mongolica*）（Matov & Kononenko, 2012）。

Lophomilia nekrasovi Kononenko & Behounek, 2009 淡清微夜蛾

Lophomilia nekrasovi Kononenko & Behounek, 2009, *Zootaxa*, 1989: 10, figs. 9, 10, 29, 44. Type locality: Russia, South Primorye terr., vic. Ryazanovka.

Distribution. China (Liaoning, Shaanxi); Russia (Siberia), Korea.

Genus *Stenbergmania* Bryk, 1949 白戚夜蛾属

Stenbergmania Bryk, 1949, *Arkiv för Zoologi*, 41 (A) 1: 142. Type species: *Herminia albomaculalis* Bremer, 1864 [Russia, Primorye terr.].

Sino-Pacific, South Siberian monotypic genus.

注：首次给出中文属名。

Stenbergmania albomaculalis (Bremer, 1864) 白戚夜蛾

Herminia albomaculalis Bremer, 1864, *Memoires de l'Académie Imperiale des Sciences de St.-Petersbourg*, (7)8, 1: 65, pl. 5: 24. Type locality: Russia, "Ussuri"[Primorye terr.].

Distribution. China (Heilongjiang, Jilin); Russia (Transbaikalia, East), Korea, Japan.

寄主：麻栎（*Quercus acutissima*）（Matov & Kononenko, 2012）。

注：首次给出中文种名。

Genus *Paragabara* Hampson, 1926 戚夜蛾属

Paragabara Hampson, 1926, *Descriptions of New Genera and Species of Lepidoptera Phalaenae of the Subfamily Noctuinae* (*Noctuidae*) *in the British Museum* (*Natural History*): 260. Type species: *Madopa flavomacula* Oberthür, 1880 [Russia, Primorye terr., Askold I.].

The genus includes 6 species, distributed in the Sino-Pacific subregion of the Palaeartctic and partly in the Oriental regions; 3 species are in the Three Provinces of Northeast China.

Paragabara flavomacula (Oberthür, 1880) 戚夜蛾

Madopa flavomacula Oberthür, 1880, *Études d'Entomologie*, 5: 87, pl. 4: 5. Type locality: Russia, Primorye terr., Askold I..

Distribution. China (Heilongjiang, Jilin, Liaoning, Hebei, Jiangsu); Russia (Siberia), Korea, Japan.

寄主：两型豆（*Amphicarpaea edgeworthii*）、野葛（*Pueraria thunbergiana*）（Matov & Kononenko, 2012）。

Paragabara curvicornuta Kononenko & Matov, 2010 弧戚夜蛾

Paragabara curvicornuta Kononenko & Matov, 2010, *Zootaxa*, 2679: 59, Figs 14, 25, 26, 35. Type locality: Russia, Primorye terr.

Distribution. China (Jilin); Russia (Siberia), Korea, Japan.

寄主：豆科新鲜或枯叶（Kononenko & Matov, 2010）。

Paragabara ochreipennis Sugi, 1962 赭戚夜蛾

Paragabara ochreipennis Sugi, 1962, *Akitu*, 10: 39: 3, 4. Type locality: Japan, Kyushu, Fukuoka Pref., Yoshii.

Distribution. China (Heilongjiang, Jilin, Liaoning, Shanghai); Russia (Siberia), Korea, Japan.

注：首次给出中文种名；由 Kononenko, Han & Matov (2010)根据 20 世纪初采自 "Manchzhuria, Ertsendjantszy station" 和上海的标本材料首次记录本种在中国具有分布。

Subfamily **Rivulinae** Grote, 1895 涓夜蛾亚科

Rivulini Grote, 1895, *Proceedings of the American Philosophical Society* 34: 419. Type genus: *Rivula* Guenée, [1845].

Synonymy: Rivulinae Beck, 1960.

Rivulinae is a small, unrevised subfamily with some tropical species and genera of uncertain position. Lödl (1998) removed the genera *Hiaspis* and *Acidon* from Rivulinae to *Mecistoptera* generic group of Hypeninae.

Holloway (2005, 2008, 2011) redefined the genus *Oglasa* as Bagisarinae and placed the genus *Bocula* to Rivulinae (Holloway, 2011), what is supported by molecular study (Zahiri *et al.*, 2011).

The subfamily contains about 10-12 unrevised genera in the Old and New Worlds mainly in subtropical and tropical areas. 1 genus with 10 species is known from the Palaearctic region; 1 species is known from the the Three Provinces of Northeast China.

References. Chen, 1982, 1999 (Noctuidae); Sugi, 1982 (Noctuidae: Ophiderinae); Kitching 1984 (Noctuidae: Rivulinae); Speidel *et al.*, 1996 (Noctuidae: Rivulinae); Kononenko *et al.*, 1998 (Noctuidae: Rivulinae); Kitching, Rawlins 1999; Sviridov, 2003 (Noctuidae: Rivulinae); Fibiger, Lafontaine, 2005 (Erebidae: Rivulinae); Fibiger, Hacker, 2005 (Erebidae: Rivulinae); Kononenko, 2005 (Noctuidae: Rivulinae), 2010 (Erebidae: Rivulinae); Lafontaine, Fibiger, 2006 (Noctuidae: Rivulinae); Kononenko, Han, 2007 (Noctuidae: Rivulinae); Matov *et al.*, 2008 (Noctuidae: Rivulinae); Holloway, 2008 (Noctuidae: Rivulinae); Fibiger *et al.*, 2010 (Noctuidae: Rivulinae); Zahiri *et al.*, 2011, 2012 (Erebidae).

Genus ***Rivula*** Guenée, [1845] 1844 **涓夜蛾属**

Rivula Guenée, [1845] 1844, *in* Duponchel, *Catalogue Mèthodique des Lépidoptères d'Europe Distribuls en Families, Tribus et Genres, etc.*, (2): 206. Type species: *Phalaena sericialis* Scopoli, 1763 [Europe, Slovenia].

Synonymy: *Motina* Walker, 1863; *Ploteia* Walker, 1863, preocc. (Walker, [1858] 1857, Lepidoptera); *Cholimma* Walker, 1864; *Rhazunda* Walker, 1866; *Pasira* Moore, 1882 preocc. (Stall, 1859, Hemiptera); *Rivulana* Bethune-Baker, 1911; *Alikangiana* Strand, 1920; *Paurosceles* Turner, 1945; *Thopelia* Nye, 1975, repl. name.

About 50 species of the genus are known over the World, distributed mainly in the tropical and subtropical areas; 1 species is in the Three Provinces of Northeast China.

Rivula sericealis **(Scopoli, 1763) 涓夜蛾**

Phalaena sericealis Scopoli, 1763, *Entomologia Carniolica Exhibens Insecta Carnioliae Indigena et Distributa in Ordines, Genera, Species, Varietates Methodo Linnaeana*: 242. Type locality: Europe, Slovenia.

Synonymy: *Phalaena munda* Hufnagel, 1766; *Pasira ochracea* Moore, 1882; *Rivula dubitatrix* Bryk, 1948.

Distribution. China (Heilongjiang, Jilin, Liaoning, Jiangsu, Yunnan); Russia (European part, Siberia), Korea, Japan, India,Vietnam; Central Asian, Europe, North Africa.

寄主：鹅观草（*Agropyron tsukushiense*）、四季青（*Agrostis stolonifera*）、羽状短柄草（*Brachypodium pinnatum*）、短柄草（*Brachypodium sylvaticum*）、鸭茅（*Dactylis glomerata*）、偃麦草（*Elytrigia repens*）、高羊茅（*Festuca arundinacea*）、牛尾草（*Festuca elatior*）、黑麦草（*Lolium perenne*）、苇状酸沼草（*Molinia caerulea*）、丝带草（*Phalaris arundinacea*）、荨麻属（*Urtica*）、黄芪属（*Astragalus*）、雀麦属（*Bromus*）、早熟禾属（*Poa*）、赤竹属（*Sasa*）、拂子茅属（*Calamagrostis*）、小冠花属（*Coronilla*）、蚕豆属（*Vicia*）、苔属（*Carex*）植物（Matov & Kononenko, 2012）。

Subfamily **Scoliopteryginae** Herrich-Schäffer, [1852] 棘翅夜蛾亚科

Scoliopteryges Herrich-Schäffer, [1852] 1843–56 (1850), *Systematische Bearbeitung der Schmetterlinge von Europa*, 6 (Systema Lepidopterorum Europae): 21. Type genus: *Scoliopteryx* Germar, 1810.

Synonymy: Gonopteridae Herrich-Schäffer, [1850] 1845; Gonopteridae Guenée, 1852; Anomiinae Grote, 1882; Anomiini Grote, 1890; Scoliopteryginae Spuler, 1907; Scoliopteryginae Börner, 1932; Scoliopterygini Forbes, 1954; Scoliopterygini Beck, 1960; Anomiini Wiltshire, 1990; Anomini Berio, 1992.

Earlier authors placed genera of Scoliopteryginae to the subfamily Catocalinae (sensu auctorum). Kitching & Rawlins (1999) and Goater *et al.* (2003) treated this group as tribe Gonopterini Herrich-Schäffer [1850] of the subfamily Calpinae. Lafontaine & Fibiger (2005) changed the name of Gonopterini to Scoliopterygini Herrich-Schäffer [1852] as the name of prevailing usage. Holloway (2005) treated the group as tribe of Catocalinae and consequently as a distinct subfamily (Holloway, 2011; Zahiri *et al.*, 2011). The genus *Anomis* was placed by Lafontaine & Fibiger (2006) in its own tribe, Anomini. However, Holloway (2005) joined *Scoliopteryx*, *Anomis* and allied genera in tribe Scoliopterygini and later (Holloway, 2011) in the subfamily Scoliopteryginae. Here we follow to the last treatment of the subfamily (Holloway, 2011). The subfamily numbers 5 genera and 7 species are in the Three Provinces of Northeast China.

References. Sugi, 1982 (Noctuidae: Catocalinae); Chen, 1982, 1999 (Noctuidae); Kononenko *et al.*, 1998 (Noctuidae: Catocalinae); Kitching, Ravlins, 1999 (Noctuidae: Calpinae, Gonopterini); Sviridov, 2003 (Noctuidae: Catocalinae); Goater *et al.*, 2003 (Catocalinae, Gonopterini); Fibiger, Lafontaine, 2005 (Erebidae: Calpinae, Scoliopterygini, Anomini); Fibiger, Hacker, 2005 (Erebidae: Calpinae, Gonopterini); Holloway, 2005 (Noctuidae: Catocalinae), 2011 (Erebidae: Scoliopteryginae); Kononenko, Pinratana, 2005 (Noctuidae: Catocalinae), 2013 (Erebidae: Scoliopteryginae); Lafontaine, Fibiger, 2006 (Noctuidae: Calpinae, Scoliopterygini, Anomini); Kononenko, Han, 2007 (Noctuidae: Calpinae); Matov *et al.*, 2008 (Noctuidae: Calpinae); Zahiri *et al.*, 2011, 2012 (Erebidae: Scoliopteryginae).

Genus ***Scoliopteryx*** Germar, 1810 棘翅夜蛾属

Scoliopteryx Germar, 1810, *Dissertatio sistems Bombycum Species*: 14. Type species: *Phalaena libatrix* Linnaeus, 1758 [Europe].

Synonymy: *Pterodonta* Reichenbach & Leipzig, 1817; *Euphemias* Hübner, [1821] 1816; *Euphais* Hübner, [1822]; *Gonoptera* Berthold, 1827.

The genus includes 2 species, one of which is widespread in the Holarctic, the other one is Central Asian; 1 species is in the Three Provinces of Northeast China.

Scoliopteryx libatrix **(Linnaeus, 1758)** 棘翅夜蛾

Phalaena libatrix Linnaeus, 1758, *Systema Naturae* (Edn. 10) 1: 507. Type locality: Europe.

Synonymy: *Noctua libatricus*: Haworth, 1809, emend..

Distribution. China (Heilongjiang, Jilin, Liaoning, Beijing, Shaanxi, Henan, Yunnan); Russia (European part, Siberia); North Africa, North America from Alaska to New Mexico.

寄主：北美白桦（*Betula papyrifera*）、柔毛桦（*Betula pubescens*）、银白杨（*Populus alba*）、大叶钻天杨（*Populus balsamifera*）、加拿大杨（*Populus canadensis*）、钻天杨（*Populus italica*）、辽杨（*Populus maximowiczi*）、欧洲黑杨（*Populus nigra*）、甜杨（*Populus suaveolens*）、欧洲山杨（*Populus tremula*）、美洲山杨（*Populus tremuloides*）、白柳（*Salix alba*）、黄华柳（*Salix atrocinerea*）、耳柳（*Salix aurita*）、*Salix australior*、黄花柳（*Salix caprea*）、灰柳（*Salix cinerea*）、粉枝柳（*Salix daphnoides*）、爆竹柳（*Salix fragilis*）、河柳（*Salix gilgiana*）、（*Salix lapponicum*）、尖果密穗柳（*Salix oxycarpa*）、五蕊柳（*Salix pentandra*）、（*Salix petiolaris*）、东陵山柳（*Salix phylicifolia*）、杞柳（*Salix purpurea*）、匍匐柳（*Salix repens*）、细叶沼柳（*Salix rosmarinifolia*）、（*Salix rubens*）、波纹柳（*Salix starkeana*）、细穗柳（*Salix tenuijulis*）、三蕊柳（*Salix triandra*）、蒿柳（*Salix viminalis*）、安多杜鹃（*Andromeda polifolia*）、黑果越橘（*Vaccinium myrtillus*）、越橘（*Vaccinium vitisidaea*）、水柏枝（*Myricaria germanica*）、欧洲木莓（*Rubus caesius*）、覆盆子（*Rubus idaeus*）、绣线菊（*Spiraea salicifolia*）、柳兰（*Chamaenerion angustifolium*）、（*Fraxinus potamophila*）及茶藨子属（*Ribes*）、栎属（*Quercus*）植物（Matov & Kononenko, 2012）。

Genus ***Cosmophila*** Boisduval, 1833 小桥夜蛾属

Cosmophila Boisduval, 1833, *Annales de la Société Entomologique de France*, 2: 242. Type species: *Cosmophila xanthindyma* Boisduval, 1833 [Mauritius] (=*Noctua flava* Fabricius, 1775).

The genus includes 15 species, distributed mainly in the tropical area of the Old World (Oriental, Ethiopian and Australian regions); among them, 1 species occurs in the Three Provinces of Northeast China.

注：首次给出中文属名。

Cosmophila flava **(Fabricius, 1775) 小桥夜蛾**

Noctua flava Fabricius, 1775, *Entomologiae Systematica emendata et Aucta. Secundum Classes, Ordines, Genera, Species Adjectis Synonimis, Locis, Observationibus, Descriptionibus*: 601. Type locality: "India Oriental".

Synonymy: *Noctua stigmatizans* Fabricius, 1775; *Xanthia fimbriago* Stephens, 1829; *Cosmophila xanthindyma* Boisduval, 1833; *Cosmophila auragoides* Guenée, 1852; *Cosmophila indica* Guenée, 1852; *Cirroedia variolosa* Walker, 1857; *Cirroedia edentata* Walker, 1857; *Cosmophila aurantiaca* Prittwitz, 1867; *Anomis serrata* Barnes et McDunnough, 1913.

Distribution. China (Jilin, Liaoning, Nei Mongol, Shandong, Henan, Fujian, Taiwan); Russia (Transbaikalia, East), Korea, Japan, India, New Zealand; North America, Africa.

寄主：棉、木槿、蜀葵、苘麻、冬苋、烟草、木耳菜等；成虫吸食柑橘、芒果、番石榴、黄皮等果汁（陈一心，1999）；黄麻（*Corchorus capsularis*）、木棉（*Bombax ceiba*）、黄葵（*Abelmoschus moschatus*）、苘麻（*Abutilon theophrasti*）、蜀葵（*Alcea rosea*）、海岛棉（*Gossypium barbadense*）、草棉（*Gossypium herbaceum*）、印度棉（*Gossypium indicum*）、大麻槿（*Hibiscus cannabinus*）、沼泽地木槿（*Hibiscus diversifolius*）、黄秋葵（*Hibiscus esculentus*）、百合朱槿（*Hibiscus liliiflorus*）、菜芙蓉（*Hibiscus manihot*）、木芙蓉叶（*Hibiscus mutabilis*）、朱槿（*Hibiscus rosasinensis*）、玫瑰茄（*Hibiscus sabdariffa*）、木槿（*Hibiscus syriacus*）、黄槿（*Hibiscus tiliaceus*）、野西瓜（*Hibiscus trionum*）、翅果麻（*Kydia calycina*）、（*Malachra capitata*）、（*Malvaviscus concinnus*）、桤叶黄花稔（*Sida alnifolia*）、心叶黄花稔（*Sida cordifolia*）、黄花稔（*Sida rhombifolia*）、地桃花（*Urena lobata*）、野苹果（*Malus sylvestris*）、菜豆（*Phaseolus vulgaris*）、黑吉豆（*Vigna mungo*）、赤小豆（*Vigna umbellata*）、豇豆（*Vigna unguiculata*）、番茄（*Lycopersicum esculentum*）、烟草（*Nicotiana tabacum*）、甘薯（*Ipomoea batatas*）（Matov & Kononenko, 2012）。

注：《中国动物志》（夜蛾科）中将本种归入“*Anomis* Hübner, [1821] 1816 桥夜蛾属”，在此予以变更。

Genus ***Gonitis*** Guenée, 1852 关桥夜蛾属

Gonitis Guenée, 1852, *in* Boisduval & Guenée, *Histoire Naturelle des Insectes. Species Général des Lépidoptères*, 6: 403. Type species: *Gonitis zeditrix* Guenée, 1852 [Haiti].

Synonymy: *Tiridata* Walker, 1865; *Gonotis* Moore, 1882, missp..

The genus includes 13 species, distributed mainly in the tropical and subtropical areas of the Old and New World; 1 species occurs in the Three Provinces of Northeast China.

注：首次给出中文属名。

Gonitis mesogona **Walker, 1858 中桥夜蛾**

Gonitis mesogona Walker, 1858, *List of the Specimens of Lepidopterous Insects in the Collection of the British Museum*, 13: 1002. Type locality: Sri Lanka.

Distribution. China (Heilongjiang, Jilin, Hebei, Shandong, Zhejiang, Hubei, Hunan, Fujian, Hainan, Guizhou, Yunnan, Taiwan); Russia (Siberia), Korea, Japan, Southeast Asia, Sri Lanka, India, Nepal, Pakistan.

寄主：红悬钩、醋栗（陈一心，1999）；木槿（*Hibiscus syriacus*）、黑醋栗（*Ribes nigrum*）、蓬藁（*Rubus hirsutus*）、（*Rubus ribesioedeus*）、（*Rubus lasiocarpus*）、多毛悬钩子（*Rubus rufus*）、炮烙莓（*Rubus sieboldii*）、日本黑莓（*Rubus trifidus*）、五色梅（*Lantana camara*）（Matov & Kononenko, 2012）。

注：《中国动物志》（夜蛾科）将此种归入“*Anomis* Hübner, [1821] 1816 桥夜蛾属”，在此予以变更。

Gonitis involuta **Walker, [1858] 1857 黄麻桥夜蛾**

Gonitis involuta Walker, [1858] 1857, *List of the Specimens of Lepidopterous Insects in the Collection of the British Museum*, 13: 1003. Type locality: Sri Lanka.

Synonymy: *Gonitis basalis* Walker, [1858] 1857; *Tiridata colligata* Walker, 1865; *Gonitis vitiensis* Butler, 1886; *Cosmophila dona* Swinhoe, 1919; *Anomis brima* Swinhoe, 1920.

Distribution. China (Jilin, Liaoning, Zhejiang, Guangdong, Xizang, Taiwan); Russia (Siberia), Korea, Japan, India, Nepal, Pakistan, Afghanistan; Southeast Asia, Oceania.

寄主：黄麻（陈一心，1999）；刺蒴麻属（*Triumfetta*）、黄麻属（*Corchorus*）、桐棉属（*Thespesia*）、梵天花属（*Urena*）、黄槿（*Hibiscus tiliaceus*）植物（Matov & Kononenko, 2012）。

注：《中国动物志》（夜蛾科）将此种归入"*Anomis* Hübner, [1821] 1816 桥夜蛾属"，在此予以变更。

Genus ***Rusicada*** Walker, [1858] 1857 仿桥夜蛾属

Rusicada Walker, [1858] 1857, *List Specimens lepid. Insects CollnBr. Mus.*, 13: 984. Type species: *Rusicada nigritarsis* Walker, [1858] 1857 [Sri Lanka].

Synonymy: *Gonopteronia* Bethune-Baker, 1906.

The genus includes 17 species, distributed mainly in tropical and subtropical areas of the Old and New World; 1 species occurs in the Three Provinces of Northeast China.

注：首次给出中文属名。

Rusicada privata **(Walker, 1865) 坎仿桥夜蛾**

Gonitis privata Walker, 1865, *List of the Specimens of Lepidopterous Insects in the Collection of the British Museum,* 33: 863. Type locality: China, Shanghai.

Synonymy: *Anomis commoda* Butler, 1878; *Rusicada subfulvida* Warren, 1913; *Rusicada griseolineata* Warren, 1913; *Rusicada fulvida* auct..

Distribution. China (Liaoning, Hunan, Taiwan); Russia (Siberia), Korea, Japan, Sri Lanka, India, Australia; Southeast Asia, North America (introduced by man).

寄主：木槿属（*Hibiscus*）植物（Matov & Kononenko, 2012）。

注：《中国动物志》（夜蛾科）中引用本种同物异名"*Anomis commoda* Butler, 1878"，在此予以变更。

Rusicada leucolopha **(Prout, 1928) 巨仿桥夜蛾**

Anomis metaxantha leucolopha Prout, 1928, *Bulletin of the Hill Museum*, 2: 266. Type locality: Indonesia, Sumatra.

Synonymy: *Anomis maxima* Berio, 1956; *Anomis longipennis* Sugi 1982.

Distribution. China (Liaoning, Jiangsu, Zhejiang, Guangdong, Taiwan); Russia (Siberia), Korea, Japan; Southeast Asia, North America.

寄主：华东椴（*Tilia japonica*）、木槿属（*Hibiscus*）植物（Matov & Kononenko, 2012）。

注：《中国动物志》（夜蛾科）中引用本种同物异名"*Anomis maxima* Berio, 1956"，在此予以变更。

Genus *Dinumma* Walker, 1858 双衲夜蛾属

Dinumma Walker, 1858, *List of the Specimens of Lepidopterous Insects in the Collection of the British Museum*, 15:1805. Type species: *Dinumma placens* Walker, 1858 [Sri Lanka].

Synonymy: *Ortheaga* Walker, 1865.

The genus includes 3 species are in the world; 1 species is in the Three Provinces of Northeast China.

Dinumma deponens **Walker, 1858 曲带双衲夜蛾**

Dinumma deponens Walker, 1858, *List of the Specimens of Lepidopterous Insects in the Collection of the British Museum*, 15: 1806. Type locality: India, Hindostan.

Synonymy: *Dinumma bipunctata* Motschulsky, [1861] 1860; *Amphipyra largeteaui* Oberthur, 1884.

Distribution. China (Liaoning, Shandong, Henan, Jiangsu, Zhejiang, Hunan, Fujian, Jiangxi, Guangdong, Guangxi, Yunnan, Taiwan); Korea, Japan, India, Nepal.

寄主：樱桃、大叶合欢（陈一心，1999）。

Subfamily **Calpinae** Boisduval, 1840 壶夜蛾亚科

Calpides Boisduval, 1840, *Genera et Index Methodicus Europaeorum Lepidopterorum*:, 156. Type genus: *Calpe* Treitscke, 1825 (=*Calyptra* Ochsenheimer, 1816).

Synonymy: Calpidi Guenée, 1841; Calpides Herrich-Schäffer, [1851] 1845; Ophideridae Guenée, 1852; Othreinae Berio, 1955.

Goater *at al.* (2003) listed many genera in Calpinae (sensu Fibiger, 2003), later most of them were moved to other subfamilies and tribes (Holloway, 2005; Fibiger & Lafontaine, 2005; Lafontaine & Fibiger, 2006). The exact number of genera and limits of Calpinae are still disputed. Fibiger & Lafontaine (2005) and Lafontaine & Fibiger (2006) treated Calpinae with three tribes: Anomini, Calpini and Scoliopterygini. Lafontaine & Schmidt (2010), Holloway (2011) and Zahiri *et al.* (2012a) separated Calpinae (sensu Lafontaine & Fibiger, 2006) into two subfamilies of Erebidae: Scoliopteryginae and Calpinae. In present treatment the subfamily Calpinae comprised by the tribes Calpini, Ophiderini and Phyllodini (Holloway, 2011) supported by molecular study (Zahiri *et al.*, 2012a, 2012b). The worldwide checklist of tribe Calpini was published by Zaspel & Branham (2008). The check list of the genus *Eudocima* presented by Zilli & Hogenes (2002). The revision of the genus *Calyptra* and fruit piercing and blood-sucking behavior of *Calyptra* species published by Bänziger (1983, 1989).

The subfamily includes approximately 200 species world wide and is most diverse in the tropics and subtropics, mainly in areas of forest. It includes in the regional fauna two tribes: Calpini and Ophiderini. The subfamily is represented in the Three Provinces of Northeast China by 9 species of 4 genera in 2 tribes.

References. Chen, 1982, 1999 (Noctuidae); Sugi, 1982 (Noctuidae: Catocalinae); Bänziger, 1983, 1989; Kononenko *et al.*, 1998 (Noctuidae: Catocalinae); Kitching, Ravlins, 1999 (Noctuidae: Calpinae); Sviridov, 2003

(Noctuidae: Catocalinae); Goater *et al.*, 2003 (Noctuidae: Catocalinae, Calpini); Fibiger, Lafontaine, 2005 (Erebidae: Calpinae, Calpini); Fibiger, Hacker, 2005 (Erebidae: Calpinae, Calpini); Kononenko, 2005 (Erebidae: Calpinae), 2010 (Noctuidae: Calpinae, Calpini); Holloway, 2005 (Noctuidae: Catocalinae, Calpini), 2011 (Erebidae: Calpinae, Calpini); Kononenko, Pinratana, 2005 (Noctuidae: Catocalinae), 2013 (Erebidae: Calpinae, Calpini); Lafontaine, Fibiger, 2006 (Noctuidae: Calpinae, Calpini); Zaspel *et al.*, 2007 (Noctuidae: Calpinae); Zaspel, Branham, 2008 (Noctuidae: Calpinae, Calpini); Kononenko, Han, 2007 (Noctuidae: Calpinae); Matov *et al.*, 2008 (Noctuidae: Calpinae); Kononenko, 2010 (Noctuidae: Calpinae); Zahiri *et al.*, 2012 (Erebidae); Zaspel *et al.*, 2012 (Erebidae: Calpinae, Calpini).

注：首次给出中文亚科名。

Tribe **Calpini** Boisduval, 1840 壶夜蛾族

Calpides Boisduval, 1840, *Genera et Index Methodicus Europaeorum Lepidopterorum*: 156. Type genus: *Calpe* Treitscke, 1825 (=*Calyptra* Ochsenheimer, 1816).

Synonymy: Calpidi Guenée, 1841; Calpides Herrich-Schäffer, [1851] 1845.

Tropical origin tribe containing genera *Calyptra*, *Plusiodonta*, *Oraesia* and some other. Widely distributed and most diverse in the Old and New Worlds tropics and subtropics; some species represented in the south of the Palaearctic region. Three genera are represented in the Three Provinces of Northeast China.

Refrerences. Chen, 1982, 1999 (Noctuidae: Ophiderinae); Sugi, 1982 (Noctuidae: Catocalinae); Bänziger, 1983 (Noctuidae: Catocalinae); Kononenko *et al.*, 1998 (Noctuidae: Catocalinae); Kitching, Ravlins, 1999 (Noctuidae: Calpinae); Sviridov, 2003 (Noctuidae: Catocalinae); Goater *et al.*, 2003 (Noctuidae: Catocalinae, Calpini); Fibiger, Hacker, 2005 (Erebidae: Calpinae, Calpini); Fibiger, Lafontaine, 2005 (Erebidae: Calpinae, Calpini); Kononenko, 2005 (Erebidae: Calpinae), 2010 (Noctuidae: Calpinae, Calpini); Holloway, 2005 (Noctuidae: Catocalinae, Calpini), 2011 (Erebidae: Calpinae, Calpini); Kononenko, Pinratana, 2005 (Noctuidae: Catocalinae), 2013 (Erebidae: Calpinae, Calpini); Lafontaine, Fibiger, 2006 (Noctuidae: Calpinae, Calpini); Kononenko, Han, 2007 (Noctuidae: Calpinae); Zaspel *et al.*, 2007 (Noctuidae: Calpinae); Matov *et al.*, 2008 (Noctuidae: Calpinae); Zaspel, Branham, 2008 (Noctuidae: Calpinae, Calpini); Kononenko, 2009, 2010 (Noctuidae: Calpinae); Zahiri *et al.*, 2012 (Erebidae); Zaspel *et al.*, 2012 (Erebidae: Calpinae, Calpini).

注：首次给出中文族名。

Genus ***Calyptra*** Ochsenheimer, 1816 壶夜蛾属

Calyptra Ochsenheimer, 1816, *Die Schmetterlinge von Europa*, 4: 78. Type species: *Phalaena thalictri* Borkhausen, 1790 [Europe].

Synonymy: *Culasta* Moore, 1881; *Hypocalpe* Butler, 1883; *Percalpe* Berio, 1956; *Calpe* Treitschke, 1825, unnecessary repl. name.

The genus includes 19 species, distributed mainly in the Eastern Palaearctic and Oriental regions, 1 species is known from North America; 4 species are in the Three Provinces of Northeast China.

Calyptra thalictri (Borkhausen, 1790) 壶夜蛾

Phalaena thalictri Borkhausen, 1790, *Naturgeschichte der Europaischen Schmetterlinge nach Systematischer Ordnung*, 3: 425. Type locality: not stated.

Synonymy: *Phalaena* (*Noctua*) *capucina* Esper, 1789; *Calpe sodalis* Butler, 1878; *Calpe capucina centralitalica* Dannehl, 1925 (var.); *Calpe thalictri pallida* Schwingenschuss, 1938 (ssp.).

Distribution. China (Heilongjiang, Jilin, Liaoning, Xinjiang, Shandong, Henan, Zhejiang, Fujian, Sichuan, Yunnan); Russia (European part, Ural, Siberia), Mongolia, Korea, Japan; Central Asia, Europe.

寄主：唐松草（陈一心，1999）；千金藤（*Stephania japonica*）、唐松草（*Thalictrum aquilegifolium*）、（*Thalictrum bauhini*）、黄唐松草（*Thalictrum flavum*）、亚欧唐松草（*Thalictrum minus*）、耧斗菜属（*Aquilegia*）植物（Matov & Kononenko, 2012）。

Calyptra hokkaida (Wileman, 1922) 北海道壶夜蛾

Calpe hokkaida Wileman, 1922, *Entomologist*, 55: 198. Type locality: Japan, Hokkaido.

Synonymy: *Calpe hoenei* Berio, 1956.

Distribution. China (Jilin, Beijing, Zhejiang); Russia (Siberia), Korea, Japan.

寄主：黄紫堇（*Corydalis ochotensis*）（Park *et al.*, 2006）；亚欧唐松草（*Thalictrum minus*）、蝙蝠葛（*Menispermum dahuricum*）、异果黄堇（*Corydalis heterocarpa*）、刻叶紫堇（*Corydalis incisa*）（Matov & Kononenko, 2012）。

Calyptra lata (Butler, 1881) 平嘴壶夜蛾

Calpe lata Butler, 1881, *Transactions of the Entomological Society of London*, 1881: 21. Type locality: Japan, Tokyo.

Synonymy: *Calpe aureola* Graeser, 1889 [1890].

Distribution. China (Heilongjiang, Jilin, Hebei, Shandong, Fujian, Yunnan); Russia (Siberia), Korea, Japan.

寄主：柑桔、唐松草、紫堇（陈一心，1999）；蝙蝠葛（*Menispermum dahuricum*）、木防已（*Cocculus trilobus*）（Matov & Kononenko, 2012）。

Calyptra gruesa (Draudt, 1950) 翎壶夜蛾

Calpe gruesa Draudt, 1950, *Mitteilungen der MünchnerEntomologischen Gesellschaft*, 40: 168. Type locality: China, Prov. Zhejiang [=Chekiang], Prov. Shaanxi [West Tien-mu-shan, Tai-pei-shan].

Distribution. China (Jilin, Liaoning, Shaanxi, Zhejiang, Hubei, Hunan); South Korea, Japan.

寄主：防已属植物（陈一心，1999）；黄紫堇（*Corydalis ochotensis*）（Park *et al.*, 2006）。

Genus *Oraesia* Guenée, 1852 嘴壶夜蛾属

Oraesia Guenée, 1852, *in* Boisduval & Guenée, *Histoire Naturelle des Insectes. Species Général des Lépidoptères*, 6: 362. Type species: *Noctua emarginata* Fabricius, 1794 [India].

The genus includes 17 species distributed in the tropical area of the Old World, most diverse in the Neotropics; 2 species are in the Three Provinces of Northeast China.

Oraesia emarginata (Fabricius, 1794) 嘴壶夜蛾

Noctua emarginata Fabricius, 1794, *Entomologiae Systematica emendata et Aucta. Secundum Classes, Ordines, Genera, Species Adjectis Synonimis, Locis, Observationibus, Descriptionibus*, 3 (2): 82. Type locality: India, Transquebariae.

Synonymy: *Oraesia metallescens* Guenée, 1852; *Oraesia tetans* Walker, [1858] 1857; *Oraesia allicias* Walker, [1858] 1857.

Distribution. China (Jilin, Liaoning, Shandong, Jiangsu, Zhejiang, Fujian, Guangdong, Hainan, Guangxi, Yunnan, Taiwan); Russia (Siberia), Korea, Japan, India, Nepal, Pakistan, Oman; Southeast Asia, East Africa.

寄主：锡生藤（*Cissampelos pareira*）、毛木防己（*Cocculus orbiculatus*）、木防己（*Cocculus trilobus*）、番石榴（*Psidium guajava*）、龙眼属（*Dimocarpus*）、番薯属（*Ipomoea*）植物（Matov & Kononenko, 2012）。

Oraesia excavata (Butler, 1878) 鸟嘴壶夜蛾

Calpe excavata Butler, 1878, *Annals and Magazine of Natural History*, (5)1: 202. Type locality: Japan, Yokohama.

Distribution. China (Jilin, Shandong, Zhejiang, Jiangsu, Fujian, Hunan, Guangdong, Guangxi, Yunnan, Taiwan); Korea, Japan, Philippines.

寄主：成虫取食柑橘、苹果、梨、葡萄、无花果、桃、芒果、黄皮等的果汁（陈一心，1999）；木防己（*Cocculus trilobus*）（Matov & Kononenko, 2012）。

Genus *Plusiodonta* Guenée, 1852 肖金翅夜蛾属

Plusiodonta Guenée, 1852, *in* Boisduval & Guenée, *Histoire Naturelle des Insectes. Species Général des Lépidoptères*, 6: 359. Type species: *Plusiodonta chalsytoides* Guenée, 1852 [Indonesia, Java].

Synonymy: *Deva* Walker, [1858] 1857; *Gadera* Walker, [1858] 1857; *Tafalla* Walker, 1869; *Odontina* Guenée, 1862, preocc. (Zborzhewski, 1834 [Mollusca]); *Tinnodoa* Nye, 1975 unnecessary repl. name.

The genus includes 37 species, distributed in the tropical area over the World; 1 species is in the Three Provinces of Northeast China.

Plusiodonta casta **(Butler, 1878) 纯肖金翅夜蛾**

Platydia casta Butler, 1878, *Illustrations of Typical Specimens of Lepidoptera Heterocera in the Collection of the British Museum*, 2: 54, pl. 38: 1. Type locality: Japan, Yokohama.

Distribution. China (Heilongjiang, Jilin, Liaoning, Beijing, Shandong, Zhejiang, Jiangsu, Hubei, Hunan, Fujian, Taiwan); Russia (Siberia), Korea, Japan.

寄主：蝙蝠葛（陈一心，1999）；木通（*Akebia quinata*）、木防己（*Cocculus trilobus*）、蝙蝠葛（*Menispermum dahuricum*）（Matov & Kononenko, 2012）。

Tribe **Ophiderini** Guenée, 1852 强喙夜蛾族

Ophideridae Guenée,1852, *in* Boisduval & Guenée, *Histoire Naturelle des Insectes. Species Général des Lépidoptères*, 7: 108. Type genus: *Ophideres* Boisduval, 1832.

The tribe is most diverse in the tropical area of the Old and New Worlds. Many species have high migratory ability and migrate to south part of the Palaearctic region. The tribe includes 1 genus in the Three Provinces of Northeast China.

References. Chen, 1982, 1999 (Noctuidae: Ophiderinae); Sugi, 1982 (Noctuidae: Ophiderinae); Kononenko *et al.*, 1998 (Noctuidae: Ophiderinae); Zilli, Hogenes, 2002 (Noctuidae: Ophiderinae); Sviridov, 2003 (Noctuidae: Catocalinae); Holloway, 2005 (Noctuidae: Catocalinae, Calpini); Kononenko, 2005 (Erebidae: Calpinae), 2010 (Noctuidae: Calpinae, Calpini); Kononenko, Pinratana, 2005 (Noctuidae: Catocalinae), 2013 (Erebidae: Calpinae, Ophiderini); Kononenko, Han, 2007 (Noctuidae: Calpinae, Calpini); Matov *et al.*, 2008 (Noctuidae: Calpinae); Zahiri *et al.*, 2011 (Erebidae: Calpinae, Ophiderini), 2012 (Erebidae: Calpinae, Ophiderini).

Genus ***Eudocima*** Billberg, 1820 艳叶夜蛾属

Eudocima Billberg, 1820, *Enumeratio Insectorum in Museo G.J. Billberg*: 85. Type species: *Phalaena salaminia* Cramer, 1777 ["Surinam"].

Synonymy: *Othreis* Hübner, [1823] 1816; *Corycia* Hübner, [1823] 1816; *Elygea* Billberg, 1820; *Leptophara* Billberg, 1820; *Trissophaes* Hübner, [1823]; *Maenas* Hübner, [1823]; *Rhytia* Hübner, [1823]; *Acacallis* Hübner, [1823]; *Ophideres* Boisduval, 1832; *Acacalis* Agassiz, 1846, emend.; *Ophioderes*: Agassiz, 1846, emend.; *Othryis*: Agassiz, 1846, emend.; *Moenas*: Walker, [1858], missp.; *Khadira* Moore, 1881; *Adris* Moore, 1881; *Purbia* Moore, 1881; *Vandana* Moore, 1881; *Argadesa* Moore, 1881; *Halastus* Butler, 1892; *Eumaenas* Hampson, 1924.

The genus includes 45 species, distributed in the tropical area over the World; 2 species are in the Three Provinces of Northeast China.

Eudocima tyrannus **(Guenée, 1852) 枯艳叶夜蛾**

Ophideres tyrannus Guenée, 1852, *in* Boisduval & Guenée, *Histoire Naturelle des Insectes. Species Général des Lépidoptères*, 7: 110. Type locality: Central India.

Synonymy: *Ophideres tyrannus amurensis* Staudinger, 1892 (var.).

Distribution. China (Liaoning, Hebei, Shandong, Zhejiang, Jiangsu, Fujian, Hubei, Guangxi, Hainan, Sichuan, Taiwan); Russia (Siberia), Korea, Japan, India(North), Nepal; Southeast Asia.

寄主：通草，成虫取食多种水果果汁（陈一心，1999）；木通（*Akebia quinata*）、五指那藤（*Stauntonia hexaphylla*）、木防已（*Cocculus trilobus*）、蝙蝠葛（*Menispermum dahuricum*）、台湾十大功劳（*Mahonia japonica*）、南天竹（*Nandina domestica*）、柊树（*Osmanthus heterophyllus*）（Matov & Kononenko, 2012）。

***Eudocima phalonia* (Linnaeus, 1763) 凡艳叶夜蛾**

Phalaena (*Noctua*) *phalonia* Linnaeus, 1763, *Centuria Insect. rariorum*: 28. Type locality: not stated.

Synonymy: *Phalaena fullonia* Clerk, 1764; *Phalaena* (*Noctua*) *fullionica* Linnaeus, 1767; *Noctua pomona* Cramer, 1775; *Noctua dioscoreae* Fabricius, 1781; *Ophideres princeps* Boisduval, 1832; *Ophideres obliterans* Walker, [1858] 1857; *Ophideres dioscorae* Moore; 1877, missp.; *Noctua dioscoriae* Hampson, 1894, missp.; *Noctua discoreae* Swinhoe, 1900, missp.; *Noctua discoreae* Warren, 1913; *Noctua discoriae* Zhu & Chen, 1963; *Ophideres fullionica* Pinhey, 1975, missp..

Distribution. China (Heilongjiang, Jilin, Liaoning, Shandong, Zhejiang, Jiangsu, Fujian, Hunan, Guangdong, Guangxi, Hainan, Sichuan, Yunnan, Taiwan); Russia (Siberia), Korea, Japan, India, Nepal, New Zealand, Australia; Southeast Asia, Central Africa.

寄主：木通（陈一心，1999）；木通（*Akebia quinata*）、（*Albertisia ferruginea*）、印防已（*Anamirta cocculus*）、（*Arcangelisia flava*）、（*Carronia multisepalea*）、（*Cissampelos hispida*）、锡生藤（*Cissampelos pareira*）、（*Cocculus hirsutus*）、樟叶木防已（*Cocculus laurifolia*）、木防已（*Cocculus trilobus*）、（*Coscinium blumeanum*）、（*Cyclea atjehensis*）、毛叶轮环藤（*Cyclea barbata*）、盾形轮环藤（*Cyclea peltata*）、铁藤（*Cyclea polypetala*）、（*Cyclea varians*）、（*Dioscoreophyllum volkensis*）、苍白秤钩风（*Diploclisia glaucescens*）、刺桐（*Erythrina variegata*）、绿白天仙藤（*Fibraurea chloroleuca*）、（*Hypserpa decumbens*）、（*Legnephora moorei*）、连蕊藤（*Parabaena sagittata*）、细圆藤（*Pericampylus glaucus*）、（*Pleogyne australis*）、（*Sarcopetalum harveyanum*）、青藤（*Sinomenium acutum*）、（*Stephania aculeata*）、（*Stephania brevipes*）、（*Stephania capitata*）、（*Stephania capitata*）、丁克拉千金藤（*Stephania dinklagei*）、雅丽千金藤（*Stephania elegans*）、西藏地不容（*Stephania glabra*）、千金藤（*Stephania japonica*）、（*Stephania kerrii*）、福曼氏千金藤（*Stephania suberosa*）、斯普林氏千金藤（*Stephania venosa*）、（*Tiliacora acuminata*）、（*Tiliacora funifera*）、（*Tiliacora triandra*）、大叶藤（*Tinomiscium petiolare*）、本齐青牛胆（*Tinospora baenzigeri*）、心叶青牛胆（*Tinospora cordifolia*）、民间药发冷藤（*Tinospora crispa*）、（*Tinospora homosepala*）、中华青牛胆（*Tinospora sinensis*）、菝葜青牛胆（*Tinospora smilacina*）、似雪三莪青牛胆（*Tinospora tinosporoides*）、可可（*Theobroma cacao*）、蝙蝠葛属（*Menispermum*）、小檗属（*Berberis*）、十大功劳属（*Mahonia*）植物（Matov & Kononenko, 2012）。

注：《中国动物志》（夜蛾科）中引用本种同物异名"*Phalaena fullonia* Clerk, 1764"，在此予以变更。

Subfamily **Hypocalinae** Guenée, 1852 鹰夜蛾亚科

Hypocalidae Guenée, 1852, *in* Boisduval & Guenée, *Histoire Naturelle des Insectes. Species Général des Lépidoptères*, 7: 73. Type genus: *Hypocala* Guenée, 1852.

Hypocalinae is a small tropical subfamily included single genus *Hypocala*. This group is treated in the subfamily rank of Erebinae by Lafontaine & Schmidt (2010) and Holloway (2011). The subfamily numbers 1 genus and 4 species are in the Three Provinces of Northeast China.

References. Chen, 1982, 1999 (Noctuidae: Ophiderinae); Sugi, 1982 (Noctuidae: Catocalinae); Sviridov, 2003 (Noctuidae: Catocalinae); Fibiger, Lafontaine, 2005 (Erebidae: Catocalinae, Hypocalini); Kononenko, 2005 (Erebidae: Catocalinae), 2010 (Noctuidae: Catocalinae, Hypocalini); Holloway, 2005 (Noctuidae: Catocalinae, Hypocalini), 2013 (Erebidae: Hypocalinae); Kononenko, Pinratana, 2005 (Noctuidae Catocalinae), 2013 (Erebidae: Hypocalinae); Kononenko, Han, 2007 (Noctuidae: Hypocalini); Matov *et al.*, 2008 (Noctuidae: Catocalinae); Zahiri *et al.*, 2012 (Erebidae: Hypocalinae).

注：首次给出中文亚科名。

Genus ***Hypocala*** Guenée, 1852 鹰夜蛾属

Hypocala Guenée, 1852, *in* Boisduval & Guenée, *Histoire Naturelle des Insectes. Species Général des Lépidoptères*, 7: 73. Type species: *Hyblaea deflorata* Fabricius, 1794 [India].

The genus contains 16 species, widespread mainly in the tropical area of the Old World, 1 species known from the Neotropical region; 4 species are in the Three Provinces of Northeast China.

***Hypocala deflorata* (Fabricius, 1794) 鹰夜蛾**

Hyblaea deflorata Fabricius, 1794, *Entomologiae Systematica emendata et Aucta. Secundum Classes, Ordines, Genera, Species Adjectis Synonimis, Locis, Observationibus, Descriptionibus*, 3 (2): 127. Type locality: "India oriental".

Synonymy: *Hypocala angulipalps* Guenée, 1852; *Hypocala moorei* Butler, 1892; *Hypocala australiae* Butler, 1892.

Distribution. China (Liaoning, Hebei, Shandong, Fujian, Guangdong, Hainan, Guizhou, Sichuan, Taiwan); Russia (Siberia), Korea, Japan, Indonesia, Sri Lanka, India, Nepal, New Zealand, Australia, USA(Hawaii); Central and South Africa.

寄主：柿（陈一心，1999）；柿（*Diospyros kaki*）、山柿（*Diospyros montana*）、美果榄（*Pouteria sapota*）、山榄属（*Planchonella*）植物（Matov & Kononenko, 2012）。

***Hypocala subsatura* Guenée, 1852 苹梢鹰夜蛾**

Hypocala subsatura Guenée, 1852, *in* Boisduval & Guenée, *Histoire Naturelle des Insectes. Species Général des Lépidoptères*, 7: 75. Type locality: Bangladesh, Silhet.

Synonymy: *Hypocala aspersa* Butler, 1883; *Hypocala limbata* Butler, 1889; *Pseudophia tungusa* Graeser, 1890.

Distribution. China (Liaoning, Nei Mongol, Beijing, Shaanxi, Henan, Shandong, Zhejiang, Jiangsu, Fujian, Guangdong, Guangxi, Hainan, Yunnan, Xizang, Taiwan); Russia (Siberia), Korea, Japan, India, Nepal, Pakistan, Philippines, Indonesia.

寄主：苹果、栎（陈一心，1999）；麻栎（*Quercus acutissima*）、竹叶青冈（*Quercus bambusifolia*）、膨大栎（*Quercus dilatata*）、灰色栎树（*Quercus incana*）、柿（*Diospyros kaki*）、印度乌木（*Diospyros melanoxylon*）、山柿（*Diospyros montana*）、苹果（Malus pumila）、（*Malus sylvestris*）、苏木（*Caesalpinia sappan*）、马鲁古苦油楝（*Carapa moluccensis*）、（*Xylocarpus moluccensis*）、藤黄属（*Garcinia*）和柑橘属（*Citrus*）植物（Matov & Kononenko, 2012）。

Hypocala violacea Butler, 1879 红褐鹰夜蛾

Hypocala violacea Butler, 1879, *Transactions of the Entomological Society of London*, 1879: 6. Type locality: India, Cachar.

Synonymy: *Hypocala clarissima* Butler, 1892; *Hypocala kebaea* Bethune-Baker, 1906.

Distribution. China (Liaoning, Yunnan, Taiwan); Russia (Siberia), Korea, Japan, Philippines, Indonesia, India, Nepal.

Hypocala rostrata (Fabricius, 1794) 钩鹰夜蛾

Noctua rostrata Fabricius, 1794, *Entomologiae Systematica emendate et Aucta. Secundum Classes, Ordines, Genera, Species Adjectis Synonimis, Locis, Observationibus, Descriptionibus*, 3 (2): 127. Type locality: "India oriental."

Synonymy: *Hyblaea plumicornis* Guenée, 1852; *Hypocala efflorescens* Guenée, 1852; *Hyblaea genuiai* Wallengren, 1856; *Hyblaea bohemani* Wallengren, 1856.

Distribution. China (Liaoning, Sichuan, Taiwan); Korea, Japan, India, Nepal, Philippines, Indonesia, Pakistan, Afghanistan; Central and South Africa.

Subfamily **Hypenodinae** Forbes, 1954 拟髯须夜蛾亚科

Hypenodinae Forbes, 1954, *Memoirs of the Cornell University agriculture Experimental Station*, 329: 381. Type genus: *Hypenodes* Doubleday, 1850.

This subfamily includes small sized moths distributed mainly in the tropical area. It is especially diverse in South China and Indonesia. Kitching & Rawlins (1998) used the name Strepsimaninae for this group of genera, but Fibiger & Lafontaine (2005) excluded *Strepsimanes* from the Hypenodinae and considered Hypenodinae and Strepsimaniinae as two distinct subfamilies.

The classification of the subfamily follows to Holloway (2011) and Zahiri *et al.* (2012). Hypenodinae includes 2 tribes Hypenodini and Micronoctuini. The later was described as a family of Noctuoidea in Fibiger & Lafontaine (2005) but associated with Erebidae and downgraded to the tribe of the Hypenodinae by Holloway (2011) what is strongly supported by molecular study (Zahiri *et al.*, 2012). Correspondingly all subfamilies of Micronoctuidae described by Fibiger (2007, 2008, 2010, 2011) were downgraded to subtribes of Micronoctuini.

The subfamily Hypenodinae numbers 2 tribes, 2 genera and 3 species are in the Three Provinces of Northeast China.

References. Inoue, 1979 (Noctuidae); Sugi, 1982 (Noctuidae: Ophiderinae); Kitching, 1984; Speidel *et al.,* 1996; Kononenko *et al.*, 1998 (Noctuidae: Catocalinae); Kitching, Rawlins, 1999 [1998] (Noctuidae: Strepsimaninae); Sviridov, 2003 (Noctuidae: Catocalinae); Fibiger, Hacker, 2005 (Erebidae: Hypenodinae); Fibiger, Lafontaine, 2005 (Erebidae: Hypenodinae); Lafontaine, Fibiger, 2006 (Noctuidae: Hypenodinae); Kononenko, Han, 2007 (Noctuidae: Hypenodinae, part); Matov *et al.*, 2008 (Noctuidae: Hypenodinae); Holloway, 2008 (Noctuidae: Hypenodinae), 2011 (Erebidae: Hypenodinae); Fibiger *et al.*, 2010 (Noctuidae: Hypenodinae); Fibiger *et al.*, 2011 (Erebidae: Hypenodinae); Zahiri *et al.*, 2012 (Erebidae: Hypenodinae).

注：首次给出中文亚科名。

Tribe **Micronoctuini** Fibiger, 2005 小夜蛾族

Micronoctuidae Fibiger, 2005. *Esperiana*, 11: 33. Type genus *Micronoctua* Fibiger, 1997.

The tribe Micronoctuini is highly diverse in the tropics of the Old World, especially in the Oriental region and contains many still not yet described species. There are very few information about life circle and life style of Micronoctuini. Presumingly, due small size and weak flight of the adults, they are species with low migratory ability and their high diversity could be explain by precence many local species in tropical forests. The known larvae are feeding on fungi (Fibiger, 2007, 2011). The East Asian taxa of Micronoctuini were revised and described in series of publications by Fibiger (2007, 2008, 2010, 2011). Those of the Rurrian（Siberia） and China were revised by Fibiger & Kononenko (2008a) and Fibiger *et al.* (2011). 1 genus and 1 species occurs in the Three Provinces of Northeast China.

References. Sugi, 1982 (Noctuidae: Ophiderini); Fibiger 1997 (Noctuidae); Sviridov, 2003 (Noctuidae: Catocalinae); Fibiger, Lafontaine, 2005 (Micronoctuidae); Fibiger, Hacker, 2005 (Micronoctuidae); Lafontaine, Fibiger, 2006 (Micronoctuidae); Fibiger, 2007, 2008, 2010, 2011 (Micronoctuidae); Kononenko, Han, 2007 (Micronoctuidae); Fibiger, Kononenko, 2008a (Micronoctuidae); Matov *et al.*, 2008 (Micronoctuidae); Fibiger *et al.*, 2010 (Micronoctuidae); Fibiger *et al.*, 2011 (Micronoctuidae); Fibiger *et al.*, 2011 (Micronoctuidae); Fibiger *et al.*, 2011 (Erebidae: Hypenodinae, Micronoctuini); Holloway, 2011 (Erebidae: Hypenodinae, Micronoctuini); Zahiri *et al.*, 2012 (Erebidae: Hypenodinae, Micronoctuini); Kononenko, Pinratana, 2013 (Erebidae: Hypenodinae, Micronoctuini).

Genus ***Parens*** Fibiger, 2011 帕小夜蛾属

Parens Fibiger, 2011, *Zootaxa*, 2842: 18. Type species: *Parens paraocci* Fibiger, 2011 [Taiwan].

The genus includes 4 species, mainly distributed in China; 1 species is in the Three Provinces of Northeast China.

注：首次给出中文属名。

Parens occi **(Fibiger & Kononenko, 2008) 枕帕小夜蛾**

Micronoctua occi Fibiger & Kononenko, 2008, *Zootaxa*, 1890: 52, figs 1, 2, 7, 10. Type locality: Russia, Primorye terr., Gornotaezhnoe.

Distribution. China (Liaoning, Shaanxi, Fujian, Zhejiang, Guangdong, Taiwan); Russia (Siberia), South Korea, Japan.

注：首次给出中文种名。

Tribe **Hypenodini** Forbes, 1954 拟髯须夜蛾族

Hypenodinae Forbes, 1954, *Memoirs of the Cornell University agriculture Experimental Station*, 329: 381. Type genus: *Hypenodes* Doubleday, 1850.

Similarly to Micronoctuidae the tribe distributed mainly in the tropical regions, while several species of the tribe occurs in the temperate zone. Poorly studied group with many yet not described species. The tribe includes 1 genus and 2 species in the Three Provinces of Northeast China.

References. Inoue, 1979 (Noctuidae: Ophiderinae); Sugi, 1982 (Noctuidae: Ophiderinae); Kononenko *et al.*, 1998 (Noctuidae: Hypenodinae); Sviridov, 2003 (Noctuidae: Catocalinae); Fibiger, Hacker, 2005 (Erebidae: Hypenodinae); Fibiger, Lafontaine, 2005 (Erebidae: Hypenodinae); Lafontaine, Fibiger, 2006 (Noctuidae: Hypenodinae); Kononenko, Han, 2007 (Erebidae: Hypenodinae); Matov *et al.*, 2008 (Noctuidae: Hypenodinae); Holloway, 2008, 2011 (Erebidae: Hypenodinae, Hypenodini); Fibiger *at al.*, 2010 (Noctuidae: Hypenodinae), 2011 (Erebidae: Hypenodinae); Fibiger *et al.*, 2011 (Erebidae: Hypenodinae, Hypenodini); Zahiri *et al.*, 2012 (Erebidae: Hypenodinae, Hypenodini); Kononenko, Pinratana, 2013 (Erebidae: Hypenodinae, Hypenodini).

注：首次给出中文族名。

Genus ***Schrankia*** Hübner, [1825] 1816 施夜蛾属

Schrankia Hübner, [1825] 1816, *Verzeichniss bekannter Schmettlinge* [sic!]: 345. Type species: *Pyralis taenialis* Hübner [1809] [Europe].

Synonymy: *Hypenodes* Guenée, 1854, preocc. (Doubleday, 1850 [Lepidoptera, Noctuidae]); *Costankia* Beck, 1996.

The genus includes 33 species, distributed in the Old and New Worlds, mainly in the tropical area; 2 species are in the Three Provinces of Northeast China.

Schrankia costaestrigalis **(Stephens, 1834) 窄缘施夜蛾**

Cledeobia costaestrigalis Stephens, 1834, *Illustrations of British Entomology* (*Haustellata*), 4: 20, 45. Type locality: Great Britain.

Synonymy: *Hypenodes costistrigalis lugubris* Dannehl, 1925 (var.).

Distribution. China (Liaoning, Guangxi, Taiwan); Russia (European part, Siberia), Korea, Japan, Australia, New Zealand; Central Asia, Near East, Europe, Africa.

寄主：帚石楠（*Calluna vulgaris*）、水薄荷（*Mentha aquatica*）、牛至（*Origanum vulgare*）、百里香（*Thymus serpyllum*）、蚕豆（*Vicia faba*）、匍匐风铃草（*Campanula rapunculoides*）、加拿大一枝黄花（*Solidago canadensis*）、铁线莲属（*Clematis*）、柳属（*Salix*）、蔷薇属（*Rosa*）、倒挂金钟属（*Fuchsia*）、鼬瓣花属（*Galeopsis*）、莴苣属（*Lactuca*）、山罗花属（*Melampyrum*）植物（Matov & Kononenko, 2012）。

Schrankia separatalis **(Herz, 1904) 晕施夜蛾**

Hypenodes separatalis Herz, 1904, *Annuaire du Musée Zoologique de l'Académie Imperiale des Sciences de St.-Petersbourg*, 9: 331, pl. 1: 7. Type locality: North Korea, 38°5′N, 128° E.

Synonymy: *Hypenodes squalida* Wileman & South, 1917.

Distribution. China (Jilin, Liaoning); Russia (Siberia), Korea, Japan.

注：首次给出中文种名。

Subfamily **Boletobiinae** Guenée [1858] 1857 菌夜蛾亚科

Boletobidae Guenée, [1858] 1857, *in* Boisduval & Guenée, *Histoire Naturelle des Insectes. Species Général des Lépidoptères*, 9: 328. Type genus: *Boletobia* Boisduval, 1840 (=*Parascotia* Hübner, [1825]).

Synonymy: Boletobiini Grote, 1895.

The subfamily contains genera placed by earlier authors to subfamilies Catocalinae (sensu auctorum), Erastriinae (sensu auctorum) [Erastrianae (*sensu* Hampson, 1910)], Acontiinae (sensu auctorum) (Sugi, 1982, 1992; Kononenko, 1990; Kononenko *et al.*, 1998; Goater *et al.*, 2003) and part of Calpinae (sensu Fibiger 2003 in Goater *et al.*, 2003). In current treatment it includes five tribes (Boletobiini, Araeopteronini, Aventiini, Eublemmini and Phytometrini). All those tribes were treated by Fibiger & Lafontaine (2005) and Lafontaine & Fibiger (2006) as distinct subfamilies of Erebidae, but later they were downgraded to tribes and associated with the subfamily Boletobiinae (Lafontaine & Schmidt, 2010; Holloway, 2011) on the basis of morphological and molecular study (Zahiri *et al.*, 2012). These groups are highly diverse in tropical areas. Among them the

Araeopteronini is mainly unrevised group. The revision of the East Siberiaern species of the genus *Araeopteron* and complete checklist all described species has been published by Fibiger & Kononenko (2008). Holloway (2009) placed several tropical genera (*Acidaliodes*, *Hyriodes*, *Pseudocraspedia*, *Sigela* and *Niaccaba*) in Araeopteroninae. Many genera of Boletobiinae, are also requires global revision. The subfamily numbers 6 tribes, 23 genera and 39 species are in the Three Provinces of Northeast China.

References. Seitz, [1907] –1914, 1931–1938 (Noctuidae: Acontiinae, Erastriinae); Hampson, 1910b, 1912 (Noctuidae: Erastrianae, part, Acontianae, part.); Chen, 1982, 1999 (Noctuidae: Ophiderinae part; Acontiinae, part); Sugi, 1982 (Noctuidae: Acontiinae); Kononenko *et al.*, 1998 (Noctuidae: Catocalinae, part, Acontiinae, part); Kononenko, 2003 (Acontiinae, part); Goater *et al.*, 2003 (Noctuidae: Catocalinae, part), 2010 (Boletobiinae, Araeopteroninae, Eublemminae, Phytometrinae), 2011 (Erebidae: Boletobiinae); Sviridov, 2003 (Noctuidae: Catocalinae, part); Fibiger, Lafontaine, 2005 (Erebidae: Boletobiinae); Fibiger, Hacker, 2005 (Erebidae: Boletobiinae); Kononenko, 2005 (Erebidae: Boletobiinae, Phytometrinae, Aventiinae, Araeopteroninae, Eublemminae), 2010 (Noctuidae: Boletobinae, Phytometrinae, Aventiinae, Araeopteroninae, Eublemminae); Holloway, 2005 (Noctuidae: Catocalinae, Boletobiini), 2008 (Noctuidae: Araeopteroninae, Aventiinae, Eublemminae); Lafontaine, Fibiger, 2006 (Noctuidae: Boletobiinae); Lafontaine, Schmidt, 2010 (Erebidae: Boletobiinae, Aventiinae, Eublemminae, Phytometrinae); Holloway, 2011 (Erebidae: Boletobiinae); Zahiri *et al.*, 2012 (Erebidae: Boletobiinae); Kononenko, Pinratana, 2013 (Erebidae: Boletobiinae).

注：首次给出中文亚科名。

Tribe **Boletobiini** Guenée [1858] 1857 菌夜蛾族

Boletobidae Guenée, [1858] 1857, *in* Boisduval & Guenée, *Histoire Naturelle des Insectes. Species Général des Lépidoptères*, 9: 328. Type genus: *Boletobia* Boisduval, 1840 (=*Parascotia* Hübner, [1825]).

Synonymy: Boletobiini Grote, 1895.

Rather characteristic tropical tribe with several genera extending their range to the temperate zone. The known larvae of many species are tropically connected with fungi. 4 genera are belong to the tribe in the Three Provinces of Northeast China.

References. Chen, 1982, 1999 (Noctuidae: Ophiderinae); Sugi, 1982 (Noctuidae: Catocalinae, part, Acontiinae, part); Kononenko *et al.*, 1998 (Noctuidae: Catocalinae, part, Acontiinae, part); Sviridov, 2003 (Noctuidae: Catocalinae, part); Fibiger, Lafontaine, 2005 (Erebidae: Boletobiinae); Fibiger, Hacker, 2005 (Erebidae: Boletobiinae); Kononenko, 2005 (Erebidae: Boletobiinae), 2010 (Noctuidae: Boletobiinae); Lafontaine, Fibiger, 2006 (Noctuidae: Boletobiinae); Matov *et al.*, 2008 (Noctuidae: Boletobiinae); Fibiger *et al.*, 2010 (Noctuidae: Boletobiinae), 2011 (Erebidae: Boletobiinae, Boletobiini); Lafontaine, Schmidt, 2010 (Erebidae: Boletobiinae); Holloway, 2011 (Erebidae: Boletobiinae, Boletobiini); Lafontaine, Schmidt, 2013 (Erebidae: Boletobiinae, Boletobiini); Zahiri *et al.*, 2012 (Erebidae: Boletobiinae, Boletobiini).

注：首次给出中文族名。

Genus *Diomea* Walker, [1858] 1857 狄夜蛾属

Diomea Walker, [1858] 1857, *List of the Specimens of Lepidopterous Insects in the Collection of the British Museum*, 13: 1079, 1109. Type species: *Diomea rotundata* Walker, [1858] 1857 [Sri Lanka].

Synonymy: *Corsa* Walker, 1862; *Zigera* Walker, 1862; *Heteroscotia* Bryk, 1948.

The genus includes about 20 species, distributed in the Chinese subregion of the Palaearctic, Oriental and Ethiopian regions; 2 species are in the Three Provinces of Northeast China.

Diomea cremata (Butler, 1878) 星狄夜蛾

Capnodes cremata Butler, 1878, *Annals and Magazine of Natural History*, (5)1: 295. Type locality: Japan, Yokohama.

Synonymy: *Heteroscotia stygia* Bryk, 1948.

Distribution. China (Heilongjiang, Jilin, Liaoning, Hebei, Yunnan); Russia (Siberia), Korea, Japan, India.

寄主：香菇（*Lentinula edodes*）、云芝属（*Polystictus*）植物（Matov & Kononenko, 2012）。

Diomea jankowskii (Oberthür, 1880) 黑斑狄夜蛾

Capnodes jankowskii Oberthür, 1880, *Études d'Entomologie*, 5: 87, pl. 9: 1. Type locality: Russia: Primorye terr., Askold I..

Distribution. China (Heilongjiang, Jilin, Taiwan); Russia (Siberia), Korea, Japan, Thailand.

Genus *Hypostrotia* Hampson, 1926 熏夜蛾属

Hypostrotia Hampson, 1926, *Descriptions of New Genera and Species of Lepidoptera Phalaenae of the Subfamily Noctuinae (Noctuidae) in the British Museum (Natural History)*: 293. Type species: *Capnodes cinerea* Butler, 1878 [Japan].

Synonymy: *Capnodes* auct..

Monotypis genus, distributed in the Chinese subregion of the Palaearctic region; 1 species is in the Three Provinces of Northeast China.

Hypostrotia cinerea (Butler, 1878) 熏夜蛾

Capnodes cinerea Butler, 1878, *Annals and Magazine of Natural History*, (5)1: 294. Type locality: Japan, Yokohama.

Distribution. China (Heilongjiang, Jilin); Russia (Transbaikalia, Siberia), Korea, Japan.

寄主：香菇（Lentinula edodes）、云芝属（Polystictus）植物（Matov & Kononenko, 2012）。

Genus *Anatatha* Hampson, 1926 曼夜蛾属

Anatatha Hampson, 1926, *Descriptions of New Genera and Species of Lepidoptera Phalaenae of the Subfamily Noctuinae* (*Noctuidae*) *in the British Museum* (*Natural History*): 291. Type species: *Catada nigrisigna* Hampson, 1895 [India].

The genus includes 7 species distributed in the Chinese subregion of the Palaearctic and Oriental regions; 1 species is in the Three Provinces of Northeast China.

Anatatha lignea (Butler, 1879) 黑斑曼夜蛾

Bleptina lignea Butler, 1879, *Illustrations of Typical Specimens of Lepidoptera Heterocera in the Collection of the British Museum*, 3: 64, pl. 56: 14. Type locality: Japan.

Distribution. China (Liaoning, Hebei); Russia (Siberia), Korea, Japan.

Genus *Paragona* Staudinger, 1892 仿曼夜蛾属

Paragona Staudinger, 1892, *in* N.M. Romanoff (ed.), *Mémoires sur les Lépidoptères*, 6: 5. Type species: *Acidalia multisignata* Christoph, 1881 [Russia, Amur reg., Pompeevka].

The genus includes 4 species distributed in the Chinese subregion of the Palaearctic and partly Oriental regions; 2 species are in the Three Provinces of Northeast China.

注：首次给出中文属名。

Paragona multisignata (Christoph, 1881) 远东仿曼夜蛾

Acidalia multisignata Christoph, 1881, *Bulletin de la Société Impériale des Naturalistes de Moscou*, 55 (3): 50. Type locality: Russia, Amur reg., [Pompeevka].

Distribution. China (Heilongjiang, Jilin); Russia (Siberia, Transbaikalia), Mongolia, Korea, Japan.

注：首次给出中文种名。

Paragona cognata (Staudinger, 1892) 凡仿曼夜蛾

Boletobia cognata Staudinger, 1892, *in* N.M. Romanoff (ed.), *Mémoires sur les Lépidoptères*, 6: 606. Type locality: Russia, Primorye terr.: Askold I., Suifun [Razdolnaya river].

Distribution. China (Jilin); Russia (Siberia, Transbaikalia), Mongolia, Korea.

注：首次给出中文种名。

Tribe **Araeopteronini** Fibiger, 2005 纤翅夜蛾族

Araeopteroninae Fibiger, 2005. *Esperiana*, 11: 25. Type genus: *Araeopteron* Hampson, 1893.

Purely studied tropical origin tribe including the genus *Araeopteron* and probably some related ones. Many species distributed in the tropical regions are still undescribed. Japan fauna of the genus *Araeopteron* described by Inoue (1958, 1965); those of the Russia, South of the Far East and Korea by Fibiger & Kononenko (2008). 1 genus belongs to the tribe in the Three Provinces of Northeast China.

References. Inoue, 1958, 1965; Fibiger & Lafontaine, 2005; Fibiger & Kononenko, 2008.

注：首次给出中文族名。

Genus ***Araeopteron*** Hampson, 1893 纤翅夜蛾属

Araeopteron Hampson, 1893, *Illustrations of Typical Specimens of Lepidoptera Heterocera in the Collection of the British Museum*, 9:33, 136. Type species: *Araeopteron pictale* Hampson, 1893 [Sri Lanka].

Synonymy: *Araeopterum* Hampson, 1895, emend.; *Thelxinoa* Turner, 1902; *Essonistis* Meyrick, 1902; *Araeoptera* Hampson, 1910, emend.; *Araeopterella* Fibiger & Hacker, 2001.

The genus includes 40 described species and many undescribed, distributed mainly in the Oriental and Ethiopian regions, and in the eastern Palaearctic region. 4 species are known in Korea and Japan; 3 species are in the Three Provinces of Northeast China.

注：首次给出中文属名。

***Araeopteron amoena* Inoue, 1958 变纤翅夜蛾**

Araeopteron amoena Inoue, 1958, *Tinea*, 4: 230: 2. Type locality: Japan, Kanagawa Pref., Chigasaki.

Distribution. China (Jilin, Liaoning); Russia (Siberia), Korea, Japan.

注：首次给出中文种名。

***Araeopteron fragmenta* Inoue, 1965 碎纤翅夜蛾**

Araeopteron fragmenta Inoue, 1965, *Tinea*, 7: 81, pl., 15: 5A, 5B. Type locality: Japan, Kanagawa Pref., Fujisawa.

Distribution. China (Liaoning); Russia (Siberia), South Korea, Japan.

注：首次给出中文种名。

***Araeopteron nebulosa* Inoue, 1965 棕纤翅夜蛾**

Araeopteron nebulosa Inoue, 1965, *Tinea* 7: 82, pl. 15: 4A, 4B. Type locality: Japan, Shizouka Pref., Odaru

Spa.

Distribution. China (Jilin, Liaoning); Russia (Siberia), South Korea, Japan.

注：首次给出中文种名。

Tribe **Aventiini** Tutt, 1896 勒夜蛾族

Aventiidae Tutt, 1896, *British Moths*: 226. Type genus: *Aventia* Duponchel, 1829 (=*Laspeyria* Germar, 1810).

Synonymy: Trisatelini Beck, 1999.

Predominantly tropical tribe with many genera and species distributed in tropical area of the Old and New Worlds, especially diverse in the Southeast Asia. Known larvae feed on lichens. 16 genera are recorded in the Three Provinces of Northeast China.

References. Hampson, 1910 (Noctuidae: Acontianae); Chen, 1982, 1999 (Noctuidae: Ophiderinae, part; Acontiinae, part); Sugi, 1982 (Noctuiae: Acontiinae); Kononenko *et al.*, 1998 (Noctuidae: Acontiinae); Kononenko, 2003 (Noctuidae: Acontiinae, part); Goater *et al.*, 2003 (Noctuidae: Catocalinae, Catocalini, Aventiina, part); Fibiger, Lafontaine, 2005 (Erebidae: Aventiinae); Fibiger, Hacker, 2005 (Erebidae: Aventiinae); Kononenko, 2005 (Erebidae: Aventiinae), 2010 (Noctuidae: Aventiinae); Lafontaine, Fibiger, 2006 (Noctuidae: Aventiinae); Kononenko, Han, 2007 (Noctuidae: Aventiinae); Matov *et al.*, 2008 (Noctuidae: Aventiinae); Holloway, 2009 (Noctuidae: Aventiinae), 2011 (Erebidae: Boletobiinae, Aventiini); Lafontaine, Schmidt, 2010 (Erebidae: Aventiinae); Fibiger *et al.*, 2010 (Noctuidae: Phytometrinae, part), 2011 (Erebidae: Boletobiinae, Aventiini); Zahiri *et al.*, 2012 (Erebidae: Boletobiinae, Aventiini); Lafontaine, Schmidt, 2013 (Erebidae: Boletobiinae, Aventiini); Kononenko, Pinratana, 2013 (Erebidae: Boletobiinae, Aventiini).

注：首次给出中文族名。

Genus ***Laspeyria*** Germar, 1810 勒夜蛾属

Laspeyria Germar, 1810, *Dissertatio sistems Bombycum Species*: 13. Type species: *Bombyx flexula* [Denis & Schiffermüller], 1775 [Austria].

Synonymy: *Colposia* Hübner, 1816; *Laspeyresia*: Reichenbach & Leipzig, 1817, emend.; *Aventia* Duponchel, 1829; *Laspeyresia* Herrich-Schäffer, 1839, emend.; *Euteles* Gistl, 1849, unnecessary repl. name; *Sophta* Walker, [1863] 1862; *Nacerasa* Walker, 1866; *Perynea* Hampson, 1910; *Trogatha* Hampson, 1910.

The genus includes 7 species, distributed in the Palaearctic, Oriental, Australian and Ethiopian regions, of them *L. flexula* is widely distributed in the Palaearctic, *L. subrosea* known from the Chinese subregion of the Palaearctic and *L. ruficeps* is widely distributed in the Oriental region; 3 species are in the Three Provinces of Northeast China.

Laspeyria flexula ([Denis & Schiffermüller], 1775) 勒夜蛾

Bombyx flexula [Denis & Schiffermüller], 1775, *Ankündung eines Systematisches Werkes von den Schmetterlinge der Wiener Gegend. Wien*: 64. Type locality: Austria, Vienna district.

Synonymy: *Phalaena sinuata* Fabricius, 1775; *Geometra flexularia*: Hübner, 1799, emend..

Distribution. China (Heilongjiang, Hebei, Jiangsu); Russia (European part, Siberia), Korea, Japan, Kazakhstan; Near East, Europe.

寄主：蜈蚣衣（*Physcia stellaria*）、石黄衣（*Xanthoria parietina*）、鼓球藻属（*Desmococcus*）、原球藻属（Protococcus）、梅衣属（*Parmelia*）、地卷属（*Peltigera*）、文字衣属（*Graphis*）植物（Matov & Kononenko, 2012）。

Laspeyria subrosea (Butler, 1881) 亚勒夜蛾

Rivula subrosea Butler, 1881, *Transactions of the Entomological Society of London*, 1881: 580. Type locality: Japan, Tokyo.

Synonymy: *Corgatha nawai* Nagano, 1918; *Corgatha gifuensis* Nagano, 1918.

Distribution. China (Jilin, Liaoning); Russia (Siberia), Korea, Japan.

注：首次给出中文种名。

Laspeyria ruficeps (Walker, 1864) 赭灰勒夜蛾

Thermesia ruficeps Walker, 1864, *Journal of the Proceedings of the Linnean Society of London, Zoology*, 7: 187. Type locality: Borneo, Sarawak [UM, Oxford].

Synonymy: *Thermesia sparsa* Walker, 1864; *Corgatha castaneiceps* Hampson, 1896.

Distribution. China (Heilongjiang, Jilin, Liaoning, Sichuan, Taiwan); South Korea, Japan, Sri Lanka, India, Nepal, Indonesia, Malaysia.

Genus *Enispa* Walker, [1866] 1865 点孔夜蛾属

Enispa Walker, [1866] 1865, *List of the Specimens of Lepidopterous Insects in the Collection of the British Museum*, 34: 1275. Type species: *Enispa eosarialis* Walker, [1866] 1865 [Borneo, Sarawak].

Synonymy: *Micraeschus* Butler, 1878; *Chara* Staudinger,1892; *Trogacontia* Hampson, 1892; *Penisa* Warren, 1911.

The genus includes approximately 60 species, distributed in the Old World, mainly in the Oriental and Ethiopian regions; 3 species are in the Three Provinces of Northeast China.

注：首次给出中文属名。

***Enispa lutefascialis* (Leech, 1889) 黑线点孔夜蛾**

Mestleta lutefascialis Leech, 1889, *Entomologist*, 22: 65, pl. 2: 15. Type locality: Japan, Satsuma.

Synonymy: *Aventiola solitaria* Staudinger, 1892.

Distribution. China (Liaoning); Russia (Siberia), South Korea, Japan.

寄主：*Lichen* sp.（Park *et al.*, 2006）

注：首次给出中文种名。

***Enispa albosignata* (Staudinger, 1892) 多点孔夜蛾**

Chara albosignata Staudinger, 1892, *in* N.M. Romanoff (ed.), *Mémoires sur les Lépidoptères*, 6: 559, pl. 14: 2. Type locality: Russia, Primorye terr., Askold I..

Distribution. China (Jilin, Liaoning); Russia (Siberia), Korea.

注：首次给出中文种名。

***Enispa bimaculata* (Staudinger, 1892) 白条点孔夜蛾**

Chara bimaculata Staudinger, 1892, *in* N.M. Romanoff (ed.), *Mémoires sur les Lépidoptères*, 6: 561. Type locality: Russia, Primorye terr., Vladivostok.

Synonymy: *Aventiola leucosticta* Hampson, 1910.

Distribution. China (Jilin); Russia (Siberia), Korea, Japan.

注：首次给出中文种名。

Genus ***Corgatha*** Walker, [1859] 1858 孔夜蛾属

Corgatha Walker, [1859] 1858, *List of the Specimens of Lepidopterous Insects in the Collection of the British Museum*, 16: 215. Type species: *Corgatha zonalis* Walker, [1859] 1858 [Sri Lanka].

Synonymy: *Palura* Walker, 1861; *Guriauna* Walker, 1861; *Ausinza* Walker, 1864; *Zitna* Walker, [1866]; *Nacerasa* Walker, 1866; *Celeopsyche* Butler, 1879; *Pseudephyra* Butler, 1886; *Callipyris* Meyrick, 1891; *Aventina* Staudinger, 1892; *Penisa* Warren, 1911; *Trichogatha* Warren, 1913.

The genus includes about 140 species (including recently described ones from Africa by Hacker in 2019), distributed in the Old World, mainly in Oriental, Ethiopian and Australian regions. 14 species are known from Chinese subregion of the Palaearctic region; 6 species are in the Three Provinces of Northeast China.

***Corgatha obsoleta* Marumo, 1932 旧孔夜蛾**

Corgatha obsoleta Marumo, 1932, *Kontyû*, 6: 12, fig. 3. Type locality: Japan.

Distribution. China (Jilin); Russia (Siberia), Korea, Japan.

注：首次给出中文种名。

Corgatha costimacula (Staudinger, 1892) 白边孔夜蛾

Aventiola costimacula Staudinger, 1892, *in* N.M. Romanoff (ed.), *Mémoires sur les Lépidoptères*, 6: 604, pl. 14: 4. Type locality: Russia, Primorye terr., Baranovsky, Partizansk ["Sutschan"].

Distribution. China (Jilin); Russia (Siberia), Korea, Japan.

Corgatha nitens (Butler, 1879) 昭孔夜蛾

Celeopsyche nitens Butler, 1879, *Illustrations of Typical Specimens of Lepidoptera Heterocera in the Collection of the British Museum*, 3: 68, pl. 57: 10. Type locality: Japan, Yokohama.

Synonymy: *Corgatha fasciola* Leech, 1900.

Distribution. China (Jilin, Liaoning, Jiangsu); Korea, Japan.

寄主：地衣（陈一心，1999）。

Corgatha argillacea (Butler, 1879) 土孔夜蛾

Egnasia argillacea Butler, 1879, *Annals and Magazine of Natural History*, (5)4: 450. Type locality: Japan, Yokohama.

Distribution. China (Jilin, Liaoning, Hebei, Henan); Russia (Siberia), Korea, Japan.

Corgatha dictaria (Walker, 1861) 柑橘孔夜蛾

Acidalia dictaria Walker, 1861, *List of the Specimens of Lepidopterous Insects in the Collection of the British Museum*, 23: 759. Type locality: China, Shanghai.

Synonymy: *Oruza liberata* Walker, 1864; *Acantholipes metalligera* Butler, 1881; *Corgatha tana* Swinhoe, 1900; *Acidalia sasakii* Matsumura, 1917.

Distribution. China (Liaoning, Jiangsu, Zhejiang, Sichuan); Korea, Japan.

寄主：柑橘（陈一心，1999）。

Corgatha pygmaea Wileman, 1911 竹孔夜蛾

Corgatha pygmaea Wileman, 1911 *Transactions of the Entomological Society of London*, 1911: 228. Type locality: Japan, Yamato, Yoshino.

Synonymy: *Corgatha rosacea* Marumo, 1932.

Distribution. China (Liaoning); Korea, Japan.

注：首次给出中文种名。

Genus ***Oruza*** Walker, 1861 巧夜蛾属

Oruza Walker, 1861, *List of the Specimens of Lepidopterous Insects in the Collection of the British Museum*, 23: 947. Type species: *Oruza costata* Walker, 1862 [India].

Synonymy: *Curvatula* Staudinger, 1892; *Vittappressa* Bethune-Baker, 1906; *Vittapressa* auct., missp..

The genus includes 58 species (including recently described by Hacker from Africa in 2019), distributed in the Old World, mainly in the Oriental, Ethiopian and Australian regions; 6 species are known from Chinese subregion of the Palaearctic region; 2 species are in the Three Provinces of Northeast China.

Oruza mira **(Butler, 1879) 奇巧夜蛾**

Selenis mira Butler, 1879, *Illustrations of Typical Specimens of Lepidoptera Heterocera in the Collection of the British Museum*, 3: 29, pl. 47: 6. Type locality: Japan, Hakodate.

Synonymy: *Curvatula pallicostata* Staudinger, 1892.

Distribution. China (Heilongjiang, Jilin, Liaoning, Anhui, Hubei); Russia (Siberia), Korea, Japan.

寄主：青冈 (*Quercus glauca*)、盐肤木（*Rhus chinensis*）（Matov & Kononenko, 2012）。

Oruza yoshinoensis **(Wileman, 1911) 日巧夜蛾**

Corgatha yoshinoensis Wileman, 1911, *Transactions of the Entomological Society of London*, 1911: 228, pl. 31: 14. Type locality: Japan, Yamato, Yoshino.

Distribution. China (Liaoning); Russia (Siberia), South Korea, Japan.

注：首次给出中文种名。

Genus ***Lophoruza*** Hampson, 1910 蝠夜蛾属

Lophoruza Hampson, 1910, *Catalogue of the Lepidoptera Phalaenae in the British Museum*, 10: 225. Type species: *Lophoruza vacillatrix* Hampson, 1910 [India, Sikkim].

Synonymy: *Eugnathia* Warren, 1912; *Lophograpta* Warren, 1913.

The genus includes about 23 species, distributed in the Old World, mainly in the Oriental, Ethiopian, Australian and Palaearctic regions; 1 species is in the Three Provinces of Northeast China.

Lophoruza pulcherrima **(Butler, 1879) 美蝠夜蛾**

Egnasia pulcherrima Butler, 1879, *Illustration of Typical Specimens of Lepidoptera Heterocera in the Collection of the British Museum*, 3: 67, pl. 57: 8. Type locality: Japan, Yokohama.

Distribution. China (Jilin, Liaoning, Hebei, Guangxi, Sichuan); South Korea, Japan.

寄主：菝葜、牛尾菜（陈一心，1999）。

Genus *Naganoella* Sugi 1982 尺夜蛾属

Naganoella Sugi 1982, *Moths of Japan*, 1: 886; 2: 400. Type species: *Dierna timandra* Alphéraky, 1897 [Korea].

Synonymy: *Dierna* auct..

Monotypic genus, distributed in Chinese subregion of the Palaearctic and partly Oriental regions.

Naganoella timandra (Alphéraky, 1897) 红尺夜蛾

Dierna timandra Alphéraky, 1897, *in* N.M. Romanoff (ed.), *Mémoires sur les Lépidoptères*, 9: 179, pl. 11: 7. Type locality: Korea.

Synonymy: *Perynea pvilcherina* [sic!] Nagano, 1918.

Distribution. China (Heilongjiang, Jilin, Liaoning, Hebei, Henan, Zhejiang, Hunan); Russia (Siberia), Korea, Japan, Thailand.

寄主：麻栎（*Quercus acutissima*）（Matov & Kononenko, 2012）。

Genus *Hyposada* Hampson, 1910 次孔夜蛾属

Hyposada Hampson, 1910, *Zoological collections from Northern Rhodesia and adjacent territoties: Lepidoptera Phalaenae,* 1910: 412. Type species: *Phibalapteryx hydrocampata* Guenée, 1857 [India].

Synonymy: *Hyposada* Hampson, 1910 [December].

The genus includes about 23 species, distributed in the Old World, mainly in the Oriental, Ethiopian, Australian and Palaearctic regions; 1 species is in the Three Provinces of Northeast China.

注：首次给出中文属名。

Hyposada brunnea (Leech, 1900) 波带次孔夜蛾

Thermesia brunnea Leech, 1900, *Transactions of the Entomological Society of London*, 1900: 571. Type locality: Japan, Oiwake.

Synonymy: *Hyposada mesophaea* Hampson, 1910.

Distribution. China (Liaoning); Korea, Japan.

注：首次给出中文种名。

Genus *Aventiola* Staudinger, 1892 燕夜蛾属

Aventiola Staudinger, 1892, *in* N.M. Romanoff (ed.), *Mémoires sur les Lépidoptères*, 6: 602. Type species: *Aventiola maculifera* Staudinger, 1892 [Russia, Primorye terr.] (=*Egnasia pusilla* Butler. 1879).

The genus contains 6 species, distributed in the Oriental and Australian regions, 1 species is in the Three Provinces of Northeast China.

Aventiola pusilla **(Butler, 1879) 燕夜蛾**

Egnasia pusilla Butler, 1879, *Illustrations of Typical Specimens of Lepidoptera Heterocera in the Collection of the British Museum*, 3: 67, pl. 57: 9. Type locality: Japan, Yokohama.

Synonymy: *Aventiola maculifera* Staudinger, 1892; *Saraca nigripalpis* Hirayama, 1937.

Distribution. China (Heilongjiang, Jilin, Liaoning, Hebei, Jiangsu, Sichuan); Russia (Transbaikalia, Siberia), Korea, Japan.

寄主：地衣（陈一心，1999）。

Genus ***Hemipsectra*** Hampson, 1891 半夜蛾属

Hemipsectra Hampson, 1891, *Illustrations of Typical Speciemens of Lepidoptera Heterocera in the Collection of the British Museum. Part VIII. The Lepidoptera Heterocera of the Nilgiri District*, 1891: 84. Type species: *Hemipsectra plumipars* Hampson, 1891 [India].

The genus includes 2 species, distributed in India and the Far Eastern countries; 1 species occurs in the Three Provinces of Northeast China.

注：首次给出中文属名。

Hemipsectra fallax **(Butler, 1789) 淡半夜蛾**

Egnasia fallax Butler, 1789, *Annals and Magazine of Natural History*, (5)4: 450. Type locality: Japan, Yokohama.

Synonymy: *Gelastocera insignata* Wileman, 1911.

Distribution. China (Jilin, Liaoning); Korea, Japan.

注：首次给出中文种名。

Genus ***Holocryptis*** Lucas, 1892 全夜蛾属

Holocryptis Lucas, 1892, *Proceedings of the Linnean Society of New South Wales*, (2)7: 256. Type species: *Holocryptis phasianura* Lucas, 1892 [Australia].

The genus includes 14 species, distributed in the Chinese subregion of the Palaearctic, Oriental, Ethiopian and Australian regions; 2 species are in the Three Provinces of Northeast China.

注：首次给出中文属名。

Holocryptis nymphula **(Rebel, 1909) 黑斑全夜蛾**

Troctoptera nymphula Rebel, 1909, *Verhandlungen der k.k. Zoologisch-Botanischen Gesellschaft in Wien*, 59: 375. Type locality: Russia, Khabarovsk.

Distribution. China (Jilin, Liaoning); Russia (Siberia), Korea, Japan.

注：首次给出中文种名。

***Holocryptis ussuriensis* (Rebel, 1901) 乌苏里全夜蛾**

Troctoptera ussuriensis Rebel, 1901, *in* Staudinger & Rebel, *Catalog der Lepidopteren des Palaearctischen Faunengebietes*, 2: 231. Type locality: Russia, Vladivostok.

Distribution. China (Liaoning); Russia (Siberia), Korea, Japan.

注：首次给出中文种名。

Genus ***Remmigabara*** Kononenko, Han & Matov 2010 雷夜蛾属

Remmigabara Kononenko, Han & Matov, 2010, *Zootaxa*, 2679: 57. Type species: *Paragabara secunda* Remm, 1983 [Russia, Primorye terr.].

Monotypis genus, distributed in the Chinese subregion of the Palaearctic region.

注：首次给出中文属名。

***Remmigabara secunda* (Remm, 1983) 雷夜蛾**

Paragabara secunda Remm, 1983, *Entomologisckoe Obozrenie,* 62 (3): 596: 7, 8. Type locality: Russia, Primorye terr., Tigrovoi.

Distribution. China (Liaoning, Beijing); Russia (Siberia), Korea.

注：首次给出中文种名。

Subfamily Boletobiinae, *incertae seids* 菌夜蛾亚科内隶属未明确

Holloway (2011) placed the genera *Prolophota*, *Ectogonia* and some other to the group of miscellaneous taxa in the subfamily Boletobiinae.

Genus ***Prolophota*** Hampson, 1896 斑夜蛾属

Prolophota Hampson, 1896, *The Fauna of British India, including Ceylon and Burma*, 4: 546. Type species: *Prolophota trigonifera* Hampson, 1896 [Sri Lanka].

The genus includes 4 species distributed in the Palaearctic and Oriental regions; 1 species is in the Three Provinces of Northeast China.

注：首次给出中文属名。

Prolophota trigonifera Hampson, 1896 三角缘斑夜蛾

Prolophota trigonifera Hampson, 1896, *The Fauna of British India, including Ceylon and Burma*, 4: 547, fig. 284. Type locality: Sri Lanka, Hambantota.

Distribution. China (Liaoning, Taiwan); South Korea, Japan, Sri Lanka.

注：首次给出中文种名。

Genus *Ectogonia* Hampson, 1896 角衣夜蛾属

Ectogonia Hampson, 1896, *The Fauna of British India, including Ceylon and Burma*, 4: 543. Type species: *Camptochilus viola* Hampson, 1895 [Sikkim].

Synonymy: *Camptochilus* Hampson, 1895.

The genus includes 3 species, distributed in the world; 1 species is in the Three Provinces of Northeast China.

Ectogonia butleri (Leech, 1900) 褐角衣夜蛾

Camptochilus butleri Leech, 1900, *Transactions of the Entomological Society of London*, 1900: 647. Type locality: Japan, Nagasaki, Kyushu.

Distribution. China (Jilin, Liaoning); South Korea, Japan.

注：首次给出中文种名。

Tribe **Trisatelini** Beck, 1999 三线夜蛾族

Trisatelini Beck, 1999, *Entomologist*, 72: 141. Type genus: *Trisateles* Tams, 1939.

The monotypic tribe with single species distributed in the temperate zone of the Palaearctic. The larva feed on dry leaves of deciduous trees. 1 genus of the tribe occurs in the Three Provinces of Northeast China.

References. Chen, 1982, 1999 (Noctuidae: Ophiderinae); Sugi, 1982 (Noctuidae: Acontiinae, part); Kononenko *et al.*, 1998 (Noctuidae: Acontiinae, part); Fibiger, Hacker, 2005 (Erebidae: Phytometrinae, part); Kononenko, 2003 (Noctuidae: Acontiinae, part), 2005 (Erebidae: Eublemminae), 2010 (Noctuidae: Aventiinae); Kononenko, Han, 2007 (Noctuidae: Aventiinae); Matov *et al.*, 2008 (Noctuidae: Aventiinae); Fibiger *et al.*, 2010 (Noctuidae: Phytometrinae, part); Zahiri *et al.* 2012 (Erebidae: Boletobiinae, Trisatelini).

注：首次给出中文族名。

Genus *Trisateles* Tams, 1939 三线夜蛾属

Trisateles Tams, 1939, *Entomologist*, 72: 141. Type species: *Pyralis emortualis* [Denis & Schiffermuller], 1775 [Austria].

Synonymy: *Aethia* Hübner, 1816, preocc. (Merren, 1788 [Aves]); *Sophronia* Duponchel, 1845, preocc. (Hübner, [1825] 1816 [Lepidoptera, Gelechiidae]); *Standfussia* Spuler, 1907, preocc. (Tutt, 1900 [Lepidoptera, Psychidae]).

Monotypic Palaearctic genus; 1 species is in the Three Provinces of Northeast China.

注：首次给出中文属名。

Trisateles emortualis **([Denis & Schiffermüller], 1775) 三线夜蛾**

Pyralis emortualis [Denis & Schiffermüller], 1775, *Ankündung eines Systematisches Werkes von den Schmetterlinge der Wiener Gegend. Wien*: 120. Type locality: Austria, Vienna district.

Synonymy: *Crambus emortuatus*: Haworth, 1809, emend..

Distribution. China (Jilin); Russia (European part, Siberia), Korea, Japan; Near East, Europe.

寄主：夏栎（*Quercus robur*, *Quercus* spp.）、（*Acer campestris*）、山毛榉属（*Fagus*）、桦木属（*Betula*）、鹅耳枥属（*Carpinus*）、柳属（*Salix*）、椴树属（*Tilia*）植物（Matov & Kononenko, 2012）。

注：首次给出中文种名。

Tribe **Eublemmini** Forbes, 1954 猎夜蛾族

Eublemmini Forbes, 1954, *Memoirs of the Cornell University agriculture Experimental Station*, 329: 271. Type genus: *Eublemma* Hübner, [1821].

Synonymy: Anthophilidae Duponchel, [1845] 1844 unavailable (art. 39 ICZN); Micradi Stephens, 1850; Eublemmini Franclemont & Todd, 1983.

The tribe distributed over the World mainly in the tropical zone and warm arid steppe and semidesert areas. Moths are often with daytime activity. The known larvae often feed on flowers or seeds of their foodplants. Over 500 species are in the world fauna, many tropical species are not described or not revised yet; 2 genera and 3 species are in the Three Provinces of Northeast China.

References. Hampson, 1910 (Noctuidae: Acontianae); Chen, 1982, 1999 (Noctuidae: Acontiinae); Sugi, 1982 (Noctuiae: Acontiinae); Kononenko *et al.*, 1998 (Noctuidae: Acontiinae); Kononenko, 2003 (Noctuidae: Acontiinae, part), 2005 (Erebidae: Eublemminae), 2010 (Noctuidae: Eublemminae); Fibiger, Lafontaine, 2005 (Erebidae: Eublemminae); Fibiger, Hacker, 2005 (Erebidae: Eublemminae); Lafontaine, Fibiger, 2006 (Noctuidae: Eublemminae, Eublemmini); Kononenko, Han, 2007 (Noctuidae: Eublemminae); Matov *et al.*, 2008 (Noctuidae: Eublemminae); Kononenko, Matov, 2009 (Noctuidae: Eublemminae); Holloway, 2009 (Noctuidae: Eublemminae), 2011 (Erebidae: Boletobiinae, Eublemmini); Fibiger *et al.*, 2010 (Noctuidae: Eublemminae), 2011 (Erebidae: Boletobiinae, Eublemmini); Lafontaine, Schmidt, 2010 (Erebidae: Eublemminae); Zahiri *et al.*, 2012 (Erebidae: Boletobiinae, Eublemmini); Zahiri *et al.*, 2012 (Erebidae: Boletobiinae, Eublemmini); Lafontaine, Schmidt, 2013 (Erebidae: Boletobiinae, Eublemmini); Kononenko, Pinratana, 2013 (Erebidae: Boletobiinae, Eublemmini).

注：首次给出中文族名。

Genus *Odice* Hübner, [1823] 1816 灰猎夜蛾属

Odice Hübner, [1823] 1816, *Verzeichniss bekannter Schmettlinge* [sic!]: [258]. Type species: *Noctua inamoena* Hübner, [1803] [Europe] (=*Phalaena arcuinna* [Hübner, 1790]).

Synonymy: *Glossodice* Berio, 1991.

The genus includes 11 species, distributed in the Southwest Palaearctic region and North Africa; 1 species is in the Three Provinces of Northeast China.

注：首次给出中文属名。

Odice arcuinna (Hübner, 1790) 灰猎夜蛾

Phalaena arcuinna Hübner, 1790, *Beiträge zur Geschichte der Schmetterlinge*, 2 (4): 93, pl. 4: O. Type locality: Germany, Augsburg.

Synonymy: *Noctua inamoena* Hübner, [1803]; *Noctua argillacea* Tauscher, 1809; *Noctua ingrata* Herrich-Schaffer, 1845; *Cucullia pergrata* Rambur, 1858.

Distribution. China (Liaoning, Nei Mongol, Xinjiang, Hebei, Shaanxi, Shandong); Russia (European part, Ural, Altai), Mongolia, Korea; Near East, Europe.

寄主：昭苏滇紫草（*Onosma echioides*）（Matov & Kononenko, 2012）。

Genus *Eublemma* Hübner, [1821] 1816 猎夜蛾属

Eublemma Hübner, [1821] 1816, *Verzeichniss bekannter Schmettlinge* [sic!]: 256. Type species: *Noctua amoena* Hübner, [1803] [Europe] (=*Phalaena respersa* Hübner, 1790).

Synonymy: *Anthophilae* Hübner, [1806], suppr.; *Anthophila* Ochsenheimer, 1816, preocc. (Haworth, 1822 [Lepidoptera, Glyphipterigidae]); *Porphyrinia* Hübner, [1821] 1816; *Eromene* Hübner, [1821] 1816; *Trothisa* Hübner, [1821] 1816; *Ecthetis* Hübner, [1821] 1816; *Odice* Hübner, [1821] 1816; *Heliomanes* Sodoffsky, 1837, repl. name; *Microphisa* Boisduval, 1840; *Micra* Guenée, 1841; *Microphysa*: Guenée, 1841, emend., preocc. (Westwood, 1834 [Hemiptera]); *Microphysa*: Agassiz, 1846, emend.; preocc. (Westwood, 1834 [Hemiptera]); *Glaphyra* Guenée, 1841, preocc. (Newman, 1840 [Coleoptera]); *Thalpochares* Lederer, 1853, repl. name; *Mixocharis* Lederer, 1853; *Autoba* Walker, [1863] 1864; *Silda* Walker, 1863; *Vescisa* Walker, 1864; *Zalaca* Walker, [1866] 1865; *Mestleta* Walker, 1865; *Eumestleta* Butler, 1892; *Eublemmoides* Bethune-Baker, 1906; *Thalomicra* Spuller, 1907; *Coccidiphaga* Spuler, 1907; *Zonesthiousa* Thierry-Mieg, 1907; *Polyorycta* Warren, 1911; *Gyophora* Warren, 1913; *Smicroloba* Warren, 1913; *Anthophya*: Duponchel, 1929, missp.; *Eumicremma* Berio, 1954; *Eupsoropsis* Berio, 1969; *Honeyania* Berio, 1989; *Rhyplaga* Nye, 1975; *Roseoblemma* Beck, 1996; *Panoblemma* Beck, 1996; *Parvablemma* Beck, 1996.

The genus includes about 460 species (including recently described by Hacker from Africa in 2019), distributed in the Old and New Worlds, mainly in arid, subtropical and tropical areas, in the Palaearctic region approximately 50 species; 2 species are in the Three Provinces of Northeast China.

Eublemma ostrina **(Hufnagel, 1808) 小紫猎夜蛾**

Noctua ostrina Hübner, 1808, *Sammlung Europaischer Schmetterlinge. Eulen*, 2, 3, pl. 85: 399. Type locality: Europe.

Synonymy: *Anthophila numida* Lukas, 1849; *Micra carthami* Herrich-Schäffer, 1851; *Micra ostrina aestivalis* Guenée, 1852 (var.); *Porphyrinia thasia* Koutsaftikis, 1973.

Distribution. China (Jilin); Russia (Southeast European part, Ural, Altai), Mongolia, Kazakhstan; Central Asia, Near East, South Europe, North Africa.

寄主：刺苞菊（*Carlina corymbosa*）、欧洲刺苞菊（Carlina vulgaris）、硬叶蓝刺头（*Echinops ritro*）、蓝刺头（*Echinops sphaerocephalus*）、白叶蜡菊（*Helichrysum angustifolium*）、沙生蜡菊（Helichrysum arenarium）、飞廉属（*Carduus*）、蓟属（*Cirsium*）、菜蓟属（*Cynara*）、大翅蓟属（*Onopordon*）植物（Matov & Kononenko, 2012）。

注：首次给出中文种名。

Eublemma amasina **(Eversmann, 1842) 桃红猎夜蛾**

Anthophila amasina Eversmann, 1842, *Bulletin de la Société Impériale des Naturalistes de Moscou*, 1842 (3): 555. Type locality: Russia, Orenburg.

Synonymy: *Anthophila paradisea* Butler, 1878.

Distribution. China (Heilongjiang, Jilin, Hebei, Shaanxi, Jiangsu, Hubei); Russia (Ural, Altai, Tuva, Transbaikalia, East), Korea, Japan; Central Asia.

寄主：日本蓟 (*Cirsium nipponicum*)（Matov & Kononenko, 2012）。

Tribe **Phytometrini** Hampson, 1913 植夜蛾族

Phytometrinae Hampson, 1913, *Catalogue of the Lepidoptera Phalaenae in the British Museum*, 12: 401. Type genus: *Phytometra* Haworth, 1809, type species of *Phytometra* Haworth, 1809 incorrectly identified as *Phalaena festucae* Linnaeus, 1758, now in *Plusia* Ochsenheimer, 1816.

Synonymy: Phytometrini Wiltshire, 1990.

Fibiger & Lafontaine (2005) listed 14 genera in the subfamily, mainly distributed in the Old World and partially in the New Worlds, predominantly in the tropical and subtropical areas with a total of 196 species in the World fauna. 2 species of 2 genera are in the Three Provinces of Northeast China.

References. Chen, 1982, 1999 (Noctuidae); Sugi, 1982 (Noctuidae: Catocalinae); Kononenko *et al.*, 1998 (Noctuidae: Catocalinae); Sviridov, 2003 (Noctuidae: Catocalinae, part); Fibiger, Hacker, 2005 (Erebidae: Phytometrinae); Fibiger, Lafontaine, 2005 (Erebidae: Phytometrinae); Kononenko, 2005 (Erebidae: Phytometrinae), 2010 (Noctuidae: Phytometrinae); Lafontaine, Fibiger, 2006 (Noctuidae: Phytometrinae);

Kononenko, Han, 2007 (Noctuidae: Phytometrinae); Matov *et al.*, 2008 (Noctuidae: Phytometrinae); Holloway, 2009 (Noctuidae: Phytometrinae), 2011 (Erebidae: Boletobiinae, Phytometrini); Fibiger *et al.*, 2010 (Noctuidae: Phytometrinae), 2011 (Erebidae: Boletobiinae, Phytometrini); Zahiri *et al.*, 2012 (Erebidae: Boletobiinae, Phytometrini); Kononenko, Pinratana, 2013 (Erebidae: Boletobiinae, Eublemmini).

注：首次给出中文族名。

Genus ***Phytometra*** Haworth, 1809 植夜蛾属

Phytometra Haworth, 1809, *Lepidoptera Britannica sistens Digestionem novam Insectorum Lepidopterorum que in Magna Britannia Reperiuntur, Larvarum pabulo, Temporeque Pascendi; Expansione Alarum; Mensibusque Voland; Synonymis Atque Locis Observationibusque Varris*, Pt. 2: 254. Type species: *Noctua acarnea* [Denis & Schiffermüller], 1775 [Austria].

Synonymy: *Antarchaea* Hübner, [1821] 1816; *Prothymia* Hübner, [1823]; *Nanthilda* Blanchard, 1840; *Pyralidesthes* Warren, 1913.

The genus includes 62 species, mainly distributed in the subtropical and tropical of the Palaearctic, Oriental, Ephiopean and Nearctic regions; among them 1 species occurs in the Three Provinces of Northeast China.

注：首次给出中文属名。

Phytometra amata **(Butler, 1879) 红线植夜蛾**

Rhodaria amata Butler, 1879, *Illustrations of Typical Specimens of Lepidoptera Heterocera in the Collection of the British Museum*, 3: 72, pl. 57: 11. Type locality: Japan, Yokohama.

Synonymy: *Pyralidestes inamoena* Filipjev, 1925.

Distribution. China (Liaoning); Russia (Siberia), Korea, Japan.

寄主：麻栎（*Quercus acutissima*）（Matov & Kononenko, 2012）。

注：首次给出中文种名。

Genus ***Colobochyla*** Hübner, [1825] 1816 残夜蛾属

Colobochyla Hübner, [1825] 1816, *Verzeichniss bekannter Schmettlinge* [sic!]: 344. Type species: *Pyralis salicalis* [Denis & Schiffermüller], 1775 [Austria].

Synonymy: *Salia* Hübner, 1806, rejected name; *Cholobochyla*: Hübner, 1816, missp.; *Madopa* Stephens, 1829; *Colobochila* Agassiz, 1846, emend.; *Calobochila*: Walker, [1859] 1858, missp..

The genus includes 9 species, distributed in the Old and New Worlds, mainly in the tropical and subtropical areas, 2 species in the Palaearctic region; only *C. salicalis* ([Denis & Schiffermüller], 1775) is in the Three Provinces of Northeast China.

***Colobochyla salicalis* ([Denis & Schiffermüller], 1775) 残夜蛾**

Pyralis salicalis [Denis & Schiffermüller], 1775, *Ankündung eines Systematisches Werkes von den Schmetterlinge der Wiener Gegend. Wien*: 122. Type locality: Austria, Vienna district.

Synonymy: *Phalaena obliquata* Fabricius, 1794; *Crambus salicatus* Haworth, 1809; *Amblygoes cinerea* Butler, 1879, emend.; *Madopa salicatus laetalis* Staudinger, 1892 (var.).

Distribution. China (Heilongjiang, Jilin, Liaoning, Xinjiang, Hebei, Shandong, Henan, Fujian); Russia (European part, Ural, Siberia), Korea, Japan, Kazakhstan; Near East, Europe.

寄主：新疆杨（*Populus pyramidalis*）、欧洲山杨（*Populus tremula*）、白柳（*Salix alba*）、黄华柳（*Salix atrocinerea*）、黄花柳（*Salix caprea*）、灰柳（*Salix cinerea*）、（*Salix divers*）（Matov & Kononenko, 2012）。

Subfamily **Erebinae** Leach, [1815] 1830 目夜蛾亚科

Erebida Leach, [1815] 1830, *in* Brewster, *The American Edition of the New Edinburgh Encyclopedia, Conducted by David Brewster — Assisted by upwards of One Hundred Gentlemen in Europe*, 9: 134. Type genus: *Erebus* Latreille, 1810.

Synonymy: Erebidae Guenée, 1852; Erebinae Forbes, 1954; Erebini Forbes, 1954; Thysaniini Grote, 1895.

Erebinae is biggest subfamily in Erebidae and Noctuoidea and comprises approx 10 000 species world wide. Initially it was treated as a distinct family with subfamilies Rivulinae, Boletobiinae, Araeopteroninae, Hypenodinae, Eublemminae, Herminiinae, Hypeninae, Phytometrinae, Aventiinae, Erebinae, Calpinae and Catocalinae (Fibiger & Lafontaine, 2005) and in subsequent faunal publications (Hacker & Fibiger, 2005; Kononenko, 2005). Further Lafontaine & Fibiger (2006) changed the status of the Erebinae to subfamily and listed tribes as the subfamilies of the family Noctuidae. Then it was raised again to the family (Lafontaine & Schmidt, 2010; Holloway, 2011; Zahiri *et al.,* 2012) and expanded by including above listed subfamilies and Lymantriinae, Arctiinae, Pangraptinae, Scoliopteryginae, Hypocalinae, Scolecocampinae, Anobinae and Eulepidotinae. Currently it is s treated as a subfamily of the family Erebidae (sensu Lafontaine & Fibiger, 2006) with tribes Acantolipini, Sypnini, Erebini, Catocalini, Cocytini, Euclidini, Toxocampini, Ophiusini, Poaphilini, Pericimini, Ommatophorini, Hulodini and Hypopyrini (Holloway, 2011; Zahiri *et al.,* 2012). The genera of unclear systematic position are placed to Erebinae of incertae sedis. The genera *Arcte* and *Cyclodes* (tribe Arcteini) were replaced by Holloway (2009) to the subfamily Pantheinae, then to subfamily Dyopsinae of the Noctuidae (Zahiri *et al.,* 2013b); the tribe Toxocampini rised to subfamily (Fibiger *et al.*, 2010).

The subfamily Erebinae numbers 12 tribes, 24 genera and 68 species are in the Three Provinces of Northeast China.

References. Hampson, 1913a, 1913b (Noctuidae: Catocalinae); Seitz, [1907] –1914 (Noctuidae: Catocalinae); Boursin, 1955 (Noctuidae: Catocalinae); Berio, Fletcher, 1958 (Noctuidae: Catocalinae); Chen, 1982, 1999

(Noctuidae: Catocalinae, Ophiderinae, part); Sugi, 1982 (Noctuidae: Catocalinae, Ophiderinae, part); Kononenko, 1990 (Noctuidae: Catocalinae, part, Ophiderinae, part); Kononenko *et al.*, 1998 (Noctuidae: Catocalinae, part, Ophiderinae, part); Goater *et al.*, 2003 (Noctuidae: Catocalinae); Sviridov, 2003 (Noctuidae: Catocalinae); Goater *et al.*, 2003 (Noctuidae, Catocalinae); Fibiger, Hacker, 2005 (Erebidae); Fibiger, Lafontaine, 2005 (Erebidae); Holloway, 2005 (Noctuidae: Catocalinae), 2011 (Erebidae: Erebinae); Kononenko, 2005 (Noctuidae: Erebinae, Catocalinae), 2010 (Erebidae: Catocalinae); Kononenko, Pinratana, 2005 (Noctuidae: Catocalinae), 2013 (Erebidae: Erebinae); Lafontaine, Fibiger, 2006 (Noctuidae: Erebinae); Kononenko, Han, 2007 (Noctuidae: Catocalinae); Matov *et al.*, 2008 (Noctuidae: Erebinae, Catocalinae); Holloway, 2008 (Noctuidae: Catocalinae), 2011 (Erebidae: Erebinae, Catocalinae); Fibiger *et al.*, 2011 (Noctuidae: Erebinae); Zahiri *et al.*, 2012 (Erebidae: Erebinae).

注：首次给出中文亚科名。

Tribe **Arytrurini** Fibiger & Lafontaine, 2005 大棱夜蛾族

Arytrurini Fibiger & Lafontaine, 2005, *Esperiana*, 11: 31. Type genus: *Aryrtrura* John, 1912.

Synonymy: Arytrurini Goater, Ronkay & Fibiger 2003, not available (art. 16.1 ICZN).

The tribe includes 1 genus with 2 species, distributed in the Palaearctic temperate zone. Fibiger *et al.* (2011) placed the genus to the tribe Pericymini, while do not synonymised that tribes. 1 genus with 2 species occurs in the Three Provinces of Northeast China.

References. Chen, 1982, 1999 (Noctuidae: Ophiderinae, part); Sugi, 1982 (Noctuidae: Ophiderinae, part); Kononenko, 1990 (Othreinae, part); Kononenko *et al.*, 1998 (Noctuidae: Catocalinae, part); Goater *et al.*, 2003 (Noctuidae: Catocalinae, Catocalini, Arytrurina); Sviridov, 2003 (Noctuidae: Catocalinae, part); Fibiger, Hacker, 2005 (Erebidae: Catocalinae, Arytrurini); Fibiger, Lafontaine, 2005 (Erebidae: Catocalinae, Arytrurini); Kononenko, 2005 (Erebidae: Catocalinae, Arytrurini), 2010 (Noctuidae: Catocalinae, Arytrurini); Lafontaine, Fibiger, 2006 (Noctuidae: Catocalinae, Arytrurini); Fibiger *et al.*, 2011 (Pericymini, part).

注：首次给出中文族名。

Genus ***Arytrura*** John, 1912 大棱夜蛾属

Arytrura John, 1912, *Revue Russe d'Entomologie*, 12: 133. Type species: *Zethes musculus* Ménétriès, 1859 [Russia, Amur; Japan].

Synonymy: *Megazethes* Warren, 1913; *Diapolia* Hampson, 1926.

The genus includes 2 species, *A. musculus* and *A. subfalcata* in the Three Provinces of Northeast China.

Arytrura musculus **(Ménétriès, 1859) 大棱夜蛾**

Zethes musculus Ménétriès, 1859, *Bulletin de la Classe Physico-Mathématique de l'Académie Impériale des Sciences de St.-Pétersbourg*, 17: 220. Type locality: Russia, [Primorye terr.] "l'Oussuri"; [Khabarovsk reg.] "l'Amour"; Japan.

Distribution. China (Heilongjiang, Jilin, Fujian, Guizhou; East Part); Russia (Ural, Siberia), Korea, Japan; Europe.

寄主：欧洲山杨（*Populus tremula*）、黄花柳（*Salix caprea*）、灰柳（*Salix cinerea*）、蒿柳（*Salix viminalis*）（Matov & Kononenko, 2012）。

***Arytrura subfalcata* (Ménétriès, 1859) 镰大棱夜蛾**

Zethes subfalcata Ménétriès, 1859, *Bulletin de la Classe Physico-Mathématique de l'Académie Impériale des Sciences de St.-Pétersbourg*, 17: 220. Type locality: Russia, "Amur".

Synonymy: *Zethes limbalis* Swinhoe, 1917.

Distribution. China (Heilongjiang, Jilin, Liaoning); Russia (Siberia), Korea.

Tribe **Catocalini** Boisduval, [1828] 裳夜蛾族

Catocalini Boisduval, [1828], *Europaeorum Lepidopterorum Index Methodicus*: 97, incorrect original spelling. Type genus: *Catocala* Schrank, 1802.

Synonymy: Catocalides Boisduval, 1840; Audeini Wiltshire, 1990; Exophylini Beck, 1996.

The European and north Palaearctic fauna of Catocalinae (in old sense) has been divided into 12 tribes (Goater *et al.*, 2003). At present 7 tribes are recognised within the European fauna (Fibiger & Lafontaine, 2005; Lafontaine & Fibiger, 2006), and more tribes from the tropical Southeast Asia fauna were recognised by Holloway (2005). In current treatment the tribe Catocalini contains only the genus *Catocala* and several related genera of Afrotropic fauna.

1 genus is in the Three Provinces of Northeast China.

References. Hampson, 1913 (Noctuidae: Catocalinae, part); Mell, 1936 (Noctuidae: Catocalinae, part); Berio, Fletcher, 1958 (Noctuidae: Catocalinae, part); Chen, 1982, 1999 (Noctuidae: Catocalinae, part); Sugi, 1982 (Noctuidae: Catocalinae, part); Kononenko, 1990 (Catocalinae, part); Kononenko *et al.*, 1998 (Noctuidae: Catocalinae, part); Goater *et al.*, 2003 (Noctuidae: Catocalinae, Catocalini, Catocalina); Sviridov, 2003 (Noctuidae: Catocalinae, part); Fibiger, Hacker, 2005 (Erebidae: Catocalinae); Fibiger, Lafontaine, 2005 (Erebidae: Catocalinae); Holloway, 2005 (Erebidae: Catocalinae, Catocalini); Kononenko, 2005 (Erebidae: Catocalinae, Catocalini), 2010 (Noctuidae: Catocalinae, Catocalini); Lafontaine, Fibiger, 2006 (Noctuidae: Catocalinae, Catocalini); Fibiger *et al.*, 2011 (Erebidae: Erebinae, Catocalini); Zahiri *et al.*, 2012 (Erebidae: Erebinae, Catocalini).

Genus ***Catocala*** Schrank, 1802 裳夜蛾属

Catocala Schrank, 1802, *Fauna Boica. Durchgedachte Geschichte der in Baiera Einheimischen und Zahmen Thiere*, 2 (2): 158. Type species: *Phalaena nupta* Linnaeus, 1767 [Germany].

Synonymy: *Blepharum* Hübner, [1806], rejected name; *Blepharum* Hübner, 1808, rejected name;

Hemigeometra Haworth, 1809; *Blephara* Ochsenheimer, 1816, unavailable name; *Ephesia* Hübner, 1818; *Lamprosia* Hübner, [1821] 1806; *Blepharidia* Hübner, [1822]; *Astiotes* Hübner, [1823] 1816; *Corsice* Hübner, [1823]; *Eucora* Hübner, [1823] 1816; *Eunetis* Hübner, [1823]; *Mormonia* Hübner, [1823]; *Blepharonia* Hübner, [1823], unavailable name; *Blepharonia* Hübner, [1825]; *Blepharidia* Hübner, 1828, unavailable name; *Mormosia* Walker, [1858] 1857; *Corisee* Walker, [1858], missp.; *Andrewsia* Grote, 1882; *Alura* Möschler, 1884; *Catabapta* Hulst, 1884; *Koraïa* Herz, 1904, incorrect original spelling; *Ulotrichopus* Wallengren, 1860 (subgen.); *Simplicala* Beck, 1966; *Convercala* Beck, 1966; *Eucala* Beck, 1966; *Divercala* Beck, 1966; *Puercala* Beck, 1966; *Reticcala* Beck, 1966; *Optocala* Beck, 1966; *Metacala* Beck, 1966; *Promonia* Beck, 1966; *Bihemena* Beck, 1966; *Koraia* Nye, 1975; *Catocola*: Oken, 1815, missp.; *Catocalla*: Hübner, [1823], missp.; *Andreusia*: Hampson, 1913, emend.; *Koraia*: Nye, 1975, emend..

The genus includes more than 240 species, distributed in the cool temperate zone of the Holarctic, with almost equal number of species are in the Palaearctic (130 species) and Nearctic (113 species) regions. In the Palaearctic, the genus is most diverse in the broad-leaved and mixed forest zone in Central and West China; 33 species are in the Three Provinces of Northeast China.

Catocala fulminea (Scopoli, 1763) 光裳夜蛾

Phalaena fulminea Scopoli, 1763, *Entomologia Carniolica Exhibens Insecta Carnioliae Indigena et Distributa in Ordines, Genera, Species, Varietates Methodo Linnaeana*: 206. Type locality: Austria, Vienna district.

Synonymy: *Phalaena maturna* Hufnagel, 1766; *Phalaena paranympha* Linnaeus, 1767; *Catocala protonympha* Boisduval, 1840; *Catocala fulminea xarippe* Butler, 1877 (ssp.); *Ephesia fulminea chekiangensis* Mell, 1933 (ssp.).

Distribution. China (Heilongjiang, Jilin, Liaoning); Russia (European part, Ural, Siberia), Japan, Korea; Europe.

寄主：槲树（*Quercus dentata*）、夏栎（*Quercus robur*）、锐刺山楂（*Crataegus oxyacantha*）、山楂（*Crataegus pinnatifida*）、辽宁山楂（*Crataegus sanguinea*）、苹果（Malus pumila）、三叶海棠（*Malus sieboldii*）、北亚稠李（*Padus racemosa*）、白桃（*Persica vulgaris*）、欧洲李（*Prunus domestica*）、梅花（*Prunus mume*）、李（*Prunus salicina*）、黑刺李（*Prunus spinosa*）、樱属（*Cerasus*）、梨属（*Pyrus*）植物（Matov & Kononenko, 2012）。

Catocala obscena Alphéraky, 1879 奥裳夜蛾

Catocala obscena Alphéraky, 1879, *Deutsche Entomologische Zeitschrift Iris*, 8: 196. Type locality: Korea.

Synonymy: *Catocala obscena baihi* Ishiizuka, 2003 (ssp.).

Distribution. China (Jilin, Liaoning, Hebei, Sichuan, Yunnan); Russia (Siberia), Korea, Japan.

Catocala abamita Bremer & Grey, 1853 晦刺裳夜蛾

Catocala abamita Bremer & Grey, 1853, *Beiträge zur Schmetterlings-fiauna des NoerdlicheNorth China's*: 19. Type locality: North China, Beijing.

Synonymy: *Catocala scortum* Christoph, 1893.

Distribution. China (Jilin, Liaoning, Beijing, Hebei, Shandong, Jiangsu, Jiangxi, Fujian); Russia (Siberia), Korea.

Catocala doerriesi Staudinger, 1888 茂裳夜蛾

Catocala doerriesi Staudinger, 1888, *Entomologische Zeitung. Entomologischen Vereine zu Stettin*, 49: 271. Type locality: Russia, "Radeffka" [Khabarovsk reg., Judish Autonomy, Radde]; "Ussuri" [Primorye terr.].

Synonymy: *Catocala honrathi* Graeser, 1888 [1889]; *Catocala hampsoni* Leech, 1900.

Distribution. China (Heilongjiang, Henan, Hubei); Russia, Korea.

Catocala eminens Staudinger, 1892 兴光裳夜蛾

Catocala eminens Staudinger, 1892, *in* N.M. Romanoff (ed.), *Mémoires sur les Lépidoptères*, 6: 584, pl. 12: 5. Type locality: "Amur" [most probably South of the Primorye terr.].

Distribution. China (Heilongjiang, Liaoning, Zhejiang, Hunan); Russia (Siberia), Korea.

Catocala separans Leech, 1889 基暗裳夜蛾

Catocala separans Leech, 1889, *Proceedings of the Zoological Society of London*, 1889: 552, pl. 53: 6. Type locality: Japan, Nagahama, Fushiki.

Synonymy: *Catocala hetaera* Staudinger, 1892.

Distribution. China (Jilin); Russia (Siberia), Korea, Japan.

寄主：麻栎（*Quercus acutissima*）、栓皮栎（*Quercus variabilis*）（Matov & Kononenko, 2012）。

注：首次给出中文种名。

Catocala duplicata Butler, 1885 纹光裳夜蛾

Catocala duplicata Butler, 1885, *Cistula Entomologica*, 3: 135. Type locality: Japan.

Synonymy: *Catocala suzukii* Matsumura, 1911.

Distribution. China (Heilongjiang, Liaoning); Russia (Siberia), Korea, Japan.

寄主：麻栎（*Quercus acutissima*）、大叶栎（*Quercus crispula*）、青冈（*Quercus glauca*）、短柄枹栎（*Quercus glandulifera*）、大桴栎（*Quercus grosseserrata*）、蒙古栎（*Quercus mongolica*）、枹栎（*Quercus serrata*）、栓皮栎（*Quercus variabilis*）（Matov & Kononenko, 2012）。

Catocala helena Eversmann, 1856 珀光裳夜蛾

Catocala helena Eversmann, 1856, *Bulletin de la Société Impériale des Naturalistes de Moscou*, 30 (4): 378, pl. 1: 8. Type locality: Russia, Southeast Siberia, Transbaikalia, Kiachta.

Synonymy: *Ephasia* [sic!] *helena kurenzovi* Moltrecht, 1927 (ssp.); *Ephesia helena beicki* Mell, 1936 (ssp.).

Distribution. China (Heilongjiang, Jilin, Liaoning); Russia (Transbaikalia, Siberia), Mongolia, Korea.

Catocala nymphaeoides Herrich-Schäffer, 1845 宁裳夜蛾

Catocala nymphaeoides Herrich–Schäffer, 1845, *Systematische Bearbeitung der Schmetterlinge von Europa*, 6: 90: 599, 600. Type locality: Russia [most probably South Transbaikalia].

Synonymy: *Catocala davidi* Oberthür, 1881; *Catocala nymphula* Staudinger, 1892.

Distribution. China (Heilongjiang, Jilin); Russia (Transbaikalia, Siberia), Mongolia.

Catocala ella Butler, 1877 意光裳夜蛾

Catocala ella Butler, 1877, *Cistula Entomologica*, 2: 242. Type locality: Japan, Yokohama.

Synonymy: *Catocala nutrix* Graeser, 1888 [1899].

Distribution. China (Heilongjiang, Sichuan, Taiwan); Russia (Siberia), Korea, Japan.

寄主：辽东桤木（*Alnus hirsuta*）、日本桤木（*Alnus japonica*）、三叶海棠（*Malus sieboldii*）、梨属（*Pyrus*）植物（Matov & Kononenko, 2012）。

Catocala puella Leech, 1889 朝鲜裳夜蛾

Catocala puella Leech, 1889, *Proceedings of the Zoological Society of London*, 1889: 551, pl. 53: 5. Type locality: North Korea,Wonsan [Gensan].

Distribution. China (Heilongjiang, Liaoning, Ningxia); Korea.

Catocala deuteronympha Staudinger, 1861 显裳夜蛾

Catocala deuteronympha Staudinger, 1861, *Entomologische Zeitung. Entomologischen Vereine zu Stettin*, 22: 290. Type locality: Russia, East Siberia, Transbaikalia, Kiachta.

Synonymy: *Catocala omphale* Butler, 1881; *Catocala greyi* Staudinger, 1888; *Catocala thomsoni* Prout, 1924; *Catocala tschiliensis* Bang-Haas, 1927; *Catocala dahurica* Kljutschko, 1992.

Distribution. China (Heilongjiang, Jilin, Hebei, Fujian); Russia (Transbaikalia, Siberia), Mongolia, Korea, Japan.

寄主：春榆（*Ulmus japonica*）（Matov & Kononenko, 2012）。

Catocala praegnax Walker, 1858 前光裳夜蛾

Catocala praegnax Walker, 1858, *List of the Specimens of Lepidopterous Insects in the Collection of the British Museum*, 13: 1213.

Synonymy: *Catocala obliterata* Ménétriès, 1863; *Catocala esther* Butler, 1877; *Catocala praegnax sakaii* Kishida, 1981 (ssp.).

Distribution. China (Heilongjiang, Jiangsu, Jiangxi, Sichuan, Taiwan); Russia (Siberia), Korea, Japan.

寄主：蒙古栎（陈一心，1999）；槲栎（*Quercus aliena*）、大籽栎（*Quercus grosseserrata*）、蒙古栎（*Quercus mongolica*）、胡枝子属（*Lespedeza*）植物（Matov & Kononenko, 2012）。

Catocala agitatrix Graeser, [1889] 1888 白肾裳夜蛾

Catocala agitatrix Graeser [1889]1888, *Berliner entomologische Zeitschrift*, 32, II: 272. Type locality: Russia, [Khabarovsk].

Synonymy: *Catocala mabella* Holland, 1889.

Distribution. China (Heilongjiang, Liaoning, Henan); Russia (Transbaikalia, Siberia), Korea, Japan.

寄主：山荆子（*Malus baccata*）、毛山荆子（*Malus manshurica*）、苹果（Malus pumila）、三叶海棠（*Malus sieboldii*）、北亚稠李（*Padus racemosa*）（Matov & Kononenko, 2012）。

Catocala bella Butler, 1877 苹刺裳夜蛾

Catocala bella Butler, 1877, *Cistula Entomologica*, 2: 242. Type locality: Japan, Yokohama.

Synonymy: *Catocala serenides* Staudinger, 1888.

Distribution. China (Heilongjiang, Jilin); Russia, Mongolia, Korea, Japan.

寄主：苹果（陈一心，1999）；寄主：山荆子（*Malus baccata*）、毛山荆子（*Malus manshurica*）、苹果（*Malus pumila*）、三叶海棠（*Malus sieboldii*）（Matov & Kononenko, 2012）。

Catocala nubila Butler, 1881 淡灰裳夜蛾

Catocala nubila Butler, 1881, *Transactions of the Entomological Society of London*, 1881: 196. Type locality: Japan, Tokyo.

Distribution. China (Jilin); Russia (Transbaikalia, Siberia), Korea, Japan.

寄主：钝齿水青冈（*Fagus crenata*）（Matov & Kononenko, 2012）。

Catocala koreana Staudinger, 1892 普裳夜蛾

Catocala koreana Staudinger, 1892, *in* N.M. Romanoff (ed.), *Mémoires sur les Lépidoptères*, 6: 587. Type locality: Korea.

Synonymy: *Ephesia hymenaea ussurica* Sheljuzhko, 1943 (ssp.); *Catocala azumiensis* Sugi, 1965; *Catocala hymenaea* auct., nec [Denis et Schiffermüller], 1775.

Distribution. China (Heilongjiang, Liaoning); Russia (Siberia), Korea, Japan.

寄主：黑刺李、李、乌荆子（陈一心，1999）；小花绣线菊（*Spiraea nipponica*）（Matov & Kononenko, 2012）。

注：《中国动物志》（夜蛾科）中引用本种同物异名"*Catocala hymenaea* [Denis et Schiffermüller], 1775"，在此予以变更。

Catocala proxeneta Alphéraky, 1895 鹿裳夜蛾

Catocala proxeneta Alphéraky, 1895, *Deutsche Entomologische Zeitschrift Iris*, 8: 197. Type locality: Mongolia, Ulaan-Baator [Urga].

Synonymy: *Catocala proxeneta sutshana* Sheljuzhko, 1943 (ssp.).

Distribution. China (Heilongjiang); Russia (Transbaikalia, Siberia), Mongolia, Korea.

Catocala streckeri Staudinger, 1888 柞光裳夜蛾

Catocala streckeri Staudinger, 1888, *Entomologische Zeitung. Entomologischen Vereine zu Stettin*, 49: 272. Type locality: Russia, Primorye terr., Askold Is., [Partizansk] "Sutschan-Gebiet", Bikin.

Synonymy: *Catocala strecceri*: Hampson, 1913, emend..

Distribution. China (Heilongjiang, Jilin); Russia (Transbaikalia, Siberia, Sakhalin, Kuriles), Korea, Japan.

寄主：蒙古栎（陈一心，1999）；麻栎（*Quercus acutissima*）、青冈（*Quercus glauca*）、大桴栎（*Quercus grosseserrata*）、蒙古栎（*Quercus mongolica*）、枹栎（*Quercus serrata*）（Matov & Kononenko, 2012）。

Catocala danilovi (Bang-Haas, 1927) 丹裳夜蛾

Ephesia danilovi Bang-Haas, 1927, *Horae Macrolepidopterologiae Regionis Palaearcticae*, 1: 117. Type locality: Russia, Primorye terr., [Partizansk] "Sutschan".

Distribution. China (Jilin); Russia (Siberia), North Korea.

注：首次给出中文种名。

Catocala moltrechti O. Bang-Haas, 1927 昌裳夜蛾

Catocala moltrechti Bang-Haas, 1927, *Horae Macrolepidopterologiae Regionis Palaearcticae*, 1: 89, pl. 11:6. Type locality: Russia, Primorye terr., "Sutschan"[Partizansk].

Distribution. China (Heilongjiang); Russia (Siberia).

注：首次给出中文种名。

Catocala dissimilis Bremer, 1861 栎裳夜蛾

Catocala dissimilis Bremer, 1861, *Bulletin de l'Academie Impériale des Sciences de St.-Petersbroug*, 3: 494. Type locality: Russia, "Bureja-Gebirge" [Amur reg., Bureinsky Range].

Synonymy: *Ephesia dissimilis nigricans* Mell, 1939 (ssp.); *Ephesia dissimilis griseata* Bryk, 1948 (ssp.); *Catocala dissimilis melli* Ishizuka, 2001 (ssp.).

Distribution. China (Heilongjiang, Jilin, Liaoning, Shaanxi, Henan, Hubei, Yunnan); Russia (Transbaikalia, Siberia), Korea, Japan.

寄主：蒙古栎（陈一心，1999）；槲树（*Quercus dentata*）、大桴栎（*Quercus grosseserrata*）、蒙古栎（*Quercus mongolica*）（Matov & Kononenko, 2012）。

Catocala nagioides Wileman, 1924 圣光裳夜蛾

Catocala nagioides Wileman, 1924, *Entomologist*, 57: 287. Type locality: Japan, Honshu [Yesso], repl. name.

Synonymy: *Catocala sancta* Butler, 1885, nec Hulst, 1884.

Distribution. China (Heilongjiang, Jilin, Yunnan, Taiwan); Russia (Siberia), Korea, Japan.

寄主：槲栎（*Quercus aliena*）、槲树（*Quercus dentata*）（Matov & Kononenko, 2012）。

Catocala actaea Felder & Rogenhofer, 1874 白缘光裳夜蛾

Catocala actaea Felder & Rogenhofer, 1874, *Reise Österreichischen Fregatte Novara um die Erde in den Jahren 1857, 1859. Zoologischer Theil*, 2 (Abt. 2): pl. 112: 22. Type locality: Japan.

Synonymy: *Ephesia actaea nigricans* Mell, 1939 (ssp.).

Distribution. China (Liaoning, Hubei); Russia (Siberia), Korea, Japan.

寄主：栎（陈一心，1999）；麻栎（*Quercus acutissima*）、蒙古栎（*Quercus mongolica*）、栓皮栎（*Quercus variabilis*）（Matov & Kononenko, 2012）。

Catocala pirata (Herz, 1904) 梨裳夜蛾

Koraïa pirata Herz, 1904, *Annuaire du Musée Zoologique de l'Académie Imperiale des Sciences de St.-Petersbourg*, 9: 314, pl. 1: 18. Type locality: North Korea, ?Pungtung village.

Distribution. China (Jilin); Russia (Siberia), Korea.

注：首次给出中文种名。

Catocala bokhaica (Kononenko, 1979) 仿梨裳夜蛾

Koraia bokhaica Kononenko, 1979, *Trudy vsesoyuznogo entomologicheskogo Obshchestva*, 61: 125: 1, 2.

Type locality: Russia, Primorye terr., Benevskoe village.

Distribution. China (Jilin, Liaoning); Russia (Siberia), Korea.

注：首次给出中文种名。

Catocala fraxini (Linnaeus, 1758) 缟裳夜蛾

Phalaena fraxini Linnaeus, 1758, *Systema Naturae* (Edn. 10) 1: 512. Type locality: Europe.

Synonymy: *Catocala fraxini gaudens* Staudinger, 1901 (var.); *Catocala fraxini latefasciata* Warnecke, 1919 (var.); *Catocala jezoensis* Matsumura, 1931; *Catocala fraxini yuennanensis* Mell, 1936 (ssp.); *Catocala fraxini legionensis* Gomez-Bustillo & Vega, 1975 (ssp.).

Distribution. China (Heilongjiang, Jilin, Yunnan); Russia (European part, Siberia), Korea, Japan, Kazakhstan; Near East, Europe.

寄主：杨、柳、槭、梣（陈一心，1999）；英国山毛榉（*Fagus silvatica*）、夏栎（*Quercus robur*）、大叶钻天杨（*Populus balsamifera*）、加拿大杨（*Populus canadensis*）、钻天杨（*Populus italica*）、香杨（*Populus koreana*）、欧洲黑杨（*Populus nigra*）、西氏杨（*Populus sieboldii*）、欧洲山杨（*Populus tremula*）、垂柳（*Salix babylonica*）、黄花柳（*Salix caprea*）、爆竹柳（*Salix fragilis*）、杞柳（*Salix purpurea*）、欧洲小叶椴（*Tilia cordata*）、高精细榆树（*Ulmus campestris*）、欧洲白榆（*Ulmus laevis*）、榆（*Ulmus pumila*）、稠李（*Padus racemosa*）、栓皮槭（*Acer campestre*）、茶条槭（*Acer ginnala*）、梣叶槭（*Acer negundo*）、鞑靼槭（*Acer tataricum*）、欧洲白蜡（*Fraxinus excelsior*）、桤木属（*Alnus*）、桦木属（*Betula*）植物（Matov & Kononenko, 2012）。

Catocala lara Bremer, 1861 椴裳夜蛾

Catocala lara Bremer, 1861, *Bulletin de l'Academie Impériale des Sciences de St.-Petersbroug*, 3: 493. Type locality: Russia, "Bureja-Gebirge" [Amur reg., Bureinsky Range].

Synonymy: *Catocala lara pallidamajor* Mell, 1939 (ssp.).

Distribution. China (Heilongjiang, Jilin, Liaoning, Hebei); Russia (Siberia), Korea, Japan.

寄主：紫椴、糠椴（陈一心，1999）；紫椴（*Tilia amurensis*）、华东椴（*Tilia japonica*）、糠椴（*Tilia manshurica*）（Matov & Kononenko, 2012）。

Catocala nivea Butler, 1877 白光裳夜蛾

Catocala nivea Butler, 1877, *Cistula Entomologica*, 2: 241. Type locality: Japan, Yokohama.

Synonymy: *Catocala nivea kurosawai* Owada, 1986 (ssp.); *Catocala nivea asahinorum* Owada, 1986 (ssp.).

Distribution. China (Heilongjiang, Jilin, Liaoning, Zhejiang, Hunan, Sichuan, Taiwan); Russia (Siberia), Korea, Japan, India, Nepal.

寄主：李属（陈一心，1999）；灰叶稠李（*Prunus grayana*）、苹果属（*Malus*）植物（Matov & Kononenko, 2012）。

Catocala nupta (Linnaeus, 1767) 裳夜蛾

Phalaena Noctua nupta Linnaeus, 1767, *Systema Naturae* (Edn. 12) 1: 84. Type locality: Germany.

Synonymy: *Catocala concubia* Walker, [1858] 1857; *Catocala unicuba* Walker, [1858] 1857; *Catocala nupta obscurata* Oberthür, 1880 (var.); *Catocala nupta nuptialis* Staudinger, 1901 (var.); *Catocala nupta centralasiae* Kuznetzov, 1903 (ssp.); *Catocala nozawae* Matsumura, 1911; *Catocala nupta kansuensis* Bang-Haas, 1927 (ssp.); *Catocala nupta clara* Osthelder, 1933 (ssp.); *Catocala nupta japonica* Mell, 1936 (ssp.); *Catocala nupta likiangensis* Mell, 1936 (ssp.); *Catocala nupta alticola* Mell, 1942 (ssp.).

Distribution. China (Heilongjiang, Jilin, Liaoning, Xinjiang, Hebei, Fujian, Sichuan, Xizang); Russia (European part, Siberia), Korea, Japan, India; from Europe through the Palaearctic region to China.

寄主：杨、柳（陈一心，1999）；银白杨（*Populus alba*）、北美杨树（*Populus balsamifera*）、美洲黑杨（*Populus deltoides*）、辽杨（*Populus maximowiczii*）、钻天杨（*Populus nigra*）、欧洲山杨（*Populus tremula*）、白柳（*Salix alba*）、垂柳（*Salix babylonica*）、黄花柳（*Salix caprea*）、灰柳（*Salix cinerea*）、爆竹柳（*Salix fragilis*）、三蕊柳（*Salix triandra*）、欧洲李（*Prunus domestica*）、椴树属（*Tilia*）、山楂属（*Crataegus*）、梣属（*Fraxinus*）植物（Matov & Kononenko, 2012）。

Catocala electa (Vieweg, 1790) 柳裳夜蛾

Noctua electa Vieweg, 1790, *Tabellarisches Verzeichniss der in der Churmark Brandenburg Einheimischen Schmetterlinge*, 2: 33. Type locality: Germany, Brandenburg.

Synonymy: *Catocala zalmunna* Butler, 1877; *Catocala electa tschiliensis* Bang-Haas, 1927 (ssp.).

Distribution. China (Heilongjiang, Jilin, Xinjiang, Shandong, Henan, Hubei); Russia (European part, Siberia), Korea, Japan; Europe.

寄主：杨、柳（陈一心，1999）；细柱柳（*Salix gracilistyla*）、垂柳（*Salix babylonica*）、大白柳（*Salix maximowiczii*）、腺柳（*Salix chaenomeloides*）（Par *et al.*, 2006）；白柳（*Salix alba*）、灰柳（*Salix cinerea*）、爆竹柳（*Salix fragilis*）、匍匐柳（*Salix repens*）、细叶沼柳（*Salix rosmarinifolia*）、黑柳（*Salix nigricans*）、杞柳（*Salix purpurea*）、三蕊柳（*Salix triandra*）、蒿柳（*Salix viminalis*）、杨属（*Populus*）植物（Matov & Kononenko, 2012）。

Catocala dula Bremer, 1861 栎刺裳夜蛾

Catocala dula Bremer, 1861, *Bulletin de l'Academie Impériale des Sciences de St.-Petersbroug*, 3: 493. Type locality: Russia, "Bureja-Gebirge" [Amur reg., Bureinsky Range].

Synonymy: *Catocala dula carminea* Mell, 1939 (ssp.).

Distribution. China (Heilongjiang, Jilin, Liaoning, Nei Mongol, Henan); Russia (Transbaikalia, Siberia), Korea, Japan.

寄主：蒙古栎、槲（陈一心，1999）；钝齿水青冈（*Fagus crenata*）、麻栎（*Quercus acutissima*）、

槲树（*Quercus dentata*）、青冈（*Quercus glauca*）、大桴栎（*Quercus grosseserrata*）、蒙古栎（*Quercus mongolica*）、枹栎（*Quercu sserrata*）（Matov & Kononenko, 2012）。

***Catocala pacta* (Linnaeus, 1758) 红腹裳夜蛾**

Phalaena pacta Linnaeus, 1758, *Systema Naturae* (Edn. 10) 1: 512. Type locality: Europe.

Synonymy: *Phalaena* (*Noctua*) *suecica* Esper, 1788; *Catocala pacta deserta* Kozhanchikov, 1925 (ssp.).

Distribution. China (Heilongjiang, Jilin, Xinjiang); Russia (European part, Siberia), Mongolia, Kazakhstan; North and Central Europe.

寄主：柳（陈一心，1999）；沼柳（*Salix brachypoda*）、黄花柳（*Salix caprea*）、灰柳（*Salix cinerea*）、匍匐柳（*Salix repens*）、细叶沼柳（*Salix rosmarinifolia*）（Matov & Kononenko, 2012）。

Tribe **Cocytiini** Boisduval, 1874 同夜蛾族

Cocytides Boisduval, 1874, *Revue et Magasin de Zoologie* (s. 3) 2: 46, incorrect original spelling, must be corrected in Cocytiidae. Type genus: *Cocytia* Boisduval, 1828.

Synonymy. Cocytiidae Butler, 1875, emend.; Cocytianae Rothschild, 1915.

Tropical origin tribe with only 1 genus in the Three Provinces of Northeast China.

References. Chen, 1982, 1999 (Noctuidae: Ophiderinae, part); Sugi, 1982 (Noctuidae: Catocalinae, part); Kononenko, 1990 (Catocalinae, part); Kononenko *et al.*, 1998 (Noctuidae: Catocalinae, part); Fibiger, Lafontaine, 2005 (Erebidae: Cocytiinae); Kononenko, 2005 (Erebidae: Catocalinae, Cocytiini), 2010 (Noctuidae: Catocalinae, Cocytiini); Lafontaine, Fibiger, 2006 (Noctuidae: Cocytiinae); Matov *et al.*, 2008 (Noctuidae: Cocytiinae); Zahiri *et al.*, 2012 (Erebidae: Erebinae, Cocytiini).

注：首次给出中文族名。

Genus ***Serrodes*** Guenée, 1852 斑翅夜蛾属

Serrodes Guenée, 1852, *in* Boisduval & Guenée, *Histoire Naturelle des Insectes. Species Général des Lépidoptères*, 7: 251. Type species: *Phalaena inara* Cramer, 1779 [India].

The genus includes 12 species, distributed in the Oriental, Ethiopian and Australian regions; 1 species is in the Three Provinces of Northeast China.

***Serrodes campana* Guenée, 1852 铃斑翅夜蛾**

Serrodes campana Guenée, *in* Boisduval & Guenée, *Histoire Naturelle des Insectes. Species Général des Lépidoptères*, 7: 252, pl. 21: 6. Type locality: Bangladesh, Shilhet.

Synonymy: *Serrodes nigha* Guenée, 1852; *Serrodes callipepla* A.E. Prout, 1929.

Distribution. China (Liaoning, Zhejiang, Guangdong, Guangxi, Hainan, Sichuan, Yunnan, Taiwan); Russia

(Siberia), Korea, Japan, India(North), Nepal, Philippines, Indonesia.

寄主：无患子（陈一心，1999）；赤才（*Lepisanthes rubiginosa*）、红毛丹（*Nephelium lappaceum*）、无患子（*Sapindus mukurossi*）、毛瓣无患子（*Sapindus rarak*）、久树（*Schleichera oleosa*）、飞蛾槭（*Acer oblongum*）（Matov & Kononenko, 2012）。

Tribe **Euclidiini** Guenée, 1852 恭夜蛾族

Euclidiidae Guenée, 1852, *in* Boisduval & Guenée, *Histoire Naturelle des Insectes. Species Général des Lépidoptères*, 7: 280. Type genus: *Euclidia* Ochsenheimer, 1816.

Synonymy: Remigiidae Guenée, 1852; Mocisini Berio, 1992; Ectypina Goater, Fibiger & Ronkay, 2005, unavailable.

The tribe includes about 20 genera, mainly distributed in the tropical and subtropical regions of the Old and New worlds. Several genera occurs in the temperate zone, mainly in arid or steppe habitats. Known larvae are connected with Fabaceae. 5 genera are in the Three Provinces of Northeast China.

References. Chen, 1982, 1999 (Noctuidae); Sugi, 1982 (Noctuidae: Catocalinae, part); Kononenko, 1990 (Catocalinae, part); Kononenko *et al.*, 1998 (Noctuidae: Catocalinae, part); Goater *et al.*, 2003 (Noctuidae: Catocalinae, Catocalini, Ectypina); Sviridov, 2003 (Noctuidae: Catocalinae, part); Fibiger, Hacker, 2005 (Erebidae: Catocalinae, Euclidini); Fibiger, Lafontaine, 2005 (Erebidae: Euclidini); Kononenko, 2005 (Erebidae: Catocalinae, Euclidini), 2010 (Noctuidae: Catocalinae, Euclidini); Lafontaine, Fibiger, 2006 (Noctuidae: Catocalinae, Euclidini); Matov *et al.*, 2008 (Noctuidae: Catocalinae, Euclidini); Fibiger *et al.*, 2011 (Erebidae: Erebinae, Euclidini); Zahiri *et al.*, 2012 (Erebidae: Erebinae, Euclidini).

注：首次给出中文族名。

Genus ***Euclidia*** Ochsenheimer, 1816 恭夜蛾属

Euclidia Ochsenheimer, 1816, *Die Schmetterlinge von Europa*, 4: 96. Type species: *Phalaena glyphica* Linnaeus, 1758 [Europe].

Synonymy: *Euclidia* Hübner, [1806], rejected name; *Gonospileia* Hübner, 1816 [1823]; *Ectypa* Billberg, 1820; *Leucomelas* Hampson, 1913; *Euclidina* McDunnough, 1937.

The genus includes 9 species, widely distributed in the Palaearctic, Nearctic, Oriental and Neotropical regions; 2 species are in the Three Provinces of Northeast China.

***Euclidia dentata* Staudinger, 1871 齿恭夜蛾**

Euclidia glyphica dentata Staudinger, 1871 (var.), *in* Staudinger & Wocke, *Catalog der Lepidopteren des Europaeischen Faunengebiets*: 135. Type locality: Russia, "Ussuri" [Primorye terr.], Vladivostok, Suifun [Razdolnaya river], Askold Is., "Sutschan-Gebiete" [Partizansk].

Synonymy: *Euclidia consors* Butler, 1878; *Euclidia glyphica* auct..

Distribution. China (Heilongjiang, Jilin, Nei Mongol); Russia (Transbaikalia, Altai, Siberia), Korea, Japan.

寄主：白车轴草（*Trifolium repens*）、山野豌豆（*Vicia amoena*）（Matov & Kononenko, 2012）。

Euclidia juvenilis (Bremer, 1861) 比恭夜蛾

Agnonomia juvenilis Bremer, 1861, *Bulletin de l'Academie Impériale des Sciences de St.-Petersbroug*, 3: 495. Type locality: Russia, Amur reg., "Bureja-Gebirge" [Bureja Range], "Ussuri" [Primorye terr.].

Distribution. China (Heilongjiang, Jilin, Shaanxi); Russia (Transbaikalia, Siberia), Korea.

注：《中国动物志》（夜蛾科）中将本种归入属的同物异名"比夜蛾属 *Leucomelas* Hampson, 1913"中，在此予以变更。

Genus *Callistege* Hübner, [1823] 1816 欧夜蛾属

Callistege Hübner, [1823] 1816, *Verzeichniss bekannter Schmettlinge* [sic!]: 281. Type species: *Phalaena mi* Clerck, 1759 [Europe].

Synonymy: *Euclidimera* Hampson, 1913; *Euclidia* auct; *Gonospileia* auct..

The genus includes 10 species, distributed in the Holarctic region; 2 species are in the Three Provinces of Northeast China.

Callistege mi (Clerck, 1759) 欧夜蛾

Phalaena mi Clerck, 1759, *Icones Insectorum Rariorum cum Nominibus*, 1: pl. 9: 5. Type locality: not stated.

Synonymy: *Phalaena literata* Cyrillo, 1787; *Euclidia futilis* Staudinger, 1897; *Callistege mi extrema* Bang-Haas, 1912 (var.); *Callistege elzei* deFreina, 1976.

Distribution. China (Heilongjiang, Jilin, Liaoning, Xinjiang); Russia (European part, Siberia), Korea; Near East, Europe.

寄主：和兰翘指、苜蓿、草木樨（陈一心，1999）；萹蓄 (*Polygonum aviculare*)、斑点金丝桃（*Hypericum maculatum*）、安多杜鹃（*Andromeda polifolia*）、帚石楠（*Calluna vulgaris*）、（*Genista sagittalis*）、金雀花（*Genista tinctoria*）、牧地山黧豆（*Lathyrus pratensis*）、百脉根（*Lotus corniculatus*）、天蓝苜蓿（*Medicago lupulina*）、黄香草木樨（*Melilotus arvensis*）、草木樨 (*Melilotus officinalis*)、红车轴草（*Trifolium pratense*）、白车轴草（*Trifolium repens*）、广布野豌豆（*Vicia cracca*）、蓬子菜（*Galium verum*）、林石蚕（*Teucrium scorodonia*）、普通剪股颖（*Agrostis canina*）、沙丘野麦（*Elymus arenarius*）、绒毛草（*Holcus lanatus*）、酸沼草（*Molinia caerulea*）、虉草（*Phalaris arundinacea*）、酸模属（*Rumex*）、杨梅属（*Myrica*）、芒柄花属（*Ononis*）、薹草属（*Carex*）、羊胡子草属（*Eriophorum*）、芦苇属（*Phragmites*）植物（Matov & Kononenko, 2012）。

Genus *Melapia* Sugi, 1968 拟三角夜蛾属

Melapia Sugi, 1968, *Kontyû*, 36: 164. Type species: *Pelamia japonica* Ogata, 1961 [Japan].

Synonymy: *Pelamia* auct..

The genus includes 5 species, mainly distributed in the Oriental region; 1 species is in the Three Provinces of Northeast China.

Melapia electaria **(Bremer, 1864) 选拟三角夜蛾**

Doryodes electaria Bremer, 1864, *Memoires de l'Académie Imperiale des Sciences de St.-Petersbourg*, (7)8, 1: 84, pl. 7: 12. Type locality: Russia, "Ema-Mündung" [Primorye terr., ?Iman, the tributary of "Ussuri" river].

Synonymy: *Pelamia macroelectaria* Ogata, 1961.

Distribution. China (Heilongjiang, Hebei, Zhejiang); Russia (Siberia), Korea, Japan.

寄主：牛筋草（*Eleusine indica*）、薹草属（*Carex*）、鸭茅属（*Dactylis*）植物（Matov & Kononenko, 2012）。

Genus ***Remigia*** Guenée, 1852 实毛胫夜蛾属

Remigia Guenée, 1852, *in* Boisduval & Guenée, *Histoire Naturelle des Insectes. Species Général des Lépidoptères*, 7: 312. Type species: *Remigia latipes* Guenée, 1852 ["West Indies"].

The genus includes 5-7 species, distributed in the tropical area of the Old and New Worlds; 1 species is in the Three Provinces of Northeast China.

注：首次给出中文属名。

Remigia frugalis **(Fabricius, 1775) 实毛胫夜蛾**

Noctua frugalis Fabricius, 1775, *Entomologiae Systematica emendata et Aucta. Secundum Classes, Ordines, Genera, Species Adjectis Synonimis, Locis, Observationibus, Descriptionibus*, 1: 601. Type locality: India.

Synonymy: *Chalciope lycopodia* Geyer, 1837; *Remigia translata* Walker, 1865.

Distribution. China (Liaoning, Fujian, Guangdong, Guangxi, Hainan, Yunnan, Taiwan); Russia (Siberia), Korea, Japan, India, Sri Lanka; Southeast Asia, Oceania.

寄主：蘘荷科植物（陈一心，1999）。

Genus ***Mocis*** Hübner, [1823] 1816 毛胫夜蛾属

Mocis Hübner, [1823] 1816, *Verzeichniss bekannter Schmettlinge* [sic!]: 267. Type species: *Phalaena virbia* Cramer, 1780 [India].

Synonymy: *Pelamia* Guenée, 1852; *Baratha* Walker, 1865; *Cauninda* Moore, 1887; *Pelomia*: Warren, 1913, missp..

The genus includes 35 species, distributed in the tropical area of the Old and New Worlds; 3 species are in the Three Provinces of Northeast China.

Mocis undata (Fabricius, 1775) 毛胫夜蛾

Noctua undata Fabricius, 1775, *Entomologiae Systematica emendata et Aucta. Secundum Classes, Ordines, Genera, Species Adjectis Synonimis, Locis, Observationibus, Descriptionibus*, 1: 600. Type locality: East India.

Synonymy: *Phalaena* (*Noctua*) *archesia* Cramer, 1780; *Phalaena virbia* Cramer, 1780; *Ophiusa velata* Walker, [1863] 1864; *Remigia bifasciata* Walker, 1865; *Remigia gregalis* Guenée, 1852; *Cauninda undata bifasciata* Warren, 1913 (ssp.).

Distribution. China (Heilongjiang, Jilin, Liaoning, Hebei, Shandong, Henan, Jiangsu, Hunan, Fujian, Jiangxi, Guangdong, Guizhou, Yunnan, Taiwan); Russia (Siberia), Korea, Japan, Sri Lanka, India, Nepal, Philippines, Indonesia.

寄主：幼虫取食鱼藤、山蚂蝗属植物，成虫取食柑橘等果子（陈一心，1999）；蓼蓝（*Persicaria tinctoria*）、娑罗树（*Shorea robusta*）、刺苹婆（*Sterculia urens*）、三叶橡胶（*Hevea brasiliensis*）、两型豆（*Amphicarpaea edgeworthii*）、花生（*Arachis hypogaea*）、紫矿（*Butea monosperma*）、木豆（*Cajanus cajan*）、虫豆（*Cajanus crassus*）、金雀儿（*Cytisus scoparius*）、阔叶黄檀（*Dalbergia latifolia*）、毛鱼藤（*Derris elliptica*）、异翊鱼藤（*Derris malaccensis*）、（*Desmodium oojeinense*）、山蚂蝗（*Desmodium oxyphyllum*）、大豆（*Glycine max*）、胡枝子（*Lespedeza bicolor*）、蒙戈豇（*Phaseolus mungo*）、野葛（*Pueraria lobata*）、（*Pueraria thunbergiana*）、小鹿藿（*Rhynchosia minima*）、白灰毛豆（*Tephrosia candida*）、绿豆（*Vigna radiata*）、三裂叶豇豆（*Vigna trilobata*）、豇豆（*Vigna unguiculata*）、多花紫藤（*Wisteria floribunda*）、红毛丹（*Nephelium lappaceum*）、阳芋（*Solanum tuberosum*）、牛筋草（*Eleusine indica*）、棉属（*Gossypium*）、毛蔓豆属（*Calopogonium*）、猪屎豆属（*Crotalaria*）、木蓝属（*Indigofera*）、黧豆属（*Mucuna*）、柑橘属（*Citrus*）植物（Matov & Kononenko, 2012）。

Mocis annetta (Butler, 1878) 懈毛胫夜蛾

Remigia annetta Butler, 1878, *Annals and Magazine of Natural History*, (5)1: 193. Type locality: Japan, Yokohama, Hakodate.

Synonymy: *Cauninda annetta arabesca* Bryk, 1948 (f.).

Distribution. China (Jilin, Shandong, Jiangsu, Zhejiang, Hubei, Hunan, Fujian, Sichuan, Taiwan); Russia (Siberia), Korea, Japan.

寄主：大豆（陈一心，1999）；两型豆（*Amphicarpaea edgeworthii*）、山蚂蝗（*Desmodium oxyphyllum*）、山黑豆（*Dumasia truncata*）、大豆（*Glycine max*）、野大豆（*Glycine soja*）、胡枝子（*Lespedeza bicolor*）、刺槐（*Robinia pseudoacacia*）、多花紫藤（*Wisteria floribunda*）（Matov & Kononenko, 2012）。

Mocis ancilla (Warren, 1913) 奚毛胫夜蛾

Cauninda ancilla Warren, 1913, *in* A.Seitz (ed.), *Die Gross-Schmetterlinge der Erde*, 3: 334, pl. 61: g. Type locality: Russian East Siberia, China, Korea, Japan.

Distribution. China (Heilongjiang, Liaoning, Hebei, Shandong, Henan, Jiangsu, Zhejiang, Hunan, Fujian, Taiwan); Russia (Siberia), Korea, Japan.

寄主：葛（陈一心，1999）；山蚂蝗（*Desmodium oxyphyllum*）（Matov & Kononenko, 2012）。

Tribe **Ophiusini** Guenée, 1837 安钮夜蛾族

Ophiusidi Guenée, 1837, *Annales de la Société Entomologique de France*, 6: 321, ibidem 10: 71 (1841). Type genus: *Ophiusa* Ochsenheimer, 1816.

Synonymy: Ophiusidae Herrich-Schäffer, [1851] 1845; Ophiusidae Guenée, 1852; Ophiusini Wiltshire, 1990; Omopterini Boisduval, 1833, suppressed senior synonymy.

The tribe is widespread in the tropical and subtropical of the Old and New Worlds. Many species are strongly migratory and appear in the southern part of the temperate zone as irregular migrants; 3 genera are in the Three Provinces of Northeast China.

References. Chen, 1982, 1999 (Noctuidae); Sugi, 1982 (Noctuidae: Ophiderinae, part); Kononenko, 1990 (Catocalinae, part); Kononenko *et al.*, 1998 (Noctuidae: Ophiderinae, part); Goater *et al.*, 2003 (Noctuidae: Catocalinae, Catocalini, Ophiusina); Sviridov, 2003 (Noctuidae: Catocalinae, part); Fibiger, Hacker, 2005 (Erebidae: Catocalinae, Ophiusini); Fibiger, Lafontaine, 2005 (Erebidae: Catocalinae, Ophiusini); Kononenko, 2005 (Erebidae: Catocalinae, Ophiusini), 2010 (Noctuidae: Catocalinae, Ophiusini); Lafontaine, Fibiger, 2006 (Noctuidae: Catocalinae, Ophiusini); Matov *et al.*, 2008 (Noctuidae: Catocalinae, part); Holloway, 2011, (Erebidae, Erebinae: Ophiusini & Poaphilini); Fibiger *et al.*, 2011 (Erebidae: Catocalinae, Ophiusini); Zahiri *et al.*, 2012 (Erebidae: Catocalinae, Ophiusini).

注：首次给出中文族名。

Genus ***Artena*** Walker, 1858 关夜蛾属

Artena Walker, 1858, *List of the Specimens of Lepidopterous Insects in the Collection of the British Museum*, 14: 1343, 1388. Type species: *Artena submira* Walker, 1858 [India].

The genus includes 12 species, mainly distributed in the Oriental region; 1 species is in the Three Provinces of Northeast China.

Artena dotata **(Fabricius, 1794)** 斜线关夜蛾

Noctua dotata Fabricius, 1794, *Entomologiae Systematica emendata et Aucta. Secundum Classes, Ordines, Genera, Species Adjectis Synonimis, Locis, Observationibus, Descriptionibus*, 3 (2): 55. Type locality: East India.

Distribution. China (Liaoning, Shaanxi, Henan, Jiangsu, Zhejiang, Hubei, Hunan, Jiangxi, Fujian, Guangdong, Sichuan, Guizhou, Yunnan, Taiwan); Russia (Siberia), Korea, Japan, India(North), Nepal, Philippines, Malaysia, Indonesia.

寄主：柑橘类植物（陈一心，1999）；青冈 (*Quercus glauca*)、阔叶风车子（*Combretum latifolium*）、使君子（*Quisqualis indica*）、毗黎勒（*Terminalia bellirica*）、小叶榄仁（*Terminalia paniculata*）、毛榄仁（*Terminalia tomentosa*）、温州蜜柑（*Citrus unshiu*）（Matov & Kononenko, 2012)。

Genus *Thyas* Hübner, 1824 肖毛翅夜蛾属

Thyas Hübner, 1824, *Zuträge zur Sammlung Exotischer Schmetterlinge*, 2: pl. [203]. Type species: *Thyas honesta* Hübner, 1824 [East India].

Synonymy: *Lagoptera* Guenée, 1852; *Dermaleipa* Saalmüller, 1891.

The genus includes 15 species, mainly distributed in the tropical area of the Old World; 1 species is in the Three Provinces of Northeast China.

***Thyas juno* (Dalman, 1823) 庸肖毛翅夜蛾**

Noctua juno Dalman, 1823, *Analecta Entomologica*: 52. Type locality: not stated.

Synonymy: *Ophideres elegans* Heven, 1840; *Lagoptera multicolor* Guenée, 1852; *Thyas bella* Bremer & Grey, 1853; *Dermaleipa juno renalis* Bryk, 1948 (f.).

Distribution. China (Heilongjiang, Jilin, Liaoning, Hebei, Shandong, Henan, Anhui, Zhejiang, Hubei, Hunan, Fujian, Jiangxi, Hainan, Sichuan, Guizhou, Yunnan, Taiwan); Russia (Siberia), Korea, Japan, Philippines, Malaysia, India, Nepal, Thailand.

寄主：桦、李、木槿；成虫吸食多种水果果汁（陈一心，1999）；日本栗（*Castanea crenata*）、欧洲栗（*Castanea sativa*）、麻栎（*Quercus acutissima*）、短柄枹栎（*Quercus glandulifera*）、青冈 (*Quercus glauca*)、枹栎（*Quercus serrata*）、（*Pasania glabra*）、黑桦（*Betula dahurica*）、胡桃楸（*Juglans mandshurica*）、水胡桃（*Pterocarya rhoifolia*）、木槿（*Hibiscus syriacus*）、（*Persica vulgaris*）、榉属（*Zelkova*）、李属（*Prunus*）、梨属（*Pyrus*）、蔷薇属（*Rosa*）、葡萄属（*Vitis*）（Matov & Kononenko, 2012)。

Genus *Ophiusa* Ochsenheimer, 1816 安钮夜蛾属

Ophiusa Ochsenheimer, 1816, *Die Schmetterlinge von Europa*, 4: 93. Type species: *Phalaena tirhaca* Cramer, 1777 [South Africa].

Synonymy: *Ophiogenes* Reichenbach & Leipzig, 1817; *Ophiussa*: Hübner, [1823] 1816; *Meropis* Hübner, [1822]; *Hemachra* Sodoffsky, 1837; *Anua* Walker, 1858; *Stenopis* Mabille, 1880; *Stenopsis*: Hampson, 1913, missp.; *Subanua* Berio, 1959; *Peranua* Berio, 1959; *Perophiusa* Berio, 1959.

The genus includes about 70 species, mainly distributed in the tropical and subtropical areas of the Old World; 1 species is in the Three Provinces of Northeast China.

Ophiusa tirhaca **(Cramer, 1777) 安钮夜蛾**

Phalaena tirhaca Cramer, 1777, *Uitlandsche Kapellen*, 2: 116, pl. 172: E. Type locality: South Africa, Cap Prov..

Synonymy: *Noctua tirrhaea* Fabricius, 1781, emend.; *Phalaena* (*Noctua*) *vesta* Esper, 1789; *Noctua olivacea* Villers, 1789; *Noctua auricularis* Hübner, [1803]; *Noctua tirrhaca*: Treitschke, 1826; *Ophiodes hottentota* Guenée, 1852; *Ophiodes separans* Walker, 1858; *Anua clementi* Swinhoe, 1918; *Ophiusa tirhaca obscura* Pinker & Bacallado, 1979 (ssp.).

Distribution. China (Liaoning, Shaanxi, Shandong, Jiangsu, Zhejiang, Hubei, Jiangxi, Fujian, Guangdong, Hainan, Guangxi, Sichuan, Guizhou, Yunnan, Taiwan); Russia (Siberia), Korea, Japan, India, Australia; Near East, Southeast Asia, Europe, Africa.

寄主：乳香、漆树，成虫吸食多种水果果汁（陈一心，1999）；娑罗树（*Shorea robusta*）、苹果（Malus pumila）、（*Rosa indica*）、番石榴（*Psidium guajava*）、使君子（*Quisqualis indica*）、毗黎勒（*Terminalia bellirica*）、毛榄仁（*Terminalia tomentosa*）、石榴（*Punica granatum*）、黄栌（*Cotinus coggygria*）、（*Heeria mucronata*）、杧果（*Mangifera indica*）、乳香黄连木（*Pistacia lentiscus*）、洋清香木（*Pistacia terebinthus*）、盐肤木（*Rhus chinensis*）、毒肤杨（*Rhus coriaria*）、罗氏盐肤木（*Rhus javanica*）、火炬树（*Rhus typhina*）、马蹄纹天竺葵（*Pelargonium zonale*）、欧石南属（*Erica*）、岩蔷薇属（*Cistus*）、瑞香属（*Daphne*）、悬钩子属（*Rubus*）、桉属（*Eucalyptus*）、薄子木属（*Leptospermum*）、香桃木属（*Myrtus*）、柑橘属（*Citrus*）、沙针属（*Osyris*）、葡萄属（*Vitis*）、荚蒾属（*Viburnum*）植物（Matov & Kononenko, 2012)。

Tribe **Poaphilini** Guenée, 1852 熟夜蛾族

Poaphilidae Guenée, 1852, *in* Boisduval & Guenée, *Histoire Naturelle des Insectes. Species Général des Lépidoptères*, 7: 295. Type genus: *Poaphila* Guenée, 1852.

Holloway (2011) treated previous and this tribes together as "Tribes Ophiusini and Poaphilini". The Poaphilini includes genera *Parallelia*, *Bastilla*, *Dysgonia*, *Buzara*, *Macaldenia*, *Grammodes*, *Chalciope*, *Oxyodes*, and *Rectipalpula* in the Old World tropics and some other in the New World. The tribe is widespread in the subtropical and tropical areas of the Old and New Worlds. It is very diverse in the Oriental region and especially in the Ephiopean region. Many species are strongly migratory and some of them appear in the southern part of the temperate zone as irregular migrants. 3 genera are in the Three Provinces of Northeast China.

References. Chen, 1982, 1999 (Noctuidae: Catocalinae, part); Sugi, 1982 (Noctuidae: Ophiderinae, part); Kononenko, 1990 (Catocalinae, part); Kononenko *et al.*, 1998 (Noctuidae: Ophiderinae, part); Goater *et al.*, 2003 (Noctuidae: Catocalinae, Catocalini, Ophiusina); Sviridov, 2003 (Noctuidae: Catocalinae, part); Fibiger, Lafontaine, 2005 (Erebidae: Catocalinae, Ophiusini); Fibiger, Hacker, 2005 (Erebidae: Catocalinae, Ophiusini); Kononenko, 2005 (Erebidae: Catocalinae, Ophiusini), 2010 (Noctuidae: Catocalinae, Ophiusini); Lafontaine, Fibiger, 2006 (Noctuidae: Catocalinae, Ophiusini); Matov *et al.*, 2008 (Noctuidae: Catocalinae, part); Holloway,

2011, (Erebidae, Erebinae: Ophiusini & Poaphilini); Fibiger *et al.*, 2011 (Erebidae: Catocalinae, Ophiusini); Zahiri *et al.*, 2012 (Erebidae: Catocalinae, Ophiusini).

注：首次给出中文族名。

Genus ***Parallelia*** Hübner, 1818 条巾夜蛾属

Parallelia Hübner, 1818, *Zuträge zur Sammlung exotischer Schmettlinge, bestehend in etundigung einzeiner Fliegmuster neuer oder rarer Nichteuropäischer Gattungen*, 1818: 15. Type species: *Parallelia bistriaris* Hübner, 1818 [USA].

Synonymy: *Dysgonia* Hübner, [1823] 1816; *Ophiusa* auct., nec Ochsenheimer, 1816.

The genus includes about 5-6 species, mainly distributed in the tropical and subtropical areas of the Old and New World (Nearctis, Oriental and Australian regions); 2 species occurs in the Three Provinces of Northeast China.

注：首次给出中文属名；《中国动物志》（夜蛾科）中将本属合并到"*Dysgonia* Hübner, [1823] 1816 巾夜蛾属"，在此予以变更。

Parallelia stuposa **(Fabricius, 1794) 石榴条巾夜蛾**

Noctua stuposa Fabricius, 1794, *Entomologiae Systematica emendata et Aucta. Secundum Classes, Ordines, Genera, Species Adjectis Synonimis, Locis, Observationibus, Descriptionibus*, 3 (2): 42. Type locality: "India oriental".

Synonymy: *Ophiusa festinata* Walker, 1858; *Ophiusa stuposa tumefacta* Warren, 1913 (ssp.); *Parallelia algira japonibia* Bryk, 1948 (ssp.); *Parallelia algira sinica* Bryk, 1948 (ssp.) ; *Ophiusa analis*: Herz, 1904, nec Guenée, 1852; *algira* auct..

Distribution. China (Liaoning, Hebei, Shandong, Jiangsu, Zhejiang, Hubei, Fujian, Jiangxi, Guangdong, Hainan, Sichuan, Yunnan, Taiwan); Russia (Siberia), Korea, Japan, Thailand, India, Nepal, Philippines, Malaysia, Indonesia.

寄主：石榴（陈一心，1999）；山桐子（*Idesia polycarpa*）、蓖麻（*Ricinus communis*）、石榴（*Punica granatum*）、悬钩子属（*Rubus*）植物（Matov & Kononenko, 2012).

注：《中国动物志》（夜蛾科）中使用"石榴巾夜蛾"，根据属名变动其中文种名在此予以变更。

Parallelia arctotaenia **(Guenée, 1852) 玫瑰条巾夜蛾**

Ophiusa arctotaenia Guenée, 1852, *in* Boisduval & Guenée, *Histoire Naturelle des Insectes. Species Général des Lépidoptères*, 7: 272. Type locality: Bangladesh, Silhet.

Distribution. China (Jilin, Liaoning, Hebei, Jiangsu, Zhejiang, Hubei, Fujian, Jiangxi, Guangdong, Guangxi, Sichuan, Guizhou, Yunnan, Taiwan); Russia (Siberia), Korea, Japan, Philippines, Indonesia, Malaysia, India, Nepal, Bangladesh, Australia.

寄主：玫瑰（陈一心，1999）；乌冈栎（*Quercus phillyraeoides*）、玫瑰（*Rosa rugosa*）、叶下珠

(*Phyllanthus urinaria*)、蓖麻(*Ricinus communis*)、悬钩子属(*Rubus*)、柳属(*Salix*)植物。

注:《中国动物志》(夜蛾科)中使用"玫瑰巾夜蛾",根据属名变动其中文种名在此予以变更。

Genus ***Bastilla*** Swinhoe, 1918 暗巾夜蛾属

Bastilla Swinhoe, 1918, *Annals and Magazine of Natural History*, (9)2: 78. Type species: *Ophiusa redunca* Swinhoe, 1900 [Australia].

Synonymy: *Naxia* Guenée, 1852; *Xiana* Nye, 1975; *Ophiusa* auct.; *Parallelia* auct.; *Dysgonia* auct..

The genus includes about 10 species, distributed in the tropical and subtropical areas of the World. 1 species is in the Three Provinces of Northeast China.

注:首次给出中文属名;《中国动物志》(夜蛾科)中将本属合并到"*Dysgonia* Hübner, [1823] 1816 巾夜蛾属"中,在此予以变更。

***Bastilla maturata* (Walker, 1858) 霉暗巾夜蛾**

Ophiusa maturata Walker, 1858, *List of the Specimens of Lepidopterous Insects in the Collection of the British Museum*, 14: 1382. Type locality: Malaysia, Penang.

Synonymy: *Ophiusa falcata* Moore, 1882.

Distribution. China (Liaoning, Shandong, Henan, Jiangsu, Zhejiang, Fujian, Jiangxi, Hainan, Sichuan, Guizhou, Yunnan, Taiwan); Korea, Japan, India, Nepal, Thailand, Philippines, Malaysia, Indonesia.

注:《中国动物志》(夜蛾科)中使用"霉巾夜蛾",根据属名变动其中文种名在此予以变更。

Genus ***Dysgonia*** Hübner, [1823] 1816 巾夜蛾属

Dysgonia Hübner, [1823] 1816, *Verzeichniss bekannter Schmettlinge* [sic!]: 269. Type species: *Phalaena algira* Linnaeus, 1767 [Algeria].

Synonymy: *Naxia* Guenée, 1852; *Pasipeda* Moore, 1882, preocc., (Walker, 1858 [Lepidoptera, Noctuidae]); *Macaldenia* Moore, [1885], repl. name; *Caranilla* Moore, [1887] 1885; *Ophiusa* auct.; *Parallelia* auct..

The genus includes about 130 species, distributed in the tropical and subtropical areas of the Old and New Worlds; 3 species are in the Three Provinces of Northeast China.

***Dysgonia mandschuriana* (Staudinger, 1892) 东北巾夜蛾**

Grammodes algira var. *mandschuriana* Staudinger, 1892, *in* N.M. Romanoff (ed.), *Mémoires sur les Lépidoptères*, 6: 578. Type locality: Russia, Primorye terr., "Sutschan-Gebiet" [Partizansk].

Synonymy: *Parallelia mimula postfusca* Bryk, 1949 (ssp.); *Ophiusa algira* auct., nec Linnaeus, 1767; *Ophiusa mandschurica* auct., nec Staudinger, 1892, missp..

Distribution. China (Jilin, Liaoning, Beijing, Shandong); Russia (Siberia), Korea, Japan.

寄主：叶底珠（*Securinega suffruticosa*）（Park *et al.*, 2006; Matov & Kononenko, 2012）。

***Dysgonia dulcis* (Butler, 1878) 小直巾夜蛾**

Ophiusa dulcis Butler, 1878, *Annals and Magazine of Natural History,* (5)1: 293. Type locality: Japan, Yokohama.

Distribution. China (Heilongjiang, Jilin, Liaoning, Hebei, Hubei, Hunan, Fujian, Taiwan); Russia (Siberia), Korea, Japan.

***Dysgonia obscura* (Bremer & Grey, 1853) 小折巾夜蛾**

Ophiusa obscura Bremer & Grey, 1853, *in* Motschulsky, *Etudes Entomologiques,* 1: 66. Type locality: China, Beijing.

Synonymy: *Leucanitis hedemanni* Staudinger, 1888; *Naxia coreana* Leech, 1889.

Distribution. China (Heilongjiang, Jilin, Liaoning, Beijing, Hebei, Shandong); Russia (Siberia), Korea, Japan.

Tribe **Ercheiini** Berio, 1992 耳夜蛾族

Ercheiini Berio, 1992, *Memorie della Society Entomologica Italiana 70*: 294, 297, 300, description by reference to "Phylum di *Ercheia*" in Berio, 1960; otherwise nomen nudum, not available (art. 13. 1 ICZN); emended in Ercheiini by Speidel & Naumann, 2005. Type genus: *Ercheia* Walker, [1858].

The tribe contains 1 genus with about 26 species, distributed mainly in the tropical and subtropical areas of the Oriental and partly Australian regions; 1 genus and 2 species are represented in the Three Provinces of Northeast China.

References. Chen, 1982, 1999 (Noctuidae: Catocalinae, part); Sugi, 1982 (Noctuidae: Ophiderinae, part); Kononenko, 1990 (Catocalinae, part), 2005 (Erebidae: Catocalinae), 2010 (Noctuidae: Catocalinae); Kononenko *et al.*, 1998 (Noctuidae: Ophiderinae, part); Sviridov, 2003 (Noctuidae: Catocalinae, part); Fibiger, Lafontaine, 2005 (Erebidae: Catocalinae/Erebinae, Ercheiini); Lafontaine, Fibiger, 2006 (Noctuidae: Catocalinae, Ercheiini) Matov *et al.*, 2008 (Noctuidae: Catocalinae); Holloway, 2009, (Catocalinae), 2011 (Erebidae: Catocalinae, Ercheiini); Zahiri *et al.*, 2012 (Erebidae: Catocalinae, Ercheiini).

Genus ***Ercheia*** Walker, [1858] 1857 耳夜蛾属

Ercheia Walker, [1858] 1857, *List of the Specimens of Lepidopterous Insects in the Collection of the British Museum*, vol. 3: 1107. Type species: *Ercheia diversipennis* Walker, [1858] 1857 [Sri Lanka].

The genus includes up to 26 species, distributed in the tropical and subtropical areas of the Old and New Worlds; 2 species are in the Three Provinces of Northeast China.

Ercheia umbrosa Butler, 1881 阴耳夜蛾

Ercheia umbrosa Butler, 1881, *Transactions of the Entomological Society of London*, 1881: 194. Type locality: Japan, Tokyo.

Synonymy: *Melipotis strigipennis* Moore, 1882.

Distribution. China (Jilin, Liaoning, Jiangxi, Guangdong, Hainan, Guizhou, Taiwan); Korea, Japan, India, Nepal, Indonesia.

Ercheia niveostrigata Warren, 1913 雪耳夜蛾

Ercheia niveostrigata Warren, 1913. *in* Seitz, *Macrolepidoptera World*, 3: 335, pl. 61: h. Type locality: China.

Distribution. China (Jilin, Liaoning, Jiangsu, Zhejiang, Hunan, Fujian, Sichuan, Taiwan); Korea, Japan.

Tribe **Sypnini** Holloway, 2005 析夜蛾族

Sypnini Holloway, 2005, *Moths of Borneo*, 15 & 16: 170. Type genus: *Sypna* Guenée, 1852.

The tribe comprise 5 genera with a number of species, distributed mainly in the Chinese and Sino-Himalainan subregions of the Palaearctic and especially diverse in the Oriental regions. The Chinese species of this group was selected as "Phyllum *Sypna*" and revised by Berio (1950). Most species occur in rich broad-leaved or tropical forests; 3 genera are represented in the Three Provinces of Northeast China.

References. Chen, 1982, 1999 (Noctuidae: Ophiderinae, part); Sugi, 1982 (Noctuidae: Ophiderinae, part); Kononenko, 1990 (Catocalinae, part); Kononenko *et al.*, 1998 (Noctuidae: Ophiderinae, part); Sviridov, 2003 (Noctuidae: Catocalinae, part); Kononenko, 2005 (Erebidae: Catocalinae, Sypnini), 2010 (Noctuidae: Catocalinae, Sypnini); Matov *et al.*, 2008 (Noctuidae: Catocalinae, part); Holloway, 2009, (Catocalinae, Sypnini), 2011 (Erebidae: Catocalinae, Sypnini); Zahiri *et al.*, 2012 (Erebidae: Catocalinae, Sypnini).

注：首次给出中文族名。

Genus ***Hypersypnoides*** Berio, 1958 朋闪夜蛾属

Hypersypnoides Berio, 1954, *Annali del Museo Civico di Storia Naturale Giacomo Doria*, 66: 341. Type species: *Hypersypnoides congoensis* Berio, 1954 [Southwest Africa].

Synonymy: *Othresypna* Berio, 1950.

The genus includes 36 species, distributed in the Ethiopian (1 species) and Oriental regions, most diverse in the Himalayan subregion and Southwest China; 1 species is in the Three Provinces of Northeast China.

Hypersypnoides astrigera (Butler, 1885) 白点朋闪夜蛾

Sypna astrigera Butler, 1885, *Cistula Entomologica*, 3: 135. Type locality: Japan, Nikko; Korea, Chiuzenji.

Distribution. China (Heilongjiang, Jilin, Liaoning, Zhejiang, Fujian, Jiangxi, Hainan, Sichuan, Yunnan, Taiwan); Russia (Siberia), Korea, Japan.

Genus ***Sypnoides*** Hampson, 1913 析夜蛾属

Sypnoides Hampson, 1913, *Catalogue of the Lepidoptera Phalaenae in the British Museum*, 12: 5. Type species: *Sypna erebina* Hampson, 1926 [China, Sichuan].

Synonymy: *Hyposypnoides* Berio, 1954 (subgen.); *Supersypnoides* Berio, 1958 (subgen.); *Pysnoides* Berio, 1959.

The genus includes 31 species, distributed in the Sino-Himalayan subregion of the Palaearctic, Oriental and Ethiopian regions, most diverse in the Himalayan subregion and Southwest China; 3 species are in the Three Provinces of Northeast China.

Subgenus ***Supersypnoides*** Berio, 1958 高析夜蛾亚属

Supersypnoides Berio, 1958, *in* Berio & Fletcher, *Annali del Museo Civico di Storia Naturale Giacomo Doria*, 70: 344. Type species: *Sypna erebina* Hampson, 1926 [China].

注：首次给出中文亚属名。

***Sypnoides* (*Supersypnoides*) *hercules* (Butler, 1881) 赫析夜蛾**

Gisira hercules Butler, 1881, *Transactions of the Entomological Society of London*, 1881: 579. Type locality: Japan, Tokyo.

Synonymy: *Sypna rectifasciata* Graeser, 1888 [1889]; *Sypna hercules albifusa* Warren, 1913 (ab.); *Sypna albimedia* Warren, 1913; *Sypna hercules gigantean* Berio, 1958 (ab.).

Distribution. China (Heilongjiang, Jilin, Liaoning, Zhejiang, Xizang); Russia (Siberia), Korea, Japan, Nepal.

寄主：蒙古栎（*Quercus mongolica*）、牛叠肚（*Rubus crataegifolius*）（Park *et al.*, 2006）；钝齿水青冈（*Fagus crenata*）、麻栎（*Quercus acutissima*）、青冈（*Quercus glauca*）、大桴栎（*Quercus grosseserrata*）（Matov & Kononenko, 2012）。

***Sypnoides* (*Supersypnoides*) *picta* (Butler, 1877) 涂析夜蛾**

Sypna picta Butler, 1877, *Cistula Entomologica*, 2: 244. Type locality: Japan, Yokohama, Hakodate.

Synonymy: *Sypna achatina* Butler, 1877.

Distribution. China (Heilongjiang, Liaoning, Zhejiang, Hunan, Yunnan); Russia (Siberia), Korea, Japan.

寄主：槲、绞股蓝、悬钩子、栓皮栎（陈一心，1999）；槲树（*Quercus dentata*）、青冈（*Quercus glauca*）、蒙古栎（*Quercus mongolica*）、三叶海棠（*Malus sieboldii*）、刺蔷薇（*Rosa acicularis*）、野蔷薇（*Rosa multiflora*）、榆属（*Ulmus*）植物（Matov & Kononenko, 2012）。

***Sypnoides* (*Supersypnoides*) *fumosa* (Butler, 1877) 异析夜蛾**

Sypna fumosa Butler, 1877, *Cistula Entomologica*, 2: 245. Type locality: Japan, Yokohama, Hakodate.

Synonymy: *Sypna fuliginosa* Butler, 1877; *Sypna picta albinigra* Warren, 1913 (ab.).

Distribution. China (Heilongjiang, Jilin, Liaoning, Hunan); Russia (Siberia), Korea, Japan.

寄主：（*Rubus oldhamii*）（Park *et al*., 2006）；槲树（*Quercus dentata*）、青冈（*Quercus glauca*）、野蔷薇（*Rosa multiflora*）、悬钩子属（*Rubus*）植物（Matov & Kononenko, 2012）。

Genus ***Daddala*** Walker, 1865 炬夜蛾属

Daddala Walker, 1865, *List of the Specimens of Lepidopterous Insects in the Collection of the British Museum*, 33: 974. Type species: *Daddala quadrisignata* Walker, 1865 [India].

Synonymy: *Elpia* Walker, 1865.

The genus includes 12 species, distributed exclusively in the Oriental region, most diverse in the Southeast Asia; 1 species is in the Three Provinces of Northeast China.

***Daddala lucilla* (Butler, 1881) 光炬夜蛾**

Sypna lucilla Butler, 1881, *Transactions of the Entomological Society of London*, 1881: 206. Type locality: India, Sikkim, Darjiling.

Synonymy: *Sypna obscurata* Butler, 1881.

Distribution. China (Jilin, Liaoning, Fujian, Guangdong, Hainan, Yunnan, Taiwan); Russia (Siberia), Korea, Japan, India(North), Nepal, Philippines, Indonesia, New Guinea.

寄主：小叶青冈（*Quercus myrsinaefolis*）（Matov & Kononenko, 2012）。

Tribe **Erebini** Leach, [1815] 目夜蛾族

Erebida Leach, [1815] 1830, *in* Brewster, *The American Edition of the New Edinburgh Encyclopedia, Conducted by David Brewster — Assisted by upwards of One Hundred Gentlemen in Europe*, 9: 134. Type genus: *Erebus* Latreille, 1810.

Synonymy: Erebidae Guenée, 1852; Thysaniini Grote, 1895; Erebinae Forbes, 1954; Erebini Forbes, 1954.

The members of the tribe have predominantly tropical distribution in the Old and New Worlds, some species appear in the temperate zone of the Eastern Palaearctic as irregular immigrants; 2 genera are in the Three Provinces of Northeast China.

References. Chen, 1982, 1999 (Noctuidae: Catocalinae, part); Sugi, 1982 (Noctuidae: Ophiderinae, part); Kononenko, 1990 (Noctuidae: Catocalinae, part); Kononenko *et al.*, 1998 (Noctuidae: Ophiderinae, part); Sviridov, 2003 (Noctuidae: Catocalinae, part); Holloway, 2005, (Catocalinae, Erebini), 2011 (Erebidae: Erebinae); Kononenko, 2005 (Erebidae: Erebinae), 2010 (Noctuidae: Erebinae, Erebini); Matov *et al.*, 2008

(Noctuidae: Catocalinae, part); Zahiri *et al.*, 2012 (Erebidae: Erebinae, Erebini).

注：首次给出中文族名。

Genus ***Erebus*** Latreille, 1810 目夜蛾属

Erebus Latreille, 1810, *Considerations generates sm l'Ordre naturel des Animaux*: 365. Type species: *Phalaena crepuscularis* Linnaeus, 1758 ["East Indies"].

Synonymy: *Byas* Billberg, 1820; *Nyctipao* Hübner, [1823] 1816; *Argiva* Hübner, [1823] 1816; *Patula* Guenée, 1852; *Bocana* Walker, [1865]; *Coria* Walker, 1866; *Eupatula* Ragonot, 1894; *Crishna* Kirby, 1897, unnecessary repl. name; *Cariona* Swinhoe, 1918; *Nyctipaon* auct..

The genus includes about 40 species, most diverse in the Oriental region through to the Indo-Australian tropics, and there are few species are in Ethiopian region; 1 species is in the Three Provinces of Northeast China.

Erebus macrops **(Linnaeus, 1768) 卷裳目夜蛾**

Attacus macrops Linnaeus, 1768, *Systema Naturae* (Edn. 12) 3: 225. Type locality: "India Oriental".

Synonymy: *Noctua bubo* Fabricius, 1775; *Patula boopis* Guenée, 1852.

Distribution. China (Jilin, Liaoning, Jiangxi, Fujian, Guangdong, Hainan, Sichuan, Yunnan, Taiwan); Russia (Siberia), Korea, Japan; Southeast Asia, Central Africa.

寄主：羽叶金合欢（*Acacia pennata*）、榼藤属（*Entada*）植物（Matov & Kononenko, 2012）。

Genus ***Metopta*** Swinhoe, 1900 蚪目夜蛾属

Metopta Swinhoe, 1900, *Catalogue of Eastern and Australian Lepidoptera Heterocera in the Collection of the Oxford Univeristy Museum*, 2: 170. Type species: *Spirama rectifasciata* Ménétriès, 1863.

Synonymy: *Gialca* Walker, 1855.

Monotypic genus distributed mainly in the Chinese subregion of the Palaearctic and partly Oriental regions; 1 species is in the Three Provinces of Northeast China.

Metopta rectifasciata **(Ménétriès, 1863) 蚪目夜蛾**

Spirama rectifasciata Ménétriès, 1863, *Enumeratio Corporum Animalium Musei Imperialis Academiae Scientiarum Petropolitanae. Classis Insectorum, Ordo Lepidopterorum*, 3: pl. 17: 6. Type locality: not stated [?Japan].

Synonymy: *Spirama japonica* Walker, 1865; *Spirama interlineata* Butler, 1871.

Distribution. China (Heilongjiang, Jilin, Liaoning, Zhejiang, Jiangsu, Fujian, Jiangxi, Hunan, Taiwan); Russia (Siberia), Korea, Japan.

寄主：菝葜、牛尾菜，成虫吸食水果果汁（陈一心，1999）；菝葜（*Smilax china*）、牛尾菜（*Smilax riparia*）（Matov & Kononenko, 2012）。

Tribe **Hypopyrini** Guenée, 1852 变色夜蛾族

Hypopyridae Guenée, 1852, *in* Boisduval & Guenée, *Histoire Naturelle des Insectes. Species Général des Lépidoptères*, 7: 192. Type genus: *Hypopyra* Guenée, 1852.

The tribe includes 6 genera, mainly distributed in the tropical area of the Old and New Worlds. 1 genus is represented in the Three Provinces of Northeast China.

References. Chen, 1982, 1999 (Noctuidae: Catocalinae, part); Sugi, 1982 (Noctuidae: Ophiderinae, part); Kononenko, 1990 (Noctuidae: Catocalinae, part); Kononenko *et al.*, 1998 (Noctuidae: Ophiderinae, part); Sviridov, 2003 (Noctuidae: Catocalinae, part); Holloway, 2005, (Catocalinae, Hypopyrini), 2011 (Erebidae: Erebinae); Kononenko, 2005 (Erebidae: Erebinae), 2010 (Noctuidae: Erebinae, Hypopyrini); Matov *et al.*, 2008 (Noctuidae: Catocalinae, part); Zahiri *et al.*, 2012 (Erebidae: Erebinae, Hypopyrini).

注：首次给出中文族名。

Genus ***Spirama*** Guenée, 1852 环夜蛾属

Spirama Guenée, 1852, *in* Boisduval & Guenée, *Histoire Naturelle des Insectes. Species Général des Lépidoptères*, 7: 194. Type species: *Phalaena retorta* Clerck, 1764.

Synonymy: *Spiramia* Walker, 1858, emend..

The genus includes 19 species, mainly distributed in the Old World tropical and subtropical areas; 2 species are known from the Chinese subregion of the Palaearctic region; 1 species is in the Three Provinces of Northeast China.

Spirama helicina **(Hübner, [1831] 1825)** 绕环夜蛾

Speiredonia helicina Hübner, [1831] 1825, *Zuträge zur Sammlung Exotischer Schmetterlinge*, 3: 14 [1824], l. c.: pl. 76, figs. 437, 438. Type locality: “Berbice”, ?Japan.

Synonymy: *Spirama japonica* Guenée, 1852; *Spirama aegrota* Butler, 1881.

Distribution. China (Heilongjiang, Jilin, Liaoning, Jiangxi, Taiwan); Russia (Siberia), Korea, Japan; Southeast Asia.

寄主：合欢花（*Albizzia julibrissin*）（Matov & Kononenko, 2012）。

Tribe **Pericymini** Wiltshire, 1976 同纹夜蛾族

Pericymatini Wiltshire, 1976, *Journal of the Bombay Natural History Society*, 73: 160. Type genus: *Pericyma* Herrich-Schäffer, [1851]. Must be emended in Pericymini (Speidel & Naumann, 2005).

The tribe has its highest specific diversity in the Ethiopian and Oriental regions; 1 genus is represented in the Three Provinces of Northeast China.

References. Chen, 1982, 1999 (Noctuidae: Catocalinae, part); Sugi, 1982 (Noctuidae: Ophiderinae, part); Kononenko, 1990 (Noctuidae: Catocalinae, part); Kononenko *et al.*, 1998 (Noctuidae: Ophiderinae, part);

Sviridov, 2003 (Noctuidae: Catocalinae, part); Holloway, 2005, (Catocalinae, Pericymini), 2011 (Erebidae: Erebinae); Kononenko, 2005 (Erebidae: Erebinae), 2010 (Noctuidae: Erebinae, Pericymini); Matov *et al.*, 2008 (Noctuidae: Catocalinae, part); Zahiri *et al.*, 2012 (Erebidae: Erebinae, Pericymini).

注：首次给出中文族名。

Genus ***Cortyta*** Walker, [1858] 1857 素纹夜蛾属

Cortyta Walker, [1858] 1857, *List of the Specimens of Lepidopterous Insects in the Collection of the British Museum*, vol. 13: 1076. Type species: *Cortyta canescens* Walker, [1858] 1857 [South Africa].

The genus includes 9 species, mainly distributed in tropical to subtropical areas of the Ethiopean region; 1 species is in the Three Provinces of Northeast China.

Cortyta grisea **(Leech, 1900) 灰素纹夜蛾**

Polydesma grisea Leech. 1900, *Transactions of the Entomological Society of London*, 1900: 552. Type locality: China, Moupin, Chia-kou-ho, Kei-chow [BMNH, London].

Synonymy: *Polydesma striata* Herz. 1904.

Distribution. China (Liaoning, Shandong, Jiangsu, Zhejiang, Hunan, Jiangxi, Sichuan, Guizhou); Korea, Japan.

Tribe **Hulodini** Guenée, 1852 木夜蛾族

Hulodides Guenée, 1852, *in* Boisduval & Guenée, *Histoire Naturelle des Insectes. Species Général des Lépidoptères* 7: 206, established as subfamily of Bendidae. Type genus: *Hulodes* Guenée, 1852.

Tropical origin tribe, containing genera *Hulodes*, *Lacera*, *Speiredonia*, *Ericeia* and some other. The tribe represented by 1 genus in the Three Provinces of Northeast China.

References. Chen, 1982, 1999 (Noctuidae: Ophiderinae, part); Sugi, 1982 (Noctuidae: Ophiderinae, part); Kononenko, 1990 (Noctuidae: Catocalinae, part); Kononenko *et al.*, 1998 (Noctuidae: Catocalinae, part); Sviridov, 2003 (Noctuidae: Catocalinae, part); Holloway, 2005, (Catocalinae, Pericymini), 2011 (Erebidae: Erebinae); Kononenko, 2005 (Erebidae: Erebinae), 2010 (Noctuidae: Erebinae, Hulodini); Matov *et al.*, 2008 (Noctuidae: Catocalinae, part); Zahiri *et al.*, 2012 (Erebidae: Erebinae, Hulodini).

Genus ***Lacera*** Guenée, 1852 戟夜蛾属

Lacera Guenée, 1852, *in* Boisduval & Guenée, *Histoire Naturelle des Insectes. Species Général des Lépidoptéres. Tome Septiéme. Noctuélites*, tome 3: 336. Type species: *Phalaena alope* Cramer, 1780 [Surinam, in error for Africa].

The genus includes 11 species, mainly distributed in the Palaearctic, Oriental and Australia regions; 1 species is in the Three Provinces of Northeast China.

注：首次给出中文属名。

***Lacera procellosa* Butler, 1879 斑戟夜蛾**

Lacera procellosa Butler, 1879, *Annals and Magazine of Natural History*, (5)4: 368. Type locality: Japan, Yokohama.

Synonymy: *Lacera alope* auct., nec Cramer, 1780.

Distribution. China (Liaoning, Hunan, Jiangxi, Hainan, Xizang, Taiwan); South Korea, Japan, Vietnam, Cambodia, Sri Lanka, India, Nepal, Philippines, Indonesia, New Guinea.

寄主：云实属植物（陈一心，1999）。

Family Erebidae, *incertae sedis* 目夜蛾科内隶属未明确

Holloway (2011) placed the genus *Blasticorhinus* and some other genera of the Oriental fauna to the group of "miscellaneous genera" (incertae sedis). This group contains all genera with unclear systematic position that have not been placed to certain subfamily (Holloway, 2011). This group numbers 1 genus and 3 species are in the Three Provinces of Northeast China.

Genus *Blasticorhinus* Butler, 1893 锉夜蛾属

Blasticorhinus Butler, 1893, *Annals and Magazine of Natural History*, (6)12: 46. Type species: *Thermesia rivulosa* Walker, 1865 [India].

Synonymy: *Carsina* Hampson, 1924.

The genus includes 17 species, distributed in the Ethiopian and Oriental regions, and in the Chinese subregion of the Palaearctic region; 3 species are in the Three Provinces of Northeast China.

***Blasticorhinus ussuriensis* (Bremer, 1861) 寒锉夜蛾**

Remigia ussuriensis Bremer, 1861, *Bulletin de la Classe Physico-Mathématique de l'Académie Impériale des Sciences de St.-Pétersbourg*, 3: 496. Type locality: Russia, "Am 'Ussuri', oberhalb der Emma-Mündung" [Primorye terr.].

Distribution. China (Heilongjiang, Jilin, Liaoning, Jiangsu, Zhejiang, Hunan, Fujian); Russia (Siberia), Korea.

寄主：多花紫藤（陈一心，1999）；酸模（*Rumex acetosa*）、两型豆（*Amphicarpaea edgeworthii*）、胡枝子（*Lespedeza bicolor*）、（*Milletia japonica*）、刺槐（*Robinia pseudoacacia*）（Matov & Kononenko, 2012）。

***Blasticorhinus unduligera* (Butler, 1878) 拟寒锉夜蛾**

Azazia unduligera Butler, 1878, *Annals and Magazine of Natural History*, (5)1: 293. Type locality: Japan, Yokohama.

Distribution. China (Heilongjiang, Jilin, Liaoning); Russia (Siberia), Korea, Japan.

注：首次给出中文种名。

***Blasticorhinus hoenei* Berio, 1956 浙锉夜蛾**

Blasticorhinus hoenei Berio, 1956, *Memorie della Societa Entomologica Italiana*, 35: 27. Type locality: China, [Zhejiang] Chekiang, Ost Tien-mu-shan.

Synonymy: *Blasticorhinus hoenei* Berio, 1956; *Blasticorhinus hoenei koraiensis* Kononenko & Ahn, 1995 (ssp.).

Distribution. China (Heilongjiang, Jilin, Zhejiang, Taiwan); Korea.

注：首次给出中文种名。

Subfamily **Toxocampinae** Guenée, 1852 影夜蛾亚科

Toxocampini Guenée, 1852, *in* Boisduval & Guenée, *Histoire Naturelle des Insectes. Species Général des Lépidoptères*, 6: 419. Type genus: *Toxocampa* Guenée, 1841.

Synonymy: Lygephilini Wiltshire, 1976; Anumetini Wiltshire, 1976; Lygephilini Berio, 1992; Apopestini Beck, 1996.

The Toxocampinae as a subfamily was upgraded from tribal to subfamily level by Fibiger *et al.* (2009). The subfamily distributed mainly in temperate and arid zones of the Palaearctic region. Most species are inhabit forest-steppe, steppe and trophically connected with Fabaceae. It includes 4 genera and 13 species are in the Three Provinces of Northeast China.

References. Chen, 1982, 1999 (Noctuidae); Sugi, 1982 (Noctuidae: Ophiderinae, part); Kononenko, 1990, Kononenko *et al.*, 1998 (Noctuidae: Ophiderinae, part); Goater *et al.*, 2003 (Noctuidae: Catocalinae, Catocalini, Toxocampina); Sviridov, 2003 (Noctuidae: Catocalinae); Fibiger, Hacker, 2005 (Erebidae: Catocalinae, Toxocampini); Fibiger, Lafontaine, 2005 (Erebidae: Catocalinae, Toxocampini); Kononenko, 2005 (Noctuidae: Erebinae, Catocalinae, Toxocampini), 2010 (Erebidae: Catocalinae, Toxocampini); Lafontaine, Fibiger, 2006 (Noctuidae: Catocalinae, Toxocampini); Kononenko, Han, 2007 (Noctuidae: Catocalinae); Matov *et al.*, 2008 (Noctuidae: Erebinae, Catocalinae); Lafontaine, Schmidt, 2010 (Erebidae, Catocalinae, Toxocampini), 2013 (Erebidae: Toxocampinae); Fibiger *et al.*, 2011 (Erebidae: Toxocampinae); Zahiri *et al.*, 2012 (Erebidae: Toxocampinae); Ronkay *et al.*, 2014 (Erebidae, Toxocampinae).

注：首次给出中文亚科名。

Genus *Chrysorithrum* Butler, 1878 客来夜蛾属

Chrysorithrum Butler, 1878, *Annals and Magazine of Natural History*, (5)1: 292. Type species: *Catocala amata* Bremer & Grey, 1853 [China].

Synonymy: *Pseudophia* auct..

The genus includes 3 species, distributed in the East Palaearctic and China; 2 species are in the Three Provinces of Northeast China.

Chrysorithrum amata (Bremer & Grey, 1853) 客来夜蛾

Catocala amata Bremer & Grey, 1853, *in* Motschulsky, *Etudes Entomologie*, 1: 66. Type locality: North China, Pekin.

Synonymy: *Chrysorithrum fuscum* Butler, 1881; *Chrysorithrum rufescens* Butler, 1881; *Chrysorithrum amata steni* Bryk, 1942 (ssp.); *Chrysorithrum amatum amorina* Bryk, 1948 (ssp.); *Chrysorithrum amatum separatum* Bryk, 1948 (f.).

Distribution. China (Heilongjiang, Jilin, Liaoning, Beijing, Nei Mongol, Hebei, Shaanxi, Shandong, Henan, Zhejiang, Fujian, Yunnan); Russia (Siberia), Korea, Japan.

寄主：胡枝子（陈一心，1999）；胡枝子（*Lespedeza bicolor*）、（*Lespedeza schindleri*）（Matov & Kononenko, 2012）。

Chrysorithrum flavomaculata (Bremer, 1861) 筱客来夜蛾

Bolina flavomaculata Bremer, 1861, *Bulletin de l'Academie Impériale des Sciences de St.-Petersbroug*, 3: 492. Type locality: Russia, [Khabarovsk terr.], Kuznetzovka.

Synonymy: *Bolina maximoviczi* Bremer, 1864; *Chrysorithrum sericeum* Butler, 1878.

Distribution. China (Heilongjiang, Jilin, Liaoning, Nei Mongol, Hebei, Shaanxi, Zhejiang, Yunnan); Russia (from South Russia and South Ural, Siberia), Mongolia, Korea, Japan.

寄主：豆科植物（陈一心，1999）；广布野豌豆（*Vicia cracca*）、落叶松属（*Larix*）、柳属（*Salix*）植物（Matov & Kononenko, 2012）。

Genus *Lygephila* Billberg, 1820 影夜蛾属

Lygephila Billberg, 1820, *Enumeratio Insectorum in Museo G.J. Billberg*: 85. Type species: *Phalaena lusoria* Linnaeus, 1758 [Germany].

Synonymy: *Asticta* Hübner, 1816; *Toxocampa* Guenée, 1841; *Eccrita* Lederer, 1857; *Tathorhynhus* Hampson, 1894 (subgen.); *Craccaphila* Berio, 1996; *Sinocampa* Kononenko & Fibiger, 2008, preocc. (Chou & Chen, 1981[*Diplura*]); *Katyusha* Kemal & Koçak, 2009 (subgen.), repl. name. for *Sinocampa*.

The genus is includes 35 species, and subdivided into three subgenera *Lygephila*, *Sinocampa* and

Tathorhynchus, mainly distributed in the Palaearctic region, 1 species known from the Nearctic region; 9 species are in the Three Provinces of Northeast China.

Lygephila ludicra (Hübner, 1790) 艺影夜蛾

Phalaena (*Noctua*) *ludicra* Hübner, 1790, *Beiträge zur Geschichte der Schmetterlinge*, 2 (4): 95, 128, pl. (4), 3: r. Type locality: Germany, Sachsen; Austria.

Synonymy: *Eccrita ludicra gracilis* Staudinger, 1879 (var.); *Toxocampa ichinosawana* Matsumura, 1925; *Eccrita ludicra major* Draudt, 1950 (f.).

Distribution. China (Heilongjiang, Jilin, Shandong, Fujian); Russia (Southeast European part, Ural, Siberia), Mongolia, Korea, Japan; Near East, Central Asia, Central and South Europe.

寄主：巢菜属（陈一心，1999）；红豆草（*Onobrychis vicifolia*）、广布野豌豆（*Vicia cracca*）、山黧豆属（*Lathyrus*）植物（Matov & Kononenko, 2012）。

Lygephila maxima (Bremer, 1861) 巨影夜蛾

Toxocampa maxima Bremer, 1861, *Bulletin de l'Academie Impériale des Sciences de St.-Petersbroug*, 3: 491. Type locality: Russia, Amur reg.; Blagoveschensk, "Ussuri" [Primorye terr. terr:], "Kengka See" [Khanka lake].

Synonymy: *Toxocampa enormis* Butler, 1878.

Distribution. China (Heilongjiang, Jilin, Liaoning, Shandong, Fujian); Russia (Siberia), Korea, Japan.

寄主：禾本科植物（陈一心，1999）；鸭茅（*Dactylis glomerata*）、薹草属（*Carex*）植物（Matov & Kononenko, 2012）。

Lygephila dubatolovi Fibiger, Kononenko & Nilsson, 2008 脉影夜蛾

Lygephila dubatolovi Fibiger, Kononenko & Nilsson, 2008, *Zootaxa*, 1922: 62, figs. 1, 2, 4, 12, 17. Type locality: Russia, Primorye terr. Territory: Arsenjev; China, Liaoning: Jianchang.

Distribution. China (Liaoning); Russia (Transbaikalia, Siberia).

注：首次给出中文种名。

Lygephila vulcanea (Butler, 1881) 焚影夜蛾

Toxocampa vulcanea Butler, 1881, *Transactions of the Entomological Society of London*, 1881: 192. Type locality: Japan, Tokyo.

Distribution. China (Heilongjiang, Jilin, Shanxi); Russia (Siberia), Korea, Japan.

寄主：麻栎（*Quercus acutissima*）、槲树（*Quercus dentata*）、胡枝子属（*Lespedeza*）、紫藤属（*Wisteria*）植物（Matov & Kononenko, 2012）。

Lygephila craccae ([Denis & Schiffermüller], 1775) 放影夜蛾

Noctua craccae [Denis & Schiffermüller], 1775, *Ankündung eines Systematisches Werkes von den Schmetterlinge der Wiener Gegend. Wien*: 94. Type locality: Austria, Vienna district.

Synonymy: *Phalaena nigricollis* Villers, 1789; *Ophiusa craccae lutosa* Warren, 1913 (ssp.); *Ophiusa craccae grisea* Warren, 1913 (ssp.); *Ophiusa craccae laevigata* Warren, 1913 (ssp.); *Toxocampa craccae caliginosa* Schawerda, 1931 (var.); *Toxocampa craccae riata* Rungs, 1952 (ssp.); *Toxocampa craccae centralasiae* Sheljuzhko, 1955 (ssp.).

Distribution. China (Heilongjiang, Jilin, Nei Mongol, Xinjiang); Russia (European part, Ural, Siberia), Mongolia, Korea, Japan; Central Asia, Near East, Europe.

寄主：巢菜属、黄芪属植物（陈一心，1999）；冷黄芪（*Astragalus glycyphyllos*）、蒙古黄芪（*Astragalus membranaceus*）、湿地黄芪（*Astragalus uliginosus*）、（*Colutea canescens*）、小冠花（*Coronilla varia*）、鱼鳔槐（*Dorycnium pentaphyllum*）、（*Dorycnium subfruticosum*）、山黧豆（*Lathyrus gmelini*）、（*Lathyrus montanus*）、大托叶山黧豆（*Lathyrus pisiformis*）、牧地山黧豆（*Lathyrus pratensis*）、春山黧豆（*Lathyrus vernus*）、棘豆（*Oxytropis campanulata*）、山野豌豆（*Vicia amoena*）、（*Vicia angustifolium*）、广布野豌豆（*Vicia cracca*）、救荒野豌豆（*Vicia sativa*）、野豌豆（*Vicia sepium*）、（*Vicia silvatica*）、细叶野豌豆（*Vicia tenuifolia*）、车前草属（*Plantago*）植物（Matov & Kononenko, 2012).

Lygephila viciae (Hübner, [1822]) 蚕豆影夜蛾

Noctua viciae Hübner, [1822], *Sammlung Europaischer Schmetterlinge. Eulen*, 4: pl. 146: 671-673. Type locality: Europe.

Synonymy: *Ophiusa viciae coronillae* Herrich-Schäffer, 1855 (var.); *Toxocampa stigmata* Wileman, 1911; *Toxocampa viciae violaceogrisea* Draudt, 1950 (ssp.).

Distribution. China (Heilongjiang, Jilin, Nei Mongol, Xinjiang, Hebei, Shanxi, Shaanxi, Shandong, Zhejiang, Yunnan); Russia (Ural, Siberia), North Korea, Japan; Europe.

寄主：萹蓄 (*Polygonum aviculare*)、高山黄芪（*Astragalus alpinus*）、（*Astragalus follicularis*）、冷黄芪（*Astragalus glycyphyllos*）、（*Astragalus propinquus*）、湿地黄芪（*Astragalus uliginosus*）、小冠花（*Coronilla varia*）、（*Faba bona*）、岩黄芪（*Hedysarum esculentum*）、林地香豌豆（*Lathyrus gmelini*）、大托叶山黧豆（*Lathyrus pisiformis*）、牧地山黧豆（*Lathyrus pratensis*）、玫红山黧豆（*Lathyrus tuberosus*）、春山黧豆（*Lathyrus vernus*）、草木樨（*Melilotus officinalis*）、（*Onobrychis sibirica*）、（*Oxytropis campanulata*）、豌豆（*Pisum sativum*）、山野豌豆（*Vicia amoena*）、窄叶野豌豆（*Vicia angustifolia*）、广布野豌豆（*Vicia cracca*）、德国野豌豆（*Vicia dumetorum*）、多茎野豌豆（*Vicia multicaulis*）、救荒野豌豆（*Vicia sativa*）、（*Vicia silvatica*）、细叶野豌豆（*Vicia tenuifolia*）（Matov & Kononenko, 2012）。

Lygephila pastinum (Treitschke, 1826) 锹影夜蛾

Ophiusa pastinum Treitschke, 1826, *Die Schmetterlinge von Europa*, 5 (3): 297. Type locality: Austria, Vienna district.

Synonymy: *Ophiusa astragali* Herrich-Schäffer, [1851]; *Toxocampa pastinum dilutior* Staudinger, 1892 (var.); *Toxocampa graciocissima* Bryk, 1948.

Distribution. China (Heilongjiang, Jilin); Russia (Ural, Siberia), North Korea, Japan, Kazakhstan; Europe.

寄主：小冠花属、巢菜属、黄芪属植物（陈一心，1999）；冷黄芪（*Astragalus glycyphyllos*）、（*Astragalus propinquus*）、湿地黄芪（*Astragalus uliginosus*）、林地香豌豆（*Lathyrus gmelini*）、欧山黧豆（*Lathyrus palustris*）、大托叶山黧豆（*Lathyrus pisiformis*）、牧地山黧豆（*Lathyrus pratensis*）、春山黧豆（*Lathyrus vernus*）、（*Oxytropis campanulata*）、红车轴草（*Trifolium pratense*）、白车轴草（*Trifolium repens*）、山野豌豆（*Vicia amoena*）、窄叶野豌豆（*Vicia angustifolia*）、广布野豌豆（*Vicia cracca*）、救荒野豌豆（*Vicia sativa*）、野豌豆（*Vicia sepium*）、（*Vicia silvatica*）、细叶野豌豆（*Vicia tenuifolia*）、小冠花属（*Coronilla*）植物（Matov & Kononenko, 2012）。

Lygephila procax (Hübner, 1813) 黑缘影夜蛾

Noctua procax Hübner, 1813, *Sammlung Europaischer Schmetterlinge. Eulen*, 4: pl. 109: 510. Type locality: Europe.

Synonymy: *Asticta proclivis* Hübner, [1823]; *Ophiusa limosa* Treitschke, 1826; *Toxocampa limosa nigricostata* Graeser, 1890 (ssp.).

Distribution. China (Heilongjiang, Jilin, Nei Mongol, Xinjiang, Hebei, Shaanxi, Sichuan, Yunnan, Xizang); Russia (South of European part, Ural, Transbaikalia, Siberia), Korea, Japan, Vietnam(North), India(North), Pakistan; Middle Asia, Near East, Southeast Europe.

寄主：鱼鳔槐 (*Colutea arborescens*)、广布野豌豆（*Vicia cracca*）、小冠花属（*Coronilla*）、山黧豆属（*Lathyrus*）植物（Matov & Kononenko, 2012）。

Lygephila recta (Bremer, 1864) 直影夜蛾

Toxocampa recta Bremer, 1864, *Memoires de l'Académie Imperiale des Sciences de St.-Petersbourg*, (7)8, 1: 98, pl. 8: 9. Type locality: Russia, Primorye terr., Port May [Ternei].

Distribution. China (Heilongjiang, Jilin, Hunan, Jiangxi, Fujian); Russia (Siberia), Korea, Japan.

寄主：胡颓子属植物（陈一心，1999）；两型豆（*Amphicarpaea edgeworthii*）、胡枝子属（*Lespedeza*）植物（Matov & Kononenko, 2012）。

Genus *Autophila* Hübner, [1823] 1816 隘夜蛾属

Autophila Hübner, [1823] 1816, *Verzeichniss bekannter Schmettlinge* [sic!]: 274. Type species: *Noctua dilucida* Hübner, [1808] [Europe].

Synonymy: *Cheirophanes* Boursin, 1955 (subgen.).

The genus includes 51 species, exclusively distributed in the South Palaearctic and partly Oriental regions; 1 species is in the Three Provinces of Northeast China.

***Autophila inconspicua* (Butler, 1881) 隐隘夜蛾**

Apopestes inconspicua Butler, 1881, *Transactions of the Entomological Society of London*, 1881: 191. Type locality: Japan, Tokyo, Yokohama.

Synonymy: *Spintherops cataphanes praeligaminosa* Staudinger, 1888 (ssp.); *Autophila inconspicua cataphanoides* Boursin, 1955 (ssp.); *Autophila inconspicua altaica* Ronkay, 1989 (ssp.).

Distribution. China (Jilin, Hunan); Russia (Siberia), Mongolia, Korea, Japan, Kazakhstan.

注：本种当前具有 3 个亚种，其中，指名亚种分布于东北亚地区；*cataphanoides* Boursin, 1955 亚种分布于中国中部地区；*altaica* Ronkay, 1989 亚种分布于哈萨克斯坦的阿尔泰地区。

Genus ***Apopestes*** Hübner, [1823] 1816 仿爱夜蛾属

Apopestes Hübner, [1823] 1816, *Verzeichniss bekannter Schmettlinge* [sic!]: 275. Type species: *Noctua spectrum* Esper, 1787 [South Italy].

Synonymy: *Spintherops* Boisduval, 1840.

The genus includes 4 species, distributed in the South Palaearctic and partly Oriental regions; 2 species are in the Three Provinces of Northeast China.

***Apopestes indica* Moore, 1883 印仿爱夜蛾**

Apopestes indica Moore, 1883, *Proceedings of the Zoological Society of London*, 1883: 22. Type locality: India, Madhya Pradesh, Manpuri.

Synonymy: *Apopestes spectrum koreana* Herz, 1904 (var.).

Distribution. China (Jilin); Russia (Siberia), Korea, Japan, India(North), Nepal.

寄主：黄芪属（*Astragalus*）植物（Matov & Kononenko, 2012）。

注：首次给出中文种名。

Family **Euteliidae** Grote, 1882 尾夜蛾科

Euteliinae Grote, 1882, *New Checklist of North American Moths*: 33. Type genus: *Eutelia* Hübner, [1823].

Synonymy: Eurhipidae Herrich-Schäffer, [1851] 1845.

Family Euteliidae (together with Stictopterinae) includes 29 genera and 520 species (Nieukerken *et al.,* 2011), mainly distributed in the tropical and subtropical areas. Traditionally the Euteliinae and Stictopterinae treated as distinct subfamilies of Noctuidae (Sugi, 1982, 1992; Holloway, 1975; Kononenko *at al.*, 1998; Kononenko & Pinratana, 2005). Fibiger & Lafontaine (2005) placed them as subfamilies of Erebidae, then as subfamilies of Noctuidae (Lafontaine & Fibiger, 2005). Subsequently Zahiri *et al.* (2011) raised the status of Euteliinae to the family level and placed Stictopterinae into subfamily of Euteliidae based on molecular study. Bornean Euteliidae (together with Stictiopterinae) have been revised by Holloway (1975), and Euteliinae of Sumatra revised by Kobes (2008).

The family Euteliidae includes 2 subfamilies, Euteliinae and Stictopterinae; the subfamily Stictopterinae is divided for tribes Stictopterini and Odontidini. The tribe Stictopterini in China has been revised by Qi *et al.* (2011).

Very distinct subfamily including 25 genera and about 300 species, mainly distributed in the tropical and subtropical ares of the Oriental, Ethiopian and Neotropical regions. 10-15 species are known from the Palaearctic region; 1 genus and 2 species are presented in the Three Provinces of Northeast China.

References. Hampson, 1912 (Euteliinae); Chen, 1982, 1999 (Noctuidae); Sugi, 1960, 1982 (Noctuidae: Euteliinae); Kononenko *et al.*, 1998 (Noctuidae: Euteliinae); Kononenko, 2003 (Noctuidae: Euteliinae); Kononenko, 2005 (Erebidae: Euteliinae), 2010 (Noctuidae: Euteliinae); Kononenko, Han, 2007 (Noctuidae: Euteliinae); Matov *et al.*, 2008 (Noctuidae: Euteliinae); Holloway, 2011 (Euteliidae: Euteliinae); Zahiri *et al.*, 2012 (Erebidae: Euteliidae).

注：本科根据 Zahiri *et al.* (2012)研究由亚科提升到科级阶元。

Genus *Eutelia* Hübner, [1823] 1816 尾夜蛾属

Eutelia Hübner, [1823] 1816, *Verzeichniss bekannter Schmettlinge* [sic!]: 259. Type species: *Noctua adulatrix* Hübner, [1813] [Europe].

Synonymy: *Eutesia* Hübner, [1826] 1816, missp.; *Eurhipia* Boisduval, 1829; *Phlegetonia* Guenée, 1852; *Ripogenus* Grote, 1865; *Zobia* Saalmüller, 1891; *Alotsa* Swinhoe, 1900; *Silacida* Swinhoe, 1900; *Entelia*: Lower, 1901, missp.; *Noctasota* Clench, 1954; *Adoraria* Beck, 1996.

The genus includes 96 species, mainly distributed in the tropical and subtropical areas of the New and Old World; 2 species are in the Three Provinces of Northeast China.

Eutelia geyeri (Felder & Rogenhofer, 1874) 漆尾夜蛾

Eurhipia geyeri Felder & Rogenhofer, 1874, *Reise Österreichischen Fregatte Novara um die Erde in den Jahren 1857, 1859. Zoologischer Theil*, 2 (Abt. 2): pl. 110: 23. Type locality: Japan.

Synonymy: *Eutelia inextricata* Moore, 1882.

Distribution. China (Heilongjiang, Jilin, Liaoning, Jiangsu, Zhejiang, Hunan, Jiangxi, Fujian, Sichuan,

Yunnan, Xizang, Taiwan); Russia (Siberia), Korea, Japan.

寄主：盐肤木（陈一心，1999）；麻栎（*Quercus acutissima*）、黄栌（*Cotinus coggygria*）、盐肤木（*Rhus chinensis*）、罗氏盐肤木（*Rhus javanica*）、野漆树（*Rhus sylvestris*）（Matov & Kononenko, 2012）。

***Eutelia hamulatrix* Draudt, 1950 钩尾夜蛾**

Eutelia hamulatrix Draudt, 1950, *Mitteilungen der MünchnerEntomologischen Gesellschaft* 40: 140. pl. 8: 31. Type locality: China, West Tien-mu-shan.

Distribution. China (Jilin, Liaoning, Zhejiang, Taiwan); Korea.

Family **Nolidae** Bruand, 1846 **瘤蛾科**

Nolites [Nolinae] Bruand, 1846. *Mémoires de la Société d' Emulation du Doubs*, 2: 113. Type genus: *Nola* Leach, [1815] 1830.

Synonymy: Roeseliidae Walker, [1865] 1864; Nolidae Stainton, 1859; Nolidae Speyer & Speyer, 1862; Nolinae Hampson, 1894.

Family Nolidae represented by 186 genera and 1,738 species (Nieukerken *et al.,* 2011), distributed worldwide with maximal diversity in the tropical area. The status and placement of Nolinae (in modern sense) has been debated for a long time. Part of it (current subfamily Nolidae) was considered by early authors as a distinct family of the Noctuoidea, but Sarrothripinae, and Chloephorinae as subfamilies of the Noctuidae (Sugi, 1982; Inoue, 1982; Kononenko *et al*., 1998). The Sarrothripinae, Chloephorinae and Nolinae were recognized and joined as members of family Nolidae by Holloway (1998, 2003). Kitching & Rawlins (1999) and Fibiger & Lafontaine (2005) treated Nolidae as a distinct family. Later Lafontaine & Fibiger (2006) treated Nolinae as a subfamily of the Noctuidae. The European fauna of Nolinae has been revised by Fibiger *et al.* (2009) who placed Nolinae as a subfamily of Noctuidae. Here we accept the last concept of Nolidae based on molecular study (Zahiri *et al.*, 2013b). The Nolidae is represented in China by 6 subfamilies (Beaninae, Nolinae, Chloephorinae, Westermanniinae, Risobinae and Eligminae); totally 20 genera and 39 species of Nolidae are known from the Three Provinces of Northeast China.

References. Chen, 1982, 1999 (Noctuidae); Inoue, 1982 (Nolidae); Inoue, 1982, 1991, 2001 (Nolidae, part); Sugi, 1982 (Noctuidae: Sarrothripinae, Chloephorinae); Speidel *et al.*, 1996 (Noctuidae: Nolinae, Chloephorinae, Sarrothripinae, Camptolominae); Holloway, 1998, 2003, 2011 (Nolidae: Nolinae); Kitching, Rawlins, 1999 (Nolidae); Oh, 2001 (Noctuidae: Nolinae); Sviridov, 2003 (Noctuidae: Sarrothripinae, Chloephorinae); Tschistjakov, 2003 (Nolidae, part); Fibiger, Hacker, 2005 (Nolidae); Fibiger, Lafontaine, 2005 (Nolidae); Kononenko, 2005 (Nolidae), 2010 (Noctuidae: Nolinae); Lafontaine, Fibiger, 2006 (Noctuidae: Nolinae);

Kononenko, Han, 2007 (Noctuidae: Nolinae); Matov *et al.*, 2008 (Noctuidae: Nolinae, Chloephorinae, Eariadinae, Eligminae); Fibiger *et al.*, 2009 (Noctuidae: Nolinae), 2011 (Nolidae); Zahiri *et al.*, 2013 (Nolidae); Kononenko, Pinratana, 2013 (Nolidae).

注：根据 Zahiri *et al.* (2013)的研究成果，瘤蛾科由豆瘤蛾亚科（Beaninae）、瘤蛾亚科（Nolinae）、科瘤蛾亚科（Collomeninae）、亮瘤蛾亚科（Diphtherinae）、丽夜蛾亚科（Chloephorinae）、俊夜蛾亚科（Westermanniinae）、长角皮夜蛾亚科（Risobinae）、旋夜蛾亚科（Eligminae）等组成。

Subfamily **Nolinae** Bruand, 1846 瘤蛾亚科

Nolites [Nolinae] Bruand, 1846, *Mémoires de la Société d' Emulation du Doubs*, 2: 113. Type genus: *Nola* Leach, [1815] 1830.

20 species of 6 genera are distributed in the Three Provinces of Northeast China.

References. Inoue, 1982 (Nolidae); Inoue, 1982, 1991, 2001 (Nolidae, part); Speidel *et al.*, 1996 (Noctuidae: Nolinae); Holloway, 1998, 2003, 2011 (Nolidae: Nolinae); Kitching, Rawlins, 1999 (Nolidae); Oh, 2001 (Noctuidae: Nolinae); Tschistjakov, 2003 (Nolidae); Fibiger, Lafontaine, 2005 (Nolidae: Nolinae); Kononenko, 2005 (Nolidae, Nolinae), 2010 (Noctuidae: Nolinae); Lafontaine, Fibiger, 2006 (Noctuidae: Nolinae); Kononenko, Han, 2007 (Noctuidae: Nolinae); Matov *et al.*, 2008 (Noctuidae: Nolinae); Fibiger *et al.*, 2009 (Noctuidae: Nolinae), 2011 (Nolidae: Nolinae); Zahiri *et al.*, 2013 (Nolidae: Nolinae); Kononenko, Pinratana, 2013 (Nolidae: Nolinae).

Genus ***Meganola*** Dyar, 1898 洛瘤蛾属

Meganola Dyar, 1898, *Journal of the New York Entomological Society*, 6: 43. Type species: *Meganola conspicua* Dyar, 1898 [USA].

Synonymy: *Antennola* de Freina & Witt, 1984; *Rhynchopalpus* auct.; *Roeselia* auct.; *Mimerastria* auct..

Large genus with more than 60-70 described species distributed all over the World; 11 species are in the Three Provinces of Northeast China.

Meganola albula **([Denis & Schiffermüller], 1775) 褐白洛瘤蛾**

Noctua albula [Denis & Schiffermüller], 1775, *Ankündung eines Systematisches Werkes von den Schmetterlinge der Wiener Gegend. Wien*: 69. Type locality: Austria, Vienna district.

Synonymy: *Pyralis albulalis* Hübner, 1796; *Roeselia albulana* Hübner, [1825]; *Roeselia formosana* Wileman et West, 1928; *Nola albula nivalis* Caradja, 1934 (var.); *Roeselia albula pacifica* Inoue, 1958 (ssp.).

Distribution. China (Heilongjiang, Jilin, Jiangxi, Taiwan); Russia (European part, South Ural, Transbaikalia, Siberia), Mongolia, Korea, Japan; Central and South Europe.

寄主：茶藨、蓝莓、草莓属、委陵菜属、百脉根属、车轴菜属、薄荷属植物（陈一心，1999）；麻栎

（*Quercus acutissima*）、黄栌（*Cotinus coggygria*）、盐肤木（*Rhus chinensis*）、罗氏盐肤木（*Rhus javanica*）、野漆树（*Rhus sylvestris*）匍匐委陵菜（*Potentilla reptans*）、秦氏悬钩子（*Rubus caesius*）、树莓（*Rubus fruticosus*）、覆盆子（*Rubus idaeus*）、水薄荷（*Mentha aquatica*）、大麻叶泽兰（*Eupatorium cannabinum*）、越橘属（*Vaccinium*）、草莓属（*Fragaria*）植物（Matov & Kononenko, 2012）。

注：本种在东亚地区有2亚种分布，分别为：*Meganola albula formosalesa*（Wileman & West, 1928）和 *M. albula pacifica*（Inoue, 1958），目前为止后者仅在日本本州岛分布。本种在国内一直使用其属同物异名 *Roeselia*，在此予以变更。

Meganola fumosa **(Butler, 1879) 烟洛瘤蛾**

Nola fumosa Butler, 1879, *Illustrations of Typical Specimens of Lepidoptera Heterocera in the Collection of the British Museum*, 3: 9, pl. 43: 2. Type locality: Japan, Yokohama.

Distribution. China (Liaoning); Russia (East Transbaikalia, Siberia), Korea, Japan.

寄主：蒙古栎、槲树（方承莱，1982）。

注：本种国内一直使用其属同物异名 *Roeselia*，在此予以变更。

Meganola basifascia **(Inoue, 1958) 基洛瘤蛾**

Roeselia basifascia Inoue, 1958, *Kontyû*, 26: 235: 2, 7. Type locality: Japan, Kushiro.

Distribution. China (Jilin); Russia (Siberia), Japan.

Meganola bryophilalis **(Staudinger, 1887) 苔藓洛瘤蛾**

Nola bryophilalis Staudinger, 1887, *in* N.M. Romanoff (ed.), *Mémoires sur les Lépidoptères*, 3: 181, pl. 10: 5. Type locality: Russia, Primorye terr., Askold I..

Distribution. China (Liaoning); Russia (Siberia), Japan.

寄主：苹果属、花楸属植物（方承莱，1982）。

Meganola costalis **(Staudinger, 1887) 缘洛瘤蛾**

Nola costalis Staudinger, 1887, *in* N.M. Romanoff (ed.), *Mémoires sur les Lépidoptères*, 3: 179, pl. 10: 3. Type locality: Russia, Primorye terr., Askold I..

Synonymy: *Roeselia melanocosta* Inoue, 1961.

Distribution. China (Heilongjiang, Jilin); Russia (Siberia), Korea, Japan.

寄主：三叶海棠、豆樱、日本花楸（方承莱，1982）。

Meganola strigulosa **(Staudinger, 1887) 短斑洛瘤蛾**

Nola strigulosa Staudinger, 1887, *in* N.M. Romanoff (ed.), *Mémoires sur les Lépidoptères*, 3: 180, pl. 11: 4. Type locality: Russia, Vladivostok, Askold I..

Synonymy: *Roeselia satoi* Inoue, 1970.

Distribution. China (Jilin); Russia (Siberia), Korea, Japan.

寄主：蒙古栎、槲树（方承莱，1982）。

Meganola shimekii (Inoue, 1970) 席氏洛瘤蛾

Roeselia shimekii Inoue, 1970, *Bulletin of the Japan Entomological Academy*, 6: 5, pl. 1: 15-18. Type locality: Japan, Mt. Akiba.

Distribution. China (Heilongjiang); Russia (Siberia), Korea, Japan.

寄主：麻栎（方承莱，1982）。

Meganola mikabo (Inoue, 1970) 美洛瘤蛾

Roeselia mikabo Inoue, 1970, *Bulletin of the Japan Entomological Academy*, 6: 4, text fig. 1, pl. 1, fig. 2. Type locality: Japan, Gumma Pref., Mikaboyama.

Distribution. China (Heilongjiang); Russia (Siberia), Japan.

寄主：槲树（方承莱，1982）。

Meganola subgigas Inoue, 1982 巨洛瘤蛾

Meganola subgigas Inoue, 1982, *Moths of Japan*, 1: 665, 2: 343, pl. 229: 45, 351: 10, 353: 8. Type locality: Japan, Hokkaido, Oshima, Nakayama Pass.

Synonymy: *Meganola gigantula* auct, nec Staudinger, 1878.

Distribution. China (Jilin, Liaoning, Jiangxi); Russia (Siberia), Korea, Japan.

注：本种在国内多使用其同物异名"*Roeselia gigantula* Staudinger, 1878"，在此予以变更。

Meganola gigas (Butler, 1884) 斑洛瘤蛾

Nola gigas Butler, 1884, *Annals and Magazine of Natural History*, (5)13: 274. Type locality: Japan, Honshu [Yesso].

Synonymy: *Nola maculata* Staudinger, 1887; *Roeselia nigromaculata* Nagano, 1918.

Distribution. China (Heilongjiang, Jilin, Liaoning, Guangxi, Yunnan); Russia (Siberia), Korea, Japan.

寄主：胡桃楸（方承莱，1982）。

注：本种在国内一直使用其同物异名"*Roeselia maculata* Staudinger, 1887"，在此予以变更。

Meganola laoningensis Han & Li, 2008 辽宁洛瘤蛾

Meganola laoningensis Han & Li, 2008, *Tinea*, 20 (3): 152-154. Type locality: China, Liaoning, Jianchang [NEFU].

Synonymy: *Meganola pekarskyi* László, Ronkay & Ronkay, 2014.

Distribution. China (Liaoning, Shandong, Shaanxi, Jiangsu, Fujian, Guangdong), North Korea.

Genus *Evonima* Walker, 1865 米瘤蛾属

Evonima Walker, 1865, *List of the Specimens of Lepidopterous Insects in the Collection of the British Museum*, 32: 505. Type species: *Evonima aperta* Walker, 1865 [Indonesia, Java].

Synonymy: *Mimerastria* Butler, 1881; *Poecilonola* Hampson, 1900.

The genus includes 13-14 species, distributed from the Chinese subregion of the Palaearctic to Oriental regions; 1 species is in the Three Provinces of Northeast China.

Evonima mandschuriana (Oberthür, 1880) 苹米瘤蛾

Erastria mandschuriana Oberthür, 1880, *Études d'Entomologie*, 5: 83, pl. 2: 9. Type locality: Russia, Primorye terr., Askold I..

Synonymy: *Mimerastria mandshuriana tzygankovi* Kozhantshikov, 1924 (ssp.).

Distribution. China (Jilin, Liaoning, Beijing, Henan, Jiangxi, Sichuan); Russia (Siberia), Mongolia, Korea, Japan.

寄主：蒙古栎、槲树、大叶早樱、苹果等植物（方承莱，1982）。

注：本种在国内一直使用其属同物异名 *Mimerastris*，在此予以变更。

Genus *Manoba* Walker, [1863] 1864 斑瘤蛾属

Manoba Walker, [1863] 1864, *Journal of the Proceedings of the Linnean Society* (*Zoology*), 7:62. Type species: *Manoba implenes* Walker, [1863].

Synonymy: *Rhynchopalpus* Hampson, 1893; *Roeselia* sensu Hampson; *Meganola* auct., nec Dyar, 1898; *Mimerastria* auct..

At present, several dozens of species are included in *Manoba*, however the placement of some is uncertain. Most species are known from Chinese and Sino-Himalayan subregions of the Palaearctic, Oriental and Australian regions; 1 species is in the Three Provinces of Northeast China.

Manoba banghaasi (West, 1925) 邦氏斑瘤蛾

Nola banghaasi West, 1925, *Annals and Magazine of Natural History*, (10)3: 194. Type locality: Russia, "Ussuri" [Primorye terr.].

Synonymy: *Nola banghaasi sumi* Inoue, 1956 (ssp.).

Distribution. China (Heilongjiang, Liaoning, Guizhou, Yunnan); Russia (Siberia), Korea, Japan.

Genus *Casminola* László, Ronkay & Witt, 2010 杂瘤蛾属

Casminola László, Ronkay & Witt, 2010, *Esperiana*, 15: 14. Type species: *Poecilonola chionobasis* Hampson, 1901 [India, Himachal Pradesh].

The genus includes 10 species listed by László *et al.* (2010), most of them are distributed in the Oriental region; 1 species occurs in the Three Provinces of Northeast China.

Casminola pulchella (Leech, 1889) 美杂瘤蛾

Acontia pulchella Leech, 1889, *Proceedings of the Zoological Society of London*, 1889: 524, pl. 53: 10. Type locality: Japan, Ohoyama.

Distribution. China (Heilongjiang, Jilin, Guangxi); Russia (Siberia), Korea, Japan.

Genus *Nolathripa* Inoue, 1970 洼皮瘤蛾属

Nolathripa Inoue, 1970, Inoue, 1970, *Bulletin of the Japan Entomological Academy*, 5: 38. Type species: *Nola lactaria* Graeser, 1892 [Russia, Amur reg.].

Monotypic genus.

Nolathripa lactaria (Graeser, 1892)洼皮瘤蛾

Nola lactaria Graeser, 1892, *Berliner entomologische Zeitschrift*, 37, II: 211. Type locality: Russia, "Ussuri", "Koslofska" [South Khabarovsk terr., Kozlovka].

Synonymy: *Asinduma korbi* Püngeler, 1908; *Dialithoptera stellata* Wileman, 1911; *Nola nakagawai* Nagano, 1918.

Distribution. China (Heilongjiang, Jilin, Hebei, Jiangxi, Sichuan); Russia (Siberia), Korea, Japan.

寄主：苹果、枇杷、日本核桃（方承莱，1982）；胡桃楸（*Juglans mandshurica*）（Matov & Kononenko, 2012）。

Genus *Nola* Leach, 1815 [1830] 瘤蛾属

Nola Leach, [1815] 1830, *in* Brewster, *The American Edition of the New Edinburgh Encyclopedia, Conducted by David Brewster – Assisted by upwards of One Hundred Gentlemen in Europe*, 9: 135. Type species: *Noctua palliola* [Denis & Schiffermüller], 1775 [Austria] (=*Phalaena* (*Tinea*) *cucullatella* Linnaeus, 1758).

Synonymy: *Lira* Billberg, 1820; *Chlamifera* Hübner, 1822; *Chlamiphora* Hübner, [1825]; *Roeselia* Hübner, [1825]; *Pisara* Walker, 1862; *Necla* Walker, 1863; *Automala* Walker, 1864; *Celama* Walker, 1865; *Aradrapha* Walker, [1866]; *Lebena* Walker, 1866; *Minnagara* Walker, 1866; *Argyrophyes* Grote, 1873; *Epizeuctis* Meyrick, 1889; *Deltapterum* Hampson, 1894; *Neonola* Hampson, 1900; *Poliothripa* Hampson, 1902; *Celamoides* van

Eecke, 1920; *Idiocyttara* Turner, 1944.

About 200 described species distributed worldwide, most diverse in the tropical forests; 5 species are in the Three Provinces of Northeast China.

Nola confusalis **(Herrich-Schäffer, [1847] 1845) 栎点瘤蛾**

Roeselia confusalis Herrich-Schäffer, [1847] 1845, *Systematische Bearbeitung der Schmetterlinge von Europa*, 2: 164. Type locality: [Europe] not stated.

Synonymy: *Celama confusalis pallida* Lempke, 1938 (ssp.); *Celama confusalis fumosensis* Daniel, 1957 (ssp.); *Celama confusalis signata* Lempke, 1960 (ssp.); *Celama confusalis bicolor* Lempke, 1960 (ssp.); *Celama confusalis suffusa* Lempke, 1960 (ssp.).

Distribution. China (Heilongjiang, Jilin, Liaoning, Beijing, Hebei, Sichuan, Xizang); Russia (European part, South Ural, Siberia), Korea, Japan; Europe.

寄主：栎属、鹅耳枥属、椴树属、李属、越橘属、薄荷属植物（方承莱，1982）；夏栎（*Quercus robur*）、欧洲桤木（*Alnus glutinosa*）、黑刺李（*Prunus spinosa*）、桦木属（*Betula*）、山毛榉属（*Fagus*）、鹅耳枥属（*Carpinus*）、椴树属（*Tilia*）、山楂属（*Crataegus*）、苹果属（*Malus*）、薄荷属（*Mentha*）植物（Matov & Kononenko, 2012）。

注：本种在国内一直沿用其属名的同物异名 *Celama*，在此予以变更。

Nola aerugula **(Hübner, 1793) 锈点瘤蛾**

Phalaena aerugula Hübner, 1793, *Sammlung Auserlesener Vögel und Schmetterlinge, mit ihren Namen Herausgegeben auf Jundert nach der Natur Ansgemalten Kupfern*: 11, pl. 61. Type locality: Europe.

Synonymy: *Pyralis centonalis* Hübner, 1796; *Roeselia centonana* Hübner, 1825; *Glaphyra atomosa* Bremer, 1861; *Nola candidalis* Staudinger, 1892; *Nola impudica* Christoph, 1893; *Nola littoralis* Paux; 1901; *Nola centonalis spitzi* Schawerda, 1921 (ab.); *Nola centonalis contrarialis* Heydemann, 1934 (ssp.); *Nola trituberculana* Heslop, 1959.

Distribution. China (Heilongjiang, Jilin, Beijing); Russia (European part, South Ural, Siberia), Mongolia, Korea, Japan; Central and South Europe.

寄主：蒙古栎、悬钩子属、桦树属、百脉根属、三叶草属、草莓属、车前属等植物（方承莱，1982）；黑果越橘（*Vaccinium myrtillus*）、黄花柳（*Salix caprea*）、蕨麻（*Potentilla anserina*）、覆盆子（*Rubus idaeus*）、驴食草（*Onobrychis viciifolia*）、栎属（*Quercus*）、桤木属（*Alnus*）、桦木属（*Betula*）、金丝桃属（*Hypericum*）、杨属（*Populus*）、草莓属（*Fragaria*）、悬钩子属（*Rubus*）、百脉根属（*Lotus*）、苜蓿属（*Medicago*）、车轴草属（*Trifolium*）、野豌豆属（*Vicia*）、蒲公英属（*Taraxacum*）及刺鳞草科（Plantaginaceae）植物（Matov & Kononenko, 2012）。

注：我国分布的种类为本种的 *Nola aerugula atomosa* Bremer, 1861 亚种；属于个体变异较大的种之一，翅颜色、斑纹可有 10~15 个形态。

***Nola minutalis* Leech, 1888 新宿瘤蛾**

Nola cristatula minutalis Leech, 1888, *Proceedings of the Zoological Society of London*, 1888: 607, pl. 330: 17. Type locality: Japan, Oiwake.

Distribution. China (Jilin); Russia (Siberia), Korea, Japan.

注：本种在华立中（2005）《中国昆虫名录》中记录中国有分布，但无具体地点；根据国外文献记载东北地区有分布，当前仅发现采自长白山区的 3 头标本；同时本种在国内一直沿用属同物异名 *Celama*，在此予以变更。

***Nola taeniata* Snellen, 1875 稻穗点瘤蛾**

Nola taeniata Snellen, 1875, *Tijdschrift voor Entomlogie*, 17: 65, pl. 6: 1. Type locality: Indonesia, Celebes.

Synonymy: *Nola candida* Butler, 1879; *Roeselia fragilis* Swinhoe, 1890; *Sorocostia mesozona* Lucas, 1890; *Celama anpingicola* Strand, 1917.

Distribution. China (Jilin, Liaoning, Zhejiang, Jiangxi, Fujian, Yunnan, Taiwan); Russia (Siberia), Korea, Japan, India, Nepal, Sri Lanka, Malaysia, Indonesia, Myanmar, Australia.

寄主：稻、棉、桑、苜蓿、问荆（方承莱，1999）。

注：本种在国内一直沿用其属同物异名 *Celama*，在此予以变更。

***Nola innocua* Butler, 1880 华瘤蛾**

Nola innocua Butler, 1880, *Proceedings of the Zoological Society of London*, 1880: 671. Type locality: Taiwan.

Synonymy: *Nola costimacula* Staudinger, 1887.

Distribution. China (Jilin, Liaoning, Taiwan); Russia (Siberia, Primorye terr.), Korea, Japan.

寄主：蚊母树（*Distylium racemosum*）（Matov & Kononenko, 2012）。

注：本种是瘤蛾族昆虫中我国记载的第一个物种，至今以来没有中文种名，在此给出中文种名"华瘤蛾"。

Subfamily **Chloephorinae** Stainton, 1859 丽夜蛾亚科

Chloephorinae Stainton, 1859, *Manual British Butterflies and Moths*, 2: 186. Type genus: *Chloephora* Stephens, 1827 (=*Pseudoips* Hübner, 1822).

Synonymy: Haliadea Snellen, 1867; Hydrophilinae Börner, 1932; Hylophilini Mell, 1943; Beninae Beck, 1960.

This subfamily includes 7 tribes. Among them, 12 genera and 17 species are in the Three Provinces of Northeast China.

References. Chen, 1982, 1999 (Noctuidae: Chloephorinae); Sugi, 1982 (Noctuidae, Chloephorini); Holloway, 1998, 2003, 2011 (Nolidae: Chloephorinae); Sviridov, 2003 (Noctuidae: Sarrothripinae, Chloephorinae); Fibiger, Hacker, 2005 (Nolidae: Chloephorinae); Fibiger, Lafontaine, 2005 (Nolidae: Chloephorinae); Kononenko, 2005

(Nolidae: Chloephorinae), 2010 (Noctuidae: Nolinae, Chloephorini, Sarrothripini); Lafontaine, Fibiger, 2006 (Noctuidae: Nolinae, Chloephorinae); Kononenko, Han, 2007 (Noctuidae: Nolinae, Chloephorini); Matov *et al.*, 2008 (Noctuidae: Nolinae, Chloephorinae); Fibiger *et al.*, 2009 (Noctuidae: Nolinae, Sarrothripini, Chloephorini); Zahiri *et al.*, 2013 (Nolidae: Chloephorinae).

Tribe **Sarrothripini** Hampson, 1894 皮夜蛾族

Sarrothripinae Hampson, 1894, *The Fauna of British India including Ceylon and Burma. Moths*, 2: 365. Type genus: *Sarrothripus* Curtis, 1824, a junior synonym of *Nycteola* Hübner, 1822.

Synonymy: Nycteolae Hübner, [1826] 1816; Nycteolides Herrich-Schäffer, [1846] 1845.

The tribe is distributed from the south temperate zone to the tropical and subtropical areas of the Old World where it is most diverse. 21 genera and approximately 100 species are known worldwide, 38 of which are known from the Palaearctic region. Only 1 genus and 2 species are occur in the Three Provinces of Northeast China.

References. Mell, 1943 (Noctuidae: Sarrothripinae); Obraztsov, 1953 (Noctuidae: Nycteolinae); Inoue, 1970 (Noctuidae: Nycteolinae); Chen, 1982, 1999 (Noctuidae: Stictopterinaw); Holloway, 1998, 2003, 2011 (Nolidae: Sarrothripinae); Sviridov, 2003 (Noctuidae: Sarrothripinae, part); Fibiger, Hacker, 2005 (Nolidae: Chloephorinae, Sarrothripini); Fibiger, Lafontaine, 2005 (Nolidae: Chloephorinae, Sarrothripini); Kononenko, 2005 (Nolidae: Sarrothripinae), 2010 (Noctuidae: Nolinae, Sarrothripini); Lafontaine, Fibiger, 2006 (Noctuidae: Nolinae, Chloephorini, Sarrothripinae); Fibiger *et al.*, 2009 (Noctuidae: Nolinae, Sarrothripini); Matov *et al.*, 2008 (Noctuidae: Nolinae, Chloephorinae); Zahiri *et al.*, 2013 (Nolidae: Nolinae).

Subtribe **Sarrothripina** Hampson, 1894 皮夜蛾亚族

Sarrothripina Hampson, 1894, *The Fauna of British India including Ceylon and Burma. Moths*, 2: 365. Type genus: *Sarrothripus* Curtis, 1824, a junior synonym of *Nycteola* Hübner, 1822.

Genus ***Nycteola*** Hübner, 1822 皮夜蛾属

Nycteola Hübner, 1822, *Systematisch-Alphabetischen Verzeichniss aller Bischer bey den Fiirbildungen zur Sammlung Europaischer Schmetterlinge angegebenen Gattungsbenennugen; mit Vormerkung auch Augburgischer Gattungen*: 60, 66. Type species: *Tortrix undulana* Hübner, [1799] [Europe] (=*Phalaena revayana* Scopoli, 1772).

Synonymy: *Nycteola* Hübner, [1806], suppr.; *Sarrothripus* Curtis, 1824; *Axia* Hübner, [1825] 1816; *Sarrothripa*: Duponchel, 1834, missp.; *Sarrothripa*: Duponchel, [1845] 1844, emend.; *Sarothripus*: Agassiz, 1846, emend.; *Subrita* Walker, 1866; *Sarotricha*: Meyrick, 1888, emend.; *Icasma* Turner, 1902; *Dufayella* Căpuse, 1972.

The genus includes 28 species, distributed over the World from temperate zone to tropical area, most diverse in the Oriental and Ethiopian regions; 2 species are in the Three Provinces of Northeast China.

***Nycteola degenerana* (Hübner, [1799]) 黑纹皮夜蛾**

Tortrix degenerana Hübner, [1799], *Sammlung Europaischer Schmetterlinge. Eulen*, 7: pl. 2: 8. Type locality: Europe.

Synonymy: *Nycteola degenerana hesperica* Dufay, 1958 (ssp.); *Nycteola degenerana eurasiatica* Dufay, 1961 (ssp.).

Distribution. China (Heilongjiang); Russia (European part, Ural, through Transbaikalia to Siberia), Mongolia, Korea, Japan; Europe.

寄主：夏栎（*Quercus robur*）、（*Salix appendiculata*）、耳柳（*Salix aurita*）、（*Salix bakko*）、黄花柳（*Salix caprea*）、灰柳（*Salix cinerea*）、（*Salix myrsinifolia*）、东陵山柳（*Salix phylicifolia*）、杞柳（*Salix purpurea*）、（*Salix reinii*）、蒿柳（*Salix viminalis*）、杨属（*Populus*）植物（Matov & Kononenko, 2012）。

***Nycteola asiatica* (Krulikowsky, 1904) 亚皮夜蛾**

Sarrothripus revayana asiatica Krulikowski, 1904 (var.), *Revue Russe d'Entomologie*, 4: 91. Type locality: Turkmenia, Ashkhabad.

Synonymy: *Nycteola populana* Patočka, 1953; *Nycteola hungarica* Kovács, 1954; *Nycteola pseudasiatica* Sugi, 1959.

Distribution. China (Jilin, Liaoning, Shandong, Hunan); Russia (European part, Ural, Siberia), Mongolia, Korea, Japan, India, Nepal; Central Asia, Near East, Europe.

寄主：细柱柳（*Salix gracilistyla*）、银白杨（*Populus alba*）（Park *et al.*, 2006）；杨属（*Populus*）、柳属（*Salix*）植物（陈一心，1999；Matov & Kononenko, 2012）。

Subtribe **Bleninina** Zahiri & Holloway, 2013 ex Blenina (1943) 癣皮夜蛾亚族

Blenina Zahiri & Holloway, 2013. Emended to Bleninia by Kononenko & Pinratana, 2013.

Blenini Mell, 1943, *Zoologische Jahrbucher. Abteilung fur Systematik, Oekologie, und Geographie der Tiere* 76: 204. Type genus: *Blenina* Walker, [1858].

Tropical origin worldwide distributed. 1 genus is in the Three Provinces of Northeast China.

Genus ***Blenina*** Walker, [1858] 1857 癣皮夜蛾属

Blenina Walker, [1858] 1857, *List of the Specimens of Lepidopterous Insects in the Collection of the British Museum*, vol. 13:1214. Type species: *Blenina accipiens* Walker, [1858] 1857 [Australia, Sri Lanka].

Synonymy: *Eliocroea* Walker, 1865; *Amrella* Moore, 1882.

The genus includes 32-35 species, distributed over the World from temperate zone to tropical area, most diverse in the Oriental regions; 1 species is in the Three Provinces of Northeast China.

***Blenina senex* (Butler, 1878) 柿癣皮夜蛾**

Dandaca senex Butler, 1878, *Annals and Magazine of Natural History*, (5)1: 82. Type locality: Japan, Yokohama.

Synonymy: *Dandaca megei* Oberthur, 1881.

Distribution. China (Liaoning, Jiangsu, Zhejiang, Hunan, Jiangxi, Fujian, Hainan, Guangxi, Sichuan, Yunnan, Taiwan); Korea, Japan, Vietnam.

寄主：柿（陈一心，1999）。

Tribe **Chloephorini** Stainton, 1859 丽夜蛾族

Chloephorinae Stainton, 1859, *Manual British Butterflies and Moths*, 2: 186. Type genus: *Chloephora* Stephens, 1827 (=*Pseudoips* Hübner, 1822).

Synonymy: Haliadea Snellen, 1867; Hydrophilinae Börner, 1932; Hylophilini Mell, 1943; Beninae Beck, 1960.

Mainly tropical tribe, represented by about 500 species, with the centre of diversity in the Southeast Asia. Only a small number of the genera occurs in the temperate zone, mainly in the Chinese subregion of the Palaearctic region. 3 genera are known from the Three Provinces of Northeast China.

References. Mell, 1943 (Noctuidae: Westermanniinae); Obraztsov, 1953 (Noctuidae: Chloephorinae); Sugi, 1970, 1982, 1990 (Noctuidae: Chloephorinae); Chen, 1982, 1999 (Noctuidae); Holloway, 1998, 2003, 2011 (Nolidae: Chloephorinae); Sviridov, 2003 (Noctuidae: Chloephorinae); Fibiger, Hacker, 2005 (Nolidae: Chloephorinae, Chloephorini); Fibiger, Lafontaine, 2005 (Nolidae: Chloephorinae, Chloephorini); Kononenko, 2005 (Nolidae: Chloephorinae), 2010 (Noctuidae: Nolinae, Chloephorinae); Lafontaine, Fibiger, 2006 (Noctuidae: Nolinae, Chloephorini, Chloephorina); Fibiger *et al.*, 2009 (Noctuidae: Nolinae, Chloephorini); Matov *et al.*, 2008 (Noctuidae: Nolinae, Chloephorinae); Zahiri *et al.*, 2013 (Nolidae: Nolinae).

Genus ***Parhylophila*** Hampson, 1912 **腾夜蛾属**

Parhylophila Hampson, 1912, *Catalogue of the Lepidoptera Phalaenae in the British Museum*, 11: xiv, 514. Type species: *Chloeophora celsiana* Staudinger, 1887 [Russia, Amur reg.].

The genus includes 2 species, mainly distributed in the Chinese subregion of the Palaearctic region; 1 species is in the Three Provinces of Northeast China.

***Parhylophila celsiana* (Staudinger, 1887) 腾夜蛾**

Chloeophora celsiana Staudinger, 1887 *in* N.M. Romanoff (ed.), *Mémoires sur les Lépidoptères*, 3: 177, pl. 10: 1. Type locality: Russia, Amur reg., Radde.

Distribution. China (Heilongjiang, Jilin, Hunan); Russia (Siberia), Korea, Japan.

Genus *Pseudoips* Hübner, 1822 饰夜蛾属

Pseudoips Hübner, 1822, *Systematisch-Alphabetischen Verzeichniss aller Bischer bey den Fiirbildungen zur Sammlung Europaischer Schmetterlinge angegebenen Gattungsbenennugen; mit Vormerkung auch Augburgischer Gattungen*: 59, 63. Type species: *Pseudoips prasinana* Linnaeus, sensu Hübner, 1822 (a misidentification of *Pyralis fagana* Fabricius, 1781 [Europe]).

Synonymy: *Hylophila* Hübner, [1825] 1816; *Chloephora* Stephens, 1827; *Halias* Treitschke, 1829; *Chloephila* Constanti, 1920; *Chloeophora* auct.; *Bena* auct..

The genus includes 6 species, two of which occur in the Palaearctic region, the others are distributed mainly in the Oriental region; 2 species are in the Three Provinces of Northeast China.

Pseudoips prasinana (Linnaeus, 1758) 饰夜蛾

Phalaena Tortrix prasinana Linnaeus, 1758, *Systema Naturae* (Edn. 10) 1: 530. Type locality: Sweden.

Synonymy: *Pyralis fagana* Fabricius, 1781; *Phalaena fagana*, auct.; *Phalaena lineata* Retzius, 1783; *Phalaena viridislienata* Retzius, 1783; *Pyralis sylvana* Fabricius, 1794; *Hylophila milleri* Capr, 1883; *Hylophila fiorii* Constantini, 1911; *Hylophila japonica* Warren, 1913; *Hylophila hongarica* Warren, 1913; *Hylophila britannica* Warren, 1913.

Distribution. China (Heilongjiang, Jilin, Liaoning, Nei Mongol); Russia (European part, Ural, Siberia), Korea, Japan; Near East, Europe.

寄主：桦、栎（陈一心，1999）；欧洲栗（*Castanea sativa*）、钝齿水青冈（*Fagus crenata*）、垂枝欧洲山毛榉（*Fagus sylvatica*）、麻栎（*Quercus acutissima*）、大桴栎（*Quercus grosseserrata*）、蒙古栎（*Quercus mongolica*）、无梗花栎（*Quercus petraea*）、枹栎（*Quercus serrata*）、夏栎（*Quercus robur*）、（*Alnus incana*）、柴桦（*Betula fruticosa*）、垂枝桦（*Betula pendula*）、（*Betula pubescens*）、桦叶鹅耳枥（*Carpinus betulus*）、欧洲榛（*Corylus avellana*）、榛（*Corylus heterophylla*）、欧洲山杨（*Populus tremula*）、黑刺李（*Prunus spinosa*）、挪威枫（*Acer platanoides*）、柳属（*Salix*）、椴树属（*Tilia*）、榆属（*Ulmus*）、梣属（*Fraxinus*）植物（Matov & Kononenko, 2012）。

注：《中国动物志》（夜蛾科）中将此种归入“*Bena* Billberg, 1820 碧夜蛾属”，在此予以变更。

Pseudoips sylpha (Butler, 1879) 希饰夜蛾

Hylophila sylpha Butler, 1879, *Illustrations of Typical Specimens of Lepidoptera Heterocera in the Collection of the British Museum*, 3: 10, pl. 43: 10. Type locality: Japan, Yokohama.

Synonymy: *Hylophila kraeffti* Graeser, 1888; *Hylophila magnifica* Leech, 1890.

Distribution. China (Heilongjiang, Jilin, Liaoning, Hebei, Hubei); Russia (Transbaikalia, Siberia), Korea, Japan.

寄主：栎（陈一心，1999）；麻栎（*Quercus acutissima*）、枹栎（*Quercus serrata*）、栓皮栎（*Quercus variabilis*）（Matov & Kononenko, 2012）。

Genus *Iragaodes* Matsumura, 1931 怒夜蛾属

Iragaodes Matsumura, 1931, *Insecta Matsumurana*, 5: 106. Type species: *Heterogenea nobilis* Staudinger, 1887 [Russia, Vladivostok].

Monotypic genus, mainly distributed in Chinese subregion of the Palaearctic.

注：首次给出中文属名。

***Iragaodes nobilis* (Staudinger, 1887) 优怒夜蛾**

Heterogenea nobilis Staudinger, 1887, *in* N.M. Romanoff (ed.), *Mémoires sur les Lépidoptères*, 3: 196, pl. 11: 8. Type locality: Russia, Vladivostok.

Distribution. China (Heilongjiang, Jilin, Liaoning, Taiwan); Russia (Siberia), Korea, Japan.

寄主：昌化鹅耳枥（*Carpinus tschonoskii*）（Matov & Kononenko, 2012）。

Tribe **Camptolomini** Mell, 1943 花布夜蛾族

Camptolominae Mell, 1943, *Zoologische Jahrbucher. Abteilung fur Systematik, Oekologie, und Geographie der Tiere*, 76: 217. Type genus: *Camptoloma* Felder, 1874.

The tribe includes 1 genus with 10 species distributed in Chinese subregion of the Palaearctic and Oriental regions. 1 genus and 1 species occurs in the Three Provinces of Northeast China.

References. Mell, 1943 (Noctuidae: Camptolominae); Chen, 1982, 1999 (Noctuidae); Speidel *et al.*, 1996 (Camptolominae); Sviridov, 2003 (Noctuidae: Chloephorinae); Fibiger, Hacker, 2005 (Nolidae: Chloephorinae, Chloephorini); Fibiger, Lafontaine, 2005 (Nolidae: Chloephorinae, Chloephorini); Kononenko, 2005 (Nolidae: Camptolomini), 2010 (Noctuidae: Nolinae, Chloephorini, Camptolomina); Lafontaine, Fibiger, 2006 (Noctuidae: Nolinae, Chloephorini, Chloephorina); Matov *et al.*, 2008 (Noctuidae: Nolinae, Chloephorinae); Holloway, 2011 (Nolidae: Chloephorinae, Camptolomini); Zahiri *et al.*, 2013 (Nolidae: Nolinae, Camptolomini).

注：首次给出中文族名。

Genus *Camptoloma* Felder, 1874 花布夜蛾属

Camptoloma Felder, 1874, *in* Felder & Rogenhofer, 1874, *Reise Österreichischen Fregatte Novara um die Erde in den Jahren 1857, 1859. Zoologischer Theil.*, 2 (Abt. 2). Type species: *Numenes interiorata* Walker, [1865] [China, Shanghai].

Synonymy: *Leucopardus* Hampson, 1894.

The genus includes 10 species distributed in the Oriental and the Chinese subregion of the Palaearctic regions; 1 species is in the Three Provinces of Northeast China.

注：本属一直以来在国内被归入灯蛾亚科中，王敏等（2011）首次将其归入瘤蛾科中。

***Camptoloma interiorata* (Walker, [1865]) 花布夜蛾**

Numenes interiorata Walker, [1865], *List of the Specimens of Lepidopterous Insects in the Collection of the British Museum*, 31: 290. Type locality: China, Shanghai.

Synonymy: *Camptoloma erythrophygum* Felder & Rogenhoffer, 1874.

Distribution. China (Heilongjiang, Jilin, Liaoning, Hebei, Shaanxi, Shandong, Anhui, Jiangsu, Shanghai, Zhejiang, Jiangxi, Fujian, Hubei, Hunan, Guangdong, Guangxi, Sichuan, Yunnan, Taiwan); Russia (Siberia), Korea, Japan, India.

寄主：枹栗、木诸、麻栎、槲树、乌桕、东北楠、柳（方承莱，2000）；麻栎（*Quercus acutissima*）、槲树（*Quercus dentata*）、栗属（*Castanea*）、（*Castaneopsis*）、柯属（*Lithocarpus*）、悬钩子属（*Rubus*）植物（Matov & Kononenko, 2012）。

Tribe **Ariolicini** Mell, 1943 爱丽夜蛾族

Ariolicini Mell, 1943, *Zoologische Jahrbucher. Abteilung fur Systematik, Oekologie, und Geographie der Tiere*, 76: 212. Type genus: *Ariolica* Walker, [1863].

The tribe comprises 18 oriental-tropical genera listed by Holloway (2003); 1 genus is in the Northeast China.

References. Chen, 1982, 1999 (Noctuidae); Sviridov, 2003 (Noctuidae: Chloephorinae); Fibiger, Lafontaine, 2005 (Nolidae: Chloephorinae, Ariolicini); Kononenko, 2005 (Nolidae: Ariolicini); Lafontaine, Fibiger, 2006 (Noctuidae: Nolinae, Chloephorini, Ariolicina); Kononenko, Han, 2007 (Noctuidae: Nolinae, Chloephorini, Ariolicina); Matov *et al.*, 2008 (Noctuidae: Nolinae, Chloephorinae); Holloway, 2011 (Nolidae: Ariolicini); Zahiri *et al.*, 2013 (Nolidae: Chloephorinae, Ariolicini).

Genus ***Sinna*** Walker, 1865 豹夜蛾属

Sinna Walker, 1865, *List of the Specimens of Lepidopterous Insects in the Collection of the British Museum*, 32: 641. Type species: *Sinna calospila* Walker, 1865 [Indonesia, Java].

Synonymy:*Teinopyga* Felder, 1874.

The genus includes 8 species, distributed in the Chinese subregion of the Palaearctic and Oriental regions; 1 species is in the Three Provinces of Northeast China.

***Sinna extrema* (Walker, 1854) 胡桃豹夜蛾**

Deiopeia extrema Walker, 1854, *List of the Specimens of Lepidopterous Insects in the Collection of the British Museum*, 2: 73. Type locality: China, Shanghai.

Synonymy: *Teinopyga reticularis* Felder et Rogenhofer, 1874; *Sinna clara* Butler, 1881; *Sinna ornatissima* Alphéraky, 1897; *Sinna extrema koresinna* Bryk, 1948 (ssp.).

Distribution. China (Heilongjiang, Jilin, Liaoning, Beijing, Shaanxi, Henan, Jiangsu, Shanghai, Zhejiang, Hubei, Hunan, Jiangxi, Fujian, Hainan, Sichuan); Russia (Siberia), Korea, Japan, Thailand.

寄主：枫杨、胡桃属（陈一心，1999）；胡桃楸（*Juglans mandshurica*）、枫杨（*Pterocarya stenoptera*）（Matov & Kononenko, 2012）。

Genus *Paracrama* Moore, [1884] 翡夜蛾属

Paracrama Moore, [1884], *The Lepidoptera of Ceylon*, vol. 3: 54. Type species: *Nolasena dulcissima* Walker, [1863] 1864 [Borneo, Sarawak].

The genus includes 3 species, distributed exclusively in the Old World, mainly in Southeast Asian and Australian regions; 1 species is in the Three Provinces of Northeast China.

***Paracrama angulata* Sugi, 1985 翡夜蛾**

Paracrama angulata Sugi, 1985, *Japanese Heterocerists' Journal*, 131: 90. Type locality: Taiwan, Fenchilu.

Synonymy: *Paracrama dulcissima*: Sugi, 1982, nec Walker, [1863] 1864, misidentification.

Distribution. China (Jilin, Liaoning, Guangdong, Guangxi, Taiwan); South Korea, Japan.

Tribe **Eariadini** Hampson, 1912 钻夜蛾族

Eariadina Hampson, 1912, *Catalogue of the Lepidoptera Phalaenae in the British Museum*, 11: 456. Type genus: *Earias* Hübner, [1825].

Synonymy: Eariini Beck, 1996.

The tribe includes 1 genus with about 50 species, mainly distributed in the Oriental and Australian regions; some species occurs in the south of Palaearctic. 2 genera and 4 species are in the Three Provinces of Northeast China.

References. Chen, 1982, 1999 (Noctuidae); Sugi, 1982 (Noctuidae, Chloephorinae); Sviridov, 2003 (Noctuidae: Chloephorinae); Fibiger, Hacker, 2005 (Nolidae: Eariadinae); Fibiger, Lafontaine, 2005 (Nolidae: Eariadinae); Kononenko, 2005 (Nolidae: Eariadinae), 2010 (Noctuidae: Nolinae, Sarrothripini, Eariadina); Lafontaine, Fibiger, 2006 (Noctuidae: Nolinae, Eariadini); Kononenko, Han, 2007 (Noctuidae: Nolinae, Sarrothripini, Eariadina); Matov *et al.*, 2008 (Noctuidae: Nolinae, Eariadinae); Fibiger *et al.*, 2009 (Noctuidae: Nolinae, Sarrothripini, Eariadina); Holloway, 2011 (Nolidae: Eariadinae); Zahiri *et al.*, 2013 (Nolidae: Eariadinae).

注：首次给出中文族名。

Genus ***Earias*** Hübner, [1825] 1816 钻夜蛾属

Earias Hübner, [1825] 1816, *Verzeichniss bekannter Schmettlinge* [sic!]: 395. Type species: *Phalaena clorana* [sic!] Linnaeus, 1761 [Sweden].

Synonymy: *Aphusia* Walker, [1858] 1857; *Digba* Walker, 1862; *Earis*: Stephens, 1834, missp.; *Aphusa* auct., missp..

The genus includes 53 species, distributed exclusively in the Old World, mainly in the tropical and subtropical areas; few species known from temperate zone; 3 species are in the Three Provinces of Northeast China.

Earias pudicana **Staudinger, 1887 粉缘钻夜蛾**

Earias pudicana Staudinger, 1887, *in* N.M. Romanoff (ed.), *Mémoires sur les Lépidoptères*, 3: 174, pl. 8: 10. Type locality: Russia, Vladivostok.

Synonymy: *Earias pudicana pupillana* Staudinger, 1887 (var.); *Earias paginalis* Bryk, 1948.

Distribution. China (Heilongjiang, Jilin, Liaoning, Heibei, Beijing, Shanxi, Ningxia, Henan, Shandong, Jiangsu, Zhejiang, Hubei, Hunan, Jiangxi); Russia (Siberia), Korea, Japan.

寄主：毛白杨、柳（陈一心，1999）；杞柳（*Salix purpurea*）、榔榆（*Ulmus parvifolia*）、胡枝子（*Lespedeza bicolor*）（Matov & Kononenko, 2012）。

Earias roseifera **Butler, 1881 玫缘钻夜蛾**

Earias roseifera Butler, 1881, *Transactions of the Entomological Society of London*, 1881: 18. Type locality: Japan, Tokyo.

Synonymy: *Earias erubescens* Staudinger, 1887; *Earias jezoensis* Sugi 1982.

Distribution. China (Heilongjiang, Jilin, Liaoning, Hebei, Zhejiang, Hubei, Hunan, Sichuan, Taiwan); Russia (Siberia), Korea, Japan, Thailand.

寄主：杜鹃花（陈一心，1999）；杜鹃（*Rhododendron simsii*）（Matov & Kononenko, 2012）。

Earias cupreoviridis **(Walker, 1862) 鼎点钻夜蛾**

Xanthoptera cupreoviridis Walker, 1862, *Transactions of the Entomological Society of London*, (3)1: 92. Type locality: West Africa.

Synonymy: *Earias chromataria* Walker, 1863; *Earias fulvidana* Wallengren, 1863; *Earias fervida* Walker, 1866; *Earias limbana* Snellen, 1879.

Distribution. China (Jilin, Hubei, Hunan, Zhejiang, Yunnan, Xizang, Taiwan); Korea, Japan, Sri Lanka, India, Nepal, Pakistan, Thailand, Andamanes, Indonesia, Philippines, Saudi Arabia; Central and South Africa.

寄主：棉、蜀葵、黄花稔、野黄麻（陈一心，1999）。

Tribe **Gelastocerini** Zahiri & Holloway, 2013 霜夜蛾族

Gelastocerini Zahiri & Holloway, 2013, *Cladistics* 1 (2012): 19. Type genus: *Gelastocera* Butler, 1877.

Recently described tribe, separated on the base of molecular study and morphological analysis (Zahiri & Holloway, 2013). Earlier it was referred by Holloway (2011) as "*Gelastocera* generic group". It includes 2 genera, *Gelastocera* and *Ptyonota* listed by Holloway (2011) and we place the genus *Macrochthonia* here by its morphological affinity with *Gelastocera*. Most species are distributed in the Oriental region and Chinese subregion of the Palaearctic. 2 genera and 3 species are in the Three Provinces of Northeast China.

References. Chen, 1982, 1999 (Noctuidae); Sugi, 1982 (Noctuidae, Chloephorinae); Sviridov, 2003 (Noctuidae: Chloephorinae); Matov *et al.*, 2008 (Noctuidae: Nolinae, Chloephorinae); Kononenko, Pinratana, 2013 (Nolidae: Gelastocerini); Zahiri *et al.*, 2013 (Nolidae: Gelastocerini).

注：首次给出中文族名。

Genus *Gelastocera* Butler, 1877 霜夜蛾属

Gelastocera Butler, 1877, *Annals and Magazine of Natural History*, (4)20: 476. Type species: *Gelastocera exusta* Butler, 1877 [Japan].

The genus includes 11 species, mainly distributed in the Oriental and Chinese subregion of the Palaearctic regions; 2 species are in the Three Provinces of Northeast China.

Gelastocera ochroleucana Staudinger, 1887 赭霜夜蛾

Gelastocera ochroleucana Staudinger, 1887, *in* N.M. Romanoff (ed.), *Mémoires sur les Lépidoptères*, 3: 176, pl. 8: 11. Type locality: Russia, Raddefka [Khabarovsk terr., Judish autonomy, Radde]; "Suifun" [Primorye terr., Razdolnaya river].

Distribution. China (Heilongjiang, Jilin); Russia (Siberia), Korea.

Gelastocera exusta Butler, 1877 霜夜蛾

Gelastocera exusta Butler, 1877, *Annals and Magazine of Natural History*, (4)20: 476. Type locality: Japan, Hakodate.

Synonymy: *Gelastocera ochroleucana designata* Bryk, 1948 (f.).

Distribution. China (Heilongjiang, Jilin, Liaoning, Hubei, Hunan, Hainan, Sichuan, Xizang, Taiwan); Russia (Siberia), Korea, Japan.

寄主：日本桤木（*Alnus firma*）、千金榆（*Carpinus cordata*）、胡桃楸（*Juglans mandshurica*）、椴树属（*Tilia*）、榆属（*Ulmus*）、紫藤属（*Wisteria*）植物（Matov & Kononenko, 2012）。

Genus *Macrochthonia* Butler, 1881 土夜蛾属

Macrochthonia Butler, 1881, *Transactions of the Entomological Society of London*, 1881: 599. Type species: *Macrochthonia fervens* Butler, 1881 [Japan].

Synonymy: *Eccopteroma* Staudinger, 1892.

Monotypic Sino-Pacific genus; 1 species is in the Three Provinces of Northeast China.

Macrochthonia fervens Butler, 1881 土夜蛾

Macrochthonia fervens Butler, 1881 *Transactions of the Entomological Society of London*, 1881: 599. Type locality: Japan, Tokyo.

Synonymy: *Calymnia falcata* Graeser, 1889 [1890]; *Calymnia pyrausta* Graeser, 1889 [1890].

Distribution. China (Heilongjiang, Jilin, Jiangsu, Zhejiang, Hubei, Jiangxi, Taiwan); Russia (Siberia), Korea, Japan.

寄主：黑榆（*Ulmus davidiana*）、榔榆（*Ulmus parvifolia*）、榉树（*Zelkova serrata*）（Matov & Kononenko, 2012）。

Subfamily **Eligminae** Mell, 1943 旋夜蛾亚科

Eligmini Mell, 1943, *Zoologische Jahrbucher. Abteilung fur Systematik, Oekologie, und Geographie der Tiere*, 76: 192, Type genus: *Eligma* Hübner, [1819] (Nolidae).

Speidel *et al.* (1996) and Kitching & Rawlins (1999) placed *Eligma* within the Nolidae. Holloway (1998, 2003) stated that the position of *Eligma* in Noctuidae is uncertain, later he (Holloway, 2011) and Zahiri *et al.* (2013b) recognized the position of Eligminae in the family Nolidae of the superfamily Noctuoidea. The subfamily comprises several East Asian genera: *Eligma* Hübner, [1819], *Triorbis* Hampson, 1894, *Gadirtha* Walker, [1858], *Plotheia* Walker, [1858], *Lamprothripa* Hampson, 1912, *Ptisciana* Walker, 1865, *Phaeothripa* Hampson, 1912, *Clettharina* Hampson, 1894 and *Negritothripa* Inoue, 1970. The genus *Iscadia* Walker, 1857 is distributed also in the Old World; some genera are known from the Ethiopean and Australian regions. 3 genera and 3 species are represented in the Three Provinces of Northeast China.

References. Chen, 1982, 1999 (Noctuidae); Sugi 1982 (Noctuidae, Chloephorinae); Sviridov, 2003 (Noctuidae: Chloephorinae); Fibiger, Lafontaine 2005 (Nolidae: Eligminae); Kononenko, 2005 (Nolidae: Eligminae), 2010 (Noctuidae: Nolinae, Eligmini); Lafontaine, Fibiger 2006 (Noctuidae: Nolinae, Eligmini); Matov *et al.*, 2008 (Noctuidae: Nolinae, Eligmini); Holloway, 2011 (Nolidae: Eligminae); Zahiri *et al.*, 2013a (Nolidae: Eligminae).

注：首次给出中文亚科名。

Genus *Eligma* Hübner, [1819] 旋夜蛾属

Eligma Hübner, [1819], *Verzeichniss bekannter Schmettlinge*, 165. Type species: *Phalaena narcissus* Cramer, 1775 [China].

Synonymy: *Heligma* Agassiz, 1846, emend.; *Panligma* Moore, 1858; *Surina* Walker, 1869.

***Eligma narcissus* (Cramer, [1775]) 旋夜蛾**

Phalaena narcissus Cramer, 1775, Uitlandsche Kapellen, 1: 116, pl. 73: E, F. Type locality: China.

Synonymy: *Eligma narcissus indica* Rothschild, 1896; *Eligma narcissus javanica* Rothschild, 1896; *Eligma narcissus philippinensis* Rothschild, 1896; *Eligma narcissus celebensis* Tams, 1935; *Eligma narcissus*: Poole, 1989, Holloway, 1998.

Distribution. China (Heilongjiang, Jilin, Liaoning, Hebei, Shanxi, Hubei, Hunan, Zhejiang, Fujian, Sichuan, Yunnan, Taiwan);Russia (Siberia), Korea, Japan, India, Philippines, Indonesia.

Genus *Gadirtha* Walker, [1858] 1857 癞皮夜蛾属

Gadirtha Walker, [1858] 1857, *List of the Specimens of Lepidopterous Insects in the Collection of the British Museum*, vol. 13: 1102. Type species: *Gadirtha impingens* Walker, [1858] 1857 [India; China, Hong Kong].

Synonymy: *Iscadia* Walker, 1857; *Scolopocneme* Felder & Felder, 1862; *Sebagena* Walker, 1865; *Libunca* Walker, 1869; *Pucialia* Walker, 1869; *Algonia* Möschler, 1886; *Encalypta* Möschler, 1890; *Lophosema* Schaus, 1910; *Lophosema* Hampson, nec Schaus, 1910; *Libunea*: Hampson, 1912, missp.; *Eucalypta*: Hampson, 1912, missp..

注：《中国动物志》（夜蛾科）中引用其同物异名“*Iscadia* Walker, 1857”，在此予以变更；当前本属亚科隶属不明确，本文中暂归入旋夜蛾亚科。

***Gadirtha impingens* Walker, [1858] 缘斑癞皮夜蛾**

Gadirtha impingens Walker, [1858], *List of the Specimens of Lepidopterous Insects in the Collection of the British Museum*, 13: 1103. Type Locality: South China, Hong Kong.

Synonymy: *Scolopocneme bufonia* C. & R. Felder, 1862; *Checupa tinctoides* Snellen, 1877; *Gadirtha inexacta papuensis* Warren, 1913 (ab.); *Gadirtha inexacta uniformis* Warren, 1913 (ab.); *Gadirtha guineana* Swinhoe, 1918; *Gadirtha inexacta buruensis* Prout, 1926; *Gadirtha papuensis* Gaede, 1937; *Gadirtha impingens uniformis* (Sugi, 1982); *Gadirtha uniformis* Sugi, 1982; *Iscadia inexacta*, misidentified/misapplied.

Distribution. China (Liaoning, Guangdong, Hainan, Taiwan, Hongkong); South Korea, Japan, India, Vietnam, Myanmar, Indonesia.

注：《中国动物志》（夜蛾科）中引用其同物异名“*Iscadia uniformis* (Sugi, 1982)”，在此予以变更。

Genus *Negritothripa* Inoue, 1970 曲缘皮夜蛾属

Negritothripa Inoue, 1970, *Bulletin of the Japan Entomological Academy*, 5: 37. Type species: *Asinduma hampsoni* Wileman, 1911 [Japan].

Synonymy: *Lamprothripa* auct..

The genus includes 5 species, distributed in the Chinese subregion of the the Palaearctic and Oriental regions; 1 species is in the Three Provinces of Northeast China.

注：当前本属亚科隶属不明确，本文中暂归入旋夜蛾亚科。

Negritothripa hampsoni **(Wileman, 1911) 曲缘皮夜蛾**

Asinduma hampsoni Wileman, 1911. *Transactions of the Entomological Society of London*, 1911: 233. Type locality: Japan.

Distribution. China (Jilin, Liaoning, Zhejiang, Taiwan); Russia (Siberia), Korea, Japan.

寄主：栎（陈一心，1999）；麻栎（*Quercus acutissima*）、枹栎（*Quercus serrata*）（Matov & Kononenko, 2012）。

Family **Nolidae**, *incertae sedis* 瘤蛾科内隶属未明确

Holloway (2011) placed the genus *Kerala* and some other to the group of "miscelaneous genera" of Nolidae. The systematic position of the genus requires clarification.

Genus *Kerala* Moore, 1881 蜡丽夜蛾属

Kerala Moore, 1881, *Proceedings of the Zoological Society of London*, 1881: 329. Type species: *Kerala punctilineata* Moore, 1881 [India].

Synonymy: *Cerala* Hampson, 1912, emend..

The genus includes 6 species, distributed from the Chinese subregion of the Palaearctic to Oriental regions, where it is most diverse; 1 species is in the Three Provinces of Northeast China.

Kerala decipiens **(Butler, 1878) 黑肾蜡丽夜蛾**

Cyana decipiens Butler, 1878, *Annals and Magazine of Natural History*, (4)5: 352. Type locality: Japan.

Synonymy: *Leptina macroptera* Oberthür, 1880; *Cymatophora fujisana* Matsumura, 1909.

Distribution. China (Heilongjiang, Jilin, Liaoning, Nei Mongol, Beijing, Henan, Hunan, Sichuan); Russia (Siberia), Korea, Japan, India(North).

寄主：辽东桤木（*Alnus hirsuta*）、桦木属（*Betula*）（Matov & Kononenko, 2012）。

Family **Noctuidae** Latreille, 1809 夜蛾科

Noctuidae Latreille, 1809, *Genera Crustaceorum Insectorum*, 4: 191, 224 (as Noctuaelites), incorrect original spelling which had to be emended to Noctuidae; official list of family-group names, opinion 450. Type genus: *Noctua* Linnaeus, 1758.

Synonymy: Noctuida Leach, [1815] 1830; Noctuides Leach, [1815] 1830; Noctuelidi Boisduval, [1828] 1829; Noctuelidae Duponchel, [1845] 1844; Noctuelites Guenée, 1852; Noctues Swinhoe, 1890; Noctuo-Bombycites Latreille, 1809.

Noctuidae is the second largest family in Lepidoptera, including 1,089 genera, 11,772 described species over the world (Nieukerken *et al.*, 2011). The traditional concept of the Noctuidae have been changed dramatically by moving most of "quadrifid Noctuidae subfamilies" to Erebidae, Nolidae and Euteliidae and segregation some groups into subfamilies and tribes (Fibiger & Lafontaine, 2005). In a restricted definition it was suggested as monophyletic only recently by Fibiger & Lafontaine (2005), Zahiri *et al.* (2011, 2013a).

The family Noctuidae numbers 20 subfamilies. Among them, 231 genera and 552 species are in the Three Provinces of Northeast China.

References. Staudinger & Rebel, 1901 (Noctuidae, Trifidae, part); Hampson 1903-1913 (Noctuidae, part); Seitz, [1907] –1914, 1938 (Noctuidae, part); Spuler 1908 (Noctuidae, part); Nye 1975 (Noctuidae, part); Chen, 1982 (Noctuidae), 1999 (Noctuidae, part); Sugi 1982 (Noctuidae, part); Sugi (ed.), 1987 (Noctuidae, part); Holloway, 1988, 1989, 1998, 2005, 2008, 2009 (Noctuidae, part); Poole, 1989 (Noctuidae, part); Hacker, 1990 (Noctuidae, part); Fibiger, 1990, 1993, 1997 (Noctuidae, part); Ronkay & Ronkay, 1994 (Noctuidae, part), 1995 (Noctuidae, part); Speidel *et al.*, 1996 (Noctuidae, part); Kononenko *et al.*, 1998 (Noctuidae, part); Kitching & Rawlins, 1999 [1998] (Noctuidae, part); Goater *et al.*, 2003 (Noctuidae, part); Sviridov, 2003 (Noctuidae, part); Kononenko, 2003, 2005, 2010 (Noctuidae, part); Sugi & Jinbo 2004 (Noctuidae, part); Fibiger & Hacker *et al.,* 2005 (Noctuidae, part); Fibiger & Hacker, 2005 (Noctuidae, part); Fibiger & Lafontaine, 2005 (Noctuidae, part); Speidel & Naumann, 2005 (Noctuidae, part); Lafontaine & Fibiger, 2006 (Noctuidae); Kononenko & Han, 2007 (Noctuidae, part); Fibiger & Hacker, 2007 (Noctuidae, part); Fibiger *et al.,* 2009 (Noctuidae, part); Ronkay *et al.*, 2011 (Noctuidae, part); Fibiger *et al.*, 2011 (Noctuidae); Varga *et al.* 2015 (Noctuinae).

Subfamily **Plusiinae** Boisduval, [1828] 1829 金翅夜蛾亚科

Plusiinae Boisduval, [1828] 1829, *Europaeorum Lepidopterorum Index Methodicus*: 91 (as Plusidi). Type genus: *Plusia* Ochsenheimer, 1816.

Synonymy: Plusiidae Herrich-Schäffer, [1851] 1845; Phytometrinae Hampson, 1913.

Plusiinae are a well defined, compact subfamily of worldwide distribution with about 500 species are inhabiting all nature zones. The subfamily is subdivided into 4 tribes: Abrostolini, Argyrogrammatini, Plusiini and

Omorphinini. The Nearctic fauna of the subfamily has been extensively revised recently (Eichlin & Cunnigham, 1978; Lafontaine & Poole, 1991) and the Palaearctic fauna by Ronkay *et al.*, 2008. For Plusiinae tribal division and sequence of genera we follow to the last revision of Eurasian and North African Plusiinae by Ronkay *et al.* (2008) and Behounek *et al.* (2010). 3 tribes, 17 genera and 48 species are known from the Three Provinces of Northeast China.

References. Hampson, 1913 (Phytometrinae); McDunnough, 1944 (Noctuidae, Plusiinae); Dufay, 1960, 1968 (Noctuidae, Plusiinae); Kostrowicki, 1961 (Noctuidae, Plusiinae); Ichinose, 1962, 1973 (Noctuidae, Plusiinae); Eichlin, Cunnigham, 1978 (Noctuidae, Plusiinae); Chen, 1982, 1999 (Noctuidae); Ronkay, 1982, 1986, 1997 (Noctuidae, Plusiinae); Sugi, 1982 (Noctuidae, Plusiinae); Kljutschko, Kononenko, 1986 (Noctuidae, Plusiinae); Kitching, 1987 (Noctuidae, Plusiinae); Lafontaine, Poole, 1991 (Noctuidae, Plusiinae); Ronkay *in* Goater *et al.*, 2003 (Noctuidae, Plusiinae); Fibiger, Hacker, 2005 (Noctuidae, Plusiinae); Fibiger, Lafontaine, 2005 (Noctuidae, Plusiinae); Kononenko, 2005, 2010 (Noctuidae, Plusiinae); Lafontaine, Fibiger, 2006 (Noctuidae, Plusiinae); Matov *et al.*, 2008 (Noctuidae, Plusiinae); Ronkay *et al.*, 2008 (Noctuidae, Plusiinae); Behounek *et al.*, 2010 (Noctuidae, Plusiinae).

Tribe **Abrostolini** Eichlin & Cunningham, 1978 **隐金翅夜蛾族**

Abrostolini Eichlin & Cunningham, 1978, *United States Department of Agriculture Technical Bulletin*, 1567: 9. Type genus: *Abrostola* Ochsenheimer, 1816.

The tribe consists of 2 genera and about 40 species over the World, most species are distributed in south of the temperate zone and subtropical area. The known larvae feed on herbaceous plants. 1 genus and 18 species occur in the Palaearctic.

注：首次给出中文族名。

Genus ***Abrostola*** Ochsenheimer, 1816 **隐金翅夜蛾属**

Abrostola Ochsenheimer, 1816, *Die Schmetterlinge von Europa*, 4: 88. Type species: *Phalaena triplasia* Linnaeus, 1758 [Europe].

Synonymy: *Unca* Oken, 1815, suppr.; *Habrostola* Sodoffsky, 1837, emend.; *Inguridia* Butler, 1879; *Unca* Lhomme [1929]; *Trigeminostola* Beck, 1991; *Asclepistola* Beck, 1991.

The genus includes 46 species distributed in the Holarctic, Oriental, Ethiopian and Neotropical regions; 14 species are known from the Palaearctic region; 4 species are in the Three Provinces of Northeast China.

Abrostola triplasia **(Linnaeus, 1758) 隐金翅夜蛾**

Phalaena triplasia Linnaeus, 1758, *Systema Naturae* (Edn. 10) 1: 517. Type locality: Europe.

Synonymy: *Phalaena tripartita* Hufnagel, 1766; *Noctua trigemina* Werneburg, 1864, unnecessary repl. name.

Distribution. China (Heilongjiang, Jilin, Hebei, Shandong, Sichuan; East China); Russia (European part, Ural, Siberia), Korea, Japan; Central Asia, Near East, Europe, North Africa.

寄主：荨麻属、葎草属、野芝麻属植物（陈一心，1982）；葎草（*Humulus japonicus*）、啤酒花（*Humulus lupulus*）、花点草（*Nanocnide japonica*）、异株荨麻（*Urtica dioica*）、野芝麻属（*Lamium*）、苎麻属（*Boehmeria*）、墙草属（*Parietaria*）植物（Matov & Kononenko, 2012）。

Abrostola tripartita (Hufnagel, 1766) 拟隐金翅夜蛾

Phalaena tripartita Hufnagel, 1766; *Berlinisches Magazin*, 3: 414. Type locality: Germany, Berlin.

Synonymy: *Noctua urticae* Hübner, 1816; *triplasia* auct..

Distribution. China (Jilin, Liaoning); Russia (European part, Ural, Siberia), Korea; Central Asia, Near East, North Africa.

寄主：葎草（*Humulus japonicus*）、异株荨麻（*Urtica dioica*）、覆盆子（*Rubus idaeus*）、白前（*Vincetoxicum officinale*）、野芝麻属（*Lamium*）植物（Matov & Kononenko, 2012）。

注：首次给出中文种名。

Abrostola ussuriensis Dufay, 1958 乌苏里隐金翅夜蛾

Abrostola ussuriensis Dufay, 1958, *Bulletin Mensuel de la Société Linneenne de Lyon*, 27: 157, pl. 2: 5, 6, 11. Type locality: Russia, "Ussuri", Nikolsk [Primorye terr., Ussuriisk].

Distribution. China (Jilin); Russia (Siberia), Korea, Japan.

Abrostola korbi Dufay, 1958 赭基隐金翅夜蛾

Abrostola korbi Dufay, 1958, *Opuscula Zoologica*, 13:1:, 1. Type locality: Russia, "Ussuri, Kazakewitsch" [vicinity of Khabarovsk].

Synonymy: *Abrostola pacifica* Dufay, 1960.

Distribution. China (Heilongjiang, Jilin); Russia (Siberia), Korea, Japan.

注：首次给出中文种名。

Tribe **Argyrogrammatini** Eichlin & Cunningham, 1978 银金翅夜蛾族

Argyrogrammatini Eichlin & Cunningham, 1978, *United States Department of Agriculture Technical Bulletin*, 1567: 12 (as Argyrogrammini). Type genus: *Argyrogramma* Hübner, 1823.

The tribe comprises a number of species are in 14 genera, distributed mainly in the tropical area over the World. Known larvae feed on herbaceous plants, mostly polyphagous. Some species are able to damage the agriculture. 9 genera and 32 species are known from the Palaearctic region, 4 genera and 6 species are in the Three Provinces of Northeast China.

注：首次给出中文族名。

Genus *Trichoplusia* McDunnough, 1944 斑金翅夜蛾属

Trichoplusia McDunnough, 1944, *Memoirs of the Southern California Academy of Sciences*, 1: 204. Type species: *Plusia brassicae* Riley, 1870 [USA] (=*Noctua ni* Hübner [1803]).

Monotypic genus in Eurasia.

注：首次给出中文属名。

Trichoplusia ni (Hübner, [1803]) 粉斑金翅夜蛾

Noctua ni Hübner [1803], *Sammlung Europaischer Schmetterlinge. Eulen*, 3: pl. 58: 284. Type locality: Europe.

Synonymy: *Plusia humilis* Walker, [1858]; *Plusia extrahens* Walker, [1858]; *Plusia significans* Walker, [1858]; *Plusia innata* Herrich-Schäffer, 1868; *Plusia brassicae Riley*, 1870; *Plusia echinocystidis* Behr, 1874; *Plusia comma* Schultz, 1907; *Plusia corama* Schultz, 1907; *Plusia deserticola* Rothschild, 1913; *Plusia florida* Dannehl, 1929; *Plusia ni florida* Dannehl, 1929 (f.).

Distribution. China (Jilin, Shaanxi, Fujian, Hubei, Guangdong); Russia (South European part, South Ural, Altai, Siberia); Europe.

寄主：荨麻、棉、茄属、莴苣、芥蓝（陈一心，1982）；圆叶椒草（*Peperomia obtusifolia*）、阔果蓟罂粟（*Argemone platyceras*）、花菱草（*Eschscholzia californica*）、罂粟（*Papaver somniferum*）、甜菜（*Beta vulgaris*）、藜（*Chenopodium album*）、虎杖（*Reynoutria japonica*）、皱叶酸模（*Rumex crispus*）、康乃馨（*Dianthus caryophyllus*）、四倍体西瓜（*Citrullus vulgaris*）、黄瓜（*Cucumis sativus*）、甜瓜（*Melo sativa*）、褐芥末（*Brassica hirta*）、大叶芥菜（*Brassica juncea*）、黑芥（*Brassica nigra*）、（*Brassica oleracea*）、菜薹（*Brassica parachinensis*）、洋基芥（*Brassica tournefortii*）、海滨芥（*Cakile maritima*）、（*Napus rapa*）、豆瓣菜（*Nasturtium officinale*）、萝卜（*Raphanus sativus*）、菠菜（*Spinacia oleracea*）、木樨草（*Reseda odorata*）、蜀葵（*Althaea rosea*）、草棉（*Gossypium herbaceum*）、陆地棉（*Gossypium hirsutum*）、秋葵（*Hibiscus esculentus*）、锦葵属（*Malva*）、异株荨麻（*Urtica dioica*）、一品红（*Euphorbia pulcherrima*）、日本木瓜（*Chaenomeles japonica*）、日本木瓜（*Chaenomeles japonica*）、猫爪相思树（*Acacia greggii*）、花生（*Arachis hypogaea*）、黄芪（*Astragalus crotalariae*）、（*Crotalaria rotundifolia*）、大豆（*Glycine max*）、香豌豆（*Lathyrus odoratus*）、羽扇豆（*Lupinus arizonicus*）、（*Lupinus micranthus* =*Lupinus hirsutus*）、紫苜蓿（*Medicago sativa*）、印度草木樨（*Melilotus indica*）、菜豆（*Phaseolus vulgaris*）、豌豆（*Pisum sativum*）、豇豆（*Vigna sinensis*）、甜橙（*Citrus sinensis*）、（*Fagonia arabica*）、骆驼蹄草（*Zygophyllum oxianum*）、芹叶龙牛儿苗（*Erodium cicutarium*）、天竺葵（*Pelargonium hortorum*）、旱金莲（*Tropaeolum majus*）、澳洲鸭脚木（*Brassaia actinophylla*）、通草（*Tetrapanax papyriferus*）、旱芹（*Apium graveolens*）、野胡萝卜（*Daucus carota*）、欧防风（*Pastinaca sativa*）、番茄（*Lycopersicon esculentum*）、光烟草（*Nicotiana glauca*）、烟草（*Nicotiana tabacum*）、沙漠烟草（*Nicotiana trigonophylla*）、碧冬茄（*Petunia hybrida*）、茄（*Solanum melongena*）、龙葵（*Solanum nigrum*）、马铃

薯（*Solanum tuberosum*）、甘薯（*Ipomoea batatas*）、圆叶牵牛（*Ipomoea purpurea*）、（*Coleus japonica*）、圆叶薄荷（*Mentha suaveolens*）、罗勒（*Ocimum basilicum*）、圣罗勒（*Ocimum sanctum*）、藿香蓟（*Ageratum conyzoides*）、金盏花（*Calendula officinalis*）、红花（*Carthamus tinctorius*）、（*Chaenactis stevioides*）、（*Geraea canescens*）、非洲菊（*Gerbera jamesonii*）、加拿大莴苣（*Lactuca canadensis*）、山莴苣（*Lactuca indica*）、莴苣（*Lactuca sativa*）、（*Pectis papposa*）、瓜叶菊（*Senecio cruentus*）、（*Senecio mikaniodide*）、药用蒲公英（*Taraxacum officinale*）、石刁柏（*Asparagus officinalis*）、千年木（*Dracaena marginata*）、圆叶菝葜（*Smilax rotundifolia*）、玉米（*Zea mays*）、心型喜林芋（*Philodendron cordatum*）、小龟蔓绿绒（*Philodendron pertusum*）、菠菜属（*Spinacia*）、石头花属（*Gypsophila*）、杜鹃花属（*Rhododendron*）、南瓜属（*Cucurbita*）、独行菜属（*Lepidium*）、岩白菜属（*Bergenia*）、草莓属（*Fragaria*）、（*Faba*）、车轴草属（*Trifolium*）、蒺藜属（*Tribulus*）、老鹳草属（*Geranium*）、峨参属（*Anthriscus*）、马利筋属（*Asclepias*）、天芥菜属（*Heliotropium*）、鼠尾草属（*Salvia*）、紫菀属（*Aster*）、菊属（*Chrysanthemum*）、瓜叶菊属（*Cineraria*）、金鸡菊属（*Coreopsis*）、还阳参属（*Crepis*）、向日葵属（*Helianthus*）、苦苣菜属（*Sonchus*）、唐菖蒲属（*Gladiolus*）、苞叶芋属（*Spathiphyllum*）植物（Matov & Kononenko, 2012）。

注：国内多用"粉斑夜蛾"予以记录，根据属变动将其中文种名予以变更。

Genus ***Thysanoplusia*** Ichinose, 1973 中金弧夜蛾属

Thysanoplusia Ichinose, 1973, *Kontyû*, 41: 137. Type species: *Phytometra intermixta* Warren, 1913.

The genus includes 19 species, mainly distributed in the Ethiopian and Oriental regions, in the Palaearctic with 6 species; 1 species is in the Three Provinces of Northeast China.

注：首次给出中文属名。

***Thysanoplusia intermixta* (Warren, 1913) 中金弧夜蛾**

Phytometra intermixta Warren, 1913, *in* A.Seitz (ed.), *Die Gross-Schmetterlinge der Erde*, 3: 357, pl. 64: g. Type locality: China, Sichuan [Setchuan].

Synonymy: *Phytometra brachycalcea* Hampson, 1913.

Distribution. China (Liaoning, Hebei, Shaanxi, Fujian, Sichuan, Taiwan); Russia (Siberia), Korea, Japan, Australia; Southeast Asia.

寄主：胡萝卜、菊、牛蒡、蓟（陈一心，1982）；野胡萝卜（*Daucus carota*）、亚麻（*Linum usitatissimum*）、台湾蒲公英（*Taraxacum platycarpum*）、草莓属（*Fragaria*）、毛蔓豆属（*Calopogonium*）、薄荷属（*Mentha*）、猪屎豆属（*Crotalaria*）、菊属（*Chrysanthemum*）植物（Matov & Kononenko, 2012）。

注：国内多将本种误归入"*Diachrysia* Hübner, [1821] 1816 紫金翅夜蛾属"，在此予以变更。

Genus *Ctenoplusia* Dufay, 1970 银纹夜蛾属

Ctenoplusia Dufay, 1970, *Faune Madagascar*, 31: 91. Type species: *Plusia limbirena* Guenée, 1852 [Madagascar].

Synonymy: *Acanthoplusia* Dufay, 1970 (subgen.).

The genus subdivided into 2 subgenera, *Ctenoplusia* and *Acanthoplusia.* It includes 72 species, mainly distributed in the Ethiopian, Oriental and Australian regions, 6 species are in Palaearctic region; 2 species of 2 subgenera in the Three Provinces of Northeast China.

Subgenus *Ctenoplusia* Dufay, 1970 银纹夜蛾亚属

***Ctenoplusia* (*Ctenoplusia*) *albostriata* (Bremer & Grey, 1853) 白条夜蛾**

Plusia albostriata Bremer & Grey, 1853, *in* Motschulsky, *Études d'Entomologie*, 1: 65. Type locality: North China.

Synonymy: *Plusia subchalybaea* Walker, 1865; *Plusia oxygramma* Hampson, 1894; *Plusia transfixa* auct., nec Walker, 1865, nec Geyer, 1832.

Distribution. China (Heilongjiang, Jilin, Liaoning, Beijing, Hebei, Shaanxi, Hubei, Guangdong, Taiwan); Russia (Siberia), Korea, Japan, Australia, New Zealand; Southeast Asia.

寄主：菊科植物（陈一心，1982）；火炭母（*Polygonum chinense*）、大花六道木（*Abelia grandiflora*）、俄罗斯西门肺草（*Symphytum uplandicum*）、翠菊（*Callistephus chinensis*）、宽叶鱼眼草（*Dichrocephala latifolia*）、一年蓬（*Erigeron annuus*）、加拿大一枝黄花（*Solidago canadensis*）、打碗花属（*Calystegia*）、紫菀属（*Aster*）、金盏花属（*Calendula*）、翠菊属（*Callistephus*）、菊属（*Chrysanthemum*）、白酒草属（*Conyza*）、大丽花属（*Dahlia*）、地胆草属（*Elephantopus*）植物（Matov & Kononenko, 2012）。

注：国内多将本种误归入"*Argyrogramma* Hübner, 1823 金翅夜蛾属"，在此予以变更。

Subgenus *Acanthoplusia* Dufay, 1970 棘金翅夜蛾亚属

Acanthoplusia Dufay, 1970, *Bulletin Mensuel de la Société Linneenne de Lyon*, 39: 104. Type species: *Phytometra tarassota* Hampson, 1913 [India].

注：首次给出中文亚属名。

***Ctenoplusia* (*Acanthoplusia*) *agnata* (Staudinger, 1892) 银纹夜蛾**

Plusia agnata Staudinger, 1892, *in* N.M. Romanoff (ed.), *Mémoires sur les Lépidoptères*, 6: 547. Type locality: China, Shanghai.

Synonymy: *Plusia chalcytes* Hampson, 1913, nec Esper, 1787; *Phytometra agnata sokutsuna* Strand, 1920 (var.).

Distribution. China(all over the country); Russia (Siberia), Korea, Japan, India, Nepal; Southeast Asia.

寄主：大豆、十字花科蔬菜（陈一心，1982）；苔（*Brassica campestris*）、萝卜（*Raphanus sativus*）、大豆（*Glycine max*）、野胡萝卜（*Daucus carota*）、野薄荷（*Mentha arvensis*）、甘薯（*Ipomoea batatas*）、牛蒡（*Arctium lappa*）、土木香（*Inula helenium*）、棉属（*Gossypium*）植物（Matov & Kononenko, 2012）。

Genus ***Anadevidia*** Kostrowicki, 1961 葫芦夜蛾属

Anadevidia Kostrowicki, 1961, *Acta Zoologica Cracoviensia*, 6 384. Type species: *Noctua peponis* Fabncius, 1775 [India].

Synonymy: *Podioplusia* Ichinose, 1962.

The genus comprises 2 species, distributed in the Oriental and the Sino-Pacific subregion of the Palaearctic regions; 2 species are in the Three Provinces of Northeast China.

Anadevidia peponis **(Fabricius, 1775) 葫芦夜蛾**

Noctua peponis Fabricius, 1775, *Entomologiae Systematica emendata et Aucta. Secundum Classes, Ordines, Genera, Species Adjectis Synonimis, Locis, Observationibus, Descriptionibus*: 606. Type locality: "East Indies".

Synonymy: *Plusia agramma* Guenée, 1852, *Plusia inchoata* Walker, 1865; *Plusia fumifera* Graeser, 1889 [1890].

Distribution. China (Northeast, Northwest, North, East, South parts); Russia (Siberia), Korea, Japan; Southeast Asia, Oceania.

寄主：葫芦科植物（陈一心，1982）；甜瓜（*Cucumis melo*）、瓠果（*Cucumis sativus*）、南瓜（*Cucurbita moschata*）、葫芦（*Lagenaria sicararia=L. leucantha*）、丝瓜（*Lufa cylindrica*）、苦瓜（*Momordica charantia*）、蛇瓜（*Trichosanthes anguina*）、瓜叶栝楼（*Trichosanthes cucumerina*）、喜马拉雅瓜蒌（*Trichosanthes himalensis*）、山靛（*Mercurialis leiocarpa*）、醉鱼草（*Buddleja lindleyana*）、香蜂花（*Melissa officinalis*）、泡桐属（*Paulownia*）植物（Matov & Kononenko, 2012）。

Anadevidia hebetata **(Butler, 1889) 长纹葫芦夜蛾**

Plusia hebetata Butler, 1889, *Illustrations of Typical Specimens of Lepidoptera Heterocera in the Collection of the British Museum*, 7: 71, pl. 131: 1. Type locality: India.

Distribution. China (Jilin, Liaoning, Taiwan); Russia (Siberia), Korea, Japan, India; Southeast Asia.

注：国内多用"瓜夜蛾"予以记录，中文种名根据属变动在此予以变更。

Tribe **Plusiini** Boisduval, [1828] 1829 金翅夜蛾族

Plusiinae Boisduval, [1828] 1829, *Europaeorum Lepidopterorum Index Methodicus*: 91 (as Plusidi). Type genus: *Plusia* Ochsenheimer, 1816.

Synonymy: Plusiidae Herrich-Schäffer, [1851] 1845; Phytometrinae Hampson, 1913; Diachrisiina Beck, 1996; Plusidiina Beck, 1996.

The tribe has predominantly Holarctic distribution, it is most diverse in southern parts of the temperate zone, mountains and in the subtropical area. Known larvae feed on herbaceous plants, exept in the genus *Syngrapha*. Some species are able to mass outbreaks and cause damage of agriculture. It includes 24 genera worldwide, 16 genera and over 160 species are in the Palaearctic region; 12 genera and 38 species are in the Three Provinces of Northeast China.

Subtribe **Autoplusiina** Kitching, 1987 丫纹夜蛾亚族

Autoplusiina Kitching, 1987, *Bulletin of the British Museum* (*Natural History*) *Entomology*, 54: 196. Type genus: *Autoplusia* McDunnough, 1944.

Genus ***Erythroplusia*** Ichinose, 1962 珠纹夜蛾属

Erythroplusia Ichinose, 1962, *Kontyû*, 30: 249. Type species: *Plusia rutilifrons* Walker, 1858 [China].

Synonymy: *Perloplusia* Chou & Lu, 1978; *Peroplusia* Poole, 1989, missp..

The genus includes 2 species distributed in the Oriental and Sino-Pacific subregion of the Palaearctic regions; 2 species are in the Three Provinces of Northeast China.

***Erythroplusia rutilifrons* (Walker, 1858)** 珠纹夜蛾

Plusia rutilifrons Walker, 1858, *List of the Specimens of Lepidopterous Insects in the Collection of the British Museum*, 15: 1785. Type locality: North China.

Synonymy: *Plusia adscripta* Staudinger, 1888; *Perloplusia neorutilifrons* Chou & Lu, 1978.

Distribution. China (Jilin, Shandong, Taiwan); Russia (Siberia), Korea, Japan, India(North), Nepal.

寄主：金莲花（*Trollius chinensis*）、车前（*Plantago asiatica*）、水芹属（*Oenanthe*）、牛蒡属（*Arctium*）植物（Matov & Kononenko, 2012）。

***Erythroplusia pyropia* (Butler, 1879)** 滴纹夜蛾

Plusia pyropia Butler, 1879, *Annals and Magazine of Natural History*, (5)4: 267. Type locality: Japan, Yokohama.

Synonymy: *Peroplusia pseudopyropia* Chou & Lu, 1978.

Distribution. China (Jilin, Liaoning, Guangdong, Guangxi, Xizang, Taiwan); Russia (Siberia), Korea, Japan, India(North), Nepal, Pakistan.

寄主：水芹（*Oenanthe javanica*）、牛蒡属（*Arctium*）、蒿属（*Artemisia*）植物（Matov & Kononenko, 2012）。

Genus ***Macdunnoughia*** Kostrowicki, 1961 银锭夜蛾属

Macdunnoughia Kostrowicki, 1961, *Acta Zoologica Cracoviensia*, 6: 402. Type species: *Plusia confusa* Stephens, 1850 [Europe].

Synonymy: *Scleroplusia* Ichinose, 1962; *Puriplusia* Chou & Lu, 1974 (subgen.).

The genus subdivided into 2 subgenera, *Macdunnoughia* (s. str.) and *Puriplusia*. It comprises 7 species, distributed in the Oriental and the Chinese subregion of the Palaearctic regions; 4 species are in the Three Provinces of Northeast China.

Subgenus ***Macdunnoughia*** Kostrowicky, 1961 银锭夜蛾亚属

Macdunnoughia* (*Macdunnoughia*) *confusa **(Stephens, 1850) 瘦银锭夜蛾**

Plusia confusa Stephens, 1850, *List of the Specimens of British Animals in the Collection of the British Museum,* 5: 291. Type locality: France.

Synonymy: *Noctua circumflexa* Esper, 1788; *Plusia gutta* Guenée, 1852; *Plusia gutta bigutta* Staudinger, 1892 (ab.); *Plusia gutta aestiva* Krulikovsky, 1908 (var.); *Phytometra confusa deangulata* Strand, 1917 (ab.); *Plusia gutta*[?] *grisea* Dannehl, 1933 (ab.); *Plusia confusa reducta* Kobes, 1959 (f.); *Macdunnoughia confusa brunnescens* Lempke, 1966 (f.); *Macdunnoughia monosigna* Chou & Lu, 1979.

Distribution. China (Heilongjiang, Jilin, Liaoning, Xinjiang, Shaanxi); Russia (European part, Ural, Siberia), Mongolia, Korea, Japan; Central Asia, Near East, Europe, North Africa.

寄主：欧耆、母菊（陈一心，1982）；葡萄叶铁线莲（*Clematis vitalba*）、甜菜（*Beta vulgaris*）、卷茎蓼（*Polygonum convolvulus*）、直穗酸模（*Rumex longifolius*）、（*Silene inflata*）、狗筋麦瓶草（*Silene vulgaris*）、（*Brassica oleracea*）、大麻（*Cannabis sativa*）、异株荨麻（*Urtica dioica*）、蓖麻（*Ricinus communis*）、大豆（*Glycine max*）、紫苜蓿（*Medicago sativa*）、豌豆（*Pisum sativum*）、东北土当归（*Aralia continentalis*）、日本川芎（*Cnidium officinale*）、野胡萝卜（*Daucus carota*）、欧洲前胡（*Peucedanum officinale*）、番茄（*Lycopersicon esculentum*）、烟草（*Nicotiana tabacum*）、大车前（*Plantago major*）、短柄野芝麻（*Lamium album*）、紫花野芝麻（*Lamium purpureum*）、红辣椒千叶蓍（*Achillea millefolium*）、（*Achillea nobilis*）、田春黄菊（*Anthemis arvensis*）、春黄菊（*Anthemis tinctoria*）、牛蒡（*Arctium lappa*）、毛头牛蒡（*Arctium tomentosum*）、中亚苦蒿（*Artemisia absinthium*）、荒野蒿（*Artemisia campestris*）、野艾蒿（*Artemisia vulgaris*）、金盏花（*Calendula officinalis*）、（*Carduus tenuiflorus*）、（*Carduus tenuiflorus*）、野菊（*Chrysanthemum indicum*）、杭菊

（*Chrysanthemum morifolium*）、栽培菊苣（*Cichorium endivia*）、刺儿菜（*Cirsium arvense*）、向日葵（*Helianthus annuus*）、西洋甘菊（*Matricaria chamomilla*）、母菊（*Matricaria recutita*）、大翅蓟（*Onopordum acanthium*）、蜂斗菜（*Petasites japonicus*）、吊钟千里光（*Senecio fuchsii*）、台湾蒲公英（*Taraxacum platycarpum*）、薄荷属（*Mentha*）、水苏属（*Stachys*）、雏菊属（*Bellis*）、还阳参属（*Crepis*）、莴苣属（*Lactuca*）、大戟属（*Euphorbia*）、菜豆属（*Phaseolus*）、车轴草属（*Trifolium*）、老鹳草属（*Geranium*）植物（Matov & Kononenko, 2012）。

***Macdunnoughia* (*Macdunnoughia*) *hybrida* Ronkay, 1986 拟瘦银锭夜蛾**

Macdunnoughia hybrida Ronkay, 1986, *Annales historico-naturalis Musei nationalis hungarici*, 78: 205, pl.: 1, 2. Type locality: North Korea, North Pyongan.

Distribution. China (Jilin); Russia (Siberia), Korea.

***Macdunnoughia* (*Macdunnoughia*) *crassisigna* (Warren, 1913) 银锭夜蛾**

Phytometra crassisigna Warren, 1913, *in* A.Seitz (ed.), *Die Gross-Schmetterlinge der Erde*, 3: 352, pl. 65: h. Type locality: Korea; Japan; India.

Synonymy: *Phytometra rhopalosema* Hampson, 1913; *Phytometra gutta*: Hampson, 1913, part, nec Guenée, 1852; *Macdunnoughia xizangensis* Chou et Lu, 1982.

Distribution. China (Jilin, Hebei, Shaanxi, Jiangxi, Taiwan); Russia (Siberia), Korea, Japan, Nepal, India(North).

寄主：菊、牛蒡、胡萝卜（陈一心，1982）；野胡萝卜（*Daucus carota*）、牛蒡（*Arctium lappa*）（Matov & Kononenko, 2012）。

Subgenus *Puriplusia* Chou & Lu, 1974 淡银锭夜蛾亚属

Puriplusia Chou & Lu, 1974, *Acta Entomologica Sinica*, 17: 71. Type species: *Plusia purissima* Butler, 1878 [Japan, Yokohama].

注：首次给出中文亚属名。

***Macdunnoughia* (*Puriplusia*) *purissima* (Butler, 1878) 淡银锭夜蛾**

Plusia purissima Butler, 1878, *Annals and Magazine of Natural History*, (5)1: 202. Type locality: Japan, Yokohama.

Synonymy: *Plusia tetragona*: Leech, 1900, nec Walker, 1857.

Distribution. China (Jilin, Liaoning, Hebei, Shaanxi, Hubei, Sichuan, Guizhou); Russia (Transbaikalia, Siberia), Korea, Japan.

寄主：魁蒿（*Artemisia princeps*）、牛蒡属（*Arctium*）植物（Matov & Kononenko, 2012）。

Genus *Antoculeora* Ichinose, 1973 铗纹夜蛾属

Antoculeora Ichinose, 1973, *Kontyû*, 41: 136. Type species: *Plusia ornatissima* Walker, 1858 [North India], misidentification of *Plusia locuples* Oberthür, 1881.

Synonymy: *Cerviplusia* Chou & Lu, 1978.

The genus includes 3 species, distributed in the Oriental and the Chinese subregion of the Palaeasrctic regions; 1 species is in the Three Provinces of Northeast China.

Antoculeora locuples (Oberthür, 1881) 庐山鹿铗纹夜蛾

Plusia locuples Oberthür, 1881, *Études d'Entomologie*, 5: 85, pl. 9: 3. Type locality: Russia, Primorye terr., Askold Is..

Synonymy: *Cerviplusia lushanensis* Chou et Lu, 1978; *Plusia ornatissima* auct., nec Walker, 1858.

Distribution. China (Jilin, Jiangxi); Russia (Transbaikalia, Siberia), Korea, Japan, Nepal.

寄主：莴苣属（*Lactuca*）、蜂斗菜属（*Petasites*）、苦苣菜属（*Sonchus*）植物（Matov & Kononenko, 2012）。

Genus *Diachrysia* Hübner, [1821] 1816 紫金翅夜蛾属

Diachrysia Hübner, [1821] 1816, *Verzeichniss bekannter Schmettlinge* [sic!]: 252. Type species: *Diachrysia orichalcea* Fabricius sensu Hübner, [1821] 1816 [Italy] (=*Phalaena* (*Noctua*) *chryson* Esper, 1789).

Synonymy: *Chrychrysia* Beck, 1996; *Zosichrysia* Beck, 1996.

Holarctic temperate genus with 13 species; 8 species are in the Three Provinces of Northeast China.

注：首次给出中文属名。

Diachrysia chryson (Esper, 1789) 紫金翅夜蛾

Phalaena Noctua chryson Esper, 1789, *Die Schmetterlinge in Abbildungen nach der Natur mit Beschreibungen*, 4 (2): 446, pl. 141, noct. 62: 1. Type locality: Central Italy.

Synonymy: *Noctua aerifera* Sowerby, 1805.

Distribution. China (Heilongjiang, Jilin, Zhejiang); Russia (European part, Siberia), Korea, Japan, Kazakhstan; Near East, Europe.

寄主：无花果、泽兰属植物（陈一心，1982）；柄野芝麻（*Lamium album*）、水薄荷（*Mentha aquatica*）、野薄荷（*Mentha arvensis*）、留兰香（*Mentha spicata*）、紫苏（*Perilla frutescens*）、胶质鼠尾草（*Salvia glutinosa*）、牛蒡（*Arctium lappa*）、野艾蒿（*Artemisia vulgaris*）、卷心菜蓟（*Cirsium oleraceum*）、大麻叶泽兰（*Eupatorium cannabinum*）、泽兰（*Eupatorium japonicum*）、林泽兰（*Eupatorium lindleyanum*）、佩兰（*Eupatorium stoechadosmum*）、荨麻属（*Urtica*）、凤仙花属（*Impatiens*）、蜜蜂花属（*Melissa*）植物（Matov & Kononenko, 2012）。

Diachrysia pales (Mell, 1939) 天山紫金翅夜蛾

Phytometra chryson var. *pales* Mell, 1939, *Deutsche Entomologische Zeitschrift Iris*, 52: 138. Type locality: China, Tien Shan, lectotype designated by Ronkay & Ronkay, 2008.

Synonymy: *Phytometra leonina coreae* Strand, 1916 (ab.) (infrasubspecific name); *Phytometra chryson coreae* Bryk, 1948 (ssp.); *Diachrysia coreae* Inoue & Sugi, 1958.

Distribution. China (Jilin, Xinjiang); Russia (Siberia), Korea, Japan.

Diachrysia leonina (Oberthür, 1884) 八纹紫金翅夜蛾

Plusia leonina Oberthür, 1884, *Études d'Entomologie*, 10: 26, pl. 3: 11. Type locality: Russia, "Sidemi" [Primorye terr., Bezverkhovo].

Synonymy: *Plusia humeralis* Butler, 1886.

Distribution. China (Heilongjiang, Jilin, Shaanxi, Xizang); Russia (Siberia), Korea, Japan.

寄主：荨麻属（*Urtica*）、野芝麻属（*Lamium*）、薄荷属（*Mentha*）、紫菀属（*Aster*）、蓟属（*Cirsium*）植物（Matov & Kononenko, 2012）。

注：国内曾用中文种名"八纹夜蛾"，根据属名在此变更为"八纹紫金翅夜蛾"。

Diachrysia witti L. Ronkay, G. Ronkay & Behounek, 2008 比八纹紫金翅夜蛾

Diachrysia witti L. Ronkay, G. Ronkay & Behounek, 2008. *The Witt catalogue. A taxonomic atlas of the Eurasian and North African Noctuoidea*, Vol. 1: 32. Type locality: Russia, Primorye terr., Zanadvorovka vill.

Synonymy: *Diachrysia bieti* auct., nec Oberthür, 1884.

Distribution. China (Jilin, Beijing, Shaanxi); Russia (Siberia), Korea, Japan.

注：本种在国内多用其同物异名"*Diachrysia bieti* Oberthür, 1884"，在此予以变更；并曾用中文种名"比八纹夜蛾"，根据属名在此变更为"比八纹紫金翅夜蛾"。

Diachrysia chrysitis (Linnaeus, 1758) 黄紫金翅夜蛾

Phalaena Noctua chrysitis Linnaeus, 1758, *Systema Naturae* (Edn. 10) 1: 513. Type locality: Europe.

Synonymy: *Phytometra plesdidior* Ferdinandez, 1929.

Distribution. China (Heilongjiang, Jilin, Shaanxi); Russia (from Ural to Siberia), Mongolia, Korea, Japan; Central Asia, Near East, Europe.

寄主：藜（*Chenopodium album*）、繁缕（*Stellaria media*）、（*Betula pubescens*）、黑果越橘（*Vaccinium myrtillus*）、异株荨麻（*Urtica dioica*）、旋果蚊子草（*Filipendula ulmaria*）、覆盆子（*Rubus idaeus*）、金雀儿（*Cytisus scoparius*）、白车轴草（*Trifolium repens*）、广布野豌豆（*Vicia cracca*）、峨参（*Anthriscus silvestris*）、圆盾状忍冬（*Lonicera periclymenum*）、红花琉璃草（*Cynoglossum officinale*）、蓝蓟（*Echium vulgare*）、聚合草（*Symphytum officinale*）、大车前

（*Plantago major*）、黑夏至草（*Ballota nigra*）、鼬瓣花根（*Galeopsis tetrahit*）、柄野芝麻（*Lamium album*）、紫花野芝麻（*Lamium maculatum*）、欧地笋（*Lycopus europaeus*）、欧夏至草（*Marrubium vulgare*）、留兰香（*Mentha spicata*）、牛至（*Origanum vulgare*）、林地水苏（*Stachys silvatica*）、林石蚕（*Teucrium scorodonia*）、红辣椒千叶蓍（*Achillea millefolium*）、小牛蒡（*Arctium minus*）、狼杷草（*Bidens tripartita*）、（*Carduus lanceolatus*）、刺儿菜（*Cirsium arvense*）、翼蓟（*Cirsium vulgare*）、大麻叶泽兰（*Eupatorium cannabinum*）、（*Hypochoeris maculata*）、吊钟千里光（*Senecio fuchsii*）、加拿大一枝黄花（*Solidago canadensis*）、虉草（*Phalaris arundinacea*）、玻璃苣属（*Borago*）、勿忘草属（*Myosotis*）、毛蕊花属（*Verbascum*）、鼠尾草属（*Salvia*）、金盏花属（*Calendula*）、风毛菊属（*Leontodon*）、蒲公英属（*Taraxacum*）、蜀葵属（*Althaea*）、羽衣草属（*Alchemilla*）、草莓属（*Fragaria*）、牛舌草属（*Anchusa*）植物（Matov & Kononenko, 2012）。

注：国内多将本种误归入“*Plusia* Ochsenheimer, 1816 金翅夜蛾属”中，给出中文种名“金翅夜蛾”，中文种名根据属变动在此予以变更。

Diachrysia stenochrysis (Warren, 1913) 青紫金翅夜蛾

Phytometra stenochrysis Warren, 1913, *in* A.Seitz (ed.), *Die Gross-Schmetterlinge der Erde*, 3: 348, pl. 64f. Type locality: Japan.

Synonymy: *Plusia juncta* Tutt, 1892; *Phytometra multauri* Bryk, 1942; *Plusia tutti* Kostrowicki, 1961; *Diachrysia chrysitis* auct., nec Linnaeus, 1758.

Distribution. China (Jilin); Russia (Ural, Siberia), Mongolia, Korea, Japan; Central and South Europe.

寄主：异株荨麻（*Urtica dioica*）、大车前（*Plantago major*）、灰白益母草（*Leonurus glaucescens*）、水苏属（*Stachys*）、蒿属（*Artemisia*）、蒲公英属（*Taraxacum*）植物（Matov & Kononenko, 2012）。

注：国内曾用中文种名“青金翅夜蛾”予以记录，中文种名根据属变动在此予以变更。

Diachrysia nadeja (Oberthür, 1880) 碧紫金翅夜蛾

Plusia nadeja Oberthür, 1880, *Études d'Entomologie*, 5: 84, pl. 3: 10. Type locality: Russia, Primorye terr., Askold Is..

Synonymy: *Plusia chrysitis*: Hampson, 1913, nec Linnaeus, 1758, part.

Distribution. China (Heilongjiang, Jilin, Liaoning, Henan, Shaanxi, Shanxi); Russia (Ural, Siberia), Mongolia, Korea, Japan; Central and South Europe.

寄主：大虎杖（*Reynoutria sachalinensis*）、异株荨麻（*Urtica dioica*）、长叶车前（*Plantago lanceolata*）、香薷（*Elsholtzia ciliata*）、沙滩黄芩（*Scutellaria strigillosa*）、滨藜属（*Atriplex*）、大豆属（*Glycine*）、车轴草属（*Trifolium*）、拉拉藤属（*Galium*）、野芝麻属（*Lamium*）、蒲公英属（*Taraxacum*）植物（Matov & Kononenko, 2012）。

注：国内曾用中文种名“碧金翅夜蛾”予以记录，根据属名变动在此予以变更。

Diachrysia zosimi **(Hübner, [1822]) 绿紫金翅夜蛾**

Noctua zosimi Hübner, [1822], *Sammlung Europaischer Schmetterlinge. Eulen*, 4: pl. 142: 651. Type locality: Europe.

Distribution. China (Heilongjiang, Jilin); Russia (Ural, Siberia), Mongolia, Korea, Japan; Central and South Europe.

寄主：梅花草（*Parnassia palustris*）、山荆子（*Malus baccata*）、小地榆（*Sanguisorba minor*）、地榆（*Sanguisorba officinali*）、乳香草（*Origanum vulgare*）、红辣椒千叶蓍（*Achillea millefolium*）、荨麻属（*Urtica*）、蔷薇属（*Rosa*）植物（Matov & Kononenko, 2012）。

注：国内曾用中文种名“绿金翅夜蛾”予以记录，根据属名其中文种名在此予以变更。

Subtribe **Euchalciina** Chou and Lu, 1979 异纹夜蛾亚族

Euchalciini Chou & Lu, 1979, *Entomotaxonomia*, 1: 76, 77. Type genus: *Euchalcia* Hübner, [1821].

Synonymy: Polychrysiina Kljutshko, 1985; Panchrysiina Beck, 1996.

注：首次给出中文亚族名。

Genus ***Euchalcia*** Hübner, [1821] 1816 异纹夜蛾属

Euchalcia Hübner, [1821] 1816, *Verzeichniss bekannter Schmettlinge* [sic!]: 250. Type species: *Noctua illustrees* Fabricius, 1787 [Slovalia] (=*Noctua variabilis* Piller & Mitterpacher, 1783).

Synonymy: *Adeva* McDunnough, 1944; *Pseudeuchalcia* Ichinose, 1985 (subgen.); *Pareuchalcia* Beck, [1992] 1991 (subgen.).

The centres of diversity of the genus are the Mediterranean and Central Asian subregions; 2 species are in the Three Provinces of Northeast China.

Euchalcia sergia **(Oberthür, 1884) 北方异纹夜蛾**

Plusia sergia Oberthür, 1884, *Études d'Entomologie*, 10: 27, pl. 3: 9. Type locality: Russia, "Sidemi" [Primorye terr., Bezverkhovo].

Distribution. China (Jilin); Russia (Altai, Transbaikalia, Siberia), Korea, Japan.

注：首次给出中文种名。

Euchalcia modestoides **Poole, 1989 静异纹夜蛾**

Euchalcia modestoides Poole, 1989, *in* Heppner, J.B. (ed.), *Lepidopterorum Catalogus* (*New Series*) *Fascicle* 118. Noctuidae 1: 396, an repl. name for. Type locality: Germany, Augsburg.

Synonymy: *Phalaena* (*Noctua*) *modesta* Hübner, 1786, nom. preocc.; *Euchalcia cuprea* Esper, 1787, part.

Distribution. China (Heilongjiang); Russia (European part, Ural, Siberia), Mongolia, Korea, Japan, Kazakhstan; West Asia, Europe.

寄主：疗肺草（陈一心，1982）；红花琉璃草（*Cynoglossum officinale*）、（*Pulmonaria azurea*）、花叶肺草（*Pulmonaria mollis*）、腺毛肺草（*Pulmonaria mollissima*）、（*Pulmonaria obscura*）、疗肺草（*Pulmonaria officinalis*）、（*Symphytum nodosum*）、聚合草（*Symphytum officinale*）、块茎聚合草（*Symphytum tuberosum*）、（*Pterotheca*）、疆南星属（*Arum*）植物（Matov & Kononenko, 2012）。

注：国内曾用中文种名"静纹夜蛾"，且使用其同物异名"*Euchalcia modesta* (Hübner, 1786)"，在此予以变更。

Genus *Polychrysia* Hübner, [1821] 1816 印铜夜蛾属

Polychrysia Hübner, [1821] 1816, *Verzeichniss bekannter Schmettlinge* [sic!]: 251. Type species: *Noctua moneta* Fabricius, 1787 [Austria].

Synonymy: *Polychrisia* Bethune – Baker, 1906, missp..

Holarctic temperate genus comprising 6 species; 4 species are in the Three Provinces of Northeast China.

Polychrysia esmeralda (Oberthür, 1880) 印铜夜蛾

Plusia esmeralda Oberthür, 1880, *Études d'Entomologie*, 5: 85. Type locality: Russia, Primorye terr., Askold Is..

Synonymy: *Deva trabea* Smith, 1895; *Polychrysia esmeralda marusiki* L. Ronkay, G. Ronkay, Behounek & Mikkola, 2008 (ssp.); *Chrysoptera moneta*: Hampson, 1913.

Distribution. China (Heilongjiang, Jilin, Nei Mongol, Hebei); Russia (Siberia), Mongolia, Korea, Kazakhstan, USA (Alaska), Canada.

寄主：乌头属、翠雀属、金莲花属植物（陈一心，1982）；弯枝乌头（*Aconitum arcuatum*）、紫花高乌头（*Aconitum excelsum*）、（*Thalictrum chinensis*）、翠雀属（*Delphinium*）、金莲花属（*Trollius*）、肺草属（*Pulmonaria*）、向日葵属（*Helianthus*）植物（Matov & Kononenko, 2012）。

注：国内多用本种的同物异名"*Polychrysia moneta* (Fabricius, 1787)"予以记录，在此予以变更。

Polychrysia aurata (Staudinger, 1888) 重印铜夜蛾

Plusia aurata Staudinger, 1888, *Entomologische Zeitung. Entomologischen Vereine zu Stettin*, 49: 260. Type locality: Russia, "Ussuri" [Primorye terr.].

Distribution. China (Jilin); Russia (Altai, Siberia), Mongolia.

Polychrysia splendida (Butler, 1878) 暗印铜夜蛾

Deva splendida Butler, 1878, *Annals and Magazine of Natural History*, (5)1: 203. Type locality: Japan, Hakodate.

Synonymy: *Plusia intractata* Staudinger, 1888; *Phytometra sachalinensis* Matsumura, 1925; *Polychrisia intracta*: Poole, 1989, missp..

Distribution. China (Jilin); Russia (South Altai, Siberia), Mongolia, Korea, Japan.

寄主：高帽乌头（*Lycoctotum longecassidatum*）（Park *et al.*, 2006）；弯枝乌头（*Aconitum arcuatum*）（Matov & Kononenko, 2012）。

***Polychrysia sica* (Graeser, 1890) 断线印铜夜蛾**

Plusia sica Graeser, 1890, *Berliner entomologische Zeitschrift*, 25: 77. Type locality: Russia, Amur reg., [Radde] "Raddefka".

Distribution. China (Heilongjiang); Russia (Siberia).

注：首次给出中文种名。

Genus ***Panchrysia*** Hübner, [1821] 1816 **银钩夜蛾属**

Panchrysia Hübner, [1821] 1816, *Verzeichniss bekannter Schmettlinge* [sic!]: 252. Type species: *Noctua aurea* Hübner, [1803] [Europe] (=*Bombyx deaurata* Esper, 1787).

The genus is subdivided into 2 subgenera: *Panchrysia* (s.str.) with 6 species and the monotypic *Hexaureia*; 2 species are in the Three Provinces of Northeast China.

Subgenus ***Panchrysia*** Hübner, [1821] 1816 **银钩夜蛾亚属**

***Panchrysia* (*Panchrysia*) *ornata* (Bremer, 1864) 艳银钩夜蛾**

Plusia ornata Bremer, 1864, *Memoires de l'Académie Imperiale des Sciences de St.-Petersbourg* VII Sèrie, 8 (1): 103; pl. 8: 15. Type locality: Russia, Transbaikalia, Kiachta.

Synonymy: *Phytometra ornata contacta* W. Kozhantschikov, 1923 (var.).

Distribution. China (Jilin, Shanxi); Russia (South Ural, Siberia), Mongolia, Korea; Central Asia.

Subgenus ***Hexaureia*** Beck, 1991 **黄裳银钩夜蛾亚属**

Hexaureia Beck, 1991, *Atalanta, Würzburg*, 22 (2-4): 226. Type species: *Plusia dives* Eversmann, 1844.

Monotypic Palaearctic subgenus.

***Panchrysia* (*Hexaureia*) *dives* (Eversmann, 1844) 黄裳银钩夜蛾**

Plusia dives Eversmann, 1844, *Bulletin de la Société Impériale des Naturalistes de Moscou*, 17: 596. Type locality: Russia, Irkutsk.

Distribution. China (Heilongjiang, Jilin, Xizang); Russia (Ural, Siberia), Mongolia, Kazakhstan.

Genus *Lamprotes* Reichenbach, Leiptzig, 1817 边金翅夜蛾属

Lamprotes Reichenbach, Leiptzig, 1817, *Jenaische Allg. Lit*; *Ztg*, l: 287. Type species: *Phalaena c-aureum* Knoch, 1781.

Synonymy: *Chrysoptera* Berthold, 1827; *Cubena* Walker, 1856.

The genus includes 2 species, distributed in the temperate zone of the Palaearctic region; 1 species is in the Three Provinces of Northeast China.

***Lamprotes mikadina* (Butler, 1878) 镶边金翅夜蛾**

Plusia mikadina Butler, 1878, *Annals and Magazine of Natural History*, (5)1: 202. Type locality: Japan, Yokohama, Hakodate.

Synonymy: *Chrysoptera micadina* Hampson, 1913, emend..

Distribution. China (Jilin, Liaoning); Russia (Siberia), Korea, Japan.

寄主：耧斗菜属（*Aquilegia*）、唐松草属（*Thalictrum*）（Matov & Kononenko, 2012）。

Genus *Plusidia* Butler, 1879 闪金夜蛾属

Plusidia Butler, 1879, *Illustrations of Typical Specimens of Lepidoptera Heterocera in the Collection of the British Museum*, 3: 27. Type species: *Plusidia abrostoloides* Butler, 1879 [Japan] (=*Noctua cheiranthi* Tauscher; 1809).

The genus is monotypic, distributed in the temperate zone of the Palaearctic region.

***Plusidia cheiranthi* (Tauscher, 1809) 闪金夜蛾**

Noctua cheiranthi Tauscher, 1809, *Mémoires de la Société Impériale des Naturalistes de Moscou*, 2: 322, pl. 20: 6. Type locality: Russia.

Synonymy: *Plusia eugenia* Eversmann, 1841; *Plusidia abrostoloides* Butler, 1879; *Plusidia separanda* Warren, 1913; *Plusidia murensis* Warnecke, 1918.

Distribution. China (Jilin); Russia (Ural, Siberia), Mongolia, Korea, Japan, Kazakhstan; Near East, Europe.

寄主：西伯利亚耧斗菜（*Aquilegia sibirica*）、欧洲耧斗菜（*Aquilegia vulgaris*）、桂竹香（*Cheiranthus cheiri*）、黄唐松草（*Thalictrum flavum*）、箭头唐松草（*Thalictrum simplex*）、乌头属（*Aconitum*）植物（Matov & Kononenko, 2012）。

Subtribe **Plusiina** Boisduval, 1829 金翅夜蛾亚族

Plusiinae Boisduval, 1829, *Europaeorum Lepidopterorum Index Methodicus*: 91 (as Plusidi). Type genus: *Plusia* Ochsenheimer, 1816.

Synonymy: Plusiidae Herrich-Schäffer, [1851] 1845; Phytometrinae Hampson, 1913.

In the Three Provinces of Northeast China, the subtribe contains 3 genera: *Autographa* (9 species), *Syngrapha* (2 species) and *Plusia* (2 species).

Genus ***Autographa*** Hübner, [1821] 1816 丫纹夜蛾属

Autographa Hübner, [1821] 1816, *Verzeichniss bekannter Schmettlinge* [sic!]: 251. Type species: *Phalaena gamma* Linnaeus, 1758 [Europe].

Holarctic genus with 44 described species. Most species are inhabit various biotopes mainly in the temperate zone; 9 species are in the Three Provinces of Northeast China.

Autographa gamma **(Linnaeus, 1758) 丫纹夜蛾**

Phalaena gamma Linnaeus, 1758, *Systema Naturae* (Edn. 10) 1: 513. Type locality: Europe.

Synonymy: *Noctua pulchrina* Haworth, 1809; *Plusia gamma gammina* Staudinger, 1901 (var.); *Plusia gamma alepica* Nitsche, 1911 (ab.); *Autographa messmeri* Schadewald, 1992; *Autographa voelkeri* Schadewald, 1992.

Distribution. China (Heilongjiang, Jilin, Xinjiang, Ningxia, Gansu); Russia (European part, Ural, Siberia), Mongolia, Korea, Japan, India(North); Central Asia, Near East, Europe, North Africa, North America.

寄主：荚果蕨（*Matteuccia struthiopteris*）、（*Athyrium distentifolium*）、北欧红松（*Pinus silvestris*）、葡萄叶铁线莲（*Clematis vitalba*）、匍枝毛茛（*Rannculus repens*）、（*Beta alba*）、甜菜（*Beta vulgaris*）、（*Suaeda vera*）、酸模叶蓼（*Polygonum lapathifolium*）、小酸模（*Rumex acetosella*）、长叶酸模（*Rumex longifolius*）、垂枝桦（*Betula pendula*）、胡榛（*Corylus avellana*）、（*Rhododendron ponticum*）、黑果越橘（*Vaccinium myrtillus*）、球尾花（*Lysimachia thyrsiflora*）、毛黄连花（*Lysimachia vulgaris*）、白柳（*Salix alba*）、黄花柳（*Salix caprea*）、黄瓜（*Cucumis sativus*）、西葫芦（*Cucurbita pepo*）、大叶芥菜（*Brassica juncea*）、（*Barbarea oleracea*）、欧洲油菜（*Brassica napus*）、芜青（*Brassica rapa*）、野萝卜（*Raphanus raphanistrum*）、萝卜（*Raphanus sativus*）、野芥（*Sinapis arvensis*）、菠菜（*Spinacia oleracea*）、药蜀葵（*Althaea officinalis*）、欧锦葵（*Malva sylvestris*）、大麻（*Cannabis sativa*）、啤酒花（*Humulus lupulus*）、异株荨麻（*Urtica dioica*）、中亚羽裂叶荠（*Mercurialis annua*）、蓖麻（*Ricinus communis*）、紫叶景天（*Sedum telephium*）、黑茶藨子（*Ribes nigrum*）、红茶藨子（*Ribes rubrum*）、旋果蚊子草（*Filipendula ulmaria*）、麝香草莓（*Fragaria moschata*）、野草莓（*Fragaria vesca*）、绿色草莓（*Fragaria viridis*）、（Malus domestica）、黑刺李（*Prunus spinosa*）、覆盆子（*Rubus idaeus*）、花生（*Arachis hypogaea*）、鹰嘴豆（*Cicer arietinum*）、（*Faba bona*）、大豆（*Glycine max*）、牧地山黧豆（*Lathyrus pratensis*）、紫苜蓿（*Medicago sativa*）、白花草木樨（*Melilotus albus*）、黄香草木樨（*Melilotus officinalis*）、豌豆（*Pisum sativum*）、埃及车轴草（*Trifolium alexandrinum*）、杂种车轴草（*Trifolium hybridum*）、红车轴草（*Trifolium pratense*）、白车轴草（*Trifolium repens*）、广布野豌豆（*Vicia cracca*）、苕菜（*Vicia hirsuta*）、救荒野豌豆（*Vicia sativa*）、亚麻（*Linum usitatissimum*）、葡萄（*Vitis vinifera*）、洋常春藤（*Hedera helix*）、羊角芹（*Aegopodium podagraria*）、林当归（*Angelica silvestris*）、峨参（*Anthriscus silvestris*）、旱芹（*Apium graveolens*）、芫荽（*Coriandrum sativum*）、野胡萝卜（*Daucus carota*）、茴香（*Foeniculum officinale*）、欧防风（*Pastinaca sativa*）、欧芹（*Petroselinum crispum*）、（*Petroselinum sativum*）、茴芹（*Pimpinella anisum*）、西洋接骨木（*Sambucus nigra*）、颠茄草（*Atropa belladonna*）、辣椒（*Capsicum annuum*）、天仙子（*Hyoscyamus niger*）、番茄（*Lycopersicon esculentum*）、马铃薯（*Solanum tuberosum*）、（*Linaria alpina*）、欧洲柳穿鱼（*Linaria vulgaris*）、（*Odontites verna*）、长叶车前（*Plantago lanceolata*）、大车前（*Plantago major*）、（*Ballota undulata*）、小野芝麻（*Galeobdolon luteum*）、鼬瓣花（*Galeopsis ladanum*）、华丽鼬瓣花（*Galeopsis speciosa*）、黄鼬瓣花（*Galeopsis tetrahit*）、神香草（*Hyssopus officinalis*）、短柄野芝麻（*Lamium album*）、野芝麻（*Lamium purpureum*）、紫花猫薄荷（*Nepeta faassenii*）、（*Odontites verna*）、（*Phlomis brachyodon*）、鼠尾草（*Salvia officinalis*）、草地鼠尾草（*Salvia pratensis*）、林生水苏（*Stachys silvatica*）、（*Knautia dipsacifolia*）、（*Succisa pratensis*）、千叶蓍（*Achillea millefolium*）、珠蓍（*Achillea ptarmica*）、毛头牛蒡（*Arctium tomentosum*）、（*Artemisia campestre*）、白草蒿（*Artemisia herba-alba*）、北艾（*Artemisia vulgaris*）、金盏花（*Calendula officinalis*）、丝毛飞廉（*Carduus crispus*）、（*Carduus tenuiflorus*）、矢车菊（*Centaurea aspera*）、法国菊（*Chrysanthemum leucanthemum*）、（*Chrysanthemum vulgare*）、刺儿菜（*Cirsium arvense*）、（*Cirsium heterophyllum*）、蓟（*Cirsium palustre*）、翼蓟（*Cirsium vulgare*）、湿生鼠麴草（*Gnaphalium uliginosum*）、向日葵（*Helianthus annuus*）、山柳菊（*Hieracium umbellatum*）、莴苣（*Lactuca sativa*）、刺毛莴苣（*Lactuca virosa*）、秋蒲公英（*Leontodon autumnalis*）、洋甘菊（*Matricaria recutita*）、大翅蓟（*Onopordum acanthium*）、（*Senecio fuchsii*）、水飞蓟（*Silybum marianum*）、苣荬菜（*Sonchus arvensis*）、（*Tripleurospermum maritimum*）、（*Tussilago farfara*）、洋葱（*Allium cepa*）、葱（*Allium fistulosum*）、蒜（*Allium sativum*）、燕麦（*Avena sativa*）、无芒雀麦（*Bromus inermis*）、偃麦草（*Elytrigia repens*）、牛尾草（*Festuca elatior*）、羊茅（*Festuca ovina*）、大麦（*Hordeum vulgare*）、黍（*Panicum miliaceum*）、梯牧草（*Phleum pratense*）、小麦（*Triticum aestivum*）、玉米（*Zea mays*）、石竹属（*Dianthus*）、藜属（*Chenopodium*）、桤木属（*Alnus*）、秋海棠属（*Begonia*）、山芥属（*Barbarea*）、糖芥属（*Erysimum*）、椴树属（*Tilia*）、棉属（*Gossypium*）、羽衣草属（*Alchemilla*）、委陵菜属（*Potentilla*）、蔷薇属（*Rosa*）、桉属（*Eucalyptus*）、锦鸡儿属（*Caragana*）、羽扇豆属（*Lupinus*）、芒柄花属（*Ononis*）、菜豆属（*Phaseolus*）、拉拉藤属（*Galium*）、烟草属（*Nicotiana*）、碧冬茄属（*Petunia*）、玄参属（*Scrophularia*）、毛蕊花属（*Verbascum*）、马鞭草属（*Verbena*）、小冠薰属（*Basilicum*）、薄荷属（*Mentha*）、菊苣属（*Cichorium*）、菊属（*Dendranthema*）、滨菊属（*Leucanthemum*）、金光菊属（*Rudbeckia*）、疆菊属（*Tanacetum*）、蒲公英属（*Taraxacum*）、早熟禾属（*Poa*）植物（Matov & Kononenko, 2012）。

***Autographa mandarina* (Freyer, 1846) 满丫纹夜蛾**

Noctua mandarina Freyer, 1846, *Neuere Beiträge zur Schmetterlingskunde mit Abbildungen nach der Natur*, 5: 164, pl. 479: 4. Type locality: Russia, Ural.

Synonymy: *Plusia interscalaris* Herrich-Schäffer, [1850]; *Plusia interscalaris* Eversmann, 1857, preocc.; *Plusia typinota* Butler, 1878; *Plusia obscura* Oberthür, 1884; *Autographa sino-occidentalis* Chou et Lu, 1979; *Autographa lehri* Kljutschko, 1984.

Distribution. China (Heilongjiang, Jilin, Hebei); Russia (North of European part, Ural, Siberia, Kazakhstan), Mongolia, Korea, Japan; Europe.

寄主：胡萝卜（陈一心，1982）；异株荨麻（*Urtica dioica*）、（*Brassica oleracea*）、豌豆（*Pisum sativum*）、白车轴草（*Trifolium repens*）、羊角芹（*Aegopodium podagraria*）、峨参（*Anthriscus silvestris*）、野胡萝卜（*Daucus carota*）、大车前（*Plantago major*）、野芝麻（*Lamium purpureum*）、刺儿菜（*Cirsium arvense*）、翼蓟（*Cirsium vulgare*）、莴苣（*Lactuca sativa*）、亚麻属（*Linum*）、驴食草属（*Onobrychis*）、蒿属 (Artemisia)、苦苣菜属（*Sonchus*）、蒲公英属（*Taraxacum*）植物（Matov & Kononenko, 2012）。

Autographa buraetica **(Staudinger, 1892) 淡折线丫纹夜蛾**

Plusia pulchrina var. *buraetica* Staudinger, 1892, *Deutsche Entomologische Zeitschrift Iris*, 5: 370. Type locality: Russia, Siberia, Irkutsk.

Synonymy: *Plusia gammoides* Speyer, 1875; *Autographa ternei* Kljutschko, 1984; *Autographa pulchrina* auct., nec Haworth, 1809.

Distribution. China (Jilin); Russia (European part, Ural, Siberia), Korea, Japan, Northern Mongolia, Kazakhstan; Europe, North America.

寄主：白菜（*Brassica pekinensis*）、（*Brassica oleracea*）、大车前（*Plantago major*）、荨麻属（*Urtica*）、杜鹃花属（*Rhododendron*）植物（Matov & Kononenko, 2012）。

注：首次给出中文种名。

Autographa amurica **(Staudinger, 1892) 北方丫纹夜蛾**

Plusia pulchrina var. *amurica* Staudinger, 1892, *in* N.M. Romanoff (ed.), *Mémoires sur les Lépidoptères*, 6: 549. Type locality: Russia, “Suifun” [Primorye terr., Razdolnaya river].

Synonymy: *Plusia jota* Linnaeus, 1758; *Plusia pulchrina* Haworth, 1802.

Distribution. China (Jilin); Russia (Siberia), Korea, Japan.

Autographa v-minus **(Oberthür, 1884) V 丫纹夜蛾**

Plusia v-minus Oberthür, 1884, *Études d’Entomologie*, 10: 27. Type locality: Russia, “Sidemi” [Primorye terr., Bezverkhovo].

Distribution. China (Jilin); Russia (Siberia), Korea.

注：国内多用“V 纹金翅夜蛾”予以记录，根据属名变更为“V 丫纹夜蛾”。

Autographa urupina (Bryk, 1942) 离丫纹夜蛾

Phytometra urupina Bryk, 1942, *Deutsche Entomologische Zeitschrift Iris* 56: 57. Type locality: Russia, Kuriles, Urup I.

Distribution. China (Jilin); Russia (Siberia), Japan.

注：首次给出中文种名。

Autographa nigrisigna (Walker, 1858) 黑点丫纹夜蛾

Plusia nigrisigna Walker, 1858, *List of the Specimens of Lepidopterous Insects in the Collection of the British Museum*, 12: 928. Type locality: North India.

Distribution. China (Jilin, Hebei, Shaanxi, Sichuan, Xizang); Russia (Siberia), Korea, Japan, India(North), Nepal, Pakistan, Afghanistan.

寄主：豌豆、白菜、甘蓝、苜蓿（陈一心，1982）；甜菜（*Beta vulgaris*）、藜（*Chenopodium album*）、芸苔（*Brassica campestris*）、（*Brassica oleracea*）、芜青（*Brassica rapa*）、大麻（*Cannabis sativa*）、木豆（*Cajanus cajan*）、鹰嘴豆（*Cicer arietinum*）、紫苜蓿（*Medicago sativa*）、豌豆（*Pisum sativum*）、埃及车轴草（*Trifolium alexandrinum*）、黑吉豆（*Vigna mungo*）、豇豆（*Vigna unguiculata*）、烟草（*Nicotiana tabacum*）、马铃薯（*Solanum tuberosum*）、牛蒡（*Arctium lappa*）、莴苣（*Lactuca sativa*）、薄荷属（*Mentha*）植物（Matov & Kononenko, 2012）。

Autographa macrogamma (Eversmann, 1842) 大丫纹夜蛾

Plusia macrogamma Eversmann, 1842, *Bulletin de la Société Impériale des Naturalistes de Moscou*, 1842 (3): 554. Type locality: Russia, Ural.

Synonymy: *Plusia sevastina* Freyer, 1842; *Phytometra nigroviolacea* Rangnow, 1935.

Distribution. China (Heilongjiang); Russia (North of European part, Siberia), Mongolia(North), Kazakhstan; North Europe.

寄主：欧洲金莲花（*Trollius europaeus*）、笃斯越橘（*Vaccinium uliginosum*）、异株荨麻（*Urtica dioica*）、欧洲花楸（*Sorbus aucuparia*）、大车前（*Plantago major*）、鼬瓣根（*Galeopsis bifida*）、短柄野芝麻（*Lamium album*）、薄荷属（*Mentha*）、蓍属（*Achillea*）、牛蒡属（*Arctium*）、莴苣属（*Lactuca*）、蒲公英属（*Taraxacum*）植物（Matov & Kononenko, 2012）。

Autographa excelsa (Kretschmar, 1862) 袜丫纹夜蛾

Plusia excelsa Kretschmar, 1862, *Berliner entomologische Zeitschrift*, 6: 135, pl. 1: 5. Type locality: Russia.

Synonymy: *Plusia metabractea* Butler, 1881; *Phytometra parabractea* Hampson, 1913; *Autographa kostjuki* Klyuchko, 1984.

Distribution. China (Heilongjiang, Jilin, Sichuan); Russia (European part, Siberia), Korea, Japan; North and

Central Europe.

寄主：峨参（*Anthriscus silvestris*）、大车前（*Plantago major*）、短柄野芝麻（*Lamium album*）、野芝麻（*Lamium purpureum*）、（*Atriplex patulum*）、刺儿菜（*Cirsium arvense*）、卷心莱蓟（*Cirsium oleraceum*）、（*Tussilago farfara*）、金莲花属（*Trollius*）、荨麻属（*Urtica*）、蒿属（*Artemisia*）、蒲公英属（*Taraxacum*）植物（Matov & Kononenko, 2012）。

Genus *Syngrapha* Hübner [1821] 1816 美金翅夜蛾属

Syngrapha Hübner, [1821] 1816, *Verzeichniss bekannter Schmettlinge* [sic!]: 250. Type species: *Noctua devergens* Hübner, [1813].

Synonymy: *Caloplusia* Smith, 1884; *Palaeographa* Kljutshko, 1983; *Diasyngrapha* Beck, 1991; *Microsyngrapha* Beck, 1991; *Parsyngrapha* Beck, 1991; *Aingrapha* Beck, 1991.

Large Holarctic genus with 38 species, distributed mainly in the Nearctic region. Moths inhabit arctic, subarctic and boreal forest zones and high mountains in more southern areas of the temperate zone; 2 species are in the Three Provinces of Northeast China.

Syngrapha ain (Hochenwarth, 1785) 北方美金翅夜蛾

Noctua ain Hochenwarth, 1785, *Schriften der Berlinischen Gesellschaft Naturforschender Freunde*, 6: 337, pl. 7: 8. Type locality: Europe.

Synonymy: *Syngrapha ain persibirica* L. Ronkay, G. Ronkay, Behounek & Mikkola, 2008 (ssp.).

Distribution. China (Heilongjiang, Jilin, Liaoning); Russia (Northeast of European part, Siberia) , Mongolia, Korea, Japan, Kazakhstan; Near East, Central Europe.

寄主：西伯利亚冷杉（*Abies sibirica*）、日本落叶松（*Larix leptolepis*）、西伯利亚落叶松（*Larix sibirica*）、云杉属（*Picea*）植物（Matov & Kononenko, 2012）。

Syngrapha interrogationis (Linnaeus, 1758) 伏锌美金翅夜蛾

Phalaena (*Noctua*) *interrogationis* Linnaeus, 1758, *Systema Naturae* (Edn. 10) 1: 513. Type locality: Europe.

Synonymy: *Phalaena* (*Noctua*) *conscripta* Hübner, 1790; *Noctua subpurpurina* Haworth, 1802; *Phalaena* (*Noctua*) *aurosignata* Donovan, 1808; *Plusia interrogationis gamma-argestina* Stephens, 1850 (var.); *Plusia interrogationis rosea* Tutt, 1892 (var.); *Plusia interrogationis transbaicalensis* Staudinger, 1892 (var.); *Plusia borealis* Reuter, 1893; *Autographa epsilon* Ottolengui, 1900; *Plusia interrogationis flammifera* H. Huese, 1901 (ab.); *Autographa zeta* Ottolengui, 1902; *Plusia interrogationis aureomaculata* Vorbodt, 1912 (ab.); *Syngrapha interrogationis orbata* Warren, 1913 (ab.); *Syngrapha interrogationis ignifera* Warren, 1913 (ab.); *Syngrapha interrogationis transbaikalensis*: Hampson, 1913, emend.; *Syngrapha pyrenaica* Hampson, 1913; *Syngrapha sachalinensis* Matsumura, 1925; *Syngrapha giljarovi* Kljutschko, 1983.

Distribution. China (Heilongjiang, Jilin); Russia (European part, Siberia), Korea, Japan, Mongolia; Central and North Europe, North America.

寄主：矮桦（*Betula nana*）、垂枝桦（*Betula pendula*）、安多杜鹃（*Andromeda polifolia*）、帚石楠（*Calluna vulgaris*）、杜香（*Ledum palustre*）、笃斯越橘（Vaccinium uliginosum）、黑果越橘（*Vaccinium myrtillus*）、岩高兰（*Empetrum nigrum*）、异株荨麻（*Urtica dioica*）、欧荨麻（*Urtica urens*）、杨梅属（*Myrica*）、欧石南属（*Erica*）植物（Matov & Kononenko, 2012）。

注：国内多用“伏锌纹夜蛾”予以记录，根据属名变动将其中文种名予以变更。

Genus ***Plusia*** Ochsenheimer, 1816 金翅夜蛾属

Plusia Ochsenheimer, 1816, *Die Schmetterlinge von Europa*, 4: 89. Type species: *Phalaena festucae* Linnaeus, 1758 [Europe].

Synonymy: *Chrysapidia* Hübner, [1821] 1816; *Palaeoplusia* Hampson, 1913.

Holarctic genus with 5 species; 2 species are in the Three Provinces of Northeast China.

Plusia festucae **(Linnaeus, 1758) 金翅夜蛾**

Phalaena festucae Linnaeus, 1758, *Systema Naturae* (Edn. 10) 1: 513. Type locality: Europe.

Synonymy: *Phytometra festucae splendida* Rangnow, 1935 (ssp.); *Phytometra festucae japonibia* Bryk, 1949 (ssp.); *Phytometra festucae kamchadala* Bryk, 1949 (ssp.); *Phytometra festucae kurilensis* Bryk, 1949 (ssp.); *Phytometra festucae kantschadala* Bryk, 1949 (ssp.); *Plusia festata yokohamensis* Bryk, 1949 (ssp.); *Autographa manchurica* Lempke, 1966.

Distribution. China (Heilongjiang, Jilin, Liaoning, Xinjiang, Jiangsu); Russia (European part, Siberia), Korea, Japan; Central Asia, Near East, Europe, North Africa.

寄主：稻、杂草（陈一心，1982）；灰柳（*Salix cinerea*）、异株荨麻（*Urtica dioica*）、旋果蚊子草（*Filipendula ulmaria*）、泽泻（*Alisma plantago-aquatica*）、雄黄兰（*Crocosmia crocosmiflora*）、黄菖蒲（*Iris pseudacorus*）、（*Juncus lacustris*）、（*Juncus schoenoplectus*）、山鸢尾（*Iris setosa*）、丛苔草（*Carex cespitosa*）、金碗苔草（*Carex elata*）、蓝苔草（*Carex flacca*）、毛苔草（*Carex lasiocarpa*）、多穗苔草（*Carex panicea*）、泽生苔草（*Carex riparia*）、（*Carex sylvatica*）、胀囊苔草（*Carex vesicaria*）、小穗发草（*Deschampsia caespitosa*）、高羊茅（*Festuca arundinacea*）、牛尾草（*Festuca elatior*）、羊茅（*Festuca ovina*）、草甸羊茅（*Festuca pratensis*）、甜茅（*Glyceria aquatica*）、浮甜茅（*Glyceria fluitans*）、两蕊甜茅（*Glyceria lithuanica*）、水甜茅（*Glyceria maxima*）、（*Glyceria spectabilis*）、稻（*Oryza sativa*）、丝带草（*Phalaris arundinacea*）、梯牧草（*Phleum pratense*）、芦苇（*Phragmites australis*）、早熟禾（*Poa annua*）、林地早熟禾（*Poa nemoralis*）、草地早熟禾（*Poa pratensis*）、三棱（*Sparganium emersum*）、直立黑三棱（*Sparganium erectum*）、黑三棱（*Sparganium ramosum*）、小黑三棱（*Sparganium simplex*）、水烛（*Typha angustifolia*）、香蒲（*Typha latifolia*）、拂子茅属（*Calamagrostis*）、芸苔属（*Brassica*）、亚麻属（*Linum*）植物（Matov & Kononenko, 2012）。

Plusia putnami Grote, 1873 稻金翅夜蛾

Plusia putnami Grote, 1873, *Bulletin of the Buffalo Society of Natural Sciences*, 1: 146, 192, pl. 4: 2. Type locality: USA, New York.

Synonymy: *Plusia putnami festata* Graeser, 1889 [1890] (ssp.); *Phytometra putnami conjuncta* Warren, 1913 (ssp.); *Phytometra festata major* Warren, 1913 (ab.) (infrasubspecific name); *Phytometra putnami punctistigma* Strand, 1917 (ab.); *Phytometra festata yokohamensis* Bryk, 1949 (var.); *Autographa gracilis* Lempke, 1966; *Autographa gracilis obscura* Lempke, 1966 (f.); *Autographa gracilis coalescens* Lempke, 1966 (f.); *Autographa gracilis disconulla* Lempke, 1966 (f.); *Phytometra major* Chou & Lu, 1979.

Distribution. China (Heilongjiang, Jilin, Ningxia); Russia (European part, Ural, Siberia), Korea, Japan, Mongolia; Near East, Central and South Europe, North Africa, North America.

寄主：稻、弯嘴薹、宽叶香蒲（陈一心，1982）；（*Juncus lacustris*）、（*Juncus schoenoplectus*）、拂子茅（*Calamagrostis epigeios*）、大叶章（*Calamagrostis lansdorfi*）、鸭茅（*Dactylis glomerata*）、发草（*Deschampsia caespitosa*）、（*Myscanthus chinensis*）、稻（*Oryza sativa*）、早熟禾（*Poa annua*）、羊茅属（*Festuca*）、苔属（*Carex*）植物（Matov & Kononenko, 2012）。

Subfamily **Bagisarinae** Crumb, 1956 笆夜蛾亚科

Bagisarinae Crumb, 1956, *United States Department of Agriculture Technical Bulletin*, 1135: 76. Type genus: *Bagisara* Walker, 1858.

The subfamily Bagisarinae was segregated on the basis of larval morphology: by the presence of appendiculate crochets on the prolegs of the larva. Larvae are semi-loopers in all instars, with prolegs absent on A3 and A4 (Crumb, 1956; Beck, 1999, 2000); each crochet on prolegs subapically with a large semicircular tooth (Crumb, 1956), an additional subapical tooth present on mandible (Holloway, 1998); three SV setae on A1. The known larvae feed predominantly on Malvaceae. The genus *Amyna* was tentatively placed in Bagisarinae by Holloway (2009), but its position is dubious. The subfamily includes 40-60 species, distributed throughout the tropical and subtropical areas of the New and Old World. 4 genera and 6 species are in the Three Provinces of Northeast China.

References. Chen, 1982, 1999 (Noctuidae); Sugi, 1982 (Amphipyrinae, Acontiinae); Holloway, 1998 (Amphipyrinae), 2009, 2011 (Bagisarinae); Kononenko *et al.*, 1998 (Amphipyrinae, Acontiinae); Kononenko, 2003 (Amphipyrinae, Acontiinae, part); Fibiger, Hacker, 2005 (Bagisarinae); Fibiger, Lafontaine, 2005 (Bagisarinae); Kononenko, 2005 (Bagisarinae, Eustrotiinae, part), 2010 (Bagisarinae); Lafontaine, Fibiger, 2006 (Bagisarinae); Kononenko, Han, 2007 (Bagisarinae, Eustrotiinae, part); Matov *et al.*, 2008 (Bagisarinae, Eustrotiinae, part); Fibiger *et al.*, 2009, 2011 (Bagisarinae); Kononenko, Pinratana, 2013 (Bagisarinae).

注：首次给出中文亚科名。

Genus *Imosca* Sugi, 1984 齐夜蛾属

Imosca Sugi, 1984, *Tyô Ga*, 34 (3): 130. Type species: *Dyrzela coreana* Matsumura, 1926 [Korea], repl. name for *Allocosmia* Sugi, 1982.

Synonymy: *Allocosmia* Sugi, 1982, preocc. (Cossoman, 1897 [Mollusca]).

Imosca was placed in Bagisarinae by Sugi & Sasaki (2001); The genus includes 5 species distributed in Chinese subregion of the Palaearctic and Oriental regions; 1 species is in the Three Provinces of Northeast China.

注：《中国动物志》（夜蛾科）中引用其同物异名“*Allocosmia* Sugi, 1982”，在此予以变更。

Imosca coreana (Matsumura, 1926) 沪齐夜蛾

Dryzela[sic!] *coreana* Matsumura, 1926, *Insecta Matsumurana*, 1: 58, pl. 1: 33. Type locality: Korea, Shakoji.

Synonymy: *Iphimorpha*[sic!] *coreana moltrechti* O. Bang-Haas, 1927 (ssp.); *Ipimorpha sutchana* O. Bang-Haas, 1927.

Distribution. China (Heilongjiang, Jilin, Jiangsu); Russia (Siberia), Korea, Japan.

寄主：华东椴（*Tilia japonica*）（Matov & Kononenko, 2012）。

Genus *Pseudocosmia* Kononenko, 1985 斑白夜蛾属

Pseudocosmia Kononenko, 1985, *Tinea*, 11: 218. Type species: *Pseudocosmia maculata* Kononenko, 1985 [Russia, Primorye terr.].

Monotypic Chinese genus; 1 species is in the Three Provinces of Northeast China.

注：首次给出中文属名。

Pseudocosmia maculata Kononenko, 1985 斑白夜蛾

Pseudocosmia maculata Kononenko, 1985, *Tinea*, 11: 219, fig. 3. Type locality: Russia, Primorye terr..

Distribution. China (Jilin, Liaoning); Russia (Siberia), Korea.

注：首次给出中文种名。

Genus *Sphragifera* Staudinger, 1892 明夜蛾属

Sphragifera Staudinger, 1892, *in* N.M. Romanoff (ed.), *Mémoires sur les Lépidoptères*, 6: 554. Type species: *Anthoecia sigillata* Ménétriès, 1859 [Russia, Amur].

Synonymy: *Sphragidifera*: Bryk, 1948, emend..

Sphragifera was placed in Bagisarinae by Sugi & Sasaki (2001); The genus includes 5 species distributed in the Chinese subregion of the Palaearctic region; 2 species are in the Three Provinces of Northeast China.

Sphragifera sigillata **(Ménétriès, 1859) 丹日明夜蛾**

Anthoecia? sigillata Ménétriès, 1859, *Bulletin de l'Academie Impériale des Sciences de St.-Petersbroug*, 17: 219. Type locality: Russia, "Amur".

Distribution. China (Heilongjiang, Jilin, Liaoning, Shaanxi, Henan, Zhejiang, Fujian, Sichuan, Yunnan, Taiwan); Russia (Siberia), Korea, Japan.

寄主：千金榆（*Carpinus cordata*）、胡桃楸（*Juglans mandshurica*）（Matov & Kononenko, 2012）。

Sphragifera biplagiata **(Walker, 1865) 日月明夜蛾**

Acontia biplagiata Walker, 1865, *List of the Specimens of Lepidopterous Insects in the Collection of the British Museum*, 33: 781. repl. name.

Synonymy: *Acontia biplaga* Walker, 1857 [1858]; *Euphasia subaicalis* Walker, 1865; *Sphragidifera biplaga hexagona* Bryk, 1948 (ssp.).

Distribution. China (Jilin, Liaoning, Hebei, Henan, Hubei, Hunan, Jiangsu, Zhejiang, Fujian, Guizhou, Taiwan); Korea, Japan.

Genus ***Amyna*** Guenée, 1852 卫翅夜蛾属

Amyna Guenée 1852, *in* Boisduval & Guenée, *Histoire Naturelle des Insectes. Species Général des Lépidoptéres*, 5: 406. Type species: *Amyna selenampha* Guenée, 1852 [Bangladesh, Philippines] (=*Noctua punctum* Fabricius, 1794).

Synonymy: *Berresa* Walker, [1859]; *Ilattia* Walker, [1859]; *Lochia* Walker, 1865; *Stridova* Walker, 1869; *Pteraetholix* Grote, 1873; *Chytoryza* Grote, 1876; *Hesperimorpha* Saalmüller, 1880; *Amynodes* Warren, 1913; *Formosamyna* Strand, 1920; *Niphosticta* Turner, 1936, preocc. (Hampson, 1926 [Lepidoptera, Noctuidae]); *Trilophia* Turner, 1943; *Hurworthia* Nye, 1975; *Chytirhisa*: Lucas, 1909, missp..

The genus was tentatively placed in Bagisarinae by Holloway (2009). The systematic position of this genus requires futher study. The genus includes 35 species distributed in tropical and subtropical areas of the Old and New Worlds; 2 species are in the Three Provinces of Northeast China.

Amyna axis **Guenée, 1852 坑卫翅夜蛾**

Amyna axis Guenée, 1852, *in* Boisduval & Guenée, *Histoire Naturelle des Insectes. Species Général des Lépidoptères*, 5: 407. Type locality: Tahiti.

Synonymy: *Perigea octo* Guenée, 1852; *Poaphila stricta* Walker, 1858; *Celaena flavigutta* Walker, 1858; *Celaena perfundens* Walker, 1858; *Ilattia cephusalis* Walker, 1859; *Amyna colon* Guenée, 1862; *Celaena obstructa* Walker, 1862; *Perigea vexabilis* Wallengren, 1863; *Miana inornata* Walker, 1865; *Perigea leucospila* Walker, 1865; *Stridova albigutta* Walker, 1869; *Erastria stigmatula* Snellen, 1872; *Erastria bavia* Felder et Rogenhofer, 1874; *Segetia orbica* Morrison, 1875; *Amyna undulifera* Butler, 1875; *Chytoryza tecta* Grote, 1876;

Botys monotretalis Mabille, 1879; *Perigea supplex* Swinhoe, 1885; *Berresa rufa* Bethune-Baker, 1906.

Distribution. China (Liaoning, Hebei, Shandong, Shanxi, Jiangsu, Hunan); Russia (Siberia), Korea, Japan; Southeast Asia, Near East, Oceania, Africa, North America.

寄主：大豆（陈一心，1999）；尖叶牛膝（*Achyranthes japonica*）、鸡冠花（*Celosia cristata*）、（*Digera arvensis*）、藜（*Chenopodium album*）、菠菜（*Spinacia oleracea*）、大麻槿（*Hibiscus cannabinus*）、蓖麻（*Ricinus communis*）、花生（*Arachis hypogaea*）、菽麻（*Crotalaria juncea*）、大豆（*Glycine max*）、紫苜蓿（*Medicago sativa*）、乌头叶大豆（*Vigna aconitifolia*）、豇豆（*Vigna unguiculata*）、马铃薯（*Solanum tuberosum*）、红薯（*Ipomoea batatas*）、向日葵（*Helianthus annuus*）、木蓝属（*Indigofera*）、苋属（*Amaranthus*）、黄麻属（*Corchorus*）、黄花棯属（*Sida*）、菜豆属（*Phaseolus*）植物（Matov & Kononenko, 2012）。

注：《中国动物志》（夜蛾科）中引用本种同物异名“*Amyna octo* Guenée, 1852”，在此予以变更。

***Amyna punctum* (Fabricius, 1794) 卫翅夜蛾**

Noctua punctum Fabricius, 1794, *Entomologiae Systematica emendata et Aucta. Secundum Classes, Ordines, Genera, Species Adjectis Synonimis, Locis, Observationibus, Descriptionibus*, 3 (2): 34. Type locality: “Ost-India”.

Synonymy: *Noctua annulata* Fabricius, 1794; *Amyna selenampha* Guenée, 1852; *Alamis spoliata* Walker, [1858]; *Hama latipennis* Wallengren, 1860; *Amyna subtracta* Walker, 1862; *Perigea trivenefica* Wallengren, 1863; *Perigea urba* Wallengren, 1863; *Hadena latipennis* Walker, 1865, preocc.; *Perigea natalensis* Wallengren, 1865; *Hesperimorpha paradoxa* Saalmüller, 1880.

Distribution. China (Jilin, Liaoning, Shandong, Jiangsu, Zhejiang, Fujian, Guangdong, Hainan, Yunnan, Xizang, Taiwan); Russia (Siberia), Korea, Japan; Southeast Asia, Near East, Oceania, Africa.

寄主：巴豆（陈一心，1999）；（*Croton aromaticum*）、巴豆（*Croton tiglium*）（Matov & Kononenko, 2012）。

Subfamily **Eustrotiinae** Grote, 1882 文夜蛾亚科

Eustrotiinae Grote, 1882, *New checklist North American Moths*: 37. Type genus: *Eustrotia* Hübner, [1821].

Synonymy: Erastridi Stephens, 1850, inval.; Erastridae Herrich-Schäffer, [1851] 1845, inval.; Cerathosiini Grote, 1890; Erastriinae Tutt, 1896; inval.; Erastriini Forbes, 1954, inval.; Lithacodiinae Crumb, 1956.

Speidel *et al.* (1996) segregated the subfamilies Acontiinae, Pseudeustrotiinae, Erastriinae (*sensu* auctorum) [Erastrianae (*sensu* Hampson, 1910)]; Eustrotiinae, Tytinae and Sinocharinae from heterogeneous assemblage Acontiinae (*sensu* auctorum). Kitching & Rawlins (1999) divided Acontiinae (*sensu* auctorum) into Acontiinae and Eustrotiinae. Later Fibiger & Lafontaine (2005) and Fibiger & Hacker (2007) confirmed the status of Eustrotiinae as separate subfamily. However the concept of the subfamily is weakly developed, the subfamily requires ground revision. The classification of the *Deltote* generic group, the core of Eustrotiinae, is accepted here according to the revision by Ueda (1984, 1986). The genus *Naranga* is placed in Eustrotiinae following to Kononenko & Han (2007) and Holloway (2009). Bornean Eustrotiinae revised by Holloway (2009).

A subfamily of worldwide distribution, with about 500-700 species in the Old and New Worlds, many of them occur in the tropical and subtropical areas. In the Palaearctic about 100-150 species. The subfamily is most diverse in East Asia, and very few genera and species occur in Europe. 12 genera and 31 species occur in the Three Provinces of Northeast China.

References. Hampson, 1910; Chen, 1982, 1999 (Noctuidae); Ueda 1984, 1987; Sugi, 1993; Speidel *et al.*, 1996; Fibiger & Lafontaine, 2005; Fibiger *et al.*, 2009; Holloway, 2009.

Genus ***Phyllophila*** Guenée, 1852 姬夜蛾属

Phyllophila Guenée, 1852, *in* Boisduval & Guenée, *Histoire Naturelle des Insectes. Species Général des Lépidoptères*, 6: 254. Type species: *Anthophila wimmerii* Treitschke, 1835 [Yugodlavia] (=*Anthophila obliterata* Rambur, 1833).

The genus includes 11 species, 3 of them occur in the Palaearctic region, the others were described from Central and South Africa and South America, and their generic position requires revision; 1 species is in the Three Provinces of Northeast China.

Phyllophila obliterata **(Rambur, 1833) 姬夜蛾**

Anthophila obliterata Rambur, 1833, *Annales de la Société Entomologique de France*, 2: 27, pl. 2: 17. Type locality: South Europe, Corsica.

Synonymy: *Anthophila wimmerii* Treitschke, 1835; *Phyllophila cretacea* Butler, 1879; *Psilomonodes venustula venerica* Bryk, 1948 (ssp.).

Distribution. China (Heilongjiang, Nei Mongol, Xinjiang, Shaanxi, Hebei, Shandong, Jiangsu, Zhejiang, Hubei, Jiangxi, Fujian); Russia (European part, Siberia), Korea, Japan; Central Asia, Near East, Europe.

寄主：除虫菊、蒿（陈一心，1999）；黄花蒿（*Artemisia annua*）、荒野蒿（*Artemisia campestris*）、（*Artemisia camphorata*）、（*Artemisia coerulescens*）、魁蒿（*Artemisia princeps*）、（*Artemisia vulgaris*）、禾本科（Poaceae）植物（Matov & Kononenko, 2012）。

Genus ***Protodeltote*** Ueda, 1984 白臀俚夜蛾属

Protodeltote Ueda, 1984, *Bulletin of Kitakyushu nature History Museum*, 5: 112. Type species: *Phalaena pygarga* Hufnagel, 1766 [Germany].

Synonymy: *Deceptria* Beck, 1996; *Lithacodia* auct.; *Erastria* auct.; *Jaspidia* auct..

The genus includes 10 species, one of them widely distributed in the Palaearctic region, the other six are Sino-Pacific and one is Nearctic region; 3 species are in the Three Provinces of Northeast China.

注：首次给出中文属名。

Protodeltote pygarga (Hufnagel, 1766) 白臀俚夜蛾

Phalaena pygarga Hufnagel, 1766, *Berlinisches Magazin*, 3: 408. Type locality: Germany, Berlin.

Synonymy: *Phalaena fasciana* Linnaeus, 1762; *Noctua fuscula* [Denis et Schiffermüller], 1775; *Phalaena* (*Noctua*) *polygramma* Esper, 1790; *Noctua albilinea* Haworth, 1809; *Noctua fusca* Haworth, 1809; *Bryophila guenei Fallou*, 1864; *Lithacodia fasciana coreana* Bryk, 1949 (ssp.).

Distribution. China (Heilongjiang, Jilin, Liaoning); Russia (European part, Siberia), Korea, Japan, Kazakhstan; Near East, Europe.

寄主：秦氏悬钩子（*Rubus caesius*）、覆盆子（*Rubus idaeus*）、西伯利亚剪股颖（*Agrostis stolonifera*）、黄花茅（*Anthoxanthum odoratum*）、燕麦草（*Arrhenatherum elatius*）、羽状短柄草（*Brachypodium pinnatum*）、短柄草（*Brachypodium sylvaticum*）、拂子茅（*Calamagrostis epigeios*）、鸭茅（*Dactylis glomerata*）、偃麦草（*Elytrigia repens*）、绒毛草（*Holcus lanatus*）、黑麦草（*Lolium perenne*）、酸沼草（*Molinia caerulea*）、丝带草（*Phalaris arundinacea*）、忍冬属（*Lonicera*）植物（Matov & Kononenko, 2012）。

Protodeltote distinguenda (Staudinger, 1888) 稻白臀俚夜蛾

Erastria distinguenda Staudinger, 1888, *Entomologische Zeitung. Entomologischen Vereine zu Stettin*, 49: 265. Type locality: Russia, "Raddefka" [Khabarovck terr., Judish autonomy: Radde]; Khabarovsk terr.: Bikin; Primorye terr., Vladivostok, "Sutschan-Gebiet" [Partizansk].

Synonymy: *Lithacodia quadriorbis* Berio, 1977.

Distribution. China (Heilongjiang, Jilin, Liaoning, Jiangxi, Fujian, Guangxi, Taiwan); Russia (European part, Siberia), Korea, Japan.

寄主：紫马唐（*Digitaria violascens*）、求米草（*Oplismenus undulatifolius*）、稻（*Oryza sativa*）（Matov & Kononenko, 2012）。

注：国内多用名"稻俚夜蛾"予以记录，根据属名变动将其中文种名予以变更。

Protodeltote wiscotti (Staudinger, 1888) 直白臀俚夜蛾

Erastria wiscotti Staudinger, 1888, *Entomologische Zeitung. Entomologischen Vereine zu Stettin*, 49: 266. Type locality: Russia, Primorye terr., Askold I.; Baranovsky; "Ussuri".

Synonymy: *Eustrotia viscotti* Hampson, 1910, emend.; *Micardia wiskotti angulata* Bryk, 1942 (ssp.); *Micardia jezoensis* Sugi, 1959.

Distribution. China (Heilongjiang, Jilin); Russia (Siberia), Korea, Japan.

寄主：禾本科（Poaceae）植物（Matov & Kononenko, 2012）。

Genus ***Koyaga*** Ueda, 1984 虚俚夜蛾属

Koyaga Ueda, 1984, *Bulletin of Kitakyushu nature History Museum,* 5: 119. Type species: *Miana falsa* Butler, 1885 [Japan].

Synonymy: *Erastria* auct. Ochsenheimer, 1816; *Lithacodia* auct. Hübner, 1818; *Jaspidia* auct. Hübner, 1812.

The genus includes 7 species, distributed in Sino-Pacific subregion of the Palaearctic and Oriental regions; 3 species are in the Three Provinces of Northeast China.

注：首次给出中文属名。

Koyaga falsa **(Butler, 1885) 虚俚夜蛾**

Miana falsa Butler, 1885, *Cistula Entomologica*, 3: 132. Type locality: Japan, Yokohama.

Distribution. China (Jilin, Jiangsu, Jiangxi, Sichuan); Russia (Siberia), Korea, Japan.

寄主：千岛赤竹（*Sasa kurilensis*）（Matov & Kononenko, 2012）。

Koyaga numisma **(Staudinger, 1888) 串纹虚俚夜蛾**

Erastria numisma Staudinger, 1888, *Entomologische Zeitung. Entomologischen Vereine zu Stettin*, 49: 265. Type locality: Russia, Primorye terr., Vladivostok, Askold I., “Suifun” [Razdolnaya river].

Synonymy: *Erastria olivacea* Leech, 1889.

Distribution. China (Heilongjiang, Jilin); Russia (Siberia), Korea, Japan.

寄主：芒（*Miscanthus chinensis*）（Matov & Kononenko, 2012）。

Genus ***Sugia*** Ueda, 1984 阴俚夜蛾属

Sugia Ueda, 1984, *Bulletin of Kitakyushu nature History Museum*, 5: 125. Type species: *Erastria stygia* Butler, 1878 [Japan].

The genus includes 5 species, distributed in Sino-Pacific subregion of the Palaearctic, and Oriental regions; 3 species are in the Three Provinces of Northeast China.

注：首次给出中文属名。

Sugia stygia **(Butler, 1878) 阴俚夜蛾**

Erastria stygia Butler, 1878, *Annals and Magazine of Natural History*, (5)1: 199. Type locality: Japan, Yokohama.

Synonymy: *Erastria fasciana*: Leech, 1889; nec Linnaeus, 1761, part.

Distribution. China (Jilin, Zhejiang, Hubei, Fujian, Sichuan); Russia (Siberia), Korea, Japan.

寄主：稻、竹（陈一心，1999）；稻（*Oryza sativa*）、刚竹属（*Phyllostachys*）植物（Matov & Kononenko, 2012）。

***Sugia idiostygia* (Sugi, 1958) 苏俚夜蛾**

Jaspidia idiostygia Sugi, 1958, *Tinea*, 4: 224, fig. 3, 4. Type locality: Japan, Tokyo, Mt. Takao.

Distribution. China (Liaoning); South Korea, Japan.

***Sugia erastroides* (Draudt, 1950) 伴阴俚夜蛾**

Chytonix erastroides Draudt, 1950. *Mitteilungen der Münchner Entomologischen Gesellschaft*, 40: 112, pl. 7: 21. Type locality: China, Ost Tien-mu-shan, Mokan-shan.

Synonymy: *Jaspidia stygioides* Sugi, 1958.

Distribution. China (Jilin, Zhejiang, Jiangxi); South Korea, Japan.

注：《中国动物志》（夜蛾科）中将本种归入"*Chytonix* Grote, 1874 流夜蛾属"，在此予以变更。

Genus ***Deltote*** Reichenbah, Leipzig, 1817 **俚夜蛾属**

Deltote Reichenbah, Leipzig, 1817, *Jenaische Allgemeine Literatur-Zeitung*, 1: 388 [Germany]. Type species: *Phalaena argentula* Hübner, [1787] (=*Pyralis bankiana* Fabricius, 1775).

Synonymy: *Erastria* Ochsenheimer, 1816, preocc. (Hübner, [1813] [Lepidoptera, Geometridae]); *Eustrotia* Hübner, [1821] 1816; *Lithacodia* Hübner, 1818; *Hemeroptera* Sodoffsky, 1837, repl. name; *Hydrelia* Guenée, 1841, preocc. (Hübner, [1825] 1816 [Lepidoptera, Geometridae]); *Hyela* Stephens, 1850; *Bankia* Guenée, 1852, preocc. (Gray, 1842 [Mollusca]).

In the present treatment, the genus includes 6 species, three of which are widely distributed in the Palaearctic region, two occurs in Chinese subregion, and one (*Deltote bellicula* Hübner, 1818) the type species of *Lithacodia*, is Nearctic region; 5 species are in the Three Provinces of Northeast China. The Chinese species of *Lithacodia* are recently revised by Han & Kononenko (2015).

***Deltote deceptoria* (Scopoli, 1763) 白带俚夜蛾**

Phalaena deceptoria Scopoli, 1763, *Entomologia Carniolica Exhibens Insecta Carnioliae Indigena et Distributa in Ordines, Genera, Species, Varietates Methodo Linnaeana:* 214: 527. Type locality: Slovenia, "Carniola, Umg. Idriam.

Synonymy: *Noctua lineodes* Hufnagel, 1766; *Noctua atratula* [Denis & Schiffermüller], 1775; *Noctua atratula* Hübner, [1800-1803], nec. Denis & Schiffermüller, 1775.

Distribution. China (Heilongjiang, Jilin, Xinjiang); Russia (Ural, Siberia), Europe.

寄主：黄芪（陈一心，1999）；梯牧草（*Phleum pratense*）、阿尔泰旱禾（*Poa altaica*）（Matov & Kononenko, 2012）。

Deltote uncula (Clerck, 1759) 文俚夜蛾

Phalaena uncula Clerck, 1759, *Icones Insectorum Rariorum cum Nominibus*, 1: pl. 3: 7. Type locality: not stated.

Synonymy: *Phalaena* (*Tortrix*) *uncana* Linnaeus, 1761; *Phalaena singularis* Hufnagel, 1766; *Tortrix unca* [Denis & Schiffermüller], 1775; *Eustrotia rufotincta* Kolb, 1930.

Distribution. China (Heilongjiang); Russia (European part, Ural, Siberia), Korea, Japan; Central Asia, Europe.

寄主：苔属（陈一心，1999）；小穗发草（*Deschampsia caespitosa*）、黑苔草（*Carex nigra*）、（*Carex sylvatica*）、莎草属（*Cyperus*）植物（Matov & Kononenko, 2012）。

注：《中国动物志》（夜蛾科）中将本种归入"*Eustrotia* Hübner, [1821] 1816 文夜蛾属"，在此予以变更。

Deltote bankiana (Fabricius, 1775) 淡文俚夜蛾

Pyralis bankiana Fabricius, 1775, *Entomologiae Systematica emendata et Aucta. Secundum Classes, Ordines, Genera, Species Adjectis Synonimis, Locis, Observationibus, Descriptionibus*: 645. Type locality: England.

Synonymy: *Tortrix olivana* [Denis & Schiffermüller], 1775; *Phalaena* (*Noctua*) *argentula* Hübner, [1787]; *Noctua olivea* Hübner, [1803], emend.; *Erastria argentula amurula* Staudinger, 1892 (var.); *Erastria bankiana amurula* Staudinger, 1892 (var.); *Eustrotia olivana albescens* Draudt, 1935 (f.); *Eustrotia olivana uniformis* Draudt, 1935 (f.).

Distribution. China (Heilongjiang, Jilin, Nei Mongol, Xinjiang); Russia (European part, Ural, Siberia), Korea, Japan; Central Asia, Near East, Europe.

寄主：早熟禾等（陈一心，1999）；拳参（*Polygonum bistorta*）、燕麦草（*Arrhenatherum elatius*）、羽状短柄草（*Brachypodium pinnatum*）、直立雀麦（*Bromus erectus*）、拂子茅（*Calamagrostis epigeios*）、鸭茅（*Dactylis glomerata*）、偃麦草（*Elytrigia repens*）、酸沼草（*Molinia caerulea*）、早熟禾（*Poa annua*）、水甜茅（*Poa aquatica*）、草地早熟禾（*Poa pratensis*）、苔属（*Carex*）、莎草属（*Cyperus*）植物（Matov & Kononenko, 2012）。

注：《中国动物志》（夜蛾科）中将本种归入"*Eustrotia* Hübner, [1821] 1816 文夜蛾属"，在此予以变更。

Deltote nemorum (Oberthür, 1880) 木俚夜蛾

Erastria nemorum Oberthür, 1880, *Études d'Entomologie*, 5: 82, pl. 4: 2. Type locality: Russia, Primorye terr., Askold I..

Synonymy: *Erastria africana* Leech, 1889, nec Felder & Rogenhofer, 1874.

Distribution. China (Heilongjiang, Xinjiang, Shanxi); Russia (Siberia), Korea, Japan.

Deltote melaleuca (Hampson, 1910) 白斑俚夜蛾

Lithacodia melaleuca Hampson, 1910, *Catalogue of the Lepidoptera Phalaenae in the British Museum*, 10: 508, pl. 164: 3. Type locality: Central China, Chusan Is..

Synonymy: *Lithacodia shansiensis* Berio, 1977.

Distribution. China (Beijing, Liaoning, Shaanxi, Zhejiang, Hubei, Sichuan).

Genus *Paraphyllophila* Kononenko, 1985 叶夜蛾属

Paraphyllophila Kononenko, 1985, *Tinea*, 11 (27): 217. Type species: *Paraphyllophila confusa* Kononenko, 1985. Type locality: Russia, Primorye terr., Barabash-Levada.

Monotypic Sino-Pacific genus; 1 species is in the Three Provinces of Northeast China.

注：首次给出中文属名。

Paraphyllophila confusa Kononenko, 1985 叶夜蛾

Paraphyllophila confusa Kononenko, 1985, *Tinea*, 11 (27): 218. Type locality: Russia, Primorye terr., Kedrovaja Pad' reserve.

Distribution. China (Liaoning); Russia (Siberia: Amur reg., Khabarovsk and Primorye terr.), Korea, Japan (Honshu).

注：首次给出中文种名。

Genus *Erastroides* Hampson, 1893 代文夜蛾属

Erastroides Hampson, 1893, *Illustrations of Typical Specimens of Lepidoptera Heterocera in the Collection of the British Museum*, 9: 20, 98. Type species: *Erastroides oliviaria* Hampson, 1893 [Sri Lanka].

The genus includes 12 species, 2 species distributed in the Chinese subregion of the Palaearctic region; 6 species are in the Oriental and 3 species are in Neotropical regions; 1 species is in the Three Provinces of Northeast China.

注：首次给出中文属名。

Erastroides fentoni (Butler, 1881) 锈色代文夜蛾

Erastria fentoni Butler, 1881, *Transactions of the Entomological Society of London*, 1881: 190. Type locality: Japan, Tokyo.

Synonymy: *Erastria versicolor* Oberthür, 1884.

Distribution. China (Heilongjiang, Jilin, Hubei, Taiwan); Russia (Siberia), Korea, Japan.

注：《中国动物志》（夜蛾科）中将其归入"*Lithacodia* Hübner, [1818] 俚夜蛾属"，在此予以变更。

Genus ***Naranga*** Moore, 1881 螟蛉夜蛾属

Naranga Moore, 1881, *Proceedings of the Zoological Society of London*, 1881: 359 Type species: *Xanthodes diffusa* Walker, 1865 [Sri Lanka].

The genus includes 3 species, distributed in the Chinese subregion of the Palaearctic, and Ethiopian regions; 1 species is in the Three Provinces of Northeast China.

Naranga aenescens **Moore, 1881 稻螟蛉夜蛾**

Naranga aenescens Moore, 1881, *Proceedings of the Zoological Society of London*, 1881: 359. Type locality: Taiwan

Synonymy: *Naranga brunnea* Hampson, 1910; *Anthophila hebescens* Butler, 1879.

Distribution. China (Heilongjiang, Jilin, Liaoning, Hebei, Shaanxi, Jiangsu, Hunan, Jiangxi, Fujian, Guangxi, Taiwan); Russia (Siberia), Korea, Japan, India(North), Nepal, Indonesia.

寄主：稻、高粱、玉米、稗、茅草、茭白等（陈一心，1999）；小叶榄仁（*Terminalia paniculata*）、毛榄仁（*Terminalia tomentosa*）、（*Echinochloa indica*）、李氏禾（*Leersia hexandra*）、稻（*Oryza sativa*）、高粱（*Sorghum bicolor*）、玉米（*Zea mays*）、茭白（*Zizania latifolia*）（Matov & Kononenko, 2012）。

Genus ***Maliattha*** Walker, 1863 瑙夜蛾属

Maliattha Walker, 1863, *List of the Specimens of Lepidopterous Insects in the Collection of the British Museum*, 27: 86. Type species: *Maliattha separata* Walker, 1863 [Borneo, Sarawak].

The genus includes 68 species, distributed in the Chinese subregion of the Palaearctic region, as well as the Oriental, Ethiopian and Nearctic regions; 7 species are in the Three Provinces of Northeast China.

Maliattha rosacea **(Leech, 1889) 桃红瑙夜蛾**

Erastria rosacea Leech, 1889, *Proceedings of the Zoological Society of London*, 1889: 527, pl. 53: 9. Type locality: Japan, Oiwake.

Distribution. China (Jilin, Liaoning, Hebei, Zhejiang, Taiwan); Russia (Siberia), Korea, Japan.

Maliattha subrosacea **Ahn, 1998 拟桃红瑙夜蛾**

Maliattha subrosacea Ahn, 1998, *Illustrated Catalogue of Noctuidae in Korea*, 1998: 166, no. 427. Type locality: Korea.

Distribution. China (Liaoning); South Korea.

Maliattha chalcogramma (Bryk, 1948) 路黄瑙夜蛾

Oruza chalcogramma Bryk, 1948, *Arkiv för Zoologi*, 41 (A)1: 112, pl. 5: 7. Type locality: North Korea, Shuotsu.

Synonymy: *Maliattha vialis*: Sugi, 1982, auct., nec Moore, 1882.

Distribution. China (Heilongjiang, Jilin, Henan, Zhejiang, Hunan, Fujian, Sichuan); Russia (Siberia), Korea, Japan.

注：首次给出中文种名。

Maliattha khasanica Zolotarenko & Dubatolov, 1996 俄瑙夜蛾

Maliattha khasanica Zolotarenko & Dubatolov, 1996, *Atalanta*, 26 (1-2): 299. Type locality: Russia, Primorye terr., Khasan disrtrict, Furugelm I.

Distribution. China (Jilin); Russia (Siberia), Korea, Japan.

注：首次给出中文种名。

Maliattha bella (Staudinger, 1888) 丽瑙夜蛾

Thalpochares bella Staudinger, 1888, *Entomologische Zeitung. Entomologischen Vereine zu Stettin*, 49: 264. Type locality: Russia, Vladivostok.

Distribution. China (Heilongjiang, Jilin, Liaoning, Zhejiang, Hunan); Russia (Siberia), Korea, Japan.

Maliattha signifera (Walker, 1857) 标瑙夜蛾

Acontia signifera Walker, 1857, *List of the Specimens of Lepidopterous Insects in the Collection of the British Museum*, 12: 793. Type locality: North India.

Synonymy: *Acontia subfixa* Walker, 1866; *Maliattha signifera rufigrisea* Warren, 1913 (ssp.).

Distribution. China (Heilongjiang, Jilin, Liaoning, Hebei, Jiangsu, Hubei, Jiangxi, Fujian, Guangdong, Guangxi, Taiwan); Korea, Japan, India, Sri Lanka, Pakistan, Nepal, Vietnam, Cambodia, Myanmar, Philippines, Malaysia, Andamanes, Micronesia, Australia.

Maliattha arefacta (Butler, 1879) 斜带瑙夜蛾

Acontia arefacta Butler, 1879, *Annals and Magazine of Natural History*, (5)4: 366. Type locality: Japan, Yokohama.

Synonymy: *Bankia basalis* Moore, 1882.

Distribution. China (Liaoning, Shanghai); South Korea, Japan, Nepal.

注：首次给出中文种名。

Genus *Inabaia* Sugi, 1982 变夜蛾属

Inabaia Sugi, 1982, *in* Inoue *et al.*, *Moths of Japan*, 1982: 819. Type species: *Narosa culta* Butler, 1879 [Japan].

Monotypic Chinese-Japanese genus.

注：首次给出中文属名。

Inabaia culta (Butler, 1879) 变夜蛾

Narosa culta Butler 1879, *Annals and Magazine of Natural History*, (5)4: 356. Type locality: Japan.

Distribution. China (Jilin, Liaoning); Korea, Japan.

注：首次给出中文种名。

Genus *Ozarba* Walker, 1865 弱夜蛾属

Ozarba Walker, 1865, *List of the Speciemens of Lepidopterous Insects in the Collection of the British Museum*, vol. 32: 684. Type species: *Ozarba punctigera* Walker, 1865 [China; Australia].

Synonymy: *Acontiola* Staudinger, 1900.

The genus inludes over 200 species are in the Old and New Worlds, mainly distributed in the North Africa and South America; 1 species is in the Three Provinces of Northeast China.

Ozarba punctigera Walker, 1865 弱夜蛾

Ozarba punctigera Walker, 1865, *List of the Specimens of Lepidopterous Insects in the Collection of the British Museum*, 32: 685. Type locality: China; Australia, Queensland, Moreton Bay.

Distribution. China (Heilongjiang, Jilin, Liaoning, Jiangsu, Zhejiang, Hubei, Guizhou, Taiwan); Korea, Japan, Australia, India, Nepal, Pakistan; Central and South Africa.

Genus *Chorsia* Walker, [1863] 1864 缘夜蛾属

Chorsia Walker, [1863] 1864, *Journal of the Proceedings of the Linnean Society* (*Zoology*), 7: 59. Type species: *Chorsia maculosa* Walker, [1863] 1864 [Indonesia, Kalimantan I., Sarawak].

Synonymy: *Bryophilina* Staudinger, 1892; *Poecilogramma* Butler, 1892, preocc. (Karsch, 1887 [Orthoptera]); *Aeologramma* Strand, 1910, unnec. repl. name; *Peuderiopus* Warren, 1913; *Neustrotia* Sugi 1982.

The genus includes 24 species, distributed in the Chinese subregion of the Palaearctic, and Oriental regions; 5 species are in the Three Provinces of Northeast China.

注：《中国动物志》（夜蛾科）中引用本属同物异名“*Bryophilina* Staudinger, 1892”，在此予以变更。

Chorsia costimacula **(Oberthür, 1880) 臂斑缘夜蛾**

Erastria costimacula Oberthür, 1880, *Études d'Entomologie*, 5: 83, pl. 4: 4. Type locality: Russia, Primorye terr., Askold I..

Synonymy: *Eustrotia costimacula serica* Warren, 1912 (ssp.).

Distribution. China (Heilongjiang, Jilin, Hubei); Russia (Siberia), Korea, Japan.

寄主：苹果（*Malus pumila*）、三叶海棠（*Malus sieboldii*）（Matov & Kononenko, 2012）。

注：《中国动物志》（夜蛾科）中将本种归入“*Eustrotia* Hübner, [1821] 1816 文夜蛾属”，在此予以变更。

Chorsia noloides **(Butler, 1879) 小缘夜蛾**

Acontia noloides Butler, 1879, *Annals and Magazine of Natural History*, (5)4: 366. Type locality: Japan: Yokohama.

Distribution. China (Jilin, Liaoning, Jiangsu, Taiwan); Russia (Siberia), Korea, Japan.

注：《中国动物志》（夜蛾科）中将本种归入“Eustrotia Hübner, [1821] 1816 文夜蛾属”，在此予以变更。

Chorsia mollicula **(Graeser, 1888[1889]) 缘夜蛾**

Erastria mollicula Graeser, 1888[1889], *Berliner entomologische Zeitschrift*, 32, II: 368. Type locality: Russia, Khabarovsk.

Synonymy: *Bryophilina blandula* Staudinger, 1892.

Distribution. China (Heilongjiang, Sichuan, Taiwan); Russia (Siberia), Korea, Japan, Nepal, India; Southeast Asia.

寄主：青冈 (*Quercus glauca*)（Matov & Kononenko, 2012）。

Chorsia japonica **(Warren, 1912) 长臂缘夜蛾**

Eustrotia costimacula japonica Warren, 1912 (ssp.), *in* Seitz, *Macrolepidoptera World*, 3: 281, pl. 52: f. Type locality: Japan, Yokohama.

Distribution. China (Jilin, Liaoning); Korea, Japan.

注：首次给出中文种名

Chorsia rectilineata **Ueda, 1987 台缘夜蛾**

Neustrotia rectilineata Ueda, 1987, *Bulletin of the Kitakyushu Museum of Natural History*, 6: 78, figs. 66, 67, 88-C. Type locality: China, Taiwan; Korea, Kyongsangbug-do, Mt. Sudo-san.

Distribution. China (Heilongjiang, Jilin, Taiwan); South Korea.

寄主：春榆（*Ulmus japonica*）（Park *et al.*, 2006）。

注：首次给出中文种名。

Subfamily **Acontiinae** Guenée, 1841 绮夜蛾亚科

Acontidi Guenée, 1841. *Annales de la Société Entomologique de France*, 10: 219. Type genus: *Acontia* Ochsenheimer, 1816.

Synonymy: Agrophilidae Duponchel, [1845]; Emmelidi Stephens, 1850; Tarachini Grote, 1890.

The concept of the subfamily is accepted according to the revision by Hacker *et al.* (2008) except the tribe Aediini which was transferred from Catocalinae to Acontiinae by Fibiger and Lafontaine (2005) and presently considered as a subfamily of Noctuidae following to Holloway (2009, 2011) and Zahiri *et al.,* 2013.

The subfamily includes approximately 500 species, distributed in the Old and New Worlds, mainly in the south of the temperate zone and in subtropical and tropical areas. The vast majority of the species are in the tribe Acontiini are members of genus *Acontia.* 1 genus and 3 species occur in the Three Provinces of Northeast China.

References. Hampson, 1910 (Acontiinae, part); Chen, 1982, 1999 (Noctuidae); Sugi, 1982, (Acontiinae, part); Kononenko *et al.*, 1998 (Acontiinae, part); Kononenko, 2003 (Acontiinae, part); Fibiger, Hacker, 2005 (Acontiinae); Fibiger, Lafontaine, 2005 (Acontiinae); Kononenko, 2005, 2010 (Acontiinae); Lafontaine, Fibiger, 2006 (Acontiinae); Kononenko, Han, 2007 (Acontiinae); Matov *et al.*, 2008 (Acontiinae); Hacker *et al.*, 2008 (Acontiinae); Holloway, 2008, 2011 (Acontiinae); Fibiger *et al.*, 2009, 2011 (Acontiinae); Kononenko, Pinratana, 2013 (Acontiinae).

Tribe **Acontiini** Guenée, 1841 绮夜蛾族

Acontidi Guenée, 1841. *Annales de la Société Entomologique de France*, 10: 219. Type genus: *Acontia* Ochsenheimer, 1816.

Synonymy: Agrophilidae Duponchel, [1845]; Emmelidi Stephens, 1850; Tarachini Grote, 1890.

The tribe includes approximately 400 species, half of which are distributed in the New World; only one genus in the Three Provinces of Northeast China.

Genus ***Acontia*** Ochsenheimer, 1816 绮夜蛾属

Acontia Ochsenheimer, 1816, *Die Schmetterlinge von Europa*, 4: 91. Type species: *Noctua solaris* [Denis & Schiffermuller], 1775 [Austria] (=*Phalaena lucida* Hufnagel, 1766).

Synonymy: *Erotyla* Hübner, [1806], suppr. name; *Emmelia* Hübner, [1821] 1816 (subgen.); *Tarache* Hübner, [1823] 1816; *Erotyla* Hübner, 1822, suppr. (ICZN, 1926 Op. 97); *Eusceptis* Hübner, 1823 (subgen.); *Desmophora* Stephens, 1829; *Euphasia* Stephens, 1830; *Heliothera* Sodoffsky, 1837, repl. name; *Porrotha* Gistl, 1848, repl. name; *Tima* Walker, [1858], preocc. (Eschscholtz, 1829 [Coelenterata]); *Eugraphia* Guenée, 1852; *Metapioplasta* Wallengren, 1865 (subgen.); *Ponometia* Herrich-Schaffer, 1868; *Pseudalypia* Edwards, 1874; *Spragueia* Grote, 1875; *Trichotarache* Grote, 1875; *Fruva* Grote, 1877; *Heliodora* Neumoegen, 1891; *Graeperia* Grote, 1895, unnecess repl. name; *Therasea* Grote, 1895; *Tarachidia* Hampson, 1898; *Conacontia* Smith, 1900; *Tornacontia* Smith,

1900; *Conochares* Smith, 1905; *Hoplotarache* Hampson, 1910; *Procriosis* Hampson, 1910; *Cardiosace* Hampson, 1910; *Aulotarache* Hampson, 1910; *Neptunia* Barnes & McDunnough, 1911, preocc.; *Ceratostrotia* Warren, 1913; *Hemispargueia* Barnes & McDunnough, 1923; *Acropserotarache* Berio, 1937; *Fredina* Brandt, 1939; *Agrophila* Boisduval, 1840; *Uniptena* Nye, 1975; *Acontarache* Berio, 1977 (subgen.); *Uracontia* Beck, 1996 (subgen.); *Emmelacontia* Beck, 2000; *Olivacontia* Hacker, Legrain & Fibiger, 2008 (subgen.).

Large genus with worldwide distribution, most diverse in arid and semi-arid areas. 20 species are known from the Palaearctic region; 2 species are in the Three Provinces of Northeast China.

Subgenus ***Acontia*** Ochsenheimer, 1816 **绮夜蛾亚属**

Acontia (_Acontia_) _martjanovi_ Tschetverikov, 1904 白斑绮夜蛾

Erastria martjanovi Tschetverikov, 1904. *Revue Russe d'Entomologie*, 4: 78. Type locality: Russia, Minusinsk.

Distribution. China (Jilin); Russia (Siberia), Mongolia, South Korea.

注：首次给出中文种名。

Subgenus ***Emmelia*** Hübner, [1821] 1816 **谐夜蛾亚属**

Emmelia Hübner, [1821] 1816, *Verzeichniss bekannter Schmettlinge* [sic!]: 254. Type species: *Phalaena sulphuralis* Linnaeus, 1767 [Germany] (=*Phalaena Pyralis trabealis* Scopoli, 1763).

Acontia (_Emmelia_) _trabealis_ (Scopoli, 1763) 谐绮夜蛾

Phalaena (*Pyralis*) *trabealis* Scopoli, 1763, *Entomologia Carniolica Exhibens Insecta Carnioliae Indigena et Distributa in Ordines, Genera, Species, Varietates Methodo Linnaeana*: 240. Type locality: Slovenia "Carniola".

Synonymy: *Phalaena arabica* Hufnagel, 1766; *Phalaena sulphuralis* Linnaeus, 1767; *Noctua sulphurea* [Denis et Schiffermüller], 1775; *Tinea arlequinetta* Geofrey, 1785; *Phalaena* (*Noctua*) *trabeata* Borkhausen, 1790; *Noctua trabeata* Scriba, 1790; *Bombyx lugubris* Fabricius, 1793; *Erastria pardalina* Walker, 1865; *Agrophila nigra* Erschoff, 1874.

Distribution. China (Heilongjiang, Jilin, Liaoning, Nei Mongol, Xinjiang, Hebei, Jiangsu, Guangzhou); Russia (European part, South Ural, Siberia), Korea, Japan; Central Asia, Near East, Europe, North Africa.

寄主：甘薯、田旋花（陈一心，1999）；卷茎蓼（*Polygonum convolvulus*）、草棉（*Gossypium herbaceum*）、紫苜蓿（*Medicago sativa*）、田旋花（*Convolvulus arvensis*）、甘薯（*Ipomoea batatas*）、锦葵属（*Malva*）、滨藜属（*Atriplex*）、藜属（*Chenopodium*）、茄属（*Solanum*）（Matov & Kononenko, 2012）。

Subfamily **Pantheinae** Smith, 1898 毛夜蛾亚科

Pantheini Smith, 1898, *in* Smiths & Dyar, *Proceeding of the United States National Museum*, 21: 4. Type genus *Panthea* Hübner, [1820].

Synonymy: Trichoseini Kobes, 1992; Thiacidinae Hacker & Zill, 2007.

Kitching and Rawlins (1999) raised Pantheinae to the family rank in Noctuoidea; however Speidel *et al.* (1996a), Fibiger & Lafontaine (2005), Lafontaine & Fibiger (2006) treat Pantheinae as subfamily of Noctuidae. Holloway (2009) presented an expanded concept of Pantheinae based on Bornean fauna and transferred genera *Cyclodes* and *Arcte* (tribe Arcteini) formerly placed in Catocalinae or Erebinae to the subfamily Pantheinae, however recently these taxa and the genus *Belciades* were placed to the subfamily Dyopsinae (Holloway, 2011; Zahiri *et al.*, 2013a). The subfamily Thiacidinae synonymised with Pantheinae by Zahiri *et al.* (2013). The concept of two subfamilies, Pantheinae and Dyopsinae is accepted here. The subfamily Pantheinae includes about 50 species are in the world, distributed mainly in the Palaearctic, Oriental and Nearctic regions. The centre of diversity of the subfamily is mountain broad-leaved and mixed forests of Himalaya, South and Southwest China. There are 13 genera and 20 species in the Palaearctic region; among them 6 genera and 8 species are in the Three Provinces of Northeast China.

References. Hampson, 1913 (Mominae); Kozhanchikov, 1950 (Momini); Chen, 1982, 1999 (Noctuidae); Sugi, 1982 (Pantheinae); Sugi, Nakamura, 1989 (Pantheinae); Speidel, Kononenko, 1998 (Pantheinae); Kitching, Rawlins, 1999 (Pantheidae); Kononenko, 2003 (Pantheinae); Fibiger, Lafontaine, 2005 (Pantheinae); Kononenko, 2005 (Pantheinae), 2010 (Pantheinae, Thiacidinae); Lafontaine, Fibiger, 2006 (Pantheinae); Hacker, Zilli, 2007 (Thiacidinae, part); Kononenko, Han, 2007 (Pantheinae); Matov *et al.*, 2008 (Pantheinae); Fibiger *et al.*, 2009 (Pantheinae); Holloway, 2009, 2011 (Pantheinae); Behounek, Kononenko, 2013 (Pantheinae).

Genus ***Panthea*** Hübner, [1820] 1816 毛夜蛾属

Panthea Hübner, [1820] 1816, *Verzeichniss bekannter Schmettlinge* [sic!]: 203. Type species: *Phalaena coenobita* Esper, 1785 [Germany].

Synonymy: *Elatina* Duponchel, 1845; *Audela* Walker, 1861; *Platycerura* Packard, 1864; *Diphthera*: Hampson, 1913, nec Hübner, 1809.

The genus includes 11 species, distributed in the Palaearctic, Oriental and Nearctic regions; 1 species is in the Three Provinces of Northeast China.

***Panthea coenobita* (Esper, 1785) 毛夜蛾**

Phalaena (*Bombyx*) *coenobita* Esper, 1785, *Die Schmetterlinge in Abbildungen nach der Natur mit Beschreibungen*, 3: 196, pl. 37: 7. Type locality: Germany, Leipzig.

Synonymy: *Panthea coenobita latefasciata* Rebel, 1910 (ssp.); *Panthea coenobita ussuriensis* Warnecke, 1917 (ssp.); *Panthea coenobita kotschubeyi* Sheljuzhko, 1920 (ssp.); *Panthea coenobita immaculata* Sheljuzhko, 1920 (ssp.); *Panthea coenobita idae* Bryk, 1948 (ssp.).

Distribution. China (Heilongjiang, Jilin, Liaoning); Russia (European part, Ural, Siberia), Korea, Japan; Asia Minor, Europe.

寄主：松（陈一心，1999）；冷杉（*Abies alba*）、（*Abies excelsa*）、日本冷杉（*Abies firma*）、欧洲落叶松（*Larix decidua*）、日本落叶松（*Larix leptolepis*）、欧洲云杉（*Picea abies*）、挪威云杉（*Picea excelsa*）、瑞士五叶松（*Pinus cembra*）、赤松（*Pinus densiflora*）、红松（*Pinus koraiensis*）、北欧红松（*Pinus silvestris*）、美国白松（*Pinus strobus*）、加州巨杉（*Sequoia gigantea*）、日本柳杉（*Cryptomeria japonica*）（Matov & Kononenko, 2012）。

Genus ***Thiacidas*** Walker, 1855 短喙夜蛾属

Thiacidas Walker, 1855, *List of the Specimens of Lepidopterous Insects in the Collection of the British Museum*, 5: 979, 1027. Type species: *Thiacidas postica* Walker, 1855, *ibid*.: 1028, by monotypy [Nepal], originally placed in the Notodontidae (=*cheituna* Brandt, 1939).

Synonymy: *Panthauma* Staudinger, 1892; *Auchensia* Hampson, 1905; *Trisulana* Bethune-Backer, 1911; *Galactomoia* Fawcett, 1916; *Pteronycta* Fawcett, 1917.

The genus has been revised by Hacker & Zill (2007). It includes 43 species mainly distributed in tropical area of the Old World; 1 species is in the Three Provinces of Northeast China.

注：《中国动物志》（夜蛾科）中引用属的同物异名“*Panthauma* Staudinger, 1892”，在此予以变更。

Thiacidas egregia **(Staudinger, 1892) 短喙夜蛾**

Panthauma egregia Staudinger, 1892, *in* N.M. Romanoff (ed.), *Mémoires sur les Lépidoptères*, 6: 404, pl. 7: 2. Type locality: Russia, Primorye terr., Baranovsky.

Synonymy: *Panthauma egregia koreothauma* Bryk, 1948 (ssp.).

Distribution. China (Heilongjiang, Jilin, Liaoning, Nei Mongol); Russia (Siberia), Korea.

寄主：糠椴（*Tilia mandshurica*）（Matov & Kononenko, 2012）。

Genus ***Trichosea*** Grote, 1875 镶夜蛾属

Trichosea Grote, 1875, *Bulletin of the Buffalo Society of Natural Sciences*, 2: 213. Type species: *Phalaena ludifica* Linnaeus, 1758 [Europe].

Synonymy: *Moma* auct., nec Hübner, [1820] 1816.

The genus includes up to 15 species distributed in the Palaearctic, and mainly in the Oriental regions; 2 species are in the Three Provinces of Northeast China.

***Trichosea ludifica* (Linnaeus, 1758) 淡镶夜蛾**

Phalaena ludifica Linnaeus, 1758, *Systema Naturae* (Edn. 10) 1: 514, Nr. 100. Type locality: Europe.

Synonymy: *Moma ludifica andropoda* Bryk, 1948 (ssp.); *Moma ludifica azugensis* Bryk, 1948 (f.).

Distribution. China (Heilongjiang, Jilin, Liaoning); Russia (European part, Ural, Siberia), Korea, Japan; Europe.

寄主：千金榆（*Carpinus cordata*）、黄花柳（*Salix caprea*）、（*Crataegus laevigata*）、（*Malus sylvestris*）、稠李（*Padus racemosa*）、欧洲李（*Prunus domestica*）、黑樱桃木（*Prunus serotina*）、黑刺李（*Prunus spinosa*）、欧洲花楸（*Sorbus aria*）、（*Prunus aucuparia*）、（*Prunus matsumurana*）、栗属（*Castanea*）、山毛榉属（*Fagus*）、栎属（*Quercus*）、桦木属（*Betula*）、椴树属（*Tilia*）、榆属（*Ulmus*）、梨属（*Pyrus*）、鼠李属（*Rhamnus*）植物（Matov & Kononenko, 2012）。

注：首次给出中文种名。

***Trichosea champa* (Moore, 1879) 镶夜蛾**

Moma champa Moore, 1879, *Proceedings of the Zoological Society of London*, 1879: 403, pl. 33: 2. Type locality: India, Himachal Pradesh, Dharmsala.

Distribution. China (Heilongjiang, Jilin, Liaoning, Henan, Shaanxi, Hubei, Hunan, Fujian, Yunnan, Taiwan); Russia (Siberia), Korea, Japan, Nepal, India(North).

寄主：柃木（陈一心，1999）；大山樱（*Prunus sargentii*）（Park *et al.*, 2006）；钝齿水青冈（*Fagus crenata*）、柃木（*Eurya japonica*）、马醉木（*Pieris japonica*）、石南（*Rhododendron metternichii*）、南烛（*Vaccinium bracteatum*）、稠李（*Padus racemosa*）、白桃（*Persica vulgaris*）、扁桃（*Prunus communis*）、（*Prunus dulcis*）、窄叶火棘（*Pyracantha angustifolia*）、西洋梨（*Pyrus communis*）、川梨（*Pyrus pashia*）、小米空木（*Stephanandra incisa*）、蔷薇属（*Rosa*）、鹅耳枥属（*Carpinus*）、柳属（*Salix*）、椴树属（*Tilia*）、石楠属（*Photinia*）、槭树属（*Acer*）、葡萄属（*Vitis*）植物（Matov & Kononenko, 2012）。

Genus *Anacronicta* Warren, 1909 钝夜蛾属

Anacronicta Warren, 1909, *in* A. Seitz (ed.), *Die Gross-Schmetterlinge der Erde*, 3: 18. Type species: *Aplectoides caliginea* Butler, 1881[Japan].

Synonymy: *Anacronycta* Bryk, 1941, emend..

The genus includes 12 species, distributed in the Oriental and Sino-Pacific subregion of the Palaearctic regions; 1 species is in the Three Provinces of Northeast China.

***Anacronicta caliginea* (Butler, 1881) 暗钝夜蛾**

Aplectoides caliginea Butler, 1881, *Transactions of the Entomological Society of London*, 1881: 185. Type locality: Japan, Tokyo.

Synonymy: *Moma nitida* Graeser, 1888 [1889].

Distribution. China (Heilongjiang, Jilin, Liaoning, Shanxi, Shaanxi, Henan, Zhejiang, Hubei, Hunan, Jiangxi, Sichuan, Guizhou, Yunnan); Russia (Siberia), Korea, Japan.

寄主：柳兰（*Epilobium angustifolium*）、中国芒（*Miscanthus sinensis*）（Matov & Kononenko, 2012）。

Genus *Xanthomantis* Warren, 1909 角后夜蛾属

Xanthomantis Warren, 1909, *in* A. Seitz (ed.), *Die Gross-Schmetterlinge der Erde*, 3: 18. Type species: *Acronycta cornelia* Staudinger, 1888 [Russia, Primorye terr.].

Synonymy: *Trisuloides* auct. Butler, 1881.

The genus includes 2 species, both distributed in the Chinese subregion of the Palaearctic; 2 species are in the Three Provinces of Northeast China.

注：《中国动物志》（夜蛾科）中引用其异名"*Trisuloides* Butler, 1881"，在此予以变更。

***Xanthomantis cornelia* (Staudinger, 1888) 角后夜蛾**

Acronycta cornelia Staudinger, 1888, *Entomologische Zeitung. Entomologischen Vereine zu Stettin*, 49: 248. Type locality: Russia, Primorye terr., Askold I., "Suifun" [Razdolnaya river], Baranovsky.

Synonymy: *Diphthera honrathii* Graeser, 1888 [1889].

Distribution. China (Heilongjiang, Jilin, Liaoning, Nei Mongol); Russia (Siberia), Korea, Japan, North Thailand.

寄主：槲树（*Quercus dentata*）、蒙古栎（*Quercus mongolica*）（Matov & Kononenko, 2012）。

***Xanthomantis contaminata* (Draudt, 1937) 污角后夜蛾**

Trisuloides contaminata Draudt, 1937, *Entomologische Rundschau*, 54: 399, pl. 4: 2f. Type locality: China, Yunnan, Zhejiang [Chekiang], Shandong.

Synonymy: *Trisuloides tamsi* Park & Lee, 1977.

Distribution. China (Heilongjiang, Jilin, Liaoning, Nei Mongol, Shaanxi, Shandong, Zhejiang, Hunan, Yunnan); Russia (Siberia), Korea, Japan.

寄主：槲树（*Quercus dentata*）、蒙古栎（*Quercus mongolica*）（Matov & Kononenko, 2012）。

Genus *Colocasia* Ochsenheimer, 1816 标夜蛾属

Colocasia Ochsenheimer, 1816, *Die Schmetterlinge von Europa*, 4: 63. Type species: *Phalaena coryli* Linnaeus, 1758 [Europe].

Synonymy: *Leptostola* Billberg, 1820; *Demas* Stephens, 1829; *Phineca* Walker, 1856; *Colocasoides* Matsumura, 1931; *Calocasia* Hampson, 1913, missp..

The genus includes 5 species, one of them with Palaearctic distribution, two are Nearctic and two occur in the Chinese subregion of the Palaearctic; 2 species are in the Three Provinces of Northeast China.

Colocasia mus **(Oberthür, 1884) 鼠色标夜蛾**

Diloba mus Oberthür, 1884, *Études d'Entomologie*, 10: 17, pl. 2: 4. Type locality: Russia, Primorye terr., "Sidemi" [Bezverkhovo].

Synonymy: *Colocasia coryli ussuriensis* Kardakoff, 1928 (ab.); *Colocasia coryli grisescens* Kardakoff, 1928 (ab.).

Distribution. China (Heilongjiang, Jilin, Liaoning); Russia (Siberia), Korea, Japan.

寄主：桤木属、桦属植物（陈一心，1999）；蒙古栎（*Quercus mongolica*）、矮桤木（*Alnus fruticosa*）、辽东桤木（*Alnus hirsuta*）、白桦（*Betula platyphylla*）、毛榛（*Corylus mandshurica*）、槭树属（*Acer*）、鹅耳枥属（*Carpinus*）植物（Matov & Kononenko, 2012）。

Subfamily **Raphiinae** Beck, 1996 莽夜蛾亚科

Raphiinae Beck, 1996, *Neue entomologische Nachrichten* 36: 12. Type genus: *Raphia* Hübner, [1821].

Small subfamily containing 4 genera and about 50 species, mainly distributed in arid and subtropical areas of Old and New Worlds. The genus *Raphia* distributed in the temperate zone. Fibiger *et al.* (2009) synonymised Raphiinae with Dilobinae, however this conclusion was not supported by molecular study (Zahiri *et al.*, 2013). 1 species of 1 genus is in the Three Provinces of Northeast China.

References. Kozhantschikov, 1950 (Momini); Chen, 1982, 1999 (Noctuidae: Pantheinae); Kononenko *et al.*, 1998 (Pantheinae); Kononenko, 2003 (Pantheinae, Raphiini); Fibiger, Lafontaine, 2005 (Raphiinae); Kononenko, 2005 (Raphiinae), 2010 (Dilobinae); Lafontaine, Fibiger, 2006 (Raphiinae); Kononenko, Han, 2007 (Pantheinae); Matov *et al.*, 2008 (Raphiinae); Fibiger *et al.*, 2009, 2011 (Dilobinae); Zahiri *et al.*, 2013a (Raphiinae).

注：首次给出中文亚科名。

Genus ***Raphia*** Hübner, [1821] 莽夜蛾属

Raphia Hübner, [1821], *Verzeichniss bekannter Schmettlinge* [sic!]: 212. Type species: *Noctua hybris* Hübner, [1813] [Europe].

Synonymy: *Anodonta* Rambur, 1858, preocc. (Lamarck, 1799 [Mollusca]); *Certila* Walker, 1865; *Saligena* Walker, 1865; *Rhaphia* Agassiz, 1846, emend..

The genus includes 13 species, distributed in the Old and New Worlds, 1 species known from Africa; 1 species is in the Three Provinces of Northeast China.

Raphia peustera **Püngeler, 1906 波莽夜蛾**

Raphia peustera Püngeler, 1906, *Deutsche Entomologische Zeitschrift Iris*, 19: 216, pl. 8: 9. Type locality: China, Qinghai[Quinghai], Kuku-noor.

Synonymy: *Raphia peustera* auct., nec Püngeler, 1906; *Raphia illarioni* Filipjev, 1937.

Distribution. China (Heilongjiang, Jilin, Liaoning, Qinghai, Xizang); Russia (Siberia), Korea; Central Asia.

Subfamily **Dyopsinae** Zahiri, 2013 封夜蛾亚科

Dyopsinae Zahiri, 2013, *Zoologica Scripta, 42*: 493. Type genus: *Dyops* Guenée, 1852.

The Dyopsinae erected and treated in the subfamily status by Zahiri *et al.* (2013). The genera *Arcte*, *Cyclodes*, *Belciana* and *Anabelcia* (as *Belciana kala*) were transferred from Pantheinae to Dyopsinae based on molecular study (Zahiri *et al.*, 2013). We tentatively place the genera close to *Belciades* to Dyopsinae, while their systematic position requires claridication. Kononenko & Pinratana (2005) placed *Donda*, *Belciades* and *Belciana* in Acronictinae. These genera have been revised by Behounek & Kononenko (2012) and Behounek *et al.* (2011) as Pantheinae. The subfamily Dyopsinae (in present concept) contains 3 genera and 3 species are in the Three Provinces of Northeast China.

References. Chen, 1982, 1999 (Noctuidae: Catocalinae); Sugi, 1982 (Catocalinae); Kononenko, 1997a (Acronictinae); Kononenko *et al.*, 1998 (Catocalinae, part, Acronictinae, part); Kononenko, 2003 (Acronictinae, part); Sviridov, 2003 (Catocalinae, part); Kononenko, 2005, 2010 (Catocalinae, part, Acronictinae, part); Behounek, Han, Kononenko, 2011 (Pantheinae); Behounek, Kononenko, 2012 (Pantheinae); Holloway, 2009 (Pantheinae, part); Holloway, 2011 (Pantheinae, part); Zahiri *et al.*, 2013a (Dyopsinae).

注：首次给出中文亚科名。

Genus ***Arcte*** Kollar, [1844] 封夜蛾属

Arcte Kollar, [1844], *in* Hügel, *Kaschmir und des Reich der Sieh*, 4: 477. Type species: *Arcte polygrapha* Kollar, [1844] [North India].

Synonymy: *Cocytodes* Guenée, 1852; *Arcta*, missp..

Large sized with strong body species, having good migratory ability. The genus includes 5 species distributed mainly in tropical area of the Old World; 2 species are known from the Chinese subregion of the Palaearctic region; 1 species is in the Three Provinces of Northeast China.

Arcte coerula **(Guenée, 1852) 苎麻封夜蛾**

Cocytodes coerula Guenée, 1852, *in* Boisduval & Guenée, *Histoire Naturelle des Insectes. Species Général des Lépidoptères*, 7: 41, pl. 13: 10. Type locality: East India.

Synonymy: *Cocytodes coerulea* Hampson, 1893, emend.; *Cocytodes caerulea* Hampson, 1894, emend.;

Cocytodes coerulea auct., missp..

Distribution. China (Heilongjiang, Jilin, Liaoning, Hebei, Shandong, Zhejiang, Hubei, Hunan, Jiangxi, Fujian, Guangdong, Hainan, Sichuan, Yunnan, Taiwan); Russia (Siberia), Korea, Japan, India, Nepal; Southeast Asia.

寄主：毛叶水苎麻（*Boehmeria holosericea*）（Park *et al.*, 2006）；苎麻、黄麻、亚麻、大豆（陈一心，199）；黄麻（*Corchorus capsularis*）、构树（*Broussonetia papyrifera*）、大麻（*Cannabis sativa*）、葎草（*Humulus japonicus*）、大叶苎麻（*Boehmeria longispica*）、（*Boehmeria niponoivea*）、苎麻（*Boehmeria nivea*）、小赤麻（*Boehmeria spicata*）、瘤冠麻（*Cypholophus moluccanus*）、（*Debregeasia hypoleuca*）、异叶蝎子草（*Girardinia heterophylla*）、异株荨麻（*Urtica dioica*）、豆（*Glycine max*）、亚麻（*Linum usitatissimum*）、落尾木属（*Pipturus*）植物（Matov & Kononenko, 2012）。

Genus ***Belciades*** I. Kozhanchikov, 1950 冷靛夜蛾属

Belciades I. Kozhanchikov, 1950, *Fauna SSSR* (*Insecta, Lepid.*), *12*: 443. Type species: *Habrostola niveola* Motschulsky, 1866.

The genus includes 5 species, distributed in the Chinese subregion of the Palaearctic and partly the Oriental regions; 1 species is in the Three Provinces of Northeast China.

Belciades niveola **(Motschulsky, 1866) 冷靛夜蛾**

Habrostola niveola Motschulsky, 1866, *Bulletin de la Société Impériale des Naturalistes de Moscou*, 39 (1): 195. Type locality: Japan.

Synonymy: *Pandesma virens* Butler, 1881; *Moma japyx* Staudinger, 1892.

Distribution. China (Heilongjiang, Jilin, Liaoning, Hebei, Xizang); Russia (Siberia), Korea, Japan.

寄主：椴树属（*Tilia*）、倒吊笔属（*Wrightia*）植物（Matov & Kononenko, 2012）。

Genus ***Anabelcia*** Behounek & Kononenko, 2012 靛夜蛾属

Anabelcia Behounek & Kononenko, 2012, *Zootaxa*, 3590: 2. Type species: *Belciana kala* Prout, 1924 [Indonesia, Sumatra].

Synonymy: *Belciana* auct., nec Walker, 1862.

The genus has recently been revised by Behounek & Kononenko (2012). Predominantly tropical species distributed in the Southeast Asia. It includes 6 species; 1 species is known from the Three Provinces of Northeast China.

注：《中国动物志》（夜蛾科）中引用异名"*Belciana* Walker, 1862"，在此予以变更。

Anabelcia staudingeri **(Leech, 1900) 新靛夜蛾**

Polydesma staudingeri Leech, 1900. *Transactions of the Entomological Society of London*, 1900: 551. Type locality: North Korea, Wonsan [Gensan].

Synonymy: *Daseochaeta trinubila* Draudt, 1937.

Distribution. China (Heilongjiang, Jilin, Liaoning, Shanxi, Zhejiang, Hunan, Xizang); Russia (Siberia), Korea.

Subfamily **Balsinae** Grote, 1896 洼夜蛾亚科

Balsini Grote, 1896, *Abhandlungen des Naturwissenschaftlichen Vereins zu Bremen*, 14: 88. Type genus: *Balsa* Walker, 1860.

The subfamily includes only the holarctic genus *Balsa* with 3 Nearctic and 1 Palaearctic species; 1 species is in the Three Provinces of Northeast China.

References. Chen, 1982, 1999 (Noctuidae); Sugi, 1982 (Amphipyrinae, part); Kononenko, 1990 (Amphipyrinae, part); Kononenko *et al.*, 1998 (Amphipyrinae, part); Kononenko, 2003 (Amphipyrinae, part); Fibiger, Lafontaine, 2005 (Balsinae); Kononenko, 2005, 2010, 2016 (Balsinae); Lafontaine, Fibiger, 2006 (Balsinae); Kononenko, Han, 2007 (Balsinae); Matov *et al.*, 2008 (Balsinae).

注：首次给出中文亚科名。

Genus ***Balsa*** Walker, 1860 洼夜蛾属

Balsa Walker, 1860, *in* D'Urban, *Canadian Naturalist and Geologist*, 5: 250. Type species: *Balsa obliquifera* Walker, 1860 [Canada].

Synonymy: *Gargaza* Walker, 1866; *Nolaphana* Grote, 1873; *Asisyra* Grote, 1873.

The genus contains 3 Nearctic and 1 East Palaearctic species; 1 species is in the Three Provinces of Northeast China.

***Balsa leodura* (Staudinger, 1887) 洼夜蛾**

Nola leodura Staudinger, 1887, *in* N.M. Romanoff (ed.), *Mémoires sur les Lépidoptères*, 3: 187, pl. 10: 2. Type locality: Russia, Primorye terr., Vladivostok, Askold I..

Synonymy: *Balsa malana* auct., nec Fitch, 1856.

Distribution. China (Heilongjiang, Jilin, Liaoning, Shaanxi); Russia (Siberia), Korea, Japan.

寄主：苹果（陈一心，1999；Matov & Kononenko, 2012）。

Subfamily **Acronictinae** Heinemann, 1859 剑纹夜蛾亚科

Acronictidae Heinemann, 1859, *Die Schmetterlinge Deutschlands und der Schweiz*, 14: 297. Type genus: *Acronycta* Treitschke, 1825, emend. of *Acronicta* Ochsenheimer, 1816.

Synonymy: Apatelinae Grote, 1883; Mominae Hampson, 1902; Cymatophoropsiini Mell, 1943;

Craniophorini Beck, 1996.

The subfamily contains more than 400 species (Speidel *et al.*, 1996) in the World fauna. It is represented in all climatic zones, except high Arctic, but most diverse in the deciduous broad-leaved forests of the cool temperate zone. The species occur in all sorts of habitat, and feed on many plant families, the woody plants are preferred.

The Acronictinae is considered here in restricted sense. The genera *Donda*, *Belciades* and *Belciana* treated by Kononenko & Pinratana (2005) in Acronictinae are removed to the subfamily Dyopsinae; the placement of *Moma*, *Cymatophoropsis* and *Nacna* requires revision. The checklist of the Acronictinae of Chinese fauna published by Han & Kononenko (2010). The subfamily in the Three Provinces of Northeast China numbers 11 genera and 44 species.

References. Kozhantschikov, 1950 (Orgyidae: Acronictini); Chen, 1982, 1999 (Noctuidae: Acronictinae); Sugi, 1982 (Acronictinae); Kononenko, 1990 (Acronictinae); Kononenko, 2003 (Acronictinae); Fibiger, Lafontaine, 2005 (Acronictinae); Kononenko, 2005, 2010 (Acronictinae); Lafontaine, Fibiger, 2006 (Acronictinae); Kononenko, Han, 2007 (Acronictinae); Kononenko, Han, 2008 (Acronictinae); Matov *et al.*, 2008 (Acronictinae); Fibiger *et al.*, 2009 (Acronictinae); Han, Kononenko, 2010 (Acronictinae); Lafontaine, Schmidt, 2010 (Acronictinae).

Genus ***Cymatophoropsis*** Hampson, 1894 斑蕊夜蛾属

Cymatophoropsis Hampson, 1894, *The Fauna of British India including Ceylon and Burma. Moths*, 2: xvii, 397. Type species: *Gluphisia sinuata* Moore, 1879 [India].

Synonymy: *Trispila* Houlbert, 1921; *Thyatirides* Kozhantschikov, 1950.

The genus includes 6 species, distributed in the Oriental and Sino-Himalaian subregion of the Palaearctic regions; 2 species are in the Three Provinces of Northeast China.

Cymatophoropsis trimaculata **(Bremer, 1861) 三斑蕊夜蛾**

Thyatira trimaculata Bremer, 1861, *Bulletin de l'Academie Impériale des Sciences de St.-Petersbroug*, 3: 483. Type locality: Russia, Primorye terr..

Synonymy: *Thyatira trimaculata tripunctata* O. Bang-Haas, 1927 (ssp.); *Cymatophoropsis trimaculata tanakai* Inoue & Sugi, 1957 (ssp.).

Distribution. China (Heilongjiang, Jilin, Liaoning, Hebei, Shandong, Hunan, Fujian, Guangxi, Yunnan); Russia (Siberia), Korea, Japan.

寄主：鼠李（*Rhamnus davurica*）（Matov & Kononenko, 2012）。

Cymatophoropsis unca **(Houlbert, 1921) 大斑蕊夜蛾**

Trispila unca Houlbert, 1921, *in* Oberthür, *Études de Lépidoptérologie Comparée*, 18 (2): 237: 61. Type locality: West China, Ta-tsien-lou.

Distribution. China (Heilongjiang, Jilin, Liaoning, Zhejiang, Hubei, Jiangxi, Sichuan, Yunnan, Xizang); Russia (Siberia), Korea, Japan.

Genus ***Nacna*** Fletcher, 1961 孔雀夜蛾属

Nacna Fletcher, 1961, *Ruwenzori Expedition*, 1952: 198. Type species: *Canna pulchripicta* Walker, 1865 [India]. Repl. name for *Canna* Walker, 1865.

Synonymy: *Canna* Walker, 1865, preocc. (Gray, 1821 [Mammalia]).

The genus includes 6 species, distributed in the Oriental and the Chinese subregion of the Palaearctic regions; 1 species is in the Three Provinces of Northeast China.

Nacna malachitis **(Oberthür, 1881) 绿孔雀夜蛾**

Telesilla malachitis Oberthür, 1880, *Études d'Entomologie*, 5: 80, pl. 3: 9. Type locality: Russia, Primorye terr., Askold I..

Synonymy: *Canna splendens* Moore, 1881.

Distribution. China (Heilongjiang, Jilin, Liaoning, Shanxi, Henan, Fujian, Sichuan, Yunnan, Xizang, Taiwan); Russia (Siberia), Korea, Japan, India(North), Nepal, Vietnam.

寄主：紫椴（*Tilia amurensis*）、（*Rabdosia trichocarpa*）、野藿香（*Teucrium japonicum*）（Matov & Kononenko, 2012）。

Genus ***Euromoia*** Staudinger, 1892 东夜蛾属

Euromoia Staudinger, 1892, *in* N.M. Romanoff (ed.), *Mémoires sur les Lépidoptères*, 6: 632. Type species: *Euromoia mixta* Staudinger, 1892 [Russia, Primorye terr.].

Synonymy: *Euromoea*: Hampson, 1908, emend..

The genus includes 2 species, both distributed in the Chinese subregion of the Palaearctic and partly the Oriental regions; 2 species are in the Three Provinces of Northeast China.

Euromoia subpulchra **(Alphéraky, 1897) 后黄东夜蛾**

Hadena subpulchra Alphéraky, 1897, *in* N.M. Romanoff (ed.), *Mémoires sur les Lépidoptères*, 9: 173, pl. 12: 11. Type locality: North Korea.

Distribution. China (Heilongjiang, Jilin, Liaoning, Shaanxi, Hubei, Fujian, Taiwan); Russia (Siberia), Korea, Japan.

***Euromoia mixta* Staudinger, 1892 天目东夜蛾**

Euromoia mixta Staudinger, 1892, *in* N.M. Romanoff (ed.), *Mémoires sur les Lépidoptères*, 6: 632, pl. 12: 10. Type locality: Russia, "Sutschan-Gebiet" [Primorye terr., Partizansk].

Synonymy: *Aucha tienmushani* Draudt, 1950.

Distribution. China (Liaoning, Zhejiang, Hunan, Fujian, Sichuan); Russia (Siberia), Korea, Japan.

注：《中国动物志》（夜蛾科）中将其归入"*Aucha* Walker, [1858] 1857 灿夜蛾属"，在此予以变更。

Genus ***Subleuconycta*** Kozhantschikov, 1950 梦夜蛾属

Subleuconycta Kozhantschikov, 1950, *Fauna SSSR* (*Insecta*) *Lepidoptera*, 12: 447. Type species: *Leuconycta palshkovi* Filipjev, 1937 [Russia, Primorye terr.].

The genus includes 2 species, distributed in the Chinese subgregion of the Palaearctic and Oriental regions; 1 species is in the Three Provinces of Northeast China.

***Subleuconycta palshkovi* (Filipjev, 1937) 梦夜蛾**

Leuconycta palshkovi Filipjev, 1937, *Lambillionea*, 37: 64: 1, 2. Type locality: Russia, "Sutschan" [Primorye terr., Partizansk].

Distribution. China (Heilongjiang, Jilin, Liaoning, Hebei, Jiangsu, Zhejiang, Hunan, Jiangxi, Yunnan, Taiwan); Russia (Siberia), Korea, Japan.

Genus ***Moma*** Hübner, [1820] 1816 缤夜蛾属

Moma Hübner, [1820] 1816, *Verzeichniss bekannter Schmettlinge* [sic!]: 203. Type species: *Noctua aprilina* Linnaeus sensu Hübner, [1803] (=*Phalaena alpium* Osbeck, 1778) [Sweden].

Synonymy: *Diphthera* Hübner, [1806], suppr.; *Diphtera* Ochsenheimer, 1816, missp.; *Dipthera* Hübner, [1825], missp.; *Diphteramoma* Berio, 1961; *Diphthera* auct.; *Daseochaeta* auct..

The genus includes 5 species, distributed in the Sino-Himalayan subgregion of the Palaearctic, and the Oriental regions; 2 species are in the Three Provinces of Northeast China.

***Moma alpium* (Osbeck, 1778) 缤夜蛾**

Phalaena alpium Osbeck, 1778, *Göthborgska Wetenskaps och Witterhets Samhallet*, 1: 52, pl. 1: 2. Type locality: Germany.

Synonymy: *Noctua orion* Sepp, 1762; *Noctua aprilina* [Denis & Schiffermüller], 1775, nec Linnaeus; *Noctua orion* Esper, 1787; *Noctua runica* Gmelhaus, 1788; *Noctua aprilina* Linnaeus, *sensu* Hübner, 1803; *Moma orion murrhina* Graeser, 1888 (var.); *Diphtera designata* Turati, 1923.

Distribution. China (Heilongjiang, Jilin, Liaoning, Hubei, Jiangxi, Fujian, Sichuan, Yunnan); Russia (European part, Siberia), Korea, Japan; Asia Minor, Europe.

寄主：山毛榉、桦、栎属植物（陈一心，1999）；日本栗（*Castanea crenata*）、欧洲栗（*Castanea sativa*）、钝齿水青冈（*Fagus crenata*）、垂枝欧洲山毛榉（*Fagus sylvatica*）、麻栎（*Quercus acutissima*）、槲树（*Quercus dentata*）、大桴栎（*Quercus grosseserrata*）、蒙古栎（*Quercus mongolica*）、柔毛栎（*Quercus pubescens*）、夏栎（*Quercus robur*）、柴桦（*Betula fruticosa*）、匍生桦（*Betula humilis*）、垂枝桦（*Betula pendula*）、（*Betula pubescens*）、鹅耳枥属（*Carpinus*）、胡榛（*Corylus avellana*）、钻天杨（*Populus nigra*）、欧洲花楸（*Sorbus aucuparia*）、欧洲七叶树（*Aesculus hippocastanum*）、樱属（*Cerasus*）、苹果属（*Malus*）、李属（*Prunus*）、梨属（*Pyrus*）、梣属（*Fraxinus*）植物（Matov & Kononenko, 2012）。

Moma kolthoffi **(Bryk, 1948) 白缤夜蛾**

Diphthera alpium kolthoffi Bryk, 1948 (ssp.), *Arkiv för Zoologi*, 41 (A) 1: 52, pl. 2: 9. Type locality: China, Jiangsu[Kiangsu], Chinkiang Mts.

Synonymy: *Diphthera fulvicollis* De Lattin, 1949.

Distribution. China (Heilongjiang, Jilin, Liaoning, Jiangsu); Russia (Siberia), Korea, Japan.

寄主：鹅耳枥属（*Carpinus*）植物（Matov & Kononenko, 2012）。

Genus ***Gerbathodes*** Warren, 1911 **斋夜蛾属**

Gerbathodes Warren, 1911, *in* A.Seitz (ed.), *Die Gross-Schmetterlinge der Erde*, 3: 175. Type species: *Gerbatha angusta* Butler, 1879 [Japan].

Synonymy: *Acronictoides* Kozhantshikov, 1950.

The genus includes 2 species, both of which occur in the Chinese subregion of the Palaearctic region; 2 species are in the Three Provinces of Northeast China.

Gerbathodes paupera **(Staudinger, 1892) 散白斋夜蛾**

Craniophora paupera Staudinger, 1892 [July], *in* N.M. Romanoff (ed.), *Mémoires sur les Lépidoptères*, 6: 394. Type locality: Russia, “Raddefka” [Khabarovsk terr., Judish autonomy, Radde].

Synonymy: *Graphiphora lichenodes* Graeser, 1892; *Gerbatha connexa* Leech, 1900; *Lithacodia ruvida* Berio, 1977.

Distribution. China (Heilongjiang, Jilin, Liaoning, Taiwan); Russia (Siberia), Korea, Japan.

寄主：蒙古栎（*Quercus mongolica*）（Matov & Kononenko, 2012）。

注：首次给出中文种名。

***Gerbathodes angusta* (Butler, 1879) 斋夜蛾**

Gerbatha angusta Butler, 1879, *Illustrations of Typical Specimens of Lepidoptera Heterocera in the Collection of the British Museum*, 3: 245, pl. 47: 2. Type locality: Japan, Yokohama.

Distribution. China (Heilongjiang, Jilin, Liaoning, Jiangxi); Korea, Japan.

Genus *Acronicta* Ochsenheimer, 1816 剑纹夜蛾属

Acronicta Ochsenheimer, 1816 *Die Schmetterlinge von Europa*, 4: 62. Type species: *Phalaena leporina* Linnaeus, 1758 [Europe].

Synonymy: *Apatele* Hübner, [1806], suppr.; *Apatele* Hübner, [1808], suppr.; *Acronyctia* Meigen, 1813, missp.; *Semaphora* Guenée, 1814; *Apatelae* Ochsenheimer, 1816, unavail.; *Jocheaera* Hübner, [1820] 1816 (subgen.); *Pharetra* Hübner, [1820] 1816, preocc. (Bolten, 1798 [Brachiopoda]); *Hyboma* Hübner, [1829] 1816 (subgen.); *Apatele* Hübner, [1818], unavail.; *Triaena* Hübner, 1818 (subgen.); *Apatele* Hübner, 1822; *Acronycta* Treitschke, 1825, emend.; *Apatela* Stephens, 1829, missp.; *Cometa* Sodoffsky, 1837, repl. name.; *Sematophora* Agassiz, [1848], emend.; *Viminia* Chapman, 1890 (subgen.); *Cuspidia* Chapman, 1890; *Chamaepora* Warren, 1909; *Subacronycta* Kozhantschikov, 1950 (subgen.); *Molybdonycta* Sugi, 1979 (subgen.); *Hylonycta* Sugi, 1979 (subgen.); *Euviminia* Beck, 1966; *Aneuviminia* Beck, 1966; *Paraviminia* Beck, 1966.

The genus includes about 160 species, mainly distributed in boreal zone of the Holarctic, with 63 species are in the Palaearctic and 95 in the Nearctic. In the Palaearctic, most species are concentrated in the zone of broad-leaved forests of the East Siberia; few species are known from the Oriental and Ethiopian regions. 24 species are in the Three Provinces of Northeast China.

Subgenus *Acronicta* Ochsenheimer, 1816 剑纹夜蛾亚属

The subgenus is most diverse in the Chinese subregion; 2 species are in the Three Provinces of Northeast China.

***Acronicta* (*Acronicta*) *vulpina* (Grote, 1883) 拟剑纹夜蛾**

Apatela vulpina Grote, 1883, *Canadian entomologist*, 15: 8. Type locality: USA, New York.

Synonymy: *Apatela sancta* Edwards, 1888; *Acronycta vulpina leporella* Staudinger, 1888 (var.); *Acronycta leporina cineracea* Graeser, 1888 [1889] (var.); *Acronycta leporina* auct., nec Linnaeus, 1758.

Distribution. China (Jilin); Russia (Siberia), Mongolia, Korea, Japan; North America.

寄主：辽东桤木（*Alnus hirsuta*）、柴桦（*Betula fruticosa*）、（*Betula manshurica*）、杨属（*Populus*）、柳属（*Salix*）植物（Matov & Kononenko, 2012）。

Acronicta (*Acronicta*) *major* (Bremer, 1861) 桑剑纹夜蛾

Acronycta major Bremer, 1861, *Bulletin de l'Academie Impériale des Sciences de St.-Petersbroug*, 3: 484. Type locality: Russia, "Ussuri, Ema" [Primorye terr.].

Synonymy: *Triaena anaedina* Butler, 1881; *Acronicta major atritaigena* Dubatolov & Zolotarenko, 1995 (ssp.).

Distribution. China (Heilongjiang, Jilin, Liaoning, Shaanxi, Henan, Hubei, Hunan, Sichuan, Yunnan); Russia (Siberia), Korea, Japan.

寄主：桑、桃、梅、李、柑橘等（陈一心，1999）；胡桃（*Juglans regia*）、朝鲜柳（*Salix koreensis*）、桑（*Morus alba*）、黑桑（*Morus nigra*）、苹果（Malus pumila）、稠李（*Padus racemosa*）、李（*Prunus salicina*）、刺槐（*Robinia pseudoacacia*）、梣叶槭（*Acer negundo*）、印度七叶树（*Aesculus indica*）、温州蜜柑（*Citrus unshiu*）、香洋椿（*Cedrela serrata*）（Matov & Kononenko, 2012）。

Subgenus *Triaena* Hübner, 1818 齿剑纹夜蛾亚属

Triaena Hübner, 1818, *Zutrage Zuträge zur Sammlung Exotischer Schmetterlinge*, 1: 21. Type species: *Triaena psi* Linnaeus, 1758 [Europe].

The subgenus is most diverse in the Chinese subregion; 6 species are in the Three Provinces of Northeast China.

注：首次给出中文亚属名。

Acronicta (*Triaena*) *tridens* ([Denis & Schiffermüller], 1775) 三齿剑纹夜蛾

Noctua tridens [Denis & Schiffermüller], 1775, *Ankündung eines Systematisches Werkes von den Schmetterlinge der Wiener Gegend. Wien*: 67. Type locality: Austria, Vienna district.

Synonymy: *Acronycta kargalica* Moore, 1878; *Acronycta tridens radoti* Le Cerf, 1924 (ssp.); *Acronicta soltowensis* Schultz, 1930; *Acronicta tridens obscurior* deLattin, 1938 (ssp.).

Distribution. China (Heilongjiang, Jilin, Liaoning, Xinjiang, Gansu); Russia (European part, Siberia), North Korea, Japan, Kazakhstan; Near East, Europe.

寄主：桦、栎、山楂（陈一心，1999）；（*Malus domestica*）（Matov & Kononenko, 2012）。

Acronicta (*Triaena*) *cuspis* (Hübner, [1813]) 尖剑纹夜蛾

Noctua cuspis Hübner, [1813], *Sammlung Europaischer Schmetterlinge. Eulen*, 4: pl. 108: 504. Type locality: Europe.

Synonymy: *Acronycta cuspis belgica* Draudt, 1931 (f.).

Distribution. China (Heilongjiang, Xinjiang); Russia (European part, Siberia), Korea, Japan; Central and West Asia, Europe, Africa.

寄主：桤木属植物（陈一心，1999）；欧洲桤木（*Alnus glutinosa*）、辽东桤木（*Alnus hirsuta*）、

（*Alnus incana*）、甸生桦（*Betula humilis*）、垂枝桦（*Betula pendula*）、（*Betula pubescens*）、胡榛（*Corylus avellana*）、欧洲花楸（*Sorbus aucuparia*）（Matov & Kononenko, 2012）。

Acronicta (_Triaena_) _intermedia_ (Warren, 1909) 桃剑纹夜蛾

Acronycta intermedia Warren, 1909 [February], *in* A.Seitz (ed.), *Die Gross-Schmetterlinge der Erde*, 3: 14, pl. 2: k., repl. name for *Acronycta increta* Butler, 1878.

Synonymy: *Acronycta increta* Butler, 1878; *Acronycta incretata* Hampson, 1909; *Acronycta jezoensis* Matsumura, 1925; *Acronycta formosana* Matsumura, 1928.

Distribution. China (Heilongjiang, Jilin, Liaoning, Nei Mongol, Hebei, Fujian, Taiwan); Russia (Siberia), Korea, Japan, Vietnam, Thailand(North).

寄主：桃、梨、樱桃、梅、苹果、杏、李、柳（陈一心，1999）；大山樱（*Prunus sargentii*）（Park *et al.*, 2006）；朝鲜柳（*Salix koreensis*）、榔榆（*Ulmus parvifolia*）、樱属（*Cerasus*）、（*Malus domestica*）、苹果（*Malus pumila*）、杏（*Prunus armeniaca*）、欧洲李（*Prunus domestica*）、青肤樱（*Prunus serrulata*）、白桃（*Persica vulgaris*）、栎属（*Quercus*）、桤木属（*Alnus*）、桦木属（*Betula*）、梨属（*Pyrus*）植物（Matov & Kononenko, 2012）。

Acronicta (_Triaena_) _leucocuspis_ (Butler, 1878) 晃剑纹夜蛾

Acronycta leucocuspis Butler, 1878, *Annals and Magazine of Natural History*, (5)1: 78. Type locality: Japan, Yokohama.

Synonymy: *Acronycta sapporensis* Matsumura, 1926; *Acronycta obsuta* Draudt, 1933.

Distribution. China (Heilongjiang, Jilin, Hebei, Shandong, Yunnan); Russia (Siberia), Korea, Japan.

寄主：日本桤木（*Alnus japonica*）（Matov & Kononenko, 2012）。

Acronicta (_Triaena_) _sugii_ (Kinoshita, 1990) 日剑纹夜蛾

Triaena sugii Kinoshita, 1990, *Tyô Ga*, 41 (3): 189, 1. Type locality: Japan, Nagano.

Distribution. China (Heilongjiang, Jilin); Russia (Siberia), Korea, Japan.

注：首次给出中文种名。

Acronicta (_Triaena_) _psi_ (Linnaeus, 1758) 赛剑纹夜蛾

Phalaena psi Linnaeus, 1758, *Systema Naturae* (Edn. 10) 1: 514. Type locality: Europe.

Synonymy: *Acronycta tridens* Hübner, 1803, nec Denis & Schiffermüller, 1775; *Acronicta psi altaica* Staudinger, 1901 (var.); *Acronycta psi batnana* Draudt, 1931 (ssp.); *Acronycta psi iliensis* Draudt, 1931 (ssp.); *Acronicta psi crassistriga* de Lattin, 1938 (ssp.); *Acronycta psi solimana* Draudt, 1938 (ssp.); *Acronycta psi tehrana* Wiltshire, 1946 (ssp.); *Apatele perisi* Calle, 1974.

Distribution. China (Heilongjiang, Jilin, Liaoning, Xinjiang, Shanxi, Zhejiang, Hunan); Russia (European part, Siberia), Mongolia; Central and West Asia, Europe, North Africa.

寄主：桃、李、蔷薇等属植物（陈一心，1999）；欧洲蕨（*Pteridium aquilinum*）、夏栎（*Quercus robur*）、欧洲桤木（*Alnus glutinosa*）、（*Alnus incana*）、垂枝桦（*Betula pendula*）、（*Betula pubescens*）、胡榛（*Corylus avellana*）、黑果越橘（*Vaccinium myrtillus*）、新疆杨（*Populus pyramidalis*）、甜杨（*Populus suaveolens*）、欧洲山杨（*Populus tremula*）、耳柳（*Salix aurita*）、黄花柳（*Salix caprea*）、灰柳（*Salix cinerea*）、爆竹柳（*Salix fragilis*）、东陵山柳（*Salix phylicifolia*）、蒿柳（*Salix viminalis*）、欧洲小叶椴（*Tilia cordata*）、红茶藨子（*Ribes rubrum*）、卵叶唐棣（*Amelanchier ovalis*）、穗状唐棣（*Amelanchier spicata*）、欧洲甜樱桃（*Cerasus avium*）、草原樱桃（*Cerasus fruticosa*）、（*Cotoneaster insignis*）、（*Cotoneaster suavis*）、（*Crataegus coccinea*）、单子山楂（*Crataegus monogyna*）、锐刺山楂（*Crataegus oxyacantha*）、（*Crataegus pontica*）、（*Crataegus turkestanica*）、（Malus domestica）、（*Malus hissarica*）、新疆野苹果（*Malus sieversii*）、稠李（*Padus racemosa*）、欧洲李（*Prunus domestica*）、野生樱桃李（*Prunus divaricata*）、乌荆子李（*Prunus insititia*）、黑刺李（*Prunus spinosa*）、欧亚火棘（*Pyracantha coccinea*）、西洋梨（*Pyrus communis*）、野玫瑰（*Rosa canina*）、密刺蔷薇（*Rosa pimpinellifolia*）、玫瑰（*Rosa rugosa*）、（*RubuSouth Americanus*）、覆盆子（*Rubus idaeus*）、欧洲花楸（*Sorbus aucuparia*）、（*Sorbus hybrida*）、（*Sorbus intermedia*）、（*Sorbus torminalis*）、绣线菊（*Spiraea salicifolia*）、挪威枫（*Acer platanoides*）、火炬树（*Rhus typhina*）、羊角芹（*Aegopodium podagraria*）、栗属（*Castanea*）、山毛榉属（*Fagus*）、鹅耳枥属（*Carpinus*）、铁木属（*Ostrya*）、杏属（*Armeniaca*）、木瓜海棠属（*Chaenomeles*）、梣属（*Fraxinus*）植物（Matov & Kononenko, 2012）。

Subgenus *Jocheaera* Hübner, [1820] 1816 桦剑纹夜蛾亚属

Jocheaera Hübner, [1820] 1816, *Verzeichniss bekannter Schmettlinge* [sic!]: 201. Type species: *Phalaena alni* Linnaeus, 1767 [Europe].

The subgenus includes 1 species is in the Three Provinces of Northeast China.

注：首次给出中文亚属名。

Acronicta (*Jocheaera*) *alni* (Linnaeus, 1767) 桦剑纹夜蛾

Phalaena alni Linnaeus, 1767, *Systema Naturae* (Edn. 10), 1: 501. Type locality: Europe.

Synonymy: *Noctua degener* [Denis et Schiffermüller], 1775; *Acronycta alni intensiva* Draudt, 1937 (ssp.); *Acronycta alni korealni* Bryk, 1948 (ssp.); *Apatele alni italica* Berio, 1961 (ssp.).

Distribution. China (Heilongjiang, Jilin, Liaoning, Shaanxi, Henan); Russia (European part, Siberia), Korea, Japan; Near East, Europe.

寄主：桤木、桦、山毛榉、梾木等属植物（陈一心，1999）；钝齿水青冈（*Fagus crenata*）、槲树（*Quercus dentata*）、蒙古栎（*Quercus mongolica*）、桤木（*Alnus cremastogyne*）、欧洲桤木（*Alnus glutinosa*）、（*Alnus incana*）、桦（*Betula alba*）、岳桦（*Betula ermani*）、柴桦（*Betula fruticosa*）、垂枝桦（*Betula pendula*）、（*Betula pubescens*）、胡榛（*Corylus avellana*）、黑果越橘（*Vaccinium myrtillus*）、北美杨树（*Populus balsamifera*）、欧洲山杨（*Populus tremula*）、耳柳（*Salix aurita*）、黄花柳（*Salix caprea*）、东陵山柳（*Salix phylicifolia*）、欧洲小叶椴（*Tilia cordata*）、光榆（*Ulmus glabra*）、（Malus domestica）、（*Malus sibirica*）、稠李（*Padus racemosa*）、欧洲李（*Prunus domestica*）、（*Prunus nipponica*）、黑刺李（*Prunus spinosa*）、黑莓（*Rosa fruticosus*）、（*Rosa saxatilis*）、欧洲花楸（*Sorbus aucuparia*）、挪威枫（*Acer platanoides*）、欧亚槭（*Acer pseudoplatanus*）、欧洲红瑞木（*Cornus sanguinea*）、鹅耳枥属（*Carpinus*）、核桃属（*Juglans*）、杜鹃花属（*Rhododendron*）、茶藨子属（*Ribes*）、樱属（*Cerasus*）、山楂属（*Crataegus*）、草莓（*Fragaria ananassa*）、梨属（*Pyrus*）、毛核木属（*Symphoricarpos*）植物（Matov & Kononenko, 2012）。

Subgenus ***Hyboma*** Hübner, [1820] 1816 **果剑纹夜蛾亚属**

Hyboma Hübner, [1820]1816, *Verzeichniss bekannter Schmettlinge* [sic!]: 200. Type species: *Noctua strigosa* [Denis & Schiffermüller], 1775 [Europe].

The subgenus includes 3 species are in the Three Provinces of Northeast China.

注：首次给出中文亚属名。

***Acronicta* (*Hyboma*) *adaucta* (Warren, 1909) 光剑纹夜蛾**

Acronycta adaucta Warren, 1909, *in* A.Seitz (ed.), *Die Gross-Schmetterlinge der Erde*, 3: 16, pl. 3: e. Type locality: Japan.

Synonymy: *Acronycta phaedra* Hampson, 1910.

Distribution. China (Heilongjiang, Jilin, Liaoning, Jiangxi); Russia (Altai, Siberia), Korea, Japan.

寄主：垂丝海棠（*Malus haliana*）（Park *et al.*, 2006）；梅花（*Prunus mume*）、青肤樱（*Prunus serrulata*）（Matov & Kononenko, 2012）。

***Acronicta* (*Hyboma*) *strigosa* ([Denis & Schiffermüller], 1775) 果剑纹夜蛾**

Noctua strigosa [Denis & Schiffermüller], 1775, *Ankündung eines Systematisches Werkes von den Schmetterlinge der Wiener Gegend. Wien*: 88. Type locality: Austria, Vienna district.

Synonymy: *Phalaena* (*Noctua*) *favillacea* Esper, 1788; *Acronycta strigosa bryophiloides* Hormuzaki, 1891 (var.); *Acronicta terrigena* Graeser, 1892; *Acronicta strigose nigrescens* Barrington, 1896 (ssp.); *Acronycta strigose casparii* Staudinger, 1897 (ssp.); *Acronycta pulverosa sachalinensis* Matsumura, 1925 (ssp.).

Distribution. China (Heilongjiang, Jilin, Liaoning, Nei Mongol, Henan, Fujian, Sichuan, Yunnan); Russia (European part, Siberia), Korea, Japan; Europe.

寄主：山楂、桃、苹果（陈一心，1999）；（*Crataegus laevigata*）、单子山楂（*Crataegus monogyna*）、锐刺山楂（*Crataegus oxyacantha*）、辽宁山楂（*Crataegus sanguinea*）、毛山荆子（*Malus manshurica*）、苹果（*Malus pumila*）、稠李（*Padus racemosa*）、白桃（*Persica vulgaris*）、欧洲李（*Prunus domestica*）、黑刺李（*Prunus spinosa*）、（*Pyrus pyraster*）、欧洲花楸（*Sorbus aucuparia*）、欧鼠李（*Frangula alnus*）、桦木属（*Betula*）、樱属（*Cerasus*）、鼠李属（*Rhamnus*）植物（Matov & Kononenko, 2012）。

Acronicta (*Hyboma*) *jozana* (Matsumura, 1926) 拙剑纹夜蛾

Acronycta jozana Matsumura, 1926, *Insecta Matsumurana*, 1: 1, pl. 1: 5. Type locality: Japan, Hokkaido, Jozankei.

Synonymy: *Acronycta phaedriola* Draudt, 1931.

Distribution. China (Heilongjiang, Jilin, Liaoning, Hebei); Russia (Siberia), Korea, Japan.

寄主：（*Malus domestica*）（Matov & Kononenko, 2012）。

Subgenus *Molybdonycta* Sugi, 1979 小剑纹夜蛾亚属

Molybdonycta Sugi, 1979, *Tinea*, 10 (24): 249. Type species: *Acronycta omorii* Matsumura, 1926. Type locality: Japan, Hokkaido.

The subgenus includes 1 species is in the Three Provinces of Northeast China.

注：首次给出中文亚属名。

Acronicta (*Molybdonycta*) *bellula* (Alphéraky, 1895) 童剑纹夜蛾

Acronycta bellula Alphéraky, 1895, *Deutsche Entomologische Zeitschrift Iris*, 10: 189. Type locality: Russia, "Sidemi" [Primorye terr., Bezverkhovo].

Synonymy: *Acronycta chingana* Draudt, 1931; *Apatele cerasi* Howarth, 1951; *Acronicta omorii* auct., nec Matsumura, 1926.

Distribution. China (Heilongjiang, Jilin, Liaoning, Hebei); Russia (Siberia), Mongolia, Korea.

Subgenus *Plataplecta* Butler, 1878 霜剑纹夜蛾亚属

Plataplecta Butler, 1878, *Annals and Magazine of Natural History*, 1 (5): 195. Type species: *Polia soluta* Walker, 1865(=*Acvonycta pruinosa* Guenée, 1852). Type locality: India, North Hindostan.

The subgenus includes 8 species; one of them occurs in the Three Provinces of Northeast China.

注：首次给出中文亚属名。

Acronicta (*Plataplecta*) *pruinosa* (Guenée, 1852) 霜剑纹夜蛾

Acronycta pruinosa Guenée 1852, *In* Boisduval & Guenée, *Histoire Naturelle des Insectes. Species Général des Lépidoptéres*, 5: 53. Type locality: Indonesia, Java; Bangladesh, Silhet.

Synonymy: *Acronycta soluta* Walker, 1865; *Acronycta consanguis* Butler, 1878; *Acronycta asiatica* Poujade,

1888; *Acronycta crenulata* Bethune-Baker, 1906; *Acronycta albior* Hampson, 1909.

Distribution. China (Heilongjiang, Jilin, Liaoning, Jiangsu, Hubei, Xizang, Taiwan); Korea (Cheju Isl.), Japan, Sri Lanka, India, Bangladesh, Nepal, Vietnam, Myanmar, Malaysia, Philippines, Indonesia.

寄主：宽叶胡颓子（陈一心，1999）。

Subgenus ***Hylonycta*** Sugi, 1979 黑化剑纹夜蛾亚属

Hylonycta Sugi, 1979, *Tinea*, 10 (24): 249. Type species: *Acronycta carbonaria* Graeser, 1889 [1890] [Russia, Khabarovsk].

The subgenus includes 4 species: *A.*(*H.*) *carbonaria*, *A.*(*H.*) *subornata*, *A.*(*H.*) *catocaloida* and *A.*(*H.*) *hercules* in the Three Provinces of Northeast China.

注：首次给出中文亚属名。

***Acronicta* (*Hylonycta*) *subornata* (Leech, 1889) 白点剑纹夜蛾**

Acronycta subornata Leech, 1889, *Proceedings of the Zoological Society of London*, 1889: 477, pl. 1: 6. Type locality: Japan, Yokohama.

Distribution. China (Jilin, Liaoning); Korea, Japan.

***Acronicta* (*Hylonycta*) *carbonaria* (Graeser, [1890] 1889) 黑化剑纹夜蛾**

Acronycta carbonaria Graeser, [1890] 1889, *Berliner entomologische Zeitschrift*, 33, II: 252. Type locality: Russia, Khabarovsk.

Distribution. China (Heilongjiang, Jilin); Russia (Siberia), Korea, Japan.

寄主：蒙古栎（*Quercus mongolica*）（Park *et al.*, 2006）；麻栎（*Quercus acutissima*）、枹栎（*Quercus serrata*）（Matov & Kononenko, 2012）。

***Acronicta* (*Hylonycta*) *catocaloida* (Graeser, [1889] 1888) 白斑剑纹夜蛾**

Acronycta catocaloida Graeser, [1889] 1888, *Berliner entomologische Zeitschrift*, 32, II: 313. Type locality: Russia, Khabarovsk.

Distribution. China (Heilongjiang, Jilin, Liaoning, Hebei, Fujian, Yunnan); Russia (Siberia), Korea, Japan.

寄主：向日葵（陈一心，1999）；蒙古栎（*Quercus mongolica*）（Matov & Kononenko, 2012）。

***Acronicta* (*Hylonycta*) *hercules* (Felder & Rogenhofer, 1874) 榆剑纹夜蛾**

Acronycta hercules Felder & Rogenhofer, 1874, *Reise Österreichischen Fregatte Novara um die Erde in den Jahren 1857, 1859. Zoologischer Theil.*, 2 (Abt. 2), pl. 109: 2. Type locality: Japan.

Synonymy: *Acronycta elongata* Oberthür, 1884.

Distribution. China (Heilongjiang, Jilin, Liaoning, Hebei, Fujian, Yunnan, Taiwan); Russia (Siberia), Korea, Japan.

寄主：榆（陈一心，1999）；春榆（*Ulmus japonica*）、榔榆（*Ulmus parvifolia*）、榉属（*Zelkova*）植物（Matov & Kononenko, 2012）。

Subgenus *Viminia* Chapman, 1890 梨剑纹夜蛾亚属

Viminia Chapman, 1890[May], *Entomologist's Record and Journal of Variation*, 1: 26. Type species: *Phalaena rumicis* Linnaeus, 1758[Europe].

The subgenus includes 5 species are in the Three Provinces of Northeast China.

注：首次给出中文亚属名。

Acronicta (*Viminia*) *dahurica* Kononenko & Han, 2008 北剑纹夜蛾

Acronicta dahurica Kononenko & Han, 2008, *Zootaxa*, 1910: 47, figs. 5, 6, 13, 16. Type locality: Russia, Transbaikalia, Burgen.

Distribution. China (Heilongjiang, Nei Mongol); Russia (Transbaikalia, Southwest of the Primorye reg.).

注：首次给出中文种名。

Acronicta (*Viminia*) *rumicis* (Linnaeus, 1758) 梨剑纹夜蛾

Phalaena rumicis Linnaeus, 1758, *Systema Naturae* (Edn. 10), 1: 516. Type locality: Europe.

Synonymy: *Acronycta salicis* Curtis, 1826; *Acronycta diffusa* Walker, 1857; *Acronycta rumicis turanica* Staudinger, 1888 (var.); *Acronycta alsinoides* Goeze, 1903; *Acronicta rumicis oriens* Strand, 1916 (ssp.); *Acronycta rumicis pallida* Rothschild, 1920 (ssp.); *Acronycta meridionalis* Dannehl, 1925; *Acronycta rumicis rumicina* Bryk, 1948 (ssp.); *Apatele rumicis oriens* Inoue & Sugi, 1958 (ssp.).

Distribution. China (Heilongjiang, Jilin, Liaoning, Xinjiang, Jiangsu, Zhejiang, Hubei, Hunan, Fujian, Sichuan, Guizhou, Yunnan); Russia (Siberia), Mongolia, Korea, Japan, India; Central and West Asia, Europe, North Africa.

寄主：梨、桃、苹果、山楂、梅、柳（陈一心，1999）；（*Persicaria thunbergii*）、羊蹄（*Rumex japonica*）、北景天（*Sedum kamtschaticum*）、水蜡（*Lingustrum obtusifolium*）、毛黄连花（*Lysimachia vulgaris*）、桃树（*Prunus persica*）、萱草（*Hemerocallis fulva*）（Park *et al.*, 2006）；（*Malus domestica*）、西洋梨（*Pyrus communis*）、欧洲李（*Prunus domestica*）、覆盆子（*Rubus idaeus*）、草莓（*Fragaria ananassa*）、（*Brassica oleracea*）、酸模属（*Rumex*）植物（Matov & Kononenko, 2012）。

Acronicta (*Viminia*) *lutea* (Bremer & Grey, 1852) 黄剑纹夜蛾

Acronycta lutea Bremer & Grey, 1852, *in* Motschulsky, *Études d'Entomologie*, 1: 65. Type locality: China, Beijing.

Synonymy: *Pharetra leucoptera* Butler, 1881; *Acronycta suigensis* Matsumura, 1926; *Acronycta metaxantha funesta* Draudt, 1950 (ssp.).

Distribution. China (Heilongjiang, Jilin, Liaoning, Beijing, Hebei, Hubei); Russia (Siberia), Mongolia, Korea, Japan.

寄主：低矮草本植物（陈一心，1999）；萹蓄 (*Polygonum aviculare*)、匍生桦（*Betula humilis*）、（*Betula pubescens*）、（*Carpinus orientalis*）、香杨梅（*Myrica gale*）、钻天杨（*Populus nigra*）、欧洲山杨（*Populus tremula*）、黄花柳（*Salix caprea*）、爆竹柳（*Salix fragilis*）、酸模属（*Rumex*）、羽衣草属（*Alchemilla*）、山楂属（*Crataegus*）、大豆属（*Glycine*）植物（Matov & Kononenko, 2012）。

***Acronicta* (*Viminia*) *digna* (Butler, 1881) 威剑纹夜蛾**

Thalpophila digna Butler, 1881, *Transactions of the Entomological Society of London*, 1881: 176. Type locality: Japan, Yokohama.

Synonymy: *Acronycta michael* Oberthür, 1884; *Acronycta metaxantha* Hampson, 1909; *Sidemia hoenei* Matsumura, 1926; *Acronycta agnata* Draudt, 1937.

Distribution. China (Heilongjiang, Jilin, Liaoning, Henan, Hubei, Hunan, Taiwan); Russia (Siberia), Korea, Japan.

***Acronicta* (*Viminia*) *raphael* (Oberthür, 1884) 苏剑纹夜蛾**

Acronycta raphael Oberthür, 1884, *Études d'Entomologie*, 10: 19, pl. 2: 9. Type locality: Russia, "Sidemi" [Primorye terr., Bezverkhovo].

Synonymy: *Acronycta fixseni* Graeser, 1888 [1889]; *Acronycta raphaelis* Hampson, 1908, emend.; *Acronicta cubitata* Warren, 1914.

Distribution. China (Heilongjiang, Jilin); Russia (Siberia), Korea, Japan.

寄主：大豆（*Glycine max*）（Matov & Kononenko, 2012）。

Subgenus ***Subacronicta*** Kozhanchikov, 1950 **首剑纹夜蛾亚属**

Subacronicta Kozhanchikov, 1950, *Fauna SSSR* (*Insecta*) *Lepidoptera*, 12: 456. Type species: *Noctua megacephala* [Denis & Schifermüller], 1775 [Austria].

The subgenus includes 1 species is in the Three Provinces of Northeast China.

注：首次给出中文亚属名。

***Acronicta* (*Subacronicta*) *concerpta* (Draudt, 1937) 首剑纹夜蛾**

Acronycta concrepta Draudt, 1937, *Entomologische Rundschau*, 30: 398, pl. 4: 2e. Type locality: China, Shaanxi[Shensi], Qinling[Tsinling].

Synonymy: *Acronicta megacephala* auct., nec Denis et Schiffermüller, 1775.

Distribution. China (Heilongjiang, Jilin, Liaoning, Shaanxi, Xinjiang); Russia (Siberia), Japan.

寄主：杨、柳（陈一心，1999；Matov & Kononenko, 2012）。

注：《中国动物志》（夜蛾科）中引用本种同物异名"*Acronicta megacephala* [Denis et Schiffermüller], 1775"，在此予以变更。

Genus *Simyra* Ochsenheimer, 1816 刀夜蛾属

Simyra Ochsenheimer, 1816, *Die Schmetterlinge von Europa*, 4: 81. Type species: *Noctua nervosa* [Denis & Schiffermüller], 1775 [Austria].

Synonymy: *Cnephata* Billberg, 1820, repl. name; *Symira* Hübner, [1822] 1821, missp.; *Asema* Sodoffsky, 1837, repl. name; *Nimya* Guenée, 1841, missp.; *Arsilonche* Lederer, 1857; *Ablepharon* Grote, 1873; *Siymra* Warren, 1912, missp.; *Ommatostolidea* Benjamin, 1933; *Parasimyra* Beck, 1996.

The genus includes 12 species, distributed in the Palaearctic, Nearctic and partly Oriental regions; 2 species are in the Three Provinces of Northeast China.

Simyra splendida Staudinger, 1888 亮刀夜蛾

Simyra splendida Staudinger, 1888, *Entomologische Zeitung. Entomologischen Vereine zu Stettin*, 49: 245. Type locality: Russia, Vladivostok.

Synonymy: *Simyra niveonitens* Graeser 1888 [1889]; *Simyra splendida albisssima* Bryk, 1941 (ssp.).

Distribution. China (Heilongjiang, Jilin); Russia (Siberia), Korea, Mongolia, Kazakhstan.

寄主：北艾（*Artemisia vulgaris*）、丝路蓟（*Cirsium arvensis*）、蒲公英属（*Taraxacum*）植物（Matov & Kononenko, 2012）。

Simyra saepestriata (Alphéraky, 1895) 仿辉刀夜蛾

Simyra albovenosa saepestriata Alphéraky, 1895, *Deutsche Entomologische Zeitschrift Iris*, 8: 188. Type locality: Mongolia, Ourga [Ulan Baator].

Synonymy: *Simyra albovenosa* auct., nec Goeze, 1781; *Simyra saepistriata* Hampson, 1909, missp..

Distribution. China (Jilin); Russia (Altai, Transbaikalia), Mongolia, Japan; Central Asia.

注：首次给出中文种名。

Genus *Craniophora* Snellen, 1867 首夜蛾属

Craniophora Snellen, 1867, *Vlinders Nederland*, 1867: 262. Type species: *Noctua ligustri* [Denis & Schiffermüller], 1775 [Austria].

Synonymy: *Bisulcia* Chapman, 1890; *Miracopa* Draudt, 1950; *Acronycta* auct..

The genus includes 19 species, distributed in Palaearctic, Oriental, Australian and Ethiopian regions; 4

species are in the Three Provinces of Northeast China.

Craniophora ligustri ([Denis & Schiffermüller], 1775) 女贞首夜蛾

Noctua ligustri [Denis & Schiffermüller], 1775, *Ankündung eines Systematisches Werkes von den Schmetterlinge der Wiener Gegend. Wien*: 70. Type locality: Austria, Vienna district.

Synonymy: *Phalaena litterata* Panzer, 1804; *Phalaena* (*Noctua*) *coronula* Haworth, 1809; *Craniophora ligustri effusior* Dannehl, 1925 (var.); *Craniophora ligustri gigantea* Draudt, 1937 (ssp.); *Craniophora carbolucana* Hartig, 1968.

Distribution. China (Heilongjiang, Jilin, Liaoning, Hebei, Henan); Russia (South European part, Southeast Sebria), Korea, Japan; Near East, Europe.

寄主：女贞、梣木、桤木等属植物（陈一心，1999）；欧洲桤木（*Alnus glutinosa*）、胡榛（*Corylus avellana*）、欧洲白蜡（*Fraxinus excelsior*）、女贞（*Ligustrum lucidum*）、普通女贞（*Ligustrum vulgare*）、桦木属（*Betula*）、槭树属（*Acer*）、沙棘属（*Hippophae*）、荚蒾属（*Viburnum*）、素馨属（*Jasminum*）、丁香属（*Syringa*）植物（Matov & Kononenko, 2012）。

Craniophora praeclara (Graeser, 1890) 亮首夜蛾

Acronycta praeclara Graeser, 1890, *Berliner entomologische Zeitschrift*, 35, I: 74. Type locality: Russia, "Sidemi" [Primorye terr., Bezverkhovo], "Raddefka" [Khabarovsk terr., Judish Autonomy, Radde].

Distribution. China (Heilongjiang, Jilin); Russia (Siberia), Korea, Japan.

寄主：水曲柳（*Fraxinus mandshurica*）、花曲柳（*Fraxinus rhynchophilla*）、暴马丁香（*Syringa amurensis*）（Matov & Kononenko, 2012）。

Craniophora pacifica Filipjev, 1927 寒首夜蛾

Craniophora pacifica Filipjev, 1927 *Annuaire du Musée Zoologique de l'Académie des Sciences d'URSS*, 28: 231, pl. 12: 8, 9. Type locality: Russia, "Sutschan" [Primorye terr., Partizansk].

Synonymy: *Craniophora pacifica kalgana* Draudt, 1931 (f); *Craniophora niveosparsa*: Kozhantshikov, 1950: 541, nec Matsumura, 1926, misident.

Distribution. China (Heilongjiang, Jilin, Liaoning); Russia (Siberia), Korea, Japan.

寄主：水蜡（*Ligustrum obtusifolium*）、暴马丁香（*Syringa amurensis*）、梣属（*Fraxinus*）植物（Matov & Kononenko, 2012）。

Craniophora fasciata (Moore, 1887) 条首夜蛾

Hyboma fasciata Moore, 1887, *The Lepidoptera of Ceylon*, 3: 5, pl. 144: 4. Type locality: Sri Lanka.

Synonymy: *Hyboma divisa* Moore, 1888; *Acronycta nigrostriata* Pagenstecher, 1888.

Distribution. China (Heilongjiang, Jilin, Liaoning, Hubei, Guangdong, Hainan, Yunnan, Taiwan); Korea, Japan, Sri Lanka, India(North), Nepal, Pakistan, Vietnam, Myanma, Indonesia, Oman(North).

Genus ***Cranionycta*** Lattin, 1949 仿首夜蛾属

Cranionycta de Lattin, 1949, *Zeitschrift der Wiener Entomologischen Gesellschaft*, 34: 108. Type species: *Cranionycta oda* Lattin, 1949 [Russia, Primorye terr.].

The genus includes 3 species, distributed in the Chinese subregion of the Palaearctic and partly Oriental regions; 3 species are in the Three Provinces of Northeast China.

注：首次给出中文属名。

Cranionycta jankowskii **(Oberthür, 1880) 俄仿首夜蛾**

Apatela jankowskii Oberthür, 1880, *Études d'Entomologie*, 5: 69, pl. 7: 1. Type locality: Russia, Primorye terr., Askold I..

Synonymy: *Acronycta jancousci*: Hampson, 1909, emend..

Distribution. China (Heilongjiang, Jilin, Liaoning); Russia (Siberia), Korea, Japan.

注：首次给出中文种名。

Cranionycta albonigra **(Herz, 1904) 白黑仿首夜蛾**

Acronycta albonigra Herz, 1904, *Annuaire du Musée Zoologique de l'Académie Imperiale des Sciences de St.-Petersbourg*, 9: 269, pl. 1: 3. Type locality: North Korea.

Distribution. China (Heilongjiang, Jilin, Hebei, Shanxi, Hubei, Sichuan); Russia (Siberia), Korea.

注：国内多用“白黑首夜蛾”予以记录，根据属级单元变动更名为“白黑仿首夜蛾”。

Cranionycta oda **Lattin, 1949 仿首夜蛾**

Cranionycta oda Lattin, 1949, *Zeitschrift der Wiener Entomologischen Gesellschaft*, 34: 108: 3. Type locality: Russia, “Sutschanskii rudnik” [Primorye terr., Partizansk].

Synonymy: *Cranionycta transversa* Kozhantshikov, 1950; *Cranionycta inquieta* Draudt, 1950.

Distribution. China (Heilongjiang, Jilin, Jiangxi, Taiwan); Russia (Siberia), Korea, Japan, Nepal, Thailand.

注：国内多用“瓯首夜蛾”予以记录，根据属级单元变动更名为“仿首夜蛾”。

Genus ***Narcotica*** Sugi, 1982 缀白剑纹夜蛾属

Narcotica Sugi, 1982, *in* Inoue *et al.*, *Moths of Japan*, 1982: 682. Type species: *Acronycta niveosparsa* Matsumura, 1926 [Japan, Honshu].

Monotypic Chinese genus; 1 species is in the Three Provinces of Northeast China.

注：首次给出中文属名。

Narcotica niveosparsa **(Matsumura, 1926) 缀白剑纹夜蛾**

Acronycta niveosparsa Matsumura, 1926, *Insecta matsumana*, 1: 4, pl. 1: 8. Type locality: Japan, Honshu, Kyoto [EIHU, Sapporo].

Synonymy: *Acronycta niveosparsa*: Draudt, 1931; *Craniophora pacifica*: Kozhantschikov, 1950: 541 (incorrect synonymy); *Apatele niveosparsa*: Ogata, 1958; *Narcotica niveosparsa*: Sugi, 1982.

Distribution. China (Heilongjiang, Jilin, Liaoning, Jiangsu); Korea, Japan.

注：《中国动物志》（夜蛾科）中将本种归入"*Acronicta* Ochsenheimer, 1816 剑纹夜蛾属"中，在此予以变更。

Subfamily **Metoponiinae** Herrich-Schaffer, [1851] 美托夜蛾亚科

Metoponidae Herrich- Schaffer, [1851], *Systematische Bearbeitung der Schmetterlinge von Europa*, 2: 386. Type genus: *Metoponia* Duponchel, [1845], a synonym of *Aegle* Hübner, [1823].

Synonymy: Tytina Beck, 1999, a nomen nudum in Beck, 1996.

Metoponiinae genera were segregated from Acontiinae and associated with the New World subfamily Stirinae by Matthews (1991), then Poole (1996) with doubt included the Old World genera in Stiriinae. Fibiger & Lafontaine (2005) re-established the subfamily Metoponiinae with 10 European genera. This Subfamily includes about 100 species distributed in the Old World, mainly in the West Palaearctic and Ethiopian regions. 1 species of 1 genus is in the Three Provinces of Northeast China.

References. Chen, 1982, 1999 (Noctuidae: Amphipyrinae); Sugi, 1982 (Noctuidae, Amphipyrinae); Kononenko *et al.*, 1998 (Amphipyrinae); Kononenko, 2003, 2005, 2010 (Amphipyrinae); Fibiger, Lafontaine, 2005 (Noctuidae: Metoponiinae); Lafontaine, Fibiger, 2006 (Noctuidae: Metoponiinae); Kononenko, Han, 2007 (Amphipyrinae); Matov *et al.*, 2008 (Amphipyrinae).

注：首次给出中文亚科名。

Genus ***Mycteroplus*** Herrich–Schäffer, [1850] 鳄夜蛾属

Mycteroplus Herrich-Schäffer, [1850], *Systematische Bearbeitung der Schmetterlinge von Europa*, 2: 191, 242. Type species: *Xanthia puniceago* Boisduval, 1840 [South Russia].

Synonymy: *Stephania* Guenée, 1852; *Usbeca* Püngeler, 1914; *Acrosphalia* Rebel, 1918.

The genus contains 4 species, distributed in the West and East Palaearctic region; 1 species is in the Three Provinces of Northeast China.

Mycteroplus cornuta **(Püngeler, 1914)** 西鳄夜蛾

Usbeca cornuti Püngeler, 1914, *Deutsche Entomologische Zeitschrift Iris*, 28: 46, pl. 2: 15. Type locality:

Kazakhstan, Syr Darja, Baigacum.

Synonymy: *Usbeca kulmburgi* Rebel, 1918.

Distribution. China (Heilongjiang, Jilin, Liaoning); Russia (Siberia), Mongolia, Korea, Japan, Kazakhstan, Uzbekistan.

注：首次给出中文种名。

Subfamily **Sinocharinae** Speidel, Fänger & Naumann, 1996 瑕夜蛾亚科

Sinocharinae Speidel, Fänger & Naumann, 1996, *Systematic Entomology*, 21: 247. Type genus: *Sinocharis* Püngeler, 1912.

Speidel *et al.* (1996) placed Sinocharinae among the higher noctuids and not with the Bagisarinae, on the basis of the presence of basal abdominal coremata and structure of tympanal organ in *Sinocharis*. The subfamily contains the monotypic genus *Sinocharis*, distributed in Chinese subregion of Palaearctic region. 1 species is in the Three Provinces of Northeast China.

References. Chen, 1982, 1999 (Noctuidae); Sugi, 1982 (Noctuidae, Acontiinae); Speidel *et al.*, 1996 (Sinocharinae); Kononenko *et al.*, 1998 (Acontiinae); Kononenko, 2003 (Acontiinae); Kononenko, 2005, 2010 (Sinocharinae); Kononenko, Han, 2007 (Sinocharinae); Matov *et al.*, 2008 (Sinocharinae).

注：首次给出中文亚科名。

Genus ***Sinocharis*** Püngeler, 1912 瑕夜蛾属

Sinocharis Püngeler, 1912, *in* Korb, *Mitteilungen der Münchner Entomologischen Gesellschaft*, 3: 7. Type species: *Sinocharis korbae* Püngeler, 1912 [Russia, Khabarovsk].

Synonymy: *Noshimea* Matsumura, 1931.

Monotypic Sino-Pacific genus. 1 species occurs in the Three Provinces of Northeast China.

Sinocharis korbae **Püngeler, 1912 瑕夜蛾**

Sinocharis korbae Püngeler, 1912, *Mitteilungen der Münchner Entomologischen Gesellschaft*, 3: 4. Type locality: Russia, Khabarovsk, Kazakevichi.

Synonymy: *Noshimea fulgularis* Matsumura, 1931.

Distribution. China (Heilongjiang, Jilin, Liaoning); Russia (Siberia), Korea, Japan.

寄主：大丽花（*Dahlia pinnata*）（陈一心，1999；Matov & Kononenko, 2012）。

Subfamily **Cuculliinae** Herrich-Schäffer, [1850] 1845 冬夜蛾亚科

Cucullidae Herrich-Schäffer, [1850] 1845, *Systematische Bearbeitung der Schmetterlinge von Europa*, 2: 306, original missp.. Type genus: *Cucullia* Schrank, 1802.

Synonymy: Cucullidi Stephens, 1850.

The Cuculliinae is well defined Holarctic subfamily easily recognised by its wing-shape and by the male and female genitalia. The subfamily exceeds more than 200 members, it is distributed worldwide with most representatives in the open landscapes of cool temperate zones. Eurasian Cuculliinae has recently revised by Ronkay & Ronkay (2009) and Ronkay *et al.* (2011).

The subfamily is represented in the Three Provinces of North China by 1 genus and 25 species.

References. Hampson, 1906 (Cuculliinae, part); Boursin, 1941; Chen, 1982, 1999 (Noctuidae); Sugi, 1982 (Cuculliinae, part); Ronkay & Ronkay, 1994, 1996 (Cuculliinae); Poole, 1995 (Cuculliinae); Kononenko *et al.*, 1998 (Cuculliinae, part); Kononenko, 2003 (Cuculliinae, part); Fibiger & Lafontaine, 2005 (Cuculliinae); Kononenko, 2005, 2010 (Cuculliinae); Lafontaine & Fibiger, 2006 (Cuculliinae); Han, 2007 (Cuculliinae); Matov *et al.*, 2008 (Cuculliinae); Ronkay & Ronkay, 2009 (Cuculliinae); Ronkay *et al.*, 2011 (Cuculliinae); Kononenko, 2016 (Cuculliinae).

Genus ***Cucullia*** Schrank, 1802 冬夜蛾属

Cucullia Schrank, 1802, *Fauna Boica. Durchgedachte Geschichte der in Baiera Einheimischen und Zahmen Thiere*, 2 (2): 157. Type species: *Phalaena umbratica* Linnaeus, 1758. Type locality: [not stated].

Synonymy: *Tribonophora* Hübner [1806]; *Tribonophorae* Ochsenheimer, 1816; *Callainia* Hübner [1826] 1816; *Argyritis* Hübner, [1821]; *Callaenia* Hübner [1821]; *Empusa* Hübner [1821]; *Eucalimia* Hübner [1821]; *Euderaea* Hübner, [1821]; *Tribunophora* Hübner, 1822; *Lophia* Sodoffsky, 1837; *Lathosea* Grote, 1881; *Nycterophaeta* Smith, 1882; *Rancora* Smith, 1892; *Copicucullia* Smith, 1894; *Dasysternum* Staudinger, 1895; *Cheligalea* Hampson, 1906; *Argyrogalea* Hampson, 1906; *Argyromata* Hampson, 1906; *Empusada* Hampson, 1906; *Pseudocopicucullia* Dumont, 1928; *Metlaouia* Dumont, 1928; *Pseudonycterophaeta* Berio, 1934.

The genus contains about 220 species, mainly distributed in the arid and semiarid areas of the Holarctic, and Ethiopian regions. Of these 110 species are known from the Palaearctic region and 25 species are known from the Three Provinces of Northeast China.

Cucullia argentina **(Fabricius, 1787) 银冬夜蛾**

Noctua argentina Fabricius, 1787, *Mantissa Insectorum sistens Species nupes Detectas, Synonymis, Observationibus Emend.ibus*, 2: 162. Type locality: South Russia.

Synonymy: *Cucullia argentina grisescens* Wagner, 1931 (ssp.), precc. Leech, 1900; *Cucullia argentina anatoliensis* Kocak, 1980 (ssp.).

Distribution. China (Jilin, Xinjiang); Russia (Southeast European part, Siberia), Mongolia, Central Asia, Middle East, Central and Southeast Europe.

寄主：蒿属植物（陈一心，1999）；中亚苦蒿（*Artemisia absinthium*）、沙蒿（*Artemisia arenaria*）、荒野蒿（*Artemisia campestris*）、猪毛蒿（*Artemisia scoparia*）（Matov & Kononenko, 2012）。

Cucullia splendida (Stoll, 1782) 银装冬夜蛾

Phalaena splendida Stoll, 1782, *in* Cramer, *De Uitlandsche Kapellen Voorkomende in Drie Waereld-Deelen Asia*, *Africae North America*, 4: 242, pl. 400: F. Type locality: South Russia.

Synonymy: *Cucullia argyrea* Freyer, 1840; *Cucullia argyrea* Boisduval, 1840, nec Freyer, 1840.

Distribution. China (Jilin, Liaoning, Nei Mongol, Gansu, Qinghai, Xizang); Russia (Southeast European part, Altai, Siberia, Transbaikalia), Mongolia; Central Asia, Central Europe.

寄主：北艾（*Artemisia vulgaris*）、菊苣（*Cichorium inthybus*）、向日葵（*Helianthus annuus*）、褐苞三肋果（*Matricaria ambigua*）、匹菊属（*Pyrethrum*）、蒲公英属（*Taraxacum*）植物（Matov & Kononenko, 2012）。

Cucullia scopariae Dorfmeister, 1853 内冬夜蛾

Cucullia scopariae Dorfmeister, 1853, *Verhandlungen des Zoologisch-botanischen Vereins in Wien*, 3: 413. Type locality: Austria, Vienna.

Distribution. China (Heilongjiang, Jilin); Russia (Southeast European part, Siberia), Mongolia, Korea, Japan, Iran; Central Asia, Central and South Europe.

寄主：扫帚艾（陈一心，1999）；荒野蒿（*Artemisia campestris*）、猪毛蒿（*Artemisia scoparia*）（Matov & Kononenko, 2012）。

Cucullia scoparioides Boursin, 1941 变带冬夜蛾

Cucullia scoparioides Boursin, 1941, *Deutsche Entomologische Zeitschrift Iris*, 55: 50, pl. 7: 1, 2. Type locality: China, Shansi[=Shanxi], Mei-shan[Meishan].

Synonymy: *Cucullia cyanorea* Boursin, 1941.

Distribution. China (Heilongjiang, Jilin, Shanxi); Russia (Siberia).

注：首次给出中文种名。

Cucullia fuchsiana Eversmann, 1842 富冬夜蛾

Cucullia fuchsiana Eversmann, 1842, *Bulletin de la Société Impériale des Naturalistes de Moscou*, 3: 552, pl. 5: 6. Type locality: Russia, South Ural, Orenburg.

Synonymy: *Cucullia fuchsii* Freyer, 1842; *Cucullia scopariae* auct..

Distribution. China (Heilongjiang, Jilin, Nei Mongol, Xinjiang, Qinghai, Hebei); Russia (European part, Siberia), Mongolia, Korea, Kazakhstan.

寄主：猪毛蒿（*Artemisia scoparia*）、魁蒿（*Artemisia princeps*）、旋覆花属（*Inula*）、苦苣菜属（*Sonchus*）植物（Matov & Kononenko, 2012）。

Cucullia fraudatrix Eversmann, 1837 蒿冬夜蛾

Cucullia fraudatrix Eversmann, 1837, *Bulletin de la Société Impériale des Naturalistes de Moscou*, 1837 (6): 43. Type locality: South Russia, Orenburg.

Synonymy: *Cucullia pontica* Boisduval, 1840, *Cucullia pyrethri* Herrich-Schäffer, 1845; *Cucullia fraudatrix nigrescens* Haeger, 1962 (f.).

Distribution. China (Jilin, Liaoning, Zhejiang); Russia (European part, Siberia), Mongolia, Korea, Japan, Kazakhstan; Europe.

寄主：莴苣（陈一心，1999）；中亚苦蒿（*Artemisia absinthium*）、荒野蒿（*Artemisia campestris*）、野艾蒿（*Artemisia vulgaris*）、蓟属（*Cirsium*）、母菊属（*Matricaria*）、蒲公英属（*Taraxacum*）植物（Matov & Kononenko, 2012）。

Cucullia argentea (Hufnagel, 1766) 碧银冬夜蛾

Phalaena argentea Hufnagel, 1766, *Berlinisches Magazine*, 3: 286. Type locality: Germany, Berlin.

Synonymy: *Cucullia artemisiae* [Denis & Schifferm ü ller], 1775, nec Hufnagel, 1766; *Cucullia argentea subcaerulea* Staudinger, 1901 (var.); *Cucullia epargyra* Boursin, 1941.

Distribution. China (Heilongjiang, Jilin, Liaoning, Nei Mongol, Hebei, Xinjiang, Xizang); Russia (Southeast European part, Siberia), Mongolia, North Korea, South Korea, Japan, Kazakhstan; Europe.

寄主：蒿属（陈一心，1999）；荒野蒿（*Artemisia campestris*）、茵陈蒿（*Artemisia capillaris*）、滨蒿（*Artemisia maritima*）、野艾蒿（*Artemisia vulgaris*）、春黄菊（*Anthemis tinctoria*）、蓍属（*Achillea*）、飞廉属（*Carduus*）植物（Matov & Kononenko, 2012）。

Cucullia jankowskii Oberthür, 1884 雪冬夜蛾

Cucullia jankowskii Oberthür, 1884, *Études d'Entomologie*, 10: 23, pl. 3: 2. Type locality: Russia, Primorye terr., [Bezverkhovo] Sidemi.

Synonymy: *Cucullia japonica* Matsumura, 1925.

Distribution. China (Heilongjiang, Jilin); Russia (Siberia), Korea, Japan.

寄主：魁蒿（*Artemisia princeps*）（Matov & Kononenko, 2012）。

Cucullia artemisiae (Hufnagel, 1766) 嗜蒿冬夜蛾

Phalaena artemisiae Hufnagel, 1766, *Berlinisches Magazine*, 3: 292. Type locality: Germany, Berlin.

Synonymy: *Noctua abrotani* [Denis & Schiffermüller], 1775; *Cucullia artemisiae obscura* Turati, 1923 (f.); *Cucullia lindei* auct..

Distribution. China (Heilongjiang, Jilin, Liaoning, Hebei, Xinjiang); Russia (European part, Siberia), Korea, Japan; Central Asia, Europe.

寄主：蒿属（陈一心，1999）；南木蒿（*Artemisia abrotanum*）、中亚苦蒿（*Artemisia absinthium*）、银蒿（*Artemisia austriaca*）、荒野蒿（*Artemisia campestris*）、滨蒿（*Artemisia maritima*）、山地蒿（*Artemisia montana*）、猪毛蒿（*Artemisia scoparia*）、北艾（*Artemisia vulgaris*）、翠菊（*Bellis perennis*）、西洋甘菊（*Matricaria chamomilla*）、洋甘菊（*Matricaria recutita*）、菊蒿（*Tanacetum vulgare*）（Matov & Kononenko, 2012）。

Cucullia humilis **Boursin, 1941 倭冬夜蛾**

Cucullia humilis Boursin, 1941, *Deutsche Entomologische Zeitschrift Iris*, 55: 58, pl. 7: 13, 14. Type locality: China, Shaanxi, Qinling[Tsinling], Taibaishan[Tapaishan].

Distribution. China (Jilin, Shaanxi, Shanxi); Russia (Siberia), Mongolia, Korea.

Cucullia maculosa **Staudinger, 1888 斑冬夜蛾**

Cucullia maculosa Staudinger, 1888, *Entomologische Zeitung. Entomologischen Vereine zu Stettin*, 49: 259. Type locality: Russia, Primorye terr. [Usssuri].

Synonymy: *Cucullia jozankeana* Matsumura, 1925; *Parastichtis shibuyae* Matsumura, 1925; *Cucullia maculosa shuotsuensis* Bryk, 1949 (ssp.); *Cucullia maculosa japonibia* Bryk, 1949 (ssp.); *Apamea shibuyoides* Poole, 1989, unnecessary repl. name.

Distribution. China (Heilongjiang, Jilin, Liaoning, Hebei); Russia (Siberia), Korea, Japan.

寄主：魁蒿（*Artemisia princeps*）（Matov & Kononenko, 2012）。

Cucullia hostilis **Boursin, 1934 散斑冬夜蛾**

Cucullia hostilis Boursin, 1934, *Revue Francaise d'Entomologie*, 1: 143, pl. 2: 3. Type locality: Russia, Pimorye terr., Valdivostok, Partizansk.

Synonymy: *Cucullia marmorea* Boursin, 1941.

Distribution. China (Heilongjiang, Jilin); Russia (Siberia), Nepal, Vietnam, Thailand(North).

注：首次给出中文种名。

Cucullia propinqua **Eversmann, 1842 亮基冬夜蛾**

Cucullia propinqua Eversmann, 1842, *Bulletin de la Société Impériale des Naturalistes de Moscou*, 1842: 553. Type locality: Russia, South Ural.

Distribution. China (Jilin); Russia (Siberia), Mongolia, Korea, Japan, Kazakhstan; Europe.

寄主：绢毛蒿（*Artemisia sericea*）（Matov & Kononenko, 2012）。

注：首次给出中文种名。

Cucullia perforata Bremer, 1861 贯冬夜蛾

Cucullia perforata Bremer, 1861, *Bulletin de l'Academie Impériale des Sciences de St.-Petersbroug*, 3: 490. Type locality: Russia, north of Primorye terr. or south of Khabarovsk terr. ["am 'Ussuri' oberhalb der Ema"].

Distribution. China (Heilongjiang, Hebei, Shandong, Fujian); Russia (Siberia), Mongolia, Korea, Japan.

寄主：毛果一枝黄花（*Solidago virgaurea*）（Matov & Kononenko, 2012）。

Cucullia mandschuriae Oberthür, 1884 白纹冬夜蛾

Cucullia mandschuriae Oberthür, 1884, *Études d'Entomologie*, 10: 23, pl. 3: 4. Type locality: Russia, Primorye terr., Bezverkhovo [Sidemi].

Synonymy: *Cucullia daisensis* Matsumura, 1931; *Cucullia manshuriae* auct., nec Oberthür, 1884.

Distribution. China (Heilongjiang, Jilin, Liaoning, Shaanxi, Hunan); Russia (Siberia), Korea, Japan.

寄主：三脉紫菀（*Aster ageratoides*）（Matov & Kononenko, 2012）。

Cucullia pustulata (Eversmann, 1842) 莴苣冬夜蛾

Cucullia lactucae pustulata Eversmann, 1842, *Bulletin de la Société Impériale des Naturalistes de Moscou*, 1842: 551. Type locality: Russia, South Ural, Spassk; lectotype male, designated by Ronkay & Ronkay, 2009).

Synonymy: *Cucullia pustulata fraterna* Butler, 1878 (ssp.).

Distribution. China (Jilin, Liaoning); Russia (Southeast European part, Siberia), Mongolia, Korea, Japan, Kazakhstan; Europe.

寄主：甜菜（*Beta vulgaris*）、全光菊（*Hololeion maximowiczii*）、山莴苣（*Lactuca indica*）、莴苣（*Lactuca sativa*）、（*Mycelis muralis*）、苣荬菜（*Sonchus arvensis*）、台湾蒲公英（*Taraxacum platycarpum*）、齿裂黄鹌菜（*Youngia denticulata*）、苦苣菜（*Sonchus oleraceus*）、驴食草属（*Onobrychis*）、牛蒡属（*Arctium*）、菊属（*Chrysanthemum*）、疆菊属（*Tanacetum*）植物（Matov & Kononenko, 2012）。

注：《中国动物志》（夜蛾科）中引用其同物异名"*Cucullia pustulata fraterna* Butler, 1878"，在此予以变更。

Cucullia lucifuga ([Denis & Schiffermüller], 1775) 鲁冬夜蛾

Noctua lucifuga [Denis & Schiffermüller], 1775, *Ankündung eines Systematisches Werkes von den Schmetterlinge der Wiener Gegend. Wien*: 74. Type locality: Austria, Vienna.

Distribution. China (Heilongjiang, Jilin, Nei Mongol, Qinghai); Russia (European part, Siberia), Mongolia, Korea, Japan; Central Asia, Europe.

寄主：大果栎（*Quercus macrocarpa*）、齿叶桤木（*Alnus rugosa*）、北美桦木（*Betula papyrifera*）、帚石楠（*Calluna vulgaris*）、美洲山杨（*Populus tremuloides*）、紫叶稠李（*Prunus virginiana*）、欧芹（*Aethusa cynapium*）、野胡萝卜（*Daucus carota*）、红辣椒千叶蓍（*Achillea millefolium*）、假发矢车菊（*Centaurea phrygia*）、刺儿菜（*Cirsium arvense*）、绿毛山柳菊（*Hieracium pilosella*）、山柳菊（*Hieracium umbellatum*）、莴苣（*Lactuca sativa*）、山莴苣（*Lactuca sibirica*）、（*Petasites paradoxus*）、（*Prenanthes purpurea*）、（*Sonchus alpinus*）、苣荬菜（*Sonchus arvensis*）、花叶滇苦菜（*Sonchus asper*）、苦苣菜（*Sonchus oleraceus*）、婆罗门参（*Tragopogon pratensis*）、款冬花（*Tussilago farfara*）、欧洲百合（*Lilium martagon*）、柳属（*Salix*）、酸模属（*Rumex*）、车前属（*Plantago*）、荆芥属（*Nepeta*）、山金车属（*Arnica*）、蒿属（*Artemisia*）、菊苣属（*Cichorium*）、苦荬菜属（*Ixeris*）、风毛菊属（*Leontodon*）、风毛菊属（*Saussurea*）、千里光属（*Senecio*）、一枝黄花属（*Solidago*）、蒲公英属（*Taraxacum*）植物（Matov & Kononenko, 2012）。

***Cucullia umbratica* (Linnaeus, 1758) 冬夜蛾**

Phalaena umbratica Linnaeus, 1758, *Systema Naturae* (Edn. 10)1: 515. Type locality: Europe.

Synonymy: *Cucullia sonchi* Heinemann, 1859; *Cucullia clarior* Fuchs, 1904; *Cucullia umbratica albida* Spuler, 1908 (ab.); *Cucullia umbratica lampra* Püngeler, 1909 (ssp.); *Empusada* (*Cucullia*) *lampra* Püngeler, 1909.

Distribution. China (Jilin, Nei Mongol, Xinjiang, Xizang); Russia (European part, Siberia), Mongolia, North Korea, South Korea; Central Asia, Europe.

寄主：莴苣、苦苣菜、山柳菊（陈一心，1999）；西葫芦（*Cucurbita pepo*）、甘蓝（*Brassica oleracea*）、蓬子菜（*Galium verum*）、田旋花（*Convolvulus arvensis*）、长叶车前（*Plantago lanceolata*）、风铃草（*Campanula media*）、荒野蒿（*Artemisia campestris*）、（*Artemisia vulgaris*）、灯心草粉苞苣（*Chondrilla juncea*）、栽培菊苣（*Cichorium endivia*）、菊苣（*Cichorium intybus*）、毛叶还阳参（*Crepis capillaris*）、绿毛山柳菊（*Hieracium pilosella*）、山柳菊（*Hieracium umbellatum*）、（*Hypochoeris glabra*）、（*Hypochoeris maculata*）、欧洲猫儿菊（*Hypochoeris radicata*）、莴苣（*Lactuca sativa*）、野莴苣（*Lactuca serriola*）、秋蒲公英（*Leontodon autumnalis*）、（*Mycelis muralis*）、欧洲千里光（*Senecio vulgaris*）、苣荬菜（*Sonchus arvensis*）、花叶滇苦菜（*Sonchus asper*）、苦苣菜（*Sonchus oleraceus*）、沼生苦苣菜（*Sonchus palustris*）、药用蒲公英（*Taraxacum officinale*）、（*Eulophia angolensis*）、柳属（*Salix*）、椴树属（*Tilia*）、车轴草属（*Trifolium*）、亚麻属（*Linum*）、蓟属（*Cirsium*）、还阳参属（*Crepis*）、飞蓬属（*Erigeron*）、大翅蓟属（*Onopordum*）、婆罗门参属（*Tragopogon*）植物（Matov & Kononenko, 2012）。

***Cucullia biornata* Fischer de Waldheim, 1840 黄条冬夜蛾**

Cucullia biornata Fischer de Waldheim, 1840, *Bulletin de la Société Impériale des Naturalistes de Moscou*, 13: 83. Type locality: Russia, Volga reg..

Synonymy: *Cucullia biornata lobnoricas* Draudt, 1934 (f.).

Distribution. China (Jilin, Liaoning, Nei Mongol, Hebei, Xinjiang, Xizang); Russia (Southeast European part, Siberia), Mongolia, Korea, Kazakhstan, Pakistan, Afghanistan; Europe.

寄主：（*Artemisia dracunculi*）、金盏花（*Calendula officinalis*）、蓍属（*Achillea*）、紫菀属

（*Aster*）、苦苣菜属（*Sonchus*）、疆菊属（*Tanacetum*）、蒲公英属（*Taraxacum*）植物（Matov & Kononenko, 2012）。

Cucullia distinguenda Staudinger, 1892 蒙冬夜蛾

Cucullia distinguenda Staudinger, 1892, *Deutsche Entomologische Zeitschrift Iris*, 5: 368, pl. 3: 7. Type locality: Mongolia, Kentei.

Distribution. China (Jilin, Liaoning, Xizang); Russia (Siberia), Mongolia, Kazakhstan, Kyrgyzstan.

注：首次给出中文种名。

Cucullia gnaphalii (Hübner, [1813]) 欧冬夜蛾

Noctua gnaphalii Hübner, [1813], *Sammlung Europaischer Schmetterlinge. Eulen*, 1: pl. 125: 582, 583. Type locality: Europe.

Synonymy: *Cucullia alpherakyi* Staudinger, 1896; *Cucullia gnaphalii occidentalis* Boursin, 1945 (ssp.).

Distribution. China (Jilin); Russia (European part. Siberia), Mongolia; Central Asia, Asia Minor, Europe.

寄主：柳兰（*Chamaenerion angustifolium*）、（*Scrophularia altaica*）、红辣椒千叶蓍（*Achillea millefolium*）、雅美紫菀（*Aster amellus*）、麻菀（*Aster linosyris*）、向日葵（*Helianthus annuus*）、毛果一枝黄花（*Solidago virgaurea*）、剪秋罗属（*Lychnis*）、女娄菜属（*Melandrium*）、蒿属（*Artemisia*）、菊属（*Chrysanthemum*）、（*Chrysocoma*）、山柳菊属（*Hieracium*）植物（Matov & Kononenko, 2012）。

Cucullia virgaureae Boisduval, 1840 黄褐冬夜蛾

Cucullia virgaureae Boisduval, 1840, *Genera et Index Methodicus Europaeorum Lepidopterorum*, 1840: 154. Type locality: South Russia.

Synonymy: *Cucullia virgaureae cinnamona* Ronkay & Ronkay, 1987 (spp.); *Cucullia virgaureae cynnamomina* Ronkay & Ronkay, 1989 (spp.); *Cucullia virgaureae tianshanica* Ronkay, Ronkay & Gyulai, 2009 (spp.).

Distribution. China (Jilin); Russia (Southeast European part, Siberia); Central Asia, Europe.

寄主：麻菀（*Aster linosyris*）、毛果一枝黄花（*Solidago virgaurea*）、蒿属（*Artemisia*）植物（Matov & Kononenko, 2012）。

注：首次给出中文种名。

Cucullia amota Alphéraky, 1887 突冬夜蛾

Cucullia amota Alphéraky, 1887, *Entomologische Zeitung. Entomologischen Vereine zu Stettin*, 48: 171. Type locality: “Tura” [Russia or China, Turkestan].

Distribution. China (Jilin, Beijing, Xizang); Russia (Southeast European part, Siberia), Mongolia; Central

Asia.

寄主：麻菀（*Aster linosyris*）、蒿属（*Artemisia*）植物（Matov & Kononenko, 2012）。

注：首次给出中文种名。

Cucullia kurilullia Bryk, 1942 棕条冬夜蛾

Cucullia kurilullia Bryk, 1942, *Deutsche Entomologische Zeitschrift Iris*, 56: 45. Type locality: Russia, South Kuriles, Kunashir I., Tomari [Golovnino].

Synonymy: *Cucullia kurilullia harutai* Ronkay & Ronkay, 1998 (ssp.); *Cucullia kurilullia expulsa* Ronkay, Ronkay & Gyulai, 2009 (ssp.); *Cucullia elongata* auct., nec Butler, 1880.

Distribution. China (Heilongjiang, Jilin, Xizang); Russia(Transbaikalia, Siberia), Mongolia, Korea, Japan, Nepal, Turkmenistan, Kazakhstan, Kyrgyzstan.

寄主：紫菀属（*Aster*）植物（Matov & Kononenko, 2012）。

Cucullia elongata Butler, 1880 长冬夜蛾

Cucullia elongata Butler, 1880, *The Annals and Magazine of Natural History*, (6)5: 67. Type locality: India, Sikkim, Darjiling.

Synonymy: *Cucullia albescens* Moore, 1881; *Cucullia atkinsoni* Moore, 1882; *Cucullia elongata acutipennis* Ronkay, Ronkay & Gyulai, 2009 (spp.); *Cucullia postera*: Staudinger, 1892.

Distribution. China (Heilongjiang, Jilin, Liaoning, Xinjiang, Qinghai, Xizang); Russia (Siberia), Mongolia, Korea.

寄主：羽裂紫菀（*Aster pinnatifidus*）、东风菜（*Aster scaber*）、紫菀（*Aster yomena*）、（*Aster glehni*）、马兰属（*Kalimeris*）植物（Matov & Kononenko, 2012）。

Subfamily **Oncocnemidinae** Forbes & Franclemont, 1954 爪冬夜蛾亚科

Oncocnemidinae Forbes & Franclemont, 1954, *in* Forbes, 1954, *Memoirs of the Cornell University Agriculture Experimental Station*, 329: 122. Type genus: *Oncocnemis* Lederer, 1853.

Synonymy: Stilbidae Guenée, 1852; Pharnacisa Forbes, 1854; Cleophanidae Heinemann, 1859; Phornacisa Walker, 1862; Xylocampidi Tutt, 1896; Copihadena Morrison, 1875; Metahadena Morrison, 1976; Calophasiina Beck, 1996; Metopocerina Beck, 1996; Omiini Beck, 1996; Omphalophanina Beck, 1996; Sympistina Beck, 1996.

The members of the subfamily Oncocnemidinae comprising small, usually robust species with relatively broad brightly coloured forewing. In the male genitalia the subfamily is characterized by very long tubular vesica with an elongate field of spines on the apical part of vesica. Earlier authors associated Oncocnemidinae with Cuculliinae or treated as a tribe of the latter (Ronkay, Ronkay, 1995). Poole (1995) defined this group as the distinct subfamily. This point of view was supported by Mitchel *et al.* (2000) based on molecular data. The subfamily status of the Oncocnemidinae was accepted by Fibiger & Lafontaine (2005) and used for the Checklist of European Noctuoidea (Fibiger *et al.*, 2011). The European fauna of the subfamily was treated by Ronkay and Ronkay (1995). The Nearctic species of the genus *Oncocnemis* were revised by Troubridge (2008). The World fauna of the Oncocnemidinae enumerates 70 genera and about 600 species, mainly distributed in the steppe and savannas of the temperate zone of northern and southern hemispheres, some species of the genus *Sympistis* have a circumpolar area, other species are characteristic for steppe zone. 5 genera and 6 species are known from the Three Provinces of Northeast China.

References. Kononenko, 1979 (Cuculliinae, part); Chen, 1982, 1999 (Noctuidae); Sugi, 1982 (Cuculliinae, part); Kononenko, 1990 (Cuculliinae, part); Ronkay, Ronkay, 1995 (Cuculliinae: Oncocnemidini); Kononenko *et al.*, 1998 (Cuculliinae, part); Troubridge, Crabo, 1998 (Oncocmenidini); Kitching, Rawlins, 1999 [1998] (Cuculliinae, part); Kononenko, 2003 (Cuculliinae, part); Fibiger, Hacker, 2005 (Oncocnemidinae); Fibiger, Lafontaine, 2005 (Oncocnemidinae); Kononenko, 2005 (Oncocnemidinae); Lafontaine, Fibiger, 2006 (Oncocnemidinae); Kononenko, Han, 2007 (Oncocnemidinae); Troubridge, 2008 (Oncocnemidinae); Matov *et al.*, 2008 (Oncocnemidinae); Lafontaine, Schmidt, 2010 (Oncocnemidinae); Fibiger *et al.*, 2011 (Oncocnemidinae); Kononenko, 2016 (Oncocnemidinae).

注：首次给出中文亚科名。

Genus ***Calophasia*** Stephens, 1829 标冬夜蛾属

Calophasia Stephens, 1829 [June], *The Nomenclature of British Insectes*: 42. Type species: *Noctua linariae* [Denis & Schiffermüller], 1775 [Austria] (=*Phalaena lunula* Hufnagel, 1766).

Synonymy: *Cleophana* Boisduval, Rambur & Graslin, 1832; *Rhabdophana* Sodoffsky, 1837, repl. name for *Cleophana* Boisduval, 1832; *Calliphasia*: Agassiz, 1846, emend..

The genus includes 10 species, mainly distributed in the Mediterranean subregion of the Palaearctic region; 1 species is in the Three Provinces of Northeast China.

***Calophasia lunula* (Hufnagel, 1766) 标冬夜蛾**

Phalaena lunula Hufnagel, 1766, *Berlinisches Magazine*, 3 (3): 394. Type locality: Germany, Berlin.

Synonymy: *Noctua linariae* [Denis & Schiffermüller], 1775; *Calophasia lunula anatolica* Draudt, 1936 (ssp.), nec Lederer, 1857; *Calophasia lunula ocellata* Ronkay & Ronkay, 1995 (ssp.), repl. name for *anatolica* Draudt, 1936.

Distribution. China (Heilongjiang, Jilin, Xinjiang); Russia (European part, Siberia), Mongolia, Korea; Central Asia, Asia Minor, Europe, North America.

寄主：柳穿鱼属植物（陈一心，1999）；笃斯越橘（*Vaccinium uliginosum*）、（*Kickxia spuria*）、（*Linaria alpina*）、（*Linaria cymbalaria*）、（*Linaria dalmatica*）、卵叶柳穿鱼（*Linaria genistifolia*）、

（*Linaria italica*）、（*Linaria minor*）、（*Linaria officinalis*）、紫花柳穿鱼（*Linaria purpurea*）、（*Linaria repens*）、（*Linaria striata*）、欧洲柳穿鱼（*Linaria vulgaris*）、杨属（*Populus*）、金鱼草属（*Antirrhinum*）植物（Matov & Kononenko, 2012）。

Genus *Sympistis* Hübner, [1823] 1816 集冬夜蛾属

Sympistis Hübner, [1823] 1816, *Verzeichniss bekannter Schmettlinge* [sic!]: 261. Type species: *Noctua melaleuca* Thunberg, 1791, (a junior primary homonym of *Noctua melaleuca* Vieweg, 1790) [North Europe: Lappland].

Synonymy: *Lepipolys* Guenée, 1852, *Oncocnemis* Lederer, 1853, *Phornacisa* Walker, 1862; *Homohadena* Grote, 1873; *Adita* Grote, 1874; *Metahadena* Morrison, 1875; *Copihadena* Morrison, 1875; *Apharetra* Grote, 1901; *Homoncocnemis* Hampson, 1906; *Hemistilbia* Barnes & Benjamin, 1929; *Sympistoides* Kozhantschikov, 1947; *Pharnacisa*: Forbes, 1954, emend.; *Funepistis* Beck, 1991; *Sinupistis* Beck, 1996.

Holarctic genus includes 117 species, distributed in Arctic, montane, arid and semiarid regions of temperate zone. 24 species are known from the Palaearctic region, among them 2 species are in the Three Provinces of Northeast China.

Sympistis senica (Eversmann, 1856) 叉集冬夜蛾

Xylina senica Eversmann, 1856, *Bulletin de la Société Impériale des Naturalistes de Moscou*, 1856: pl. 3: 7. Type locality: Russia, Ural, Altai.

Synonymy: *Acronycta literata* Bremer, 1861; *Acronycta x-signata* Staudinger, 1897; *Oncocnemis senica mysterica* Ronkay, 1988 (ssp.).

Distribution. China (Jilin, Nei Mongol); Russia (Siberia), Mongolia, Japan, Kazakhstan.

寄主：欧洲山杨（*Populus tremula*）、石蚕叶绣线菊（*Spiraea chamaedryfolia*）、欧亚绣线菊（*Spiraea media*）、绣线菊（*Spiraea salicifolia*）、桦木属（*Betula*）、柳属（*Salix*）植物（Matov & Kononenko, 2012）。

注：首次给出中文种名。

Sympistis campicola (Lederer, 1853) 野爪集冬夜蛾

Oncocnemis campicola Lederer, 1853, *Verhandlungen des Zoologisch-Botanischen vereins in Wien*, 3: 369, pl. 4: 5. Type locality: [East Kazakhstan, West Altai. [East Kazakhstan, West Altai] Ust-Kamengorsk, Ust-Buchtarminsk.

Distribution. China (Heilongjiang, Jilin, Nei Mongol, Hebei, Xinjiang, Gansu, Shandong, Fujian); Russia (Southeast European part, Siberia), Mongolia, Korea, Japan, Kazakhstan, Kyrgyzstan.

寄主：楝棠升麻（*Aruncus dioicus*）、麻叶绣线菊（*Spiraea cantoniensis*）、小花绣线菊（*Spiraea nipponica*）、绣线菊（*Spiraea salicifolia*）、柳属（*Salix*）植物（Matov & Kononenko, 2012）。

Genus *Calliergis* Hübner, [1821] 1816 冶冬夜蛾属

Calliergis Hübner, [1821] 1816, *Verzeichniss bekannter Schmettlinge* [sic!], 244. Type species: *Bombyx ramosa* Esper, 1786 [Germany].

Synonymy: *Lithocampa* Guenée, 1852; *Callierges*: Hampson, 1906, missp..

The genus contains 3 species are in Mediterranean and Chinese subregions of Palaearctic region; 1 species is in the Three Provinces of Northeast China.

Calliergis ramosula **(Staudinger, 1888) 冶冬夜蛾**

Lithocampa ramosula Staudinger, 1888, *Entomologische Zeitung. Entomologischen Vereine zu Stettin*, 49: 258. Type locality: Russia, Primorye terr., Vladivostok.

Synonymy: *Callierges ramosa*: Hampson, 1906; *Calliergis ramosula*: Kononenko, 1979; *Callierges ramosula*: Ahn, Kononenko & Ronkay, 1998.

Distribution. China (Heilongjiang, Jilin, Liaoning); Russia (Siberia), Korea.

寄主：忍冬属植物（陈一心，1999）。

注：《中国动物志》（夜蛾科）中引用同物异名“*Callierges ramosa* (Esper, 1786)”，在此予以变更。

Genus *Stilbina* Staudinger, 1892 光夜蛾属

Stilbina Staudinger, 1892, *Deutsche Entomologische Zeitschrift Iris*, 4: 290. Type species: *Stilbina hypaenides* Staudinger, 1892 [Lebanon, Israel].

The genus contains 4 species, distributed in the Mediterranean and Chinese subregions of the Palaearctic region; 1 species is in the Three Provinces of Northeast China.

Stilbina koreana **Draudt, 1934 朝光夜蛾**

Stilbina koreana Draudt, 1934, *in*: Seitz (ed.), *Die Gross-Schmetterlinge der Erde*, 3. Supplement.: 172, pl. 20: k. Type locality: Korea, Kumgan Mts [Silver Mts.].

Distribution. China (Heilongjiang, Jilin, Liaoning); Russia (Transbaikalia, Siberia), Mongolia, North Korea.

Genus *Phidrimana* Kononenko, 1989 仿介夜蛾属

Phidrimana Kononenko, 1989, *Japan Heterocerist's Journal*, 152: 28. Type species: *Dryobota amurensis* Staudinger, 1892 [July] [Primorye terr.] (=*Miselia sabulosa* Graeser, 1892 [October] [Khabarovsk]).

Synonymy: *Amphidrina* auct. Staudinger, 1891.

Monotypic Eastern Palaearctic genus; 1 species is in the Three Provinces of Northeast China.

注：首次给出中文属名。

***Phidrimana amurensis* (Staudinger, 1892) 曲肾仿介夜蛾**

Dryobota amurensis Staudinger, 1892, *in* N.M. Romanoff (ed.), *Mémoires sur les Lépidoptères*, 6: 432, pl. 7: 5. Type locality: Russia, "Ussuri" [Primorye terr.].

Synonymy: *Miselia sabulosa* Graeser, 1892.

Distribution. China (Jilin); Russia (Southeast European part, Siberia), Mongolia, North Korea.

注：《中国动物志》（夜蛾科）中将本种归入"*Amphidrina* Staudinger, 1891 介夜蛾属"，在此予以变更。

Subfamily **Amphipyrinae** Guenée, 1837 杂夜蛾亚科

Amphipyrides Guenée, 1837, *Annales de la Société Entomologique de France*, 6: 222. Type genus: *Amphipyra* Ochsenheimer, 1816.

Synonymy: Amphipyridi Guenée, 1837(incorrect subsequent spelling); Pyrophilinae Grote, 1882; Anaphiinae Kiriakoff, 1970, nomen nudum; Pyramidcampina Beck, 1991; Pyroina Beck, 1996.

In the opinion of previous authors Amphipyrinae represented a huge heterogeneous group of genera, which could be characterized by trifine wing venation, naked eyes without lashes and unarmed tarsi of all legs. As a result of the revision of the Noctuoidea, the subfamily was separated into several more or less natural groups, which were integrated into different subfamilies of the Noctuinae as tribes (Lafontaine & Fibiger, 2005).

The subfamily Amphipyrinae in the restricted sense includes the genus *Amphipyra* and some related Palaearctic genera. Lafontaine & Schmidt (2010) expanded the concept of the subfamily by including to the subfamily the tribe Psaphidini, formerly treated as a distinct subfamily (Lafontaine & Fibiger, 2005). The Eurasian Psaphidini have been revised by Ronkay *et al.* (2011) as a subfamily. The broad concept of Amphipyrinae with tribes Amphipyrini and Psaphidini has been accepted here. The subfamily in the current concept includes 45 genera and about 370 species, mainly distributed in the arboreal zone of the Holarctic, Oriental, and Ethiopian regions. Some genera, currently placed in the subfamily Amphipyrinae, require revision of clarification of their systematic position. The larvae of many Amphipyrinae feed mainly on leaves of trees and shrubs of Fabaceae, Rosaceae, Tiliaceae, Betulaceae and Ulmacea, rarely on Pinaceae. Some species are polyphagous and feed on woody and herbaceous plants.

The subfamily is represented in the Three Provinces of Northeast China by 6 genera and 20 species.

References. Hampson, 1908, 1909, 1910 (Acronyctinae, part); Chen, 1982, 1999 (Noctuidae); Sugi, 1982 (Amphipyrinae, part, Cuculliinae, part); Kononenko, 1990 (Amphipyrinae, part, Cuculliinae, part); Ronkay, Ronkay, 1995 (Cuculliinae: Psaphidini); Kononenko *et al.*, 1998 (Amphipyrinae, part Cuculliinae, part); Kitching, Rawlins, 1999 [1998] (Amphipyrinae, part); Ronkay *et al.*, 2001 (Cuculliinae: Psaphidini); Kononenko, 2003 (Amphipyrinae, part, Cuculliinae, part); Fibiger, Hacker, 2005 (Amphipyrinae, Psaphidinae); Fibiger, Lafontaine, 2005 (Amphipyrinae, Psaphidinae); Kononenko, 2005 (Amphipyrinae, Psaphidinae); Lafontaine, Fibiger, 2006 (Amphipyrinae, Psaphidinae); Kononenko, Han, 2007 (Amphipyrinae, Psaphidinae); Fibiger, Hacker, 2007 (Amphipyrinae); Matov *et al.*, 2008 (Amphipyrinae, part, Psaphidinae); Lafontaine, Schmidt, 2010

(Amphipyrinae: Amphipirini, Psaphidini, Feralini); Ronkay *et al.*, 2011 (Psaphidinae); Fibiger *et al.*, 2011 (Amphipyrinae, Psaphidinae); Kononenko, 2016 (Amphipyrinae).

Tribe **Amphipyrini** Guenée, 1837 杂夜蛾族

Amphipyrides Guenée, 1837, *Annales de la Société Entomologique de France*, 6: 222. Type genus: *Amphipyra* Ochsenheimer, 1816.

Synonymy: Amphipyridi Guenée, 1837 (incorrect subsequent spelling); Pyrophilinae Grote, 1882; Anaphiinae Kiriakoff, 1970, nomen nudum; Pyramidcampina Beck, 1991; Pyroina Beck, 1996.

The tribe includes the genus *Amphipyra* with about 60 species and many genera of uncertain systematic position. It is widely distributed in Palaearctic, Oriental, Nearctic, Australian and Ethiopian regions and especially diverse in Sino-Himalayan subregion of the Palaearctic, and Oriental regions. The tribe is represented in the Three Provinces of Northeast China by 1 genus and 9 species.

References. Chen, 1982, 1999 (Noctuidae); Sugi, 1982 (Amphipyrinae, part); Kononenko, 1990 (Amphipyrinae, part); Kitching, Rawlins, 1999 [1998] (Amphipyrinae, part); Kononenko *et al.*, 1998 (Amphipyrinae, part); Kononenko, 2003 (Amphipyrinae, part); Fibiger, Hacker, 2005 (Amphipyrinae); Fibiger, Lafontaine, 2005 (Amphipyrinae); Kononenko, 2005 (Amphipyrinae); Lafontaine, Fibiger, 2006 (Amphipyrinae); Fibiger, Hacker, 2007 (Amphipyrinae); Kononenko, Han, 2007 (Amphipyrinae); Lafontaine, Schmidt, 2010 (Amphipyrinae: Amphipyrini); Fibiger *et al.*, 2011 (Amphipyrinae); Kononenko, 2016 (Amphipyrini).

Genus ***Amphipyra*** Ochsenheimer, 1816 杂夜蛾属

Amphipyra Ochsenheimer, 1816, *Die Schmetterlinge von Europa*, 4: 70. Type species: *Phalaena tragopoginis* Clerck, 1759 [Europe].

Synonymy: *Pyrophila* Hübner, [1806], suppr. (ICZN, 1926 Op. 79; 1954 Op. 278); *Scotophila* Hübner, [1821] 1816; *Pyrophila* Stephens, 1829; *Philopyra* Guenée, 1852, repl. name; *Neocomia* Rougemont, 1901; *Pyramidcampa* Beck, [1992] 1991; *Adamphipyra* Beck, [1992] 1991; *Anpyramida* Beck, 1996; *Obtuscampa* Beck, 1996.

The genus includes more than 60 species distributed in Palaearctic, Oriental, Nearctic, Australian and Ethiopian regions. There are 28 species are in the Palaearctic region, among them 9 species are in the Three Provinces of Northeast China.

***Amphipyra pyramidea* (Linnaeus, 1758)** 果红裙杂夜蛾

Phalaena Noctua pyramidea Linnaeus, 1758, *Systema Naturae* (Edn. 10) 1: 518. Type locality: Germany, Saxonia.

Synonymy: *Phalaena navicularis* Geoffrey, 1785; *Amphipyra pyramidea obscura* Oberthür, 1880 (var.); *Amphipyra pyramidea obliquilimbata* Graeser, 1888 [1889] (var.); *Amphipyra yama* Swinhoe, 1918; *Amphipyra monolitha apyra* Bryk, 1942 (ssp.); *Amphipyra pyramidea cuprior* Fletcher, 1968 (ssp.).

Distribution. China (Heilongjiang, Jilin, Liaoning, Hebei, Hubei, Jiangxi, Guangdong, Sichuan); Russia (European part, Siberia), Korea, Japan, Kazakhstan; Europe.

寄主：栎、枫、榆、杨、柳、榛、胡桃、接骨木、乌荆子、桦、桃、梨、苹果、葡萄、樱桃等（陈一心，1999）；野蔷薇（*Rosa multiflora*）（Park *et al.*, 2006）；（*Malus domestica*）、欧洲甜樱桃（*Cerasus avium*）、欧洲李（*Prunus domestica*）、胡桃（*Juglans regia*）（Matov & Kononenko, 2012）。

Amphipyra surnia **Felder & Rogenhofer, 1874 大红裙杂夜蛾**

Amphipyra surnia Felder & Rogenhofer, 1874, *Reise Österreichischen Fregatte Novara um die Erde in den Jahren 1857, 1859. Zoologischer Theil*, 2 (2): pl. 112: 17. Type locality: Japan.

Synonymy: *Amphipyra monolitha apyra* Bryk, 1942 (ssp.); *Amphipyra monolitha* auct., nec Guenée, 1852.

Distribution. China (Heilongjiang, Jilin, Liaoning, Hebei, Henan, Hubei, Jiangxi, Fujian, Guangdong, Sichuan, Yunnan, Taiwan); Russia (Siberia), Korea, Japan.

Amphipyra acheron **Draudt, 1950 流杂夜蛾**

Amphipyra acheron Draudt, 1950, *Mitteilungen der MünchnerEntomologischen Gesellschaft*, 40: 84, pl. 6: 5. Type locality: China, Yunnan, Likiang[=Lijiang]; Shaanxi, Tai-pei-shan[=Taibaishan].

Distribution. China (Heilongjiang, Jilin, Liaoning, Shaanxi, Yunnan, Xizang, Taiwan); South Korea.

Amphipyra perflua **(Fabricius, 1787) 蔷薇杂夜蛾**

Noctua perflua Fabricius, 1787, *Mantissa Insectorum sistens Species Nupes Detectas Synonymis, Observationibus, Descriptionibus, Emend.ibus*, 2: 179. Type locality: Austria.

Synonymy: *Noctua ulminea* Schrank, 1801; *Phalaena* (*Noctua*) *pyramidina* Esper, [1803].

Distribution. China (Heilongjiang, Jilin, Liaoning, Xinjiang, Henan, Jiangsu, Hubei, Guizhou, Yunnan); Russia (European part; Siberia); Central Asia, Middle East, Europe.

寄主：蒲公英（陈一心，1999）；垂枝欧洲山毛榉（*Fagus sylvatica*）、夏栎（*Quercus robur*）、欧洲桤木（*Alnus glutinosa*）、欧洲山杨（*Populus tremula*）、黄花柳（*Salix caprea*）、爆竹柳（*Salix fragilis*）、黑茶藨子（*Ribes nigrum*）、欧洲李（*Prunus domestica*）、黑刺李（*Prunus spinosa*）、覆盆子（*Rubus idaeus*）、欧洲花楸（*Sorbus aucuparia*）、喜马拉雅凤仙花（*Impatiens glandulifera*）、欧银花（*Lonicera xylosteum*）、普通女贞（*Ligustrum vulgare*）、榛属（*Corylus*）、椴树属（*Tilia*）、榆属（*Ulmus*）、山楂属（*Crataegus*）、苹果属（*Malus*）、稠李（*Padus racemosa*）、绣线菊属（*Spiraea*）、葡萄属（*Vitis*）、山茱萸属（*Cornus*）、毛核木属（*Symphoricarpos*）植物（Matov & Kononenko, 2012）。

Amphipyra schrenckii Ménétriès, 1859 桦杂夜蛾

Amphipyra schrenckii Ménétriès, 1859, *Bulletin de la Classe Physico-Mathématique de l'Académie Impériale des Sciences de St.-Pétersbourg*, 17: 219. Type locality: Russia, "Amur".

Synonymy: *Amphipyra schrenci*: Hampson, 1908, emend..

Distribution. China (Heilongjiang, Jilin, Liaoning, Shaanxi, Henan, Hubei); Russia (Transbaikalia, Siberia), Korea, Japan.

寄主：棘皮桦（陈一心，1999）；黑桦（*Betula dahurica*）（Matov & Kononenko, 2012）。

Amphipyra livida ([Denis & Schiffermüller], 1775) 紫黑杂夜蛾

Noctua livida [Denis & Schifferm ü ller], 1775, *Ankündung eines Systematisches Werkes von den Schmetterlinge der Wiener Gegend. Wien*: 85. Type locality: Austria, Vienna.

Synonymy: *Phalaena* (*Noctua*) *scotophila* Hübner, [1788]; *Amphipyra corvina* Motschulsky, 1866; *Amphipyra livida cupreina* Bryk, 1948 (ssp.).

Distribution. China (Heilongjiang, Jilin, Liaoning, Xinjiang, Henan, Jiangsu, Hubei, Guizhou, Yunnan); Russia (European part, Siberia), Korea, Japan, India; Central Asia, Middle East, Europe.

寄主：蒲公英，矮小灌木植物（陈一心，1999）；白屈菜（*Chelidonium majus*）、药用蒲公英（*Taraxacum officinale*）（Park *et al.*, 2006）；蝙蝠葛（*Menispermum dauricum*）、兰屿木耳菜（*Reynoutria elliptica*）、酸模（*Rumex acetosa*）、小酸模（*Rumex acetosella*）、散生墙草（*Parietaria diffusa*）、药用墙草（*Parietaria officinalis*）、乌蔹莓（*Cayratia japonica*）、葡萄（*Vitis vinifera*）、水芹（*Oenanthe javanica*）、山柳菊（*Hieracium umbellatum*）、（*Lactuca ramosissima*）、阿尔泰蒲公英（*Taraxacum altaicum*）、药用蒲公英（*Taraxacum officinale*）、拉拉藤属（*Galium*）、唐松草属（*Thalictrum*）、白屈菜属（*Chelidonium*）、蓼属（*Polygonum*）、山楂属（*Crataegus*）、路边青属（*Geum*）、蔷薇属（*Rosa*）、野芝麻属（*Lamium*）、蒿属（*Artemisia*）、飞廉属（*Carduus*）、风毛菊属（*Leontodon*）、禾本科（Poaceae）植物（Matov & Kononenko, 2012）。

Amphipyra tripartita Butler, 1878 匀杂夜蛾

Amphipyra tripartita Butler, 1878, *Annals and Magazine of Natural History*, 1 (5): 288. Type locality: Japan, Yokohama.

Synonymy: *Amphipyra tripartita parvula* Bryk, 1948 (ssp.).

Distribution. China (Jilin, Hubei, Sichuan); Korea, Japan.

寄主：白背牛尾菜（*Smilax nipponica*）（Park *et al.*, 2006）。

Amphipyra jankowskii Oberthür, 1884 清基暗杂夜蛾

Amphipyra jankowskii Oberth ü r, 1884, *Études d'Entomologie*, 10: 27, pl. 2: 8. Type locality: Russia,

Primorye terr., Bezverkhovo [Sidemi].

Synonymy: *Amphipyra jancousci*: Hampson, 1908, emend..

Distribution. China (Jilin); Russia (Siberia), Korea.

注：首次给出中文种名。

***Amphipyra erebina* Butler, 1878 暗杂夜蛾**

Amphipyra erebina Butler, 1878, *Annals and Magazine of Natural History* (5)1: 287. Type locality: Japan, Yokohama.

Synonymy: *Amphipyra perflua*: Leech, 1900, nec Fabricius, 1787.

Distribution. China (Heilongjiang, Jilin, Liaoning, Hubei, Yunnan); Russia (Siberia), Korea, Japan.

寄主：青冈 (*Quercus glauca*)、甘蓝（*Brassica oleracea*）、苹果（*Malus pumila*）、白桃（*Persica vulgaris*）、欧洲李（*Prunus domestica*）、葡萄（*Vitis vinifera*）、柳属（*Salix*）、忍冬属（*Lonicera*）植物（Matov & Kononenko, 2012）。

Tribe **Psaphidini** Grote, 1896 帕夜蛾族

Psaphidini Grote, 1896, *Abhandlungen des Naturwissenschaftlichen Vereins zu Bremen*, 14: 70 (as Psaphidini, an incorrect original spelling which must be emended in Psaphidini). Type genus: *Psaphida* Walker, 1865.

Synonymy: Dicopinae Grote, 1882, replaced by Grote, 1896; Asteroscopinae Tutt, 1896; Allophyini Beck, 1996 (subtribe); Asteroscopina Beck, 1996.

Holarctis distributed tribe, most diverse in the Nearctic and Sino-Himalayan subregion of the Palaearctic. Some genera distributed in zone of broad-leaved forest across Palaearctic. Moths occur in broad leaved mountain forests, the flight period of adult is mainly autumn and early spring. This tribe includes 3 subtribes, 5 genera in the Three Provinces of Northeast China.

References. Chen, 1982, 1999 (Noctuidae: Cuculliinae); Sugi, 1982 (Cuculliinae, part); Kononenko, 1990 (Cuculliinae, part); Ronkay & Ronkay, 1995 (Cuculliinae: Psaphidini); Kononenko *et al.*, 1998 (Cuculliinae, part); Kitching & Rawlins, 1999 [1998] (Psaphidinae); Ronkay *et al.*, 2001 (Cuculliinae: Psaphidini); Kononenko, 2003 (Cuculliinae, part); Fibiger & Hacker, 2005 (Psaphidinae); Fibiger & Lafontaine, 2005 (Amphipyrinae, Psaphidinae); Kononenko, 2005 (Psaphidinae); Lafontaine & Fibiger, 2006 (Psaphidinae); Kononenko, Han, 2007 (Psaphidinae: Psaphidini, Feralini); Matov *et al.*, 2008 (Psaphidinae); Lafontaine & Schmidt, 2010 (Amphipyrinae: Psaphidini, Feralini); Ronkay *et al.*, 2011 (Psaphidinae: Allophyini, Feralini); Fibiger *et al.*, 2011 (Psaphidinae: Allophyini, Feralini); Kononenko, 2016 (Psaphidini: Allophyina, Feraliina).

注：首次给出中文族名。

Subtribe **Psaphidina** Grote, 1895 帕夜蛾亚族

Psaphidini Grote, 1895, *Abhandlungen des Naturwissenschaftlichen Vereins zu Bremen*, 14: 70 (as Psaphidiini, an incorrect original spelling which must be emended in Psaphidini). Type genus: *Psaphida* Walker, 1865.

Synonymy: Dicopinae Grote, 1882, replaced by Grote, 1896; Asteroscopinae Tutt, 1896; Asteroscopina Beck, 1996.

The subtribe Psaphidina is represented in the Three Provinces of Northeast China by 2 genera and 4 species.

Genus ***Brachionycha*** Hübner, [1819] 1816 窄毛夜蛾属

Brachionycha Hübner, [1819] 1816, *Verzeichniss bekannter Schmettlinge* [sic!]: 144. Type species: *Bombyx nubeculosa* Esper, 1785 [Germany].

Synonymy: *Brachionyx*: Meigen, 1832, missp.; *Brachyonycha*: Agassiz, 1846, emend.; *Selenoscopus* Heinemann, 1859; *Brachyonix*: Berio, 1966, missp..

The genus contains 5 species distributed in temperate zone of the Holarctic region; 4 species known from the Palaearctic region, among them 2 species are in the Three Provinces of Northeast China.

注：首次给出中文属名。

Brachionycha nubeculosa **(Esper, 1785)** 窄毛夜蛾

Bombyx nubeculosa Esper, 1785, *Die Schmetterlinge in Abbildungen nach der Natur mit Beschreibungen*, 3: 248, pl. 48: 6. Type locality: not stated.

Synonymy: *Noctua centrolinea* Fabricius, 1787; *Brachionycha nubeculosa eugraphomena* Stauder, 1924 (f.); *Brachionycha nubeculosa jezoensis* Matsumura, 1928 (ssp.); *Brachionycha nubeculosa amurensis* Draudt, 1934 (ssp.); *Brachionycha nubeculosa kullbergi* Ronkay, Ronkay & Gyulai, 2011 (ssp.).

Distribution. China (Heilongjiang, Liaoning, Nei Mongol); Russia (European part, Siberia), Korea, Japan; Europe.

寄主：欧洲赤松（*Pinus sylvestris*）、垂枝欧洲山毛榉（*Fagus sylvatica*）、大桴栎（*Quercus grosseserrata*）、垂枝桦（*Betula pendula*）、（*Betula pubescens*）、桦叶鹅耳枥（*Carpinus betulus*）、欧洲山杨（*Populus tremula*）、黄花柳（*Salix caprea*）、榔榆（*Ulmus parvifolia*）、稠李（*Padus racemosa*）、欧洲李（*Prunus domestica*）、黑刺李（*Prunus spinosa*）、欧鼠李（*Rhamnus frangula*）、蓝果忍冬（*Lonicera caerulea*）、（*Sorbus aucuparia*）、欧银花（*Lonicera xylosteum*）、欧洲白蜡（*Fraxinus excelsior*）、普通女贞（*Ligustrum vulgare*）、桤木属（*Alnus*）、椴树属（*Tilia*）、山楂属（*Crataegus*）、苹果属（*Malus*）、槭树属（*Acer*）、荚蒾属（*Viburnum*）、紫菀属（*Aster*）植物（Matov & Kononenko, 2012）。

注：首次给出中文种名。

Brachionycha sajana Draudt, 1934 淡窄毛夜蛾

Brachionycha sajana Draudt, 1934, *in* A.Seitz (ed.), *Die Gross-Schmetterlinge der Erde*, 3. Supplement: 134, pl. 17d. Type locality: Russia, Buryatia, East Sayan, Mondy.

Synonymy: *Brachionycha albicilia* Sugi, 1970.

Distribution. China (Heilongjiang, Nei Mongol); Russia (Ural, Altai, Transbaikalia, Amur reg.), Japan.

寄主：日本落叶松（*Larix leptolepis*）、西伯利亚落叶松（*Larix sibirica*）（Matov & Kononenko, 2012）。

注：首次给出中文种名。

Genus *Valeria* Stephens, 1829 鹰冬夜蛾属

Valeria Stephens, 1829, *The Nomenclature of British Insectes*: 42. Type species: *Bombyx oleagina* [Denis & Schiffermüller], 1775 [Austria].

The genus contains 8 species, all known from the Palaearctic region, among them 2 species are in the Three Provinces of Northeast China.

Valeria dilutiapicata Filipjev, 1927 北鹰冬夜蛾

Valeria dilutiapicata Filipjev, 1927, *Annuaire du Musée Zoologique de l'Académie des Sciences d'URSS*, 28: 244. Type locality: Russia, Primorye terr., Tigrovyi.

Synonymy: *Euplexia splendida* Sugi, 1958.

Distribution. China (Heilongjiang); Russia (Siberia), Korea, Japan.

Valeria tricristata Draudt, 1934 碧鹰冬夜蛾

Valeria tricristata Draudt, 1934, *in* A.Seitz (ed.), *Die Gross-Schmetterlinge der Erde*, 3. Supplement: 141, pl. 17l. Type locality: China, Nanjing[=Nanking].

Distribution. China (Jilin, Jiangsu, Hunan, Jiangxi); South Korea.

Subtribe **Allophyina** Beck, 1996 贪冬夜蛾亚族

Allophyini Beck, 1996, *Neue entomologische Nachrichten*, 36: 15. Type genus: *Allophyes* Tams, 1942.

The subtribe represented in the Three Provinces of Northeast China by 2 genera and 6 species.

注：首次给出中文亚族名。

Genus *Meganephria* Hübner, [1820] 1816 巨冬夜蛾属

Meganephria Hübner, [1820] 1816, *Verzeichniss bekannter Schmettlinge* [sic!]: 206. Type species: *Phalaena bimaculosa* Linnaeus, 1767 [Germany].

Synonymy: *Miselia* auct., nec Boisduval, 1828.

The genus includes 6 species in Sino-Himalayan and Mediterranean subregions of the Palaearctic region, among them 2 species are in the Three Provinces of Northeast China.

Subgenus ***Meganephria*** Hübner, [1820] 1816 巨冬夜蛾亚属

Meganephria Hübner, [1820] 1816, *Verzeichniss bekannter Schmettlinge* [sic!]: 206. Type species: *Phalaena bimaculosa* Linnaeus, 1767 [Germany].

***Meganephria* (*Meganephria*) *kononenkoi* Poole, 1989 远东巨冬夜蛾**

Meganephria kononenkoi Poole, 1989, *Lepidopterorum Catalogus* (*New Series*), Fascicle 118. Noctuidae 2: 634. Type locality: Russia, Primorye terr., Tigrovy, repl. name.

Synonymy: *Meganephria cinerea* Kononenko, 1978, nec Butler, 1881.

Distribution. China (Heilongjiang); Russia (Siberia), Korea.

Subgenus ***Paranephria*** Ronkay, Ronkay & Gyulai, 2011 类巨冬夜蛾亚属

Paranephria Ronkay, Ronkay & Gyulai, 2011, *Witt Catalogue*, Vol. 5: 89. Type species: *Miselia tancrei* Graeser, [1889] 1888 [Russia, Vladivostok].

注：首次给出中文亚属名。

***Meganephria* (*Paranephria*) *tancrei* (Graeser, [1889] 1888) 摊巨冬夜蛾**

Miselia tancrei Graeser, [1889] 1888, *Berliner entomologische Zeitschrift*, 32, II: 329. Type locality: Russia, Vladivostok.

Distribution. China (Heilongjiang, Shaanxi); Russia (Transbaikalia, Siberia), Korea.

Genus ***Belosticta*** Butler, 1879 仿巨冬夜蛾属

Belosticta Butler, 1879, *Annals and Magazine of Natural History*, (5)4: 357. Type species: *Belosticta extensa* Butler, 1879 [Japan].

The genus includes 7 species, distributed in Sino-Himalayan and Chinese subregion of the Palaearctic region; among them 4 species are in the Three Provinces of Northeast China.

注：首次给出中文属名。

Subgenus ***Belosticta*** Butler, 1879 仿巨冬夜蛾亚属

Belosticta Butler, 1879, *Annals and Magazine of Natural History*, (5)4: 357. Type species: *Belosticta extensa*

Butler, 1879 [Japan].

***Belosticta* (*Belosticta*) *extensa* Butler, 1879 仿巨冬夜蛾**

Belosticta extensa Butler, 1879, *Annals and Magazine of Natural History*, 4 (5): 357. Type locality: Japan, Yokohama.

Distribution. China (Heilongjiang); Russia (Siberia), Korea, Japan.

寄主：榆属、李属植物（陈一心，1999）；榉树（*Zelkova serrata*）、蔷薇科（*Prunus*）、梨属（*Pyrus*）植物（Matov & Kononenko, 2012）。

***Belosticta* (*Belosticta*) *parki* (Ronkay & Kononenko, 1998) 朴仿巨冬夜蛾**

Meganephria parki Ronkay & Kononenko, 1998, *Illustrated Catalogue of the Noctuidae in Korea. Insects of Korea*, 3: 274, 394, fig.729. Type locality: Korea, Kangwondo, Chuncheon, Soyang-Dam.

Distribution. China (Heilongjiang, Jilin); Russia (Siberia), Korea.

注：首次给出中文种名。

***Belosticta* (*Belosticta*) *funesta* (Leech, 1889) 升仿巨冬夜蛾**

Miselia funesta Leech, 1889, *Proceedings of the Zoological Society of London*, 1889: 503, pl. 51: 7. Type locality: Japan, Yokohama.

Distribution. China (Jilin, Gansu, Taiwan); Korea, Japan.

注：首次给出中文种名。

Subgenus ***Debilisticta*** Ronkay, Ronkay & Gyulai, 2011 弱巨冬夜蛾亚属

Debilisticta Ronkay, Ronkay & Gyulai, 2011, *Witt Catalogue*, Vol. 5: 89. Type species: *Miselia cinerea* Butler, 1881 [Japan].

注：首次给出中文亚属名。

***Belosticta* (*Debilisticta*) *cinerea* (Butler, 1881) 灰仿巨冬夜蛾**

Miselia cinerea Butler, 1881, *Transactions of the Entomological Society of London*, 1881:184. Type locality: Japan, Yokohama.

Synonymy: *Meganephria debilis* Warnecke, 1933; *Meganephria debilis ilan* Kobayashi & Owada, 1996 (ssp.); *Meganephria cinerea morosa* Ronkay, Gyulay, Ronkai & Hacker, 2010 (ssp.).

Distribution. China (Heilongjiang, Taiwan); Russia (Transbaikalia, Siberia), Korea, Japan.

寄主：榛、春榆（陈一心，1999）；李属（*Prunus*）植物（Matov & Kononenko, 2012）。

注：首次给出中文种名。

Subtribe **Feraliina** Poole, 1995 野冬夜蛾亚族

Feraliini Poole, 1995, *in* Dominick *et al.*, *Moths of America North of Mexico*, 26 (1): 24, established as tribe of the Psaphidinae. Type genus: *Feralia* Grote, 1874.

Synonymy: Feraliini Franclemont & Todd, 1983, nomen nudum; Adpyramidcampina Beck, 1996.

Medium sized cryptic coloured moth with green background or green elements of fore wing pattern. In the male genitalia short, setose, uncus modified, in *Feralia* it has a special structure, unique for Noctuidae; valva with harpe fused with the surface of valva, apically forming short pollex; aedeagus with vesica bearing two diverticula armed with moderate, thin spines. In the female genitalia ovipositor acute; antrum with sclerotised ring, ductus bursae and caudal part of corpus bursae sclerotised.

Feralini has been defined as a tribe by Poole (1995); presently it is considered as a tribe (Ronkay *et al.*, 2011) or subtribe Feraliina (Lafontaine & Schmidt, 2010) of the subfamily Psaphidinae. Originally it included only one genus *Feralia*. Ronkay *et al.* (2011) expanded the composition of the subtribe by including several Palaearctic genera (*Daseochaeta* Warren, [1907]; *Lamprosticta* Hübner, [1820]; *Valerietta* Draudt, 1938 and *Beshkovietta* Ronkay, Ronkay, Gyulai & Hacker, 2010. The subtribe is represented in the Three Provinces of Northeast China by 1 genus and 1 species.

Genus ***Feralia*** Grote, 1874 野冬夜蛾属

Feralia Grote, 1874, *Bulletin of the Buffalo Society of Natural Sciences*, 2: 6, 58. Type species: *Diphtera jocosa* Guenée, 1852 [North America].

Synonymy: *Momophana* Grote, 1875; *Arthrochlora* Grote, 1875; *Momaphana*: Hampson, 1906, missp.; *Arthrachlora*: Hampson, 1906, missp..

Holarctic genus mainly distributed in boreal forest zone with 6 species are in the North America and 1 species are in the Palaearctic region; 1 species is in the Three Provinces of Northeast China.

Feralia sauberi **(Graeser, 1892)** 高野冬夜蛾

Valeria sauberi Graeser, 1892, *Berliner entomologische Zeitschrift*, 37, II: 233. Type locality: Russia, Amur reg., Pokrovka ["Pokrofka"], Khabarovsk terr., Nicolaevsk ["Nicolajefsk"].

Synonymy: *Bryophila* (*Metachrostis*) *muscicolor* W. Kozhantshikov, 1923; *Valeria dimorpha* O.Bang-Haas, 1927; *Diphtherocome montana* Sugi, 1968; *Feralia montana pernigra* Sugi, 1982 (ssp.).

Distribution. China (Heilongjiang, Jilin); Russia (Ural, Siberia, Altai, Transbaikalia), Mongolia, North Korea, Japan (Honshu).

寄主：日本落叶松（*Larix leptolepis*）、西伯利亚落叶松（*Larix sibirica*）（Matov & Kononenko, 2012）。

Subfamily **Heliothinae** Boisduval, [1828] 1829 实夜蛾亚科

Heliothinae Boisduval, [1828] 1829, *Europaeorum Lepidopterorum Index Methodicus*: 94 (as Heliothidi), Official List of Family Group Names, opinion 1312. Type genus: *Heliothis* Ochsenheimer, 1816.

Synonymy: Heliothidi Boisduval, [1828] 1829, original missp.; Melicleptriinae Warren, 1911; Pyrrhiini Forbes, 1954; Heliothentinae auct..

Heliothinae is a well-defined subfamily with 26 genera and about 430 species worldwide and includes some important agricultural pests. The highest diversity of the subfamily is observed in the arid zone of tropical and subtropical regions. The subfamily has been recognised for 180 years but its monopholy was not stated conclusively before Matthews (1991). The European fauna of Heliothinae revised by Fibiger *et al.* (2009).

The subfamily comprises 7 genera and 12 species are in the Three Provinces of Northeast China.

References. Hardwick, 1965, 1970 (Heliothinae); Chen, 1982, 1999 (Noctuidae); Sugi, 1982 (Heliothinae); Kononenko, 1985 (Heliothinae); Kononenko, 1990 (Heliothinae); Matthews, 1991, 1999 (Heliothinae); Speidel *et al.*, 1996 (Heliothinae); Kononenko *et al.*, 1998 (Heliothinae); Kitching, Rawlins, 1999 [1998] (Heliothinae); Kononenko, 2003 (Heliothinae); Fibiger, Hacker, 2005 (Heliothinae); Fibiger, Lafontaine, 2005 (Heliothinae); Kononenko, 2005, 2010 (Heliothinae); Lafontaine, Fibiger, 2006 (Heliothinae); Kononenko, Han, 2007 (Heliothinae); Matov *et al.*, 2008 (Heliothinae); Fibiger *et al.*, 2009 (Heliothinae); Kononenko, 2016 (Heliothinae).

Genus ***Periphanes*** Hübner, [1821] 1816 盾胫夜蛾属

Periphanes Hübner, [1821] 1816, *Verzeichniss bekannter Schmettlinge* [sic!]: 247. Type species: *Phalaena delphinii* Linnaeus, 1758 [Europe].

Synonymy: *Chariclea* Curtis, 1825; *Chariclea* Stephens, 1827; *Periphanea*: Walker, 1857; *Chariella* Bertkau, 1889; *Periphana*: Hampson, 1903, missp..

The genus, divided into subgenera *Periphanes* and *Pyrocleptria* with 2 and 3 species respectively. All species distributed in steppes and semi-deserts of Central Asian subregion of the Palaearctic region; 2 species are in the Three Provinces of Northeast China.

Subgenus ***Pyrocleptria*** Hampson, 1903 黄盾胫夜蛾亚属

Pyrocleptria Hampson, 1903, *Catalogue of the Lepidoptera Phalaenae in the British Museum* 4: ix, 25. Type species: *Heliothis cora* Eversmann, 1837, by original designation [Russia, Orenburg, Kazan].

Synonymy: *Philareta* Moore, 1881; *Helivictoria* Beck, 1996; *Calocharia* Beck, 1996.

***Periphanes* (*Pyrocleptria*) *cora* (Eversmann, 1837) 黄盾胫夜蛾**

Heliothis cora Eversmann, 1837, *Bulletin de la Société Impériale des Naturalistes de Moscou*, 1837 (6): 47.

Type locality: Russia, Ural, Orenburg.

Synonymy: *Anthaecia coreta* Guenèe, 1852.

Distribution. China (Heilongjiang, Nei Mongol); Russia (Southeast European part, Siberia), Mongolia, Kazakhstan; Asia Minor, Southeast Europe.

寄主：唐松草属（*Thalictrum*）植物（Matov & Kononenko, 2012）。

注：首次给出中文种名。

Periphanes (_Pyrocleptria_) _scutata_ (Staudinger, 1896) 盾胫夜蛾

Heliothis scutata Staudinger, 1896, *Deutsche Entomologische Zeitschrift Iris*, 8: 361, pl. 6: 13. Type locality: Mongolia, Uliassutai.

Distribution. China (Heilongjiang, Jilin, Nei Mongol, Hebei); Russia (Tuva, Transbaikalia), Mongolia.

Genus *Pyrrhia* Hübner, [1821] 1816 焰夜蛾属

Pyrrhia Hübner, [1821] 1816, *Verzeichniss bekannter Schmettlinge* [sic!]: 233. Type species: *Noctua rutilago* [Denis & Schiffermüller], 1775 [Austria] (=*Phalaena umbra* Hufnagel, 1766).

Synonymy: *Philareta* Moore, 1881.

The genus includes 7 species, distributed in temperate zone of Holarctic region; 6 species are in the Palaearctic region, among them 3 species are in the Three Provinces of Northeast China.

***Pyrrhia umbra* (Hufnagel, 1766) 焰夜蛾**

Phalaena umbra Hufnagel, 1766, *Berlinisches Magazine*, 3: 294, pl. 51: 6. Type locality: Germany, Berlin.

Synonymy: *Noctua marginata* Fabricius, 1775; *Noctua rutilago* [Denis & Schiffermüller], 1775; *Phalaena* (*Noctua*) *conspicua* Borkhausen, 1792; *Phalaena* (*Noctua*) *umbrago* Esper, 1796; *Noctua marginago* Haworth, 1809; *Heliothis cilisca* Guenée, 1852; *Hydroecia tibetana* Moore, 1878; *Chariclea vexilliger* Christoph, 1893; *Pyrrhia aconiti* Höltzermann, 1902; *Pyrrhia umbra nepalina* Hreblay & Ronkay, 1999 (ssp.); *Pyrrhia umbra melanura* Hreblay & Ronkay, 1999 (ssp.).

Distribution. China (Heilongjiang, Jilin, Liaoning, Xinjiang, Shaanxi, Shandong, Hubei, Hunan, Zhejiang, Xizang); Russia (European part, Siberia), Mongolia, Korea, Japan, India, Nepal, Kazakhstan; Europe, North Africa.

寄主：烟草、大豆、油茶、荞麦（陈一心，1999）；（*Honkenya peploides*）、萹蓄 (*Polygonum aviculare*)、酸模叶蓼（*Polygonum lapathifolium*）、欧洲栗（*Castanea sativa*）、欧洲桤木（*Alnus glutinosa*）、胡榛（*Corylus avellana*）、树形杜鹃（*Rhododendron arboreum*）、毛黄连花（*Lysimachia vulgaris*）、黄花柳（*Salix caprea*）、东陵山柳（*Salix phylicifolia*）、蒿柳（*Salix viminalis*）、异株荨麻（*Urtica dioica*）、紫叶稠李（*Padus virginiana*）、覆盆子（*Rubus idaeus*）、（*Faba bona*）、金雀花（*Genista tinctoria*）、大豆（*Glycine max*）、白花草木樨（*Melilotus albus*）、红豆草（*Onobrychis* vicifolia）、芒柄花（*Ononis arvensis*）、（*Ononis hircina*）、匍匐芒柄草（*Ononis repens*）、刺芒柄花（*Ononis spinosa*）、小苕子（*Ornithopus sativus*）、（*Phaseolus multifloris*）、菜豆（*Phaseolus vulgaris*）、黄车轴草（*Trifolium aureum*）、中间车轴草（*Trifolium medium*）、广布野豌豆（*Vicia cracca*）、野豌豆（*Vicia sepium*）、草地老鹳草（*Geranium pratense*）、鼠掌老鹳草（*Geranium sibiricum*）、欧洲白蜡（*Fraxinus excelsior*）、天仙子（*Hyoscyamus niger*）、番茄（*Lycopersicon esculentum*）、烟草（*Nicotiana tabacum*）、（*Convolvulus soldanella*）、金鱼草（*Antirrhinum majus*）、小米草（*Euphrasia officinalis*）、（*Euphrasia stricta*）、二裂柳穿鱼（*Linaria bipartita*）、欧洲柳穿鱼（*Linaria vulgaris*）、（*Melampyrum nemorosum*）、红花钓钟柳（*Penstemon barbatus*）、粘带千里光（*Senecio viscosus*）、鼬瓣花根（*Galeopsis tetrahit*）、夏枯草（*Prunella vulgaris*）、草地鼠尾草（*Salvia pratensis*）、沼生水苏（*Stachys palustris*）、荒野蒿（*Artemisia campestris*）、金盏花（*Calendula officinalis*）、豨莶（*Sigesbeckia orientalis*）、楼斗菜属（*Aquilegia*）、翠雀属（*Delphinium*）、酸模属（*Rumex*）、鹅耳枥属（*Carpinus*）、芸苔属（*Brassica*）、白芥属（*Sinapis*）、茶藨子属（*Ribes*）、蔷薇属（*Rosa*）、金雀儿属（*Cytisus*）、山蚂蝗属（*Desmodium*）、拉拉藤属（*Galium*）、天芥菜属（*Heliotropium*）、鼻花属（*Rhinanthus*）、玄参属（*Scrophularia*）、香科科属（*Teucrium*）植物（Matov & Kononenko, 2012）。

Pyrrhia bifasciata **(Staudinger, 1888) 双纹焰夜蛾**

Grammesia bifasciata Staudinger, 1888, *Entomologische Zeitung. Entomologischen Vereine zu Stettin*, 49: 254. Type locality: Russia, Primorye terr., Razdolnaya river [Suifun].

Synonymy: *Calymnia pryeri* Leech, 1889; *Heliothis olivaria* Graeser, 1888 [1889].

Distribution. China (Heilongjiang, Jilin, Hebei); Russia (Siberia), Korea, Japan.

寄主：大麻、核桃（陈一心，1999）；胡桃楸（*Juglans mandshurica*）、毛泡桐（*Paulownia tomentosa*）（Matov & Kononenko, 2012）。

Pyrrhia hedemanni **(Staudinger, 1892) 落焰夜蛾**

Chariclea hedemanni Staudinger, 1892, *in* N.M. Romanoff (ed.), *Mémoires sur les Lépidoptères*, 6: 552. Type locality: Russia, "Amur".

Synonymy: *Chariclea erubescens* Graeser, 1892; *Pyrrhia abrasa* Draudt, 1950.

Distribution. China (Heilongjiang, Jilin); Russia (Transbaikalia, Siberia), Korea.

注：《中国动物志》（夜蛾科）中引用本种同物异名"*Pyrrhia abrasa* Draudt, 1950"，在此予以变更。

Genus ***Schinia*** Hübner, 1818 尖角夜蛾属

Schinia Hübner, 1818, *Zuträge zur Sammlung Exotischer Schmetterlinge*, 1: 8, ii. Type species: *Schinia trifascia* Hübner, 1818 [USA].

Synonymy: *Euclidia* Hübner, 1808, suppr. (ICZN, 1966 Op. 789); *Melicleptria* Hübner, [1823] 1816;

Anthoecia Boisduval, 1840; *Alaria* Duncan & Westwood, 1841, preocc. (Schrank, 1788 [Vermes]); *Trypana* Guenée, 1841; *Oria* Guenée, 1852, preocc. (Hübner, [1821] 1816 [Lepidoptera, Noctuidae]); *Rhodophora* Guenée, 1852; *Tamila* Guenée, 1852; *Eulecyptera* Grote, 1865; *Lygranthoecia* Grote & Robinson, 1873; *Tricopis* Grote, 1874; *Heliophana* Grote, 1875; *Adonisea* Grote, 1875; *Oxylos* Grote, 1875; *Pippona* Harley, 1875; *Porrima* Grote, 1877; *Rhododipsa* Grote, 1877; *Bessula* Grote, 1881; *Ula* Grote, 1881; *Dasyspoudaea* Smith, 1883; *Trileuca* Grote, 1883; *Pseudotamila* Smith, 1883; *Canidia* Grote, 1890, preocc. (Thompson, 1857 [Coleoptera]); *Eupanychis* Grote, 1890; *Trichosellus* Grote, 1890; *Incita* Grote, 1895; *Palada* Smith, 1900; *Chlorocleptria* Hampson, 1903; *Trilenca* Neave, 1940, missp.; *Tricrateriofrontia* Berio, 1941; *Uollega* Berio, 1941; *Purpurschinia* Beck, 1996.

The genus includes over 150 species, it is most diverse in the Nearctic region, inhabiting xeric open environments; 10 species are known from the Palaearctic region; among them, 1 species is in the Three Provinces of Northeast China.

注：首次给出中文属名。

Schinia bieneri (Rebel, 1926) 北尖角夜蛾

Heliothis bieneri Rebel, 1926, *Verhandlungen des Zoologisch-Botanischen vereins in Wien*, 74-75: 65, fig. Type locality: Russia, [Transbaikalia, Ulan-Ude] Verhne-Udinsk.

Distribution. China (Heilongjiang, Jilin, Nei Mongol); Russia (Tuva, Transbaikalia), Mongolia.

注：首次给出中文种名。

Genus *Protoschinia*, 1970 宽胫夜蛾属

Protoschinia Hardwick, 1970, *Memoirs of the Entomological Society of Canada*, 73: 32. Type species: *Noctua scutosa* [Denis & Schiffermüller], 1775. Type locality: Austria, Vienna district.

The genus includes 2 species distributed in the Palaearctic and Nearctic regions; 1 species is in the Three Provinces of Northeast China.

Protoschinia scutosa ([Denis & Schiffermüller], 1775) 宽胫夜蛾

Noctua scutosa [Denis & Schiffermüller], 1775, *Ankündung eines Systematisches Werkes von den Schmetterlinge der Wiener Gegend. Wien*: 89. Type locality: Austria, Vienna.

Synonymy: *Phalaena* (*Noctua*) *scutosa* Goeze, 1781, incorrect authorship.

Distribution. China (Heilongjiang, Jilin, Liaoning, Hebei, Hubei, Hunan, Jiangxi, Xizang); Russia (European part, Siberia), Korea, Japan, India, Pakistan; Europe, Africa.

寄主：西葫芦（*Cucurbita pepo*）、（*Citrullus edulis*）、鹰嘴豆（*Cicer arietinum*）、金雀儿（*Cytisus scoparius*）、紫苜蓿（*Medicago sativa*）、豌豆（*Pisum sativum*）、亚麻（*Linum usitatissimum*）、（*Anisum vulgare*）、葛缕子（*Carum carvi*）、（*Foeniculum officinalis*）、芝麻（*Sesamum orientale*）、

（*Perilla ocimoides*）、鼠尾草（*Salvia officinalis*）、千叶蓍（*Achillea millefolium*）、南木蒿（*Artemisia abrotanum*）、中亚苦蒿（*Artemisia absinthium*）、银蒿（*Artemisia austriaca*）、龙蒿（*Artemisia dracunculus*）、荒野蒿（*Artemisia campestris*）、（*Artemisia monosperma*）、猪毛蒿（*Artemisia scoparia*）、（*Artemisia scoparieformis*）、（*Artemisia lencodes*）、北艾（*Artemisia vulgaris*）、茼蒿（*Chrysanthemum coronarium*）、向日葵（*Helianthus annuus*）、西洋甘菊（*Matricaria chamomilla*）、同花母菊（*Matricaria matricarioides*）、菊蒿（*Tanacetum vulgare*）、新疆三肋果（*Tripleurospermum inodorum*）、刺苍耳（*Xanthium spinosum*）、苍耳（*Xanthium strumarium*）、滨藜属（*Atriplex*）、藜属（*Chenopodium*）、黄瓜属（*Cucumis*）、棉属（*Gossypium*）、车轴草属（*Trifolium*）、芫荽属（*Coriandrum*）、旋花属（*Convolvulus*）、红花属（*Carthamus*）植物（Matov & Kononenko, 2012）。

Genus ***Heliothis*** Ochsenheimer, 1816 **实夜蛾属**

Heliothis Ochsenheimer, 1816, *Die Schmetterlinge von Europa*., 4: 91. Type species: *Phalaena dipsacea* Linnaeus, 1767 [Sweden] (=*Phalaena viriplaca* Hufnagel, 1766).

Synonymy: *Heliothis* Hübner, 1806, (suppr. ICZN, 97 & 128); *Heliothentes* Ochsenheimer, 1816, unavail. (Code, Art. 11d); *Heliotis* Lefebvre, 1827, missp.; *Heliothisa* Meigen, 1832, emend.; *Heliotis* Sodoffsky, 1837, emend.; *Chloridea* Duncan & [Westwood], 1841; *Timora* Walker, 1856; *Chazaria* Moore, 1881; *Rhodosea* Grote, 1883; *Disocnemis* Grote, 1883; *Dysocnemis* Grote, 1890; *Heliothris* Olliff, 1890, missp.; *Neocleptria* Hampson, 1903; *Rhodocleptria* Hampson, 1903; *Nubiothis* Beck, 1996; *Pelthotis* Beck, 1996.

The genus includes 54 species distributed mainly in arid and subtropical regions of Holarctic, Oriental, Australian, Ethiopian and Neotropical regions. 7-8 species are in the Palaearctic region, among them 2 species are in the Three Provinces of Northeast China.

Heliothis adaucta **Butler, 1878 苇实夜蛾**

Heliothis adaucta Butler, 1878, *Annals and Magazine of Natural History*, 1 (5): 199. Type locality: Japan, Yokohama.

Synonymy: *Chloridea maritima angarensis* Draudt, 1938 (ssp.); *Chloridea maritima bulgarica* Draudt, 1938 (ssp.); *Chloridea maritima centralasiae* Draudt, 1938 (ssp.); *Chloridea maritima hungarica* Kovacs, 1950 (ssp.).

Distribution. China (Heilongjiang, Jilin, Liaoning, Hebei); Russia (Southeast European part, Siberia), Mongolia, Korea, Japan, India, Pakistan; Central Asia, Asia Minor, Europe.

寄主：芦苇、拟漆姑属植物（陈一心，1999）；印度棉（*Gossypium indicum*）、紫苜蓿（*Medicago sativa*）、（*Pueraria thunbergiana*）、白车轴草（*Trifolium repens*）、欧洲柳穿鱼（*Linaria vulgaris*）、大豆属（*Glycine*）、石竹科（Caryophyllaceae）、藜科（Chenopodiaceae）、葫芦科（Cucurbitaceae）、菊科（Asteraceae）植物（Matov & Kononenko, 2012）。

Heliothis ononis **([Denis & Schiffermüller], 1775) 花实夜蛾**

Noctua ononis [Denis & Schifferm ü ller], 1775, *Ankündung eines Systematisches Werkes von den Schmetterlinge der Wiener Gegend. Wien*: 89. Type locality: Austria, Vienna.

Synonymy: *Heliothis ononidis* Guenée, 1852, emend.; *Heliothis ononis sibirica* Eversmann, 1857 (var.); *Melicleptria septentrionalis* H. Edwards, 1884.

Distribution. China (Heilongjiang, Jilin, Liaoning, Nei Mongol, Qinghai, Hunan, Hubei, Sichuan); Russia (Siberia), Mongolia, Korea; Central Asia, Europe.

寄主：芒柄花属、亚麻属植物（陈一心，1999）；木纹石竹（*Dianthus sylvestris*）、欧亚蝇子草（*Silene nutans*）、蛛丝卷绢（*Sempervivum arachnoideum*）、（*Astragalus goniatus*）、（*Vaccaria vulgaris*）、白花草木樨（*Melilotus alba*）、芒柄花（*Ononis arvensis*）、刺芒柄花（*Ononis spinosa*）、蕨（*Linum lewsii*）、亚麻（*Linum usitatissimum*）、欧洲山萝卜（*Knautia arvensis*）、（*Odontites lutea*）、草地鼠尾草（*Salvia pratensis*）、驴食草属（*Onobrychis*）、鼬瓣花属（*Galeopsis*）、野芝麻属（*Lamium*）植物（Matov & Kononenko, 2012）。

Genus ***Heliocheilus*** Grote, 1865 暗实夜蛾属

Heliocheilus Grote, 1865, *Proceedings of the Entomological Society of Philadelphia*, 4: 328. Type species: *Heliocheilus paradoxus* Grote, 1865 [USA].

Synonymy: *Hebdomochondra* Staudinger, 1879; *Masalia* Moore, 1881; *Pradatta* Moore, 1881; *Raghuva* Moore, 1881; *Rhodosea* Grote, 1883; *Canthylidia* Butler, 1886; *Churubasa* Moore, 1888; *Lecerfia* Dumont, 1920.

The genus includes 46 species distributed in arid and subtropical zones of Chinese subregion of the Palaearctic, Nearctic, Australian, Oriental and Ethiopian regions. 3 species are in the Palaearctic region; 1 species is in the Three Provinces of Northeast China.

Heliocheilus fervens **(Butler, 1881) 焰实夜蛾**

Heliothis fervens Butler, 1881, *Transactions of the Entomological Society of London*, 1881: 186. Type locality: Japan, Tokyo.

Synonymy: *Heliothis foveolatus* Staudinger, 1888.

Distribution. China (Heilongjiang, Xinjiang, Hebei, Jiangsu, Yunnan, Xizang); Russia (Transbaikalia, Siberia), Korea, Japan, Pakistan.

寄主：苜蓿、柳穿鱼、矢车菊、芒柄花（陈一心，1999）。

Genus ***Helicoverpa*** Hardwick, 1965 铃夜蛾属

Helicoverpa Hardwick, 1965, *Memoirs of the Entomological Society of Canada*, 40: 9. Type species: *Noctua armigera* Hübner, [1808] [Europe].

The genus includes 17 species, distributed in subtropical and arid areas of the Palaearctic, Nearctic, Oriental, Australian, Ethiopian and Neotropical regions. 3 species are in the Palaearctic region; among them, 2 species are in the Three Provinces of Northeast China.

Helicoverpa armigera **(Hübner, [1808]) 棉铃虫**

Noctua armigera Hübner, [1808], *Sammlung Europaischer Schmetterlinge. Eulen,* 2: pl. 79: 370. Type locality: Europe.

Synonymy: *Noctua barbara* Fabricius, 1794, unused senior synonym (supprressed); *Heliothis conferta* Walker, 1857; *Heliothis pulverosa* Walker, 1857; *Heliothis uniformis* Wallengren, 1860; *Heliothis fusca* Cockerell, 1889; *Heliothis guidellii* Constanti, 1922; *Heliothis armigera commoni* Hardwick, 1965 (ssp.); *Heliothis rama* Bhattcherjee & Gupts, 1972; *Bombyx obsoleta* auct., nec Fabricius, 1793.

Distribution. China; Russia (European part, Siberia), Korea, Japan, India, Australia, New Zealand; Southeast Asia, Central Asia, Middle East, Europe.

寄主：棉、玉米、小麦、大豆、烟草、番茄、辣椒、茄、芝麻、向日葵、南瓜（陈一心，1999）；矮翠雀花（*Delphinium consolida*）、假海马齿（*Trianthema portulacastrum*）、康乃馨（*Dianthus caryophyllus*）、麦瓶草（*Silene conoidea*）、反枝苋（*Amaranthus retroflexus*）、皱果苋（*Amaranthus viridis*）、藜（*Chenopodium album*）、萹蓄 (*Polygonum aviculare*)、三球悬铃木（*Platanus orientalis*）、枫杨（*Pterocarya stenoptera*）、长叶马府油树（*Madhuca longifolia*）、异柱柽柳（*Tamarix dioica*）、胡桐（*Populus euphratica*）、钻天杨（*Populus nigra*）、四子柳（*Salix tetrasperma*）、冬瓜（*Benincasa hispida*）、甜瓜（*Cucumis melo*）、瓠果（*Cucumis sativus*）、南瓜（*Cucurbita moschata*）、西葫芦（*Cucurbita pepo*）、瓠子（*Lagenaria siceraria*）、（*Lagenaria vulgaris*）、拟南芥（*Arabidopsis thaliana*）、甘蓝（*Brassica oleracea*）、（*Moricandia arvensis*）、萝卜（*Raphanus sativus*）、莠草大麻（*Cannabis ruderalis*）、大麻（*Cannabis sativa*）、可可（*Theobroma cacao*）、榴莲（*Durio zibethinus*）、咖啡黄葵（*Abelmoschus esculentus*）、（*Abutilon avicenne*）、苘麻（*Abutilon theophrasti*）、蜀葵（*Alcea rosea*）、草棉（*Gossypium herbaceum*）、陆地棉（*Gossypium hirsutum*）、印度棉（*Gossypium indicum*）、大麻槿（*Hibiscus cannabinus*）、黄秋葵（*Hibiscus esculentus*）、大花秋葵（*Hibiscus grandiflorus*）、木芙蓉叶（*Hibiscus mutabilis*）、花葵（*Lavatera arborea*）、蓖麻（*Ricinus communis*）、杏（*Armeniaca vulgaris*）、草莓（*Fragaria ananassa*）、苹果（Malus pumila）、白桃（*Persica vulgaris*）、西洋梨（*Pyrus communis*）、儿茶（*Acacia catechu*）、菲律宾合欢（*Albizzia procera*）、骆驼刺（*Alhagi camelorum*）、花生（*Arachis hypogaea*）、蒙古黄芪（*Astragalus membranaceus*）、木豆（*Cajanus cajan*）、鹰嘴豆（*Cicer arietinum*）、菽麻（*Crotalaria juncea*）、印度黄檀（*Dalbergia sissoo*）、扁豆（*Dolichos lablab*）、大豆（*Glycine max*）、光果甘草（*Glycyrrhiza glabra*）、（*Lablab niger*）、扁豆（*Lablab purpureus*）、香豌豆（*Lathyrus odoratus*）、兵豆（*Lens culinaris*）、白羽扇豆（*Lupinus albus*）、（*Medicago denticulata*）、紫苜蓿（*Medicago sativa*）、黄香草木樨（*Melilotus officinalis*）、巴西含羞草（*Mimosa invisa*）、菜豆（*Phaseolus vulgaris*）、豌豆（*Pisum *sativum*）、水黄皮（*Pongamia pinnata*）、苦豆子（*Sophora alopecuroides*）、埃及车轴草（*Trifolium alexandrinum*）、黑吉豆（*Vigna mungo*）、绿豆（*Vigna radiata*）、豇豆（*Vigna sinensis*）、（*Vigna unguiculata*）、久树（*Schleichera oleosa*）、柚（*Citrus maxima*）、甜橙（*Citrus sinensis*）、温州蜜柑（*Citrus unshiu*）、骆驼蓬（*Peganum harmala*）、印楝（*Azadirachta indica*）、腰果（*Anacardium occidentale*）、野胡萝卜（*Daucus carota*）、亚麻（*Linum usitatissimum*）、葡萄（*Vitis vinifera*）、西洋接骨木（*Sambucus nigra*）、小粒咖啡（*Coffea arabica*）、锘利果（*Morinda citrifolia*）、印第安麻（*Apocynum cannabinum*）、颠茄草（*Atropa belladonna*）、辣椒（*Capsicum annuum*）、曼陀罗（*Datura stramonium*）、天仙子（*Hyoscyamus niger*）、中亚天仙子（*Hyoscyamus pusillus*）、枸杞（*Lycium chinense*）、番茄（*Lycopersicon esculentum*）、近缘烟草（*Nicotiana affinis*）、烟草（*Nicotiana tabacum*）、刺茄（*Solanum capsicoides*）、欧白英（*Solanum dulcamara*）、茄（*Solanum melongena*）、龙葵（*Solanum nigrum*）、（*Solanum persicum*）、马铃薯（*Solanum tuberosum*）、毛果茄（*Solanum viarum*）、印度人参（*Withania somnifera*）、田旋花（*Convolvulus arvensis*）、甘薯（*Ipomoea batatas*）、金鱼草（*Antirrhinum majus*）、长叶婆婆纳（*Veronica longifolia*）、芝麻（*Sesamum indicum*）、（*Sesamum orientale*）、北车前（*Plantago media*）、五色梅（*Lantana camara*）、（*Hypogomphia bucharica*）、紫苏（*Perilla frutescens*）、南欧丹参（*Salvia sclarea*）、黄芩（*Scutellaria baicalensis*）、红辣椒千叶蓍（*Achillea millefolium*）、臭春黄菊（*Anthemis cotula*）、（*Artemisia camphorata*）、金盏花（*Calendula officinalis*）、红花（*Carthamus tinctorius*）、茼蒿（*Chrysanthemum coronarium*）、秋英（*Cosmos bipinnatus*）、（*Gerbera aurantiaca*）、非洲菊（*Gerbera jamesonii*）、小葵子（*Guizotia abyssinica*）、（*Helianthus annuus*）、粘质旋覆花（*Inula viscosa*）、莴苣（*Lactuca sativa*）、欧洲稻槎菜（*Lapsana communis*）、白夏菊（*Matricaria maritima*）、银胶菊（*Parthenium hysterophorus*）、（*Saussurea candicans*）、花叶滇苦菜（*Sonchus asper*）、（*Steirodiscus tagetes*）、万寿菊（*Tagetes erecta*）、苍耳（*Xanthium strumarium*）、百日菊（*Zinnia violacea*）、洋葱（*Allium cepa*）、虾夷葱（*Allium schoenoprasum*）、小果野蕉（*Musa acuminata*）、燕麦（*Avena sativa*）、狗牙根（*Cynodon dactylon*）、牡竹（*Dendrocalamus strictus*）、禾参子（*Eleusine coracan*）、大麦（*Hordeum vulgare*）、稻（*Oryza sativa*）、黍（*Panicum miliaceum*）、美国狼尾草（*Pennisetum americanum*）、御谷（*Pennisetum glaucum*）、高粱（*Sorghum bicolor*）、弯头高粱（*Sorghum cernuum*）、（*Sorghum contractum*）、小麦（*Triticum aestivum*）、玉米（*Zea mays*）、芋头（*Colocasia esculenta*）、爵床属（*Justicia*）、罂粟属（*Papaver*）、青葙属（*Celosia*）、酸模属（*Rumex*）、木樨草属（*Reseda*）、堇菜属（*Viola*）、醉蝶花属（*Cleome*）、木樨草属（*Reseda*）、苎麻属（*Boehmeria*）、蜀葵属（*Althaea*）、锦葵属（*Malva*）、大戟属（*Euphorbia*）、蔷薇属（*Rosa*）、（*Faba*）、木蓝属（*Indigofera*）、荆豆属（*Ulex*）、杧果属（*Mangifera*）、老鹳草属（*Geranium*）、天竺葵属（*Pelargonium*）、新风轮属（*Calamintha*）、薰衣草属（*Lavandula*）、欧夏至草属（*Marrubium*）、薄荷属（*Mentha*）、紫菀属（*Aster*）、大丽花属（*Dahlia*）、牛膝菊属（*Galinsoga*）、千里光属（*Senecio*）、山柳菊属（*Hieracium*）、风毛菊属（*Leontodon*）、蒲公英属（*Taraxacum*）、雄黄兰属（*Crocosmia*）、唐菖蒲属（*Gladiolus*）、美人蕉属（*Canna*）植物（Matov & Kononenko, 2012）。

Helicoverpa assulta (Guenée, 1852) 烟青虫

Heliothis assulta Gueneé, 1852, *in* Boisduval & Guenée, *Histoire Naturelle des Insectes. Species Général des Lépidoptères*, 6: 178. Type locality: Tahiti [MNHN, Paris].

Synonymy: *Heliothis temperata* Walker, 1857; *Heliothis separata* Walker, 1857; *Heliothis succinea* Moore, 1881; *Heliothis assulta afra* Hardwick, 1965 (ssp.).

Distribution. China (all over the country); Russia (Siberia), Korea, Japan, Pakistan, India, Nepal, Philippines, Indonesia, Australia, New Zealand; Middle East, Southeast Asia.

寄主：烟草、棉、麻、玉米、高粱、番茄、辣椒、南瓜（陈一心，1999）；木豆（*Cajanus cajan*）、毛鱼藤（*Derris elliptica*）、豌豆（*Pisum sativum*）、豇豆（*Vigna unguiculata*）、辣椒（*Capsicum annuum*）、番茄（*Lycopersicon esculentum*）、黄花烟草（*Nicotiana rustica*）、烟草（*Nicotiana tabacum*）、苦蘵（*Physalis angulata*）、小酸浆（*Physalis minima*）、灯笼果（*Physalis peruviana*）、茄（*Solanum melongena*）、紫苏（*Perilla frutescens*）、玉米（*Zea mays*）、胡椒属（*Piper*）、醋栗属（*Grossularia*）、雄黄兰属（*Crocosmia*）植物（Matov & Kononenko, 2012）。

Subfamily **Agaristinae** Boisduval, 1833 虎蛾亚科

Agaristides Boisduval, 1833, *Annales de la Société Entomologique de France*, 2: 218. Type genus: *Agarista* Leach, 1814.

Synonymy: Agaristoidea Herrich-Schäffer, [1858] 1850–8; Agaristiden Herrich-Schäffer, [1858] 1850–8; Agaristidae Walker, [1865] 1864.

Earlier authors treated this group who in the family, however its position as a subfamily Agaristinae of the family Noctuidae is well accepted. In the treatment and sequence of genera of the subfamily we follow to the revision of Palaearctic and Oriental faunas of Agaristidae by Kiriakoff (1977).

The subfamily includes about 600 species, mainly distributed in the tropical and subtropical regions of the Old and New Worlds. About 25 species of 11 genera are known from the Palaearctic region, mainly distributed in the Sino-Himalayan and Chinese subregions; among them, 3 species of 2 genera in the Three Provinces of Northeast China.

References. Strand, 1912 (Agaristidae); Kiriakoff, 1977 (Agaristidae); Chen, 1982, 1999 (Agaristidae); Sugi, 1982 (Agaristidae); Kononenko, 1987, 2003, 2005, 2010 (Agaristinae); Kononenko *et al.*, 1998 (Agaristinae); Fibiger, Lafontaine, 2005 (Agaristinae); Kononenko, Pinratana, 2005, 2013 (Agaristinae); Lafontaine, Fibiger, 2006 (Agaristinae); Kononenko, Han, 2007 (Agaristinae); Matov *et al.*, 2008 (Agaristinae); Holloway, 2011 (Agaristinae).

Genus ***Mimeusemia*** Butler, 1875 拟彩虎蛾属

Mimeusemia Butler, 1875, *Annals and Magazine of Natural History*, (4)15: 397. Type species: *Mimeusemia persimilis* Butler, 1875 [Japan].

The genus includes 32 species, mainly distributed in the Oriental region; 1 species is in the Three Provinces of Northeast China.

Mimeusemia persimilis **Butler, 1875 拟彩虎蛾**

Mimeusemia persimilis Butler, 1875, *Annals and Magazine of Natural History*, (4)15: 397. Type locality: Japan, Hokkaido, Hakodate.

Distribution. China (Heilongjiang, Jilin, Sichuan); Russia (Siberia), Korea, Japan.

寄主：山葡萄（*Vitis amurensis*）、软枣猕猴桃（*Actinidia arguta*）、蛇葡萄属（*Ampelopsis*）、乌蔹莓属（*Cayratia*）植物（Matov & Kononenko, 2012）。

Genus ***Sarbanissa*** Walker, 1865 **修虎蛾属**

Sarbanissa Walker, 1865, *List of the Specimens of Lepidopterous Insects in the Collection of the British Museum*, 33: 746. Type species: *Sarbanissa insocia* Walker, 1865 [India].

Synonymy: *Seudyra* Stretch, 1875; *Zalissa* auct..

The genus includes 29 species, distributed in the Oriental and the Chinese subregion of the Palaearctic regions; 2 species are in the Three Provinces of Northeast China.

Sarbanissa subflava **(Moore, 1877) 葡萄修虎蛾**

Seudyra subflava Moore, 1877, *Annals and Magazine of Natural History*, (4) 20: 85. Type locality: China, Zhejiang[Chekiang].

Synonymy: *Zalissa jankowskii* Alphéraky, 1897; *Seudyra subflava japonica* Bytinsky-Salz, 1939 (ssp.).

Distribution. China (Heilongjiang, Jilin, Liaoning, Hebei, Shandong, Hubei, Zhejiang, Jiangxi, Guizhou, Taiwan); Russia (Siberia), Korea, Japan.

寄主：葡萄、爬山虎（陈一心，1999）；蛇葡萄属（*Ampelopsis*）、乌蔹莓属（*Cayratia*）、地锦属（*Parthenocissus*）、葡萄属（*Vitis*）、常春藤属（*Hedera*）植物（Matov & Kononenko, 2012）。

Sarbanissa venusta **(Leech, [1889]) 艳修虎蛾**

Seudyra venusta Leech, [1889], *Proceedings of the Zoological Society of London*, 1888: 614, pl. 31: 2. Type locality: Japan.

Distribution. China (Heilongjiang, Jilin, Jiangsu, Zhejiang, Hubei, Sichuan); Russia (Siberia), Korea, Japan.

寄主：软枣猕猴桃（*Actinidia arguta*）、乌蔹莓（*Cayratia japonica*）（Park *et al.*, 2006）山葡萄（*Vitis amurensis*）（Matov & Kononenko, 2012）。

Subfamily **Condicinae** Poole, 1995 点夜蛾亚科

Condicinae Poole, 1995, *in* Dominick *et al.*, *Moths of America North of Mexico*, 26 (1): 19. Type genus: *Condica* Walker, 1856.

Synonymy: Acosmetiini Beck, 1999; Acosmetiina Beck, 1999.

Condicinae is relatively small, distributed the worldwide, the subfamily comprises externally astonishingly different genera and species. The members of the subfamily are distributed in the open and the forested areas of the tropical or subtropical area; few genera and species are known in the temperate zone. Earlier authors associated most genera of Condicinae with Amphipyrinae (auct.). In the Old World the subfamily has been recognized since the revision of Noctuidae classification by Fibiger & Lafontaine (2005). The subfamily is subdivided into 2 tribes, Condicini and Leuconyctini (Poole, 1995). It includes 26 genera and about 460 species, distributed from the southern temperate zone of Holarctic to the tropical of the Oriental and Neotropical regions where the subfamily is especially diverse. The larvae of Condicinae feed on the leaves of herbaceous plants of families Asteraceae and Polygonaceae, many species are polyphagous. 2 tribes include 19 species of 12 genera, are known from the Three Provinces of Northeast China, among them 2 species of 2 genera are incertae sedis.

References. Chen, 1982, 1999 (Noctuidae); Sugi, 1982 (Amphipyrinae, part); Kononenko *et al.*, 1998 (Amphipyrinae, part); Kitching, Rawlins, 1999 [1998] (Condicinae); Kononenko, 2003 (Amphipyrinae, part); Fibiger, Hacker, 2005, 2007 (Condicinae); Fibiger, Lafontaine, 2005 (Condicinae); Kononenko, 2005, 2010 (Condicinae); Lafontaine, Fibiger, 2006 (Condicinae); Kononenko, Han, 2007 (Condicinae); Matov *et al.*, 2008 (Condicinae); Holloway, 2011 (Condicinae); Fibiger *et al.*, 2011 (Condicinae); Kononenko, Pinratana, 2013 (Condicinae); Kononenko, 2016 (Condicinae: Condicini, Leuconyctini).

注：首次给出中文亚科名。

Tribe **Condicini** Poole, 1995 点夜蛾族

Condicinae Poole, 1995, *in* Dominick *et al.*, *Moths of America North of Mexico*, 26 (1): 19. Type genus: *Condica* Walker, 1856.

Synonymy: Acosmetiini Beck, 1999; Acosmetiina Beck, 1999.

The tribe includes many genera of the tropical area. It most diverse in the Neotropical region. The tribe Condicini enumerates 9 genera and 12 species are in the Three Provinces of Northeast China. Most of them inhabit nemoral deciduous forests in the broadleaves forest zone.

References. Chen, 1982, 1999 (Noctuidae); Sugi, 1982 (Amphipyrinae, part); Kitching, Rawlins 1998 (Condicinae, Condicini); Kononenko *et al.*, 1998 (Amphipyrinae, part); Kononenko, 2003 (Amphipyrinae, part); Fibiger, Hacker, 2005, 2007 (Condicinae, Condicini); Fibiger, Lafontaine, 2005 (Condicinae, Condicini); Kononenko, 2005, 2010 (Condicinae, Condicini); Lafontaine, Fibiger, 2006 (Condicinae, Condicini); Kononenko, Han, 2007 (Condicinae, Condicini); Matov *et al.*, 2008 (Condicinae, Condicini); Holloway, 2011 (Condicinae,

Condicini); Fibiger *et al.*, 2011 (Condicinae, Condicini); Kononenko, Pinratana, 2013 (Condicinae, Condicini).

注：首次给出中文族名。

Genus *Condica* Walker, 1856 点夜蛾属

Condica Walker, 1856, *List of the Specimens of Lepidopterous Insects in the Collection of the British Museum*, 9: 214, 239. Type species: *Condica palpalis* Walker, 1856 ["West Indies"] (=*Phalaena cupentia* Cramer, 1779).

Synonymy: *Gaphara* Walker, 1862; *Platysenta* Grote, 1874; *Myrtale* Druce, 1891; *Bicondica* Berio, 1981; *Monocondica* Berio, 1981; *Perigea* auct..

The genus contains 77 species and some number of species with uncertain systematic position. It is most diverse in the tropical region, especially in the Neotropical region. 3 species are in the Three Provinces of Northeast China.

Condica illecta **(Walker, 1865) 易点夜蛾**

Mamestra illecta Walker, 1865, *List of the Specimens of Lepidopterous Insects in the Collection of the British Museum*, 32: 667. Type locality: India, North Hindustan.

Synonymy: *Hadena funesta* Walker, 1865; *Hadena spargens* Walker, 1865.

Distribution. China (Heilongjiang, Jilin, Liaoning, Hainan, Yunnan); Russia (Siberia), Sri Lanka, India, Nepal, Pakistan, Vietnam, Thailand, Malaysia, Indonesia, Philippines; Middle East, Oceania.

寄主：藿香蓟、金鸡菊（陈一心，1999）；鸭儿芹（*Cryptotaenia japonica*）、藿香蓟（*Ageratum conyzoides*）、宽叶鱼眼草（*Dichrocephala latifolia*）、地胆草（*Elephantopus scaber*）、一点红（*Emilia sonchifolia*）、鼠麴草（*Gnaphalium purpureum*）、向日葵（*Helianthus annuus*）、卤地菊（*Wedelia prostrata*）、咖啡属（*Coffea*）、红花属（*Carthamus*）、金鸡菊属（*Coreopsis*）、大丽花属（*Dahlia*）、假泽兰属（*Mikania*）、爵床科（Acanthaceae）植物（Matov & Kononenko, 2012）。

Condica illustrata **(Staudinger, 1888) 明点夜蛾**

Perigea illustrata Staudinger, 1888, *Entomologische Zeitung. Entomologischen Vereine zu Stettin*, 49: 256. Type locality: Russia, [Primorye terr.] "Amurgebiet", "Ussuri".

Synonymy: *Segetia albopicta* Graeser, 1892.

Distribution. China (Heilongjiang); Russia (Siberia), Korea, Japan.

Condica cyclicoides **(Draudt, 1950) 环点夜蛾**

Perigea cyclicoides Draudt, 1950, *Mitteilungen der Münchner Entomologischen Gesellschaft*, 40: 94, pl. 6: 16. Type locality: China, Hoeng-shan, Tien-mu-shan, Tai-pei-shan, Lung-tan.

Distribution. China (Heilongjiang, Jilin, Shaanxi, Hubei, Zhejiang); Russia (Siberia), Korea, Japan.

注：首次给出中文种名。

Genus *Prospalta* Walker, [1858] 1857 普夜蛾属

Prospalta Walker, [1858] 1857, *List of the Specimens of Lepidopterous Insects in the Collection of the British Museum*, 13: 1079, 1114. Type species: *Prospalta leucospila* Walker, [1858] 1857 [India].

Synonymy: *Propsalta* auct., missp..

The genus includes over 20 species distributed mostly in the tropical and subtropical areas of the Oriental region. The name *Prospalta* has often been misapplied in literature to *Condica,* taxa Holloway (1989). 1 species is in the Three Provinces of Northeast China.

***Prospalta cyclica* (Hampson, 1908) 圆点普夜蛾**

Perigea cyclica Hampson, 1908, *Catalogue of the Lepidoptera Phalaenae in the British Museum*, 7: 315, pl. 115: 29. Type locality: Japan, Yokohama.

Distribution. China (Heilongjiang, Jilin, Hebei); Russia (Siberia), Korea, Japan.

注：《中国动物志》（夜蛾科）中误将本种归入"*Condica* Walker, 1856 点夜蛾属"，在此予以变更。

Genus *Acosmetia* Stephens, 1829 圆夜蛾属

Acosmetia Stephens, 1829, *The Nomenclature of British Insectes*, 43. Type species: *Noctua caliginosa* Hübner, [1813] [Europe].

Synonymy: *Hydrilla* Biosduval, 1840.

The genus includes about 20 species, distributed in the Mediterranean, Central Asian and Chinese subregions Palaearctic, Ethiopian and Oriental regions. 9 species occur in the Palaearctic, among them 2 species are in the Three Provinces of Northeast China.

注：首次给出中文属名。《中国动物志》（夜蛾科）中将本属种类归入"*Hadjina* Staudinger, 1891 赫夜蛾属"，但赫夜蛾属的种类主要分布于地中海、埃塞俄比亚、印度亚区，在东亚地区目前暂无分布记录；在此对其包含种类中文种名予以变更。

***Acosmetia chinensis* (Wallengren, 1860) 中圆夜蛾**

Perigea chinensis Wallengren, 1860, *Wiener Entomologische Monatschrift*, 4: 169. Type locality: China.

Synonymy: *Segetia mandarina* Staudinger, 1892; *Amyna pulverea* Leech, 1900; *Acosmetia rubecula* Draudt, 1950; *Acosmetia vagans* Berio, 1977.

Distribution. China (Heilongjiang, Jilin, Liaoning, Heibei, Jiangxi, Sichuan); Russia (Siberia), Korea, Japan, India.

寄主：野艾蒿（*Artemisia vulgaris*）（Matov & Kononenko, 2012）。

Acosmetia biguttula **(Motschulsky, 1866) 白斑圆夜蛾**

Mamestra biguttula Motschulsky, 1866, *Bulletin de la Société Impériale des Naturalistes de Moscou*, 39 (1): 194. Type locality: Japan.

Distribution. China (Heilongjiang, Jilin, Liaoning); Russia (Siberia, South Kuriles), Korea, Japan.

寄主：狼拔草（*Bidens tripartita*）（Matov & Kononenko, 2012）。

Genus *Iambia* Walker, 1863 雅夜蛾属

Iambia Walker, 1863, *List of the Specimens of Lepidopterous Insects in the Collection of the British Museum*, vol. 27: 109. Type species: *Iambia inferalis* Walker, 1863 [South Africa].

The genus includes around 23 species, distributed in the Palaearctic, Ethiopian and Oriental regions, among them 1 species is in the Three Provinces of Northeast China.

Iambia transversa **(Moore, 1882) 贯雅夜蛾**

Tycracona transversa Moore, 1882, *in* Hewitson & Moore, *Descriptions of new Indian Lepidopterous Insects from the Collection of the Late Mr. W.S. Atkinson* (*Heterocera*): 95, pl. 4: 5. Type locality: India, West Bengal, Darjiling.

Synonymy: *Jambia* [sic!] *sinica* Yang, 1964.

Distribution. China (Heilongjiang, Jilin, Liaoning, Shandong, Hubei, Yunnan); Korea, Japan.

Genus *Chytonix* Grote, 1874 流夜蛾属

Chytonix Grote, 1874, *Bulletin of the Buffalo Society of Natural Sciences*, 2: 14, 66. Type species: *Apamea iaspis* Guenée, 1852 [USA].

The genus includes over 40 species are in the Chinese subregion of the Palaearctic, Nearctic, Oriental, Ethiopian and Neotropical regions, some are still not described yet. In the Palaearctic region, about 10 species; 2 species are in the Three Provinces of Northeast China.

Chytonix albonotata **(Staudinger, 1892) 白点流夜蛾**

Bryophila albonotata Staudinger, 1892, *in* N.M. Romanoff (ed.), *Mémoires sur les Lépidoptères*, 6: 396, pl. 5: 9. Type locality: Russia, Primorye terr. ["Ussuri"].

Synonymy: *Chytonix nigirbasalis* Hampson, 1908; *Chytonix olethria* Wileman & West, 1929.

Distribution. China (Heilongjiang, Jilin, Jiangsu, Zhejiang, Hubei, Hunan, Fujian); Russia (Siberia), Korea, Japan.

Genus *Niphonyx* Sugi, 1982 乏夜蛾属

Niphonyx Sugi, 1982, *Moths of Japan*, 1: 779; 2: 373, 462. Type species: *Miana segregata* Butler, 1878 [Japan].

Monotypic Chinese genus.

Niphonyx segregata (Butler, 1878) 乏夜蛾

Miana segregata Butler, 1878, *Annals and Magazine of Natural History*, (5)1: 85. Type locality: Japan, Yokohama.

Synonymy: *Telesilla placens* Staudinger, 1888.

Distribution. China (Heilongjiang, Jilin, Liaoning, Hebei, Henan, Fujian, Yunnan); Russia (Siberia), Korea, Japan.

寄主：葎草（陈一心，1999；Matov & Kononenko, 2012）。

Genus *Oligonyx* Sugi, 1982 曲线禾夜蛾属

Oligonyx Sugi, 1982, *Moths of Japan*, 1: 779, 780; 2: 373, 463. Type species: *Miana vulnerata* Butler, 1878 [Japan].

Monotypic Chinese genus.

注：首次给出中文属名。

Oligonyx vulnerata (Butler, 1878) 曲线禾夜蛾

Miana vulnerata Butler, 1878, *Annals and Magazine of Natural History*, (5)1: 84. Type locality: Japan, Tokyo.

Synonymy: *Hadena christophi* Staudinger, 1888, nec Alphéraky, 1888, preocc.; *Hadena doerriesi* Staudinger, 1892, repl. name.

Distribution. China (Heilongjiang, Hebei, Hubei, Jiangxi); Russia (Siberia), Korea, Japan.

寄主：水蓼（*Polygonum hydropiper*）、酸模叶蓼（*Polygonum lapathifolium*）、戟叶蓼（*Polygonum thunbergii*）（Matov & Kononenko, 2012）。

注：《中国动物志》（夜蛾科）中将本种归入"*Oligia* Hübner, [1821] 1816 禾夜蛾属"，在此予以变更。

Genus *Pyrrhidivalva* Sugi, 1982 污禾夜蛾属

Pyrrhidivalva Sugi, 1982, *Moths of Japan*, 1: 780; 2: 373, 463. Type species: *Glottula sordida* Butler, 1871 [Japan].

Monotypic Northest Chinese genus; 1 species is in the Three Provinces of Northeast China.

注：首次给出中文属名。

***Pyrrhidivalva sordida* (Butler, 1881) 污禾夜蛾**

Glottula sordida Butler, 1881, *Transactions of the Entomological Society of London*, 1881: 147. Type locality: Japan, Yokohama.

Synonymy: *Gerbatha pseudodyops* Butler, 1884; *Segetia variegata* Oberthür, 1884.

Distribution. China (Heilongjiang, Jilin, Liaoning, Nei Mongol); Russia (Siberia), Korea, Japan.

寄主：戟叶蓼（*Polygonum thunbergii*）、水芹（*Oenanthe javanica*）、蒿属（*Artemisia*）植物（Matov & Kononenko, 2012）。

注：《中国动物志》（夜蛾科）中将本种归入"*Oligia* Hübner, [1821] 1816 禾夜蛾属"，在此予以变更。

Genus *Dysmilichia* Speiser, 1902 井夜蛾属

Dysmilichia Speiser, 1902, *Berliner Entomologische Zeitschrift,* 47: 140 repl. name for *Milichia* Snellen, [1899]. Type species: *Perigea gemella* Leech, 1889 [Japan, Yokohama].

Synonymy: *Phalacra* Staudinger, 1892 nec Walker, 1866 (Lepidoptera, Drepanidae); *Milichia* Snellen, 1898, nec Meigen, 1830 (Diptera).

The genus includes 14 species, distributed in the Mediterranean and Chinese subregions of the Palaearctic region, among them 1 species is in the Three Provinces of Northeast China.

***Dysmilichia gemella* (Leech, 1889) 井夜蛾**

Perigea gemella Leech, 1889, *Proceedings of the Zoological Society of London*, 1889: 492, pl. 53: 12. Type locality: Japan, Yokohama; Korea, Wonsan [Gensan].

Synonymy: *Dysmilichia gemella sutchanica* Filipjev, 1927 (ssp.).

Distribution. China (Heilongjiang, Jilin, Liaoning, Hebei, Zhejiang, Fujian); Russia (East), Korea, Japan.

寄主：紫苏（*Perilla frutescens*）（Matov & Kononenko, 2012）。

Tribe **Leuconyctini** Poole, 1995 北美夜蛾族

Leuconyctini Poole, 1995, *in* Dominick *et al.*, *Moths of America North of Mexico*, 26 (1): 204. Type genus: *Leuconycta* Hampson, 1908.

Rather small group, mainly distributed in the temperate zone of Holarctic region. The subtribe represented by 1 genus and 5 species are in the Three Provinces of Northeast China.

References. Chen, 1982, 1999 (Noctuidae); Sugi, 1982 (Amphipyrinae, part); Kitching, Rawlins, 1998 (Condicinae, Leuconyctini); Kononenko *et al.*, 1998 (Amphipyrinae, part); Kononenko, 2003 (Amphipyrinae, part); Fibiger, Hacker, 2005, 2007 (Condicinae, Leuconyctini); Fibiger, Lafontaine, 2005 Condicinae, Leuconyctini); Kononenko, 2005, 2010 (Condicinae, Leuconyctini); Lafontaine, Fibiger, 2006 (Condicinae, Leuconyctini); Kononenko, Han, 2007 (Condicinae, Leuconyctini); Matov *et al.*, 2008 (Condicinae,

Leuconyctini); Holloway, 2011 (Condicinae, Leuconyctini); Fibiger *et al*., 2011 (Condicinae, Leuconyctini).

Genus ***Eucarta*** Lederer, 1857 希夜蛾属

Eucarta Lederer, 1857, *Die Noctuinen Europa's*, *mit Zuziehung einger bisher mesit dazu gezahlter Arten des Asiatischen Russland's*, *Kleinasien's*, *Syrien's*, *und Labrador's*: 41, 168. Type species: *Noctua amethystina* Hübner, 1803 [Europe].

Synonymy: *Placodes* Boisduval, 1840, preocc. (Erichson, 1834 [Coleoptera]); *Telesilla* H.– S., 1856, repl. name, preocc. (Reichenbach, 1853 [Aves]); *Callogonia* Hampson, 1908, preocc. (Dall, 1889 [Mollusca]); *Dexiadena* Filipjev, 1927; *Goonallica* Nye, 1975, repl. name.

The genus includes 8 species, distributed in the Palaearctic region, mainly in its eastern part; among them 5 species are in the Three Provinces of Northeast China.

Eucarta amethystina **(Hübner, [1803]) 希夜蛾**

Noctua amethystina Hübner, [1803], *Sammlung Europaischer Schmetterlinge. Eulen*, 2: pl. 14: 69. Type locality: Europe.

Synonymy: *Telesilla amethystina austera* Warren, 1911 (ssp.).

Distribution. China (Heilongjiang, Jilin, Hubei); Russia (European part, Siberia), Korea, Japan, Kazakhstan; Asia Minor, Europe.

寄主：野胡萝卜、前胡属（陈一心，1999）；峨参（*Anthriscus silvestris*）、野胡萝卜（*Daucus carota*）、岩风（*Libanotis intermedia*）、水芹（*Oenanthe javanea*）、欧芹（*Petroselinum crispum*）、洋香菜（*Petroselinum sativum*）、准噶尔前胡（*Peucedanum morisonii*）、欧洲前胡（*Peucedanum officinale*）、沼地前胡（*Peucedanum palustre*）、宽叶石防风（*Peucedanum terebinthaceum*）、亮叶芹（*Silaum silaus*）、独活属（*Heracleum*）、蝇子草属（*Silene*）植物（Matov & Kononenko, 2012）。

Eucarta virgo **(Treitschke, 1835) 麟角希夜蛾**

Plusia Abrostola virgo Treitschke, 1835, *Die Schmetterlinge von Europa*, 10 (2): 130. Type locality: "Shrimen".

Synonymy: *Placodes spencei* Boisduval, 1840; *Plusia virgo* Motshulsky, 1866; *Platychasma virgo* Butler, 1881 *Calogonia virgo roseonitens* Sheljuzhko, 1927 (var.); *Calogonia virgo virginalis* Bryk, 1949 (ssp.); *Calogonia virgo griseofulgens* Kovács, 1968 (ssp.); *Eucarta virgo euroargenta* Rrezbani-Reser, 1999 (ssp.).

Distribution. China (Heilongjiang, Jilin, Liaoning, Nei Mongol, Hubei); Russia (European part, Siberia), Korea, Japan, Kazakhstan; Asia Minor, Europe.

寄主：岩风（*Libanotis buchtormensis*）、兴安前胡（*Peucedanum baicalense*）、（*Artemisia campestre*）、猪毛蒿（*Artemisia scoparia*）、北艾（*Artemisia vulgaris*）、艾菊（*Chrysanthemum vulgare*）、菊蒿（*Tanacetum vulgare*）、柳属（*Salix*）、薄荷属（*Mentha*）、蒲公英属（*Taraxacum*）植

物（Matov & Kononenko, 2012）。

***Eucarta arcta* (Lederer, 1853) 直线希夜蛾**

Hadena arcta Lederer, 1853, *Verhandlungen des Zoologisch-Botanischen vereins in Wien*, 3: 370, pl. 2: 5. Type locality: East Kazakhstan, West Altai, Ust-Kamengorsk, Ust-Buchtarminsk.

Synonymy: *Miana parietum* Oberthür, 1880.

Distribution. China (Heilongjiang, Jilin, Xinjiang, Fujian); Russia (Siberia, Altai, Siberia), Korea, Japan.

寄主：蒿属（*Artemisia*）植物（Matov & Kononenko, 2012）。

***Eucarta fasciata* (Butler, 1878) 纹希夜蛾**

Raphia fasciata Butler, 1878, *Annals and Magazine of Natural History*, (45)1: 193. Type locality: Japan, Yokohama.

Synonymy: *Oligia tegularis* Bryk, 1948.

Distribution. China (Heilongjiang, Jilin); Russia (Siberia), Korea, Japan.

***Eucarta arctides* (Staudinger, 1888) 窄直希夜蛾**

Hadena arctides Staudinger, 1888, *Entomologische Zeitung. Entomologischen Vereine zu Stettin*, 49: 251. Type locality: Russia, [Primorye terr.], Askold I..

Distribution. China (Heilongjiang, Jilin, Hebei, Hubei); Russia (Siberia), Korea, Japan.

寄主：魁蒿（*Artemisia princeps*）（Matov & Kononenko, 2012）。

Subfamily Condicinae, *ircerfac sedis* 点夜蛾亚科内隶属未明确

Genus ***Plusilla*** Staudinger, 1892 朴夜蛾属

Plusilla Staudinger, 1892, *in* N.M. Romanoff (ed.), *Mémoires sur les Lépidoptères*, 6: 533. Type species: *Plusilla rosalia* Staudinger, 1892 [Russia, Vladivostok].

Monotypic Chinese genus; 1 species is in the Three Provinces of Northeast China.

***Plusilla rosalia* Staudinger, 1892 朴夜蛾**

Plusilla rosalia Staudinger, 1892, *in* N.M. Romanoff (ed.), *Mémoires sur les Lépidoptères*, 6: 533, pl. 9: 11. Type locality: Russia, Vladivostok.

Distribution. China (Heilongjiang, Jilin, Liaoning, Jiangsu, Hubei); Russia (Siberia), Korea, Japan.

寄主：水蓼（*Polygonum hydropiper*）（Matov & Kononenko, 2012）。

Genus *Prometopus* Guenée, 1852 原井夜蛾属

Prometopus Guenée, 1852, *in* Boisduval & Guenée, *Histoire Naturelle des Insectes. Species Général des Lépidoptères*, 5: 38. Type species: *Prometopus inassueta* Guenée, 1852 [Australia].

The genus includes 6 species, distributed in Chinese subregion of the Palaearctic, Oriental and Australian regions; 1 species is in the Three Provinces of Northeast China.

Prometopus flavicollis (Leech, 1889) 原井夜蛾

Erastria flavicollis Leech, 1889, *Proceedings of the Zoological Society of London*, 1889: 525, pl. 53: 4. Type locality: Japan, Oiwake.

Distribution. China (Heilongjiang, Jilin); Russia (Siberia), Korea, Japan.

寄主：红松（*Pinus koraiensis*）、（*Laryx*）等植物（Matov & Kononenko, 2012）

Subfamily **Eriopinae** Herrich-Schäffer, [1851] 1845 散纹夜蛾亚科

Eriopodides Herrich-Schäffer, [1851] 1845, *Systematische Bearbeitung der Schmetterlinge von Europa*, 2: 379, emended in Eriopidae. Type genus: *Eriopus* Treitschke, 1825 (=*Callopistria* Hübner, [1821]).

Synonymy: Eriopidae Herrich-Schäffer, [1851] 1845; Eriopides Herrich-Schäffer, [1851] 1845; Callopistriidae Butler, 1892.

This group was defined by Fibiger & Lafontaine (2005) as a subfamily of Noctuidae. It is highly enigmatic subfamily, distributed worldwide, especially diverse in the tropical region. The known larvae of many species prefer ferns as hostplant, a very rare choice among Noctuidae species. The subfamily is monophyletic because of many peculiar character states both in the larvae and the adults. Due high similarity in the external appearance and the genitalia this group is extremely difficult for identification. The Taiwan fauna of Eriopinae has been reviewed by Yen & Wu (2009), they concluded that the subfamily should probably be restricted to *Callopistria.*

There are 7 genera and about 160 species are in the World fauna. Most taxa are distributed mainly in the tropical and subtropical areas of the Ethiopian and Oriental regions; 5 species of 1 genus in the Three Provinces of Northeast China.

References. Chen, 1982, 1999 (Noctuidae: Amphipyrinae); Sugi, 1982 (Amphipyrinae, part); Kononenko, 1990 (Amphipyrinae, part); Kononenko *et al.*, 1998 (Amphipyrinae, part); Kitching, Rawlins, 1999 [1998] (Hadeninae); Kononenko, 2003 (Amphipyrinae, part); Fibiger, Hacker, 2005 (Eriopinae); Fibiger, Lafontaine, 2005 (Eriopinae); Kononenko, 2005, 2010 (Eriopinae); Lafontaine, Fibiger, 2006 (Eriopinae); Kononenko, Han, 2007 (Eriopinae); Matov *et al.*, 2008 (Eriopinae); Fibiger *et al.*, 2009 (Eriopinae); Yen, Wu, 2009 (Eriopinae); Zahiri *et al.*, 2013 (Eriopinae); Kononenko, 2016 (Eriopinae).

注：首次给出中文亚科名。

Genus *Callopistria* Hübner, [1821] 1816 散纹夜蛾属

Callopistria Hübner, [1821] 1816, *Verzeichniss bekannter Schmettlinge* [sic!]: 216. Type species: *Phalaena juventina* Stoll, 1782 ["Surinam" - by mistake; ?Europe].

Synonymy: *Lagopus* Reichenbach, 1817, preocc. (Brisson, 1760 [Aves]); *Eriopus* Treitschke, 1825, repl. name; *Callopistria* Stephens, 1850, missp.; *Agraga* Walker, 1858; *Diethusa* Walker, 1858; *Agabra* Walker, 1862; *Obana* Walker, 1863, preocc. (Walker, 1862 [Lepidoptera, Noctuidae]); *Eulepa* Walker, [1863] 1864; *Cotanada* Moore, 1881; *Methorasa* Moore, 1881; *Herrichia* Grote, 1882, preocc. (Staudinger, 1871 [Lepidoptera, Oecophoridae]); *Euherrichia* Grote, 1882, repl. name; *Dissolophus* Butler, 1891; *Gnamptocera* Butler, 1891; *Haplolophus* Butler, 1891; *Hemipachycera* Butler, 1891; *Hyperdasys* Butler, 1891; *Rhoptrotrichia* Butler, 1891; *Platydasys* Butler, 1892; *Agaba* Pagenstecher, 1909, missp.; *Miropalpa* Berio, 1955.

The genus includes over 100 species, distributed mainly in the tropical and subtropical areas of the Old and New Worlds; 5 species are known from the Three Provinces of Northeast China.

Callopistria juventina (Stoll, 1782) 散纹夜蛾

Phalaena juventina Stoll, 1782, *in* Cramer, *Uitlandsche Kapellen*, 4: 245, pl. 400: N. Type locality: "Surinam" [sic!], by mistake.

Synonymy: *Noctua purpureofasciata* Piller & Mitterpacher, 1783; *Phalaena* (*Noctua*) *lagopus* Esper, 1788; *Noctua pteridis* Fabricius, 1794; *Pyralis formosissimalis* Hübner, 1796; *Callopistria obscura* Butler, 1878; *Eriopus aetnea* Costa, 1840.

Distribution. China (Heilongjiang, Jilin, Liaoning, Henan, Jiangsu, Zhejiang, Hubei, Hunan, Jiangxi, Fujian, Guangxi, Hainan, Sichuan); Russia (European part), Korea, Japan, India, Vietnam, Thailand(North); Middle East, Europe, North Africa.

寄主：蕨类（陈一心，1999）；药蕨（*Ceterach officinarum*）、欧洲蕨（*Pteridium aquilinum*）、卷柏属（*Selaginella*）、铁线蕨属（*Adiantum*）、旱蕨属（*Pellaea*）、铁角蕨属（*Asplenium*）、骨碎补属（*Davallia*）、栗蕨属（*Histiopteris*）、肾蕨属（*Nephrolepis*）、伏石蕨属（*Lemmaphyllum*）、凤尾蕨属（*Pteris*）、海金沙属（*Lygodium*）、荚果蕨属（*Matteuccia*）植物（Matov & Kononenko, 2012）。

Callopistria repleta Walker, 1858 红晕散纹夜蛾

Callopistria repleta Walker, 1858, *List of the Specimens of Lepidopterous Insects in the Collection of the British Museum*, 12: 865. Type locality: North India.

Synonymy: *Callopistria repleta postpallida* A.E. Prout, 1928 (ssp.).

Distribution. China (Heilongjiang, Jilin, Liaoning, Shaanxi, Shanxi, Henan, Zhejiang, Hubei, Hunan, Fujian, Guangxi, Hainan, Sichuan, Yunnan); Russia (Siberia), Korea, Japan, India, Pakistan, Vietnam, Thailand, Malaysia, Indonesia.

寄主：东北蹄盖蕨（*Athyrium brevifrons*）（Park *et al.*, 2006）；全缘贯众（*Cyrtomium falcatum*）、棕鳞大耳蕨（*Polystichum polyblepharum*）、日本蹄盖蕨（*Athyrium niponicum*）、欧洲蕨（*Pteridium aquilinum*）、井栏边草（*Pteris multifida*）、普通针毛蕨（*Macrothelypteris torresiana*）（Matov & Kononenko, 2012）。

Callopistria albolineola (Graeser, [1889] 1888) 白线散纹夜蛾

Eriopus albolineola Graeser, [1889] 1888, *Berliner entomologische Zeitschrift*, 32: 337. Type locality: Russia, Amur, Pokrofka.

Synonymy: *Eriopus rivularis japonibia* Strand, 1916 (ab.).

Distribution. China (Jilin, Guangdong, Hainan); Russia (Irkutsk reg., Transbaikalia, Siberia), Korea, Japan.

寄主：卷柏（*Selaginella tamariscina*）（Matov & Kononenko, 2012）。

Callopistria argyrosticta (Butler, 1881) 显斑散纹夜蛾

?Perigea argyrosticta Butler, 1881, *Transactions of the Entomological Society of London*, 1881: 177. Type locality: Japan, Tokyo.

Synonymy: *Eriopus venus* Staudinger, 1888; *Eriopus argentata* Graeser, 1888 [1889].

Distribution. China (Jilin); Russia (Siberia), Korea, Japan.

注：首次给出中文种名。

"Callopistria" miracula Herz, 1904 斜斑散纹夜蛾

Callopistria miracula Herz, 1904, *Annuaire du Musée Zoologique de l'Académie Impériale des Sciences de St.-Petersbourg*, 9: 284, pl. 1: 10. Type locality: Korea, Pungtung village [38.5°N, 128°E].

Distribution. China (Jilin, Liaoning); Korea.

注：首次给出中文种名。

Subfamily **Bryophilinae** Guenée, 1852 苔藓夜蛾亚科

Bryophilidae Guenée, 1852, *in* Boisduval & Guenée, *Histoire Naturelle des Insectes. Species Général des Lépidoptères*, 5: 21. Type genus: *Bryophila* Treitschke, 1825.

Synonymy: Bryophilidae Guenée, 1852; Jaspidiinae Aubert, 1952; Cryphiinae Inoue & Sugi, 1958; Cryphiinae Beck, 1960; Cryphiini Beck, 1996; Bryoleucini Beck, 1996; Cryphinae auct..

Bryophilinae is a rather small, compact group, with 12 genera and about 300 species worldwide considered by some previous authors (Franclemont, Todd, 1983) as a tribe of the subfamily Acronictinae, or as a subfamily closely related to Acronictinae (Sugi, 1982). Kitching & Rawlins (1999) fixed the status of the Bryophilinae as a subfamily rank and Fibiger & Lafontaine (2005) and Zilli *et al.* (2009) define the position of the Bryophlinae. The subfamily mainly allocated on the basis of uniform morphology and peculiar type of the larvae nutrition associated with lichens. The moths are nocturnal and attracted by light. The eggs are laid on or near lichens on rocks, walls, or bark. The larvae of known species feed on lichens, moss, and algae growing on tree trunks, stones or on the rocks.

Bryophilinae are mainly distributed in the Palaearctic region with centres of diversity in the Mediterranean and Chinese subregions, a number of unique species of the genus *Stenoloba* are known from the Oriental region; the genus *Athaumasta* most richly represented in the mountains of South Siberia and Central Asia. There are 5 genera and 15 species are in the Three Provinces of Northeast China.

References. Boursin, 1951, 1954b, 1961 (Cryphinae); Sugi, 1970, 1980 (Bryophilinae); Chen, 1982, 1999 (Noctuidae); Kononenko, 1990, 2003, 2005, 2010 (Bryophilinae); Kononenko *et al.*, 1998 (Bryophilinae); Kononenko, Ronkay, 2000, 2001 (Bryophilinae); Fibiger, Lafontaine, 2005 (Bryophilinae); Fibiger, Hacker, 2005 (Bryophilinae); Lafontaine, Fibiger, 2006 (Bryophilinae); Kononenko, Han, 2007 (Bryophilinae); Matov *et al.*, 2008 (Bryophilinae); Fibiger *et al.*, 2009 (Bryophilinae).

Genus *Cryphia* Hübner, 1818 藓夜蛾属

Cryphia Hübner, 1818, *Zuträge zur Sammlung Exotischer Schmetterlinge*, 1: 14. Type species: *Noctua receptricula* Hübner, 1803 [Europe].

Synonymy: *Euthales* Hübner, [1820] 1816 (subgen.); *Jaspidia* Hübner, 1822; *Bryophila* Treitschke, 1825; *Bryonycta* Boursin, 1955 (subgen.); *Scythobrya* Boursin, 1960 (subgen.); *Hymenocryphia* Boursin, 1967 (subgen.); *Nyctobrya* Boursin, 1975 (subgen.); *Cryphiomima* Berio, 1977; *Heterocryphia* Beck, 1996; *Metachrostis* auct..

Mainly a Palaearctic genus containing 44 species with the centre of diversity in Mediterranean subregion, some species are known from the Ethiopian and Nearctic regions. 1 subgenus and 6 species are in the Three Provinces of Northeast China.

注：《中国动物志》（夜蛾科）中将本属中文属名定为“苔藓夜蛾属”，但考虑到“Bryophilinae Guenée, 1852 苔藓夜蛾亚科”的模式属 *Bryophila* Treitschke, 1825 应称为苔藓夜蛾属，为了更好与“*Cryphia* Hübner, 1818 苔藓夜蛾属”相区别，在此将本属“*Cryphia* Hübner, 1818”中文属名变更为“藓夜蛾属”。

Subgenus *Cryphia* Hübner, 1818 藓夜蛾亚属

Cryphia Hübner, 1818, *Zuträge zur Sammlung Exotischer Schmetterlinge*, 1: 14. Type species: *Noctua receptricula* Hübner, [1803] [Europe].

注：首次更改中文亚属名。

***Cryphia* (*Cryphia*) *mediofusca* (Sugi, 1959) 短圆藓夜蛾**

Metachrostis mediofusca Sugi, 1959, *Tinea*, 5 (1): 186, fig. 2. Type locality: Japan, Hokkaido, Kushiro, Shibecha.

Distribution. China (Jilin, Liaoning); Russia (Siberia), Korea, Japan.

寄主：（Lichenophyta）（Matov & Kononenko, 2012）。

注：首次给出中文种名。

Cryphia (_Cryphia_) _bryophasma_ (Boursin, 1951) 北藓夜蛾

Bryophila (*Cryphia*) *bryophasma* Boursin, 1951, *Zeitschrift der Wiener Entomologischen Gesellschaft*, 36: 155, pl. 11: 1-3. Type locality: Russia, Primorye terr., [Partizansk] Sutschanskii rudnik".

Distribution. China (Heilongjiang, Jilin, Liaoning); Russia (Siberia), Korea, Japan.

寄主：（Lichenophyta）（Matov & Kononenko, 2012）。

注：首次给出中文种名。

Cryphia (_Cryphia_) _mitsuhashi_ (Marumo, 1917) 晦藓夜蛾

Bryophila mitsuhashi Marumo, 1917, *in* Nagano *et al.*, *A Collection of Essays for Mr. Yasushi Nawa, Written in Commeration of his Sixtieth Birthday*: 27, fig. 4. Type locality: Japan, Tokyo.

Synonymy: *Metachrostis obscura* Warren, 1909.

Distribution. China (Jilin); Korea, Japan.

Cryphia (_Cryphia_) _minutissima_ (Draudt, 1950) 小藓夜蛾

Bryophila minutissima Draudt, 1950, *Mitteilungen der Münchner Entomologischen Gesellschaft*, 40: 15, pl. 3: 5. Type locality: China, Zhejiang, Tianmushan[West Tien-mu-shan].

Synonymy: *Bryophila minutissina hunana* Draudt, 1950 (ssp.).

Distribution. China (Jilin, Liaoning, Hunan, Zhejiang), Korea, Japan.

Cryphia (_Cryphia_) _griseola_ (Nagano, 1918) 彩藓夜蛾

Bryophila griseola Nagano, 1918, *Insect World*, 22: 450, pl. 12: 4. Type locality: Japan.

Synonymy: *Metachrostis splendida* O.Bang-Haas, 1927; *Metachrostis algae korealgae* Bryk, 1948 (ssp.).

Distribution. China (Jilin, Liaoning); Russia (Siberia), Korea, Japan.

寄主：（Lichenophyta）（Matov & Kononenko, 2012）。

注：首次给出中文种名。

Cryphia (_Cryphia_) _sugitanii_ Boursin, 1961 白缘藓夜蛾

Cryphia sugitanii Boursin, 1961, *Zeitschrift der Wiener Entomologischen Gesellschaft*, 46: 139, pl. 10: 1. Type locality: Japan, Mt. Daisen.

Distribution. China (Liaoning); Russia (Siberia), Korea, Japan.

寄主：（Lichenophyta）（Matov & Kononenko, 2012）。

注：首次给出中文种名。

Genus *Bryophila* Treitschke, 1825 苔藓夜蛾属

Bryophila Treitschke, 1825, *Die Schmetterlinge von Europa*, 5 (1): 57. Type species: *Noctua perla* [Denis & Schiffermüller], 1775 [Austria] (=*Phalaena domestica* Hufnagel, 1766), repl. name for *Poecilia* Schrank, 1802, nec Schneider, 1801 (Pisces).

Synonymy: *Poecilia* Schrank, 1802, preocc. (Schneider, 1801 [Pisces]); *Jaspidia* Hübner, [1806], suppr. (ICZN, 1926 Op. 97; 1954 Op. 278); *Bryoleuca* Hampson, 1908 (subgen.); *Jaspidia* Ochsenheimer, 1816, unavail.; *Transbryoleuca* Beck, 1996.

The genus subdivided into 2 subgenera, *Bryophila* (s. str.) and *Bryoleuca*. It includes about 50 species distributed in the Mediterranean, Central Asian and Chinese subregions of the Palaearctic region. One subgenus, 3 species are in the Three Provinces of Northeast China.

注：首次给出中文属名。

Subgenus *Bryoleuca* Hampson, 1908 苔藓夜蛾亚属

Bryoleuca Hampson, 1908, *Catalogue of the Lepidoptera Phalaenae in the British Museum*, 7: xv, 686. Type species: *Miana trilinea* Bethune-Baker, 1894 [Egypt].

注：首次给出中文亚属名。

***Bryophila* (*Bryoleuca*) *raptricula* ([Denis & Schiffermüller], 1775) 乔苔藓夜蛾**

Noctua raptricula [Denis & Schiffermüller], 1775, *Ankündung eines Systematisches Werkes von den Schmetterlinge der Wiener Gegend. Wien*: 89. Type locality: Austria, Vienna.

Synonymy: *Phalaena* (*Noctua*) *divisa* Esper, [1791]; *Phalaena* (*Noctua*) *palliola* Borkhausen, 1792; *Phalaena* (*Noctua*) *pomula* Borkhausen, 1792; *Noctua deceptricula* Hübner, 1803; *Bryophila carbonis* Freyer, 1849; *Bryophila oxybienisis* Milliere, 1875; *Bryophila striata* Staudinger, 1879; *Bryophila raptricula eretina* Caberla, 1888 (var.); *Bryophila raptriculoides* Turati, 1912; *Bryophila divisa persica* Strand, 1915 (ab.); *Bryophila raptricula palaestinica* Strand, 1915 (f.); *Bryophila raptricula grisea* Dannehl, 1929 (f.); *Cryphia raptricula cretica* Reisser, 1962 (ssp.).

Distribution. China (Heilongjiang, Hebei, Shandong); Russia (European part, Ural, Altai, Transbaikalia, Siberia); Central Asia, Middle East, Europe, North Africa.

寄主：兜藓（陈一心，1999）；藻纹梅花衣（*Parmelia saxatilis*）、（*Parmelia centrifuga*）、（*Lecanora saxicola*）、（*Sticta pulmonaria*）、原球藻（*Protococcus viridis*）（Matov & Kononenko, 2012）。

注：《中国动物志》（夜蛾科）中将本种归入“*Cryphia* Hübner, 1818 藓夜蛾属”，在此予以变更。

Bryophila* (*Bryoleuca*) *granitalis **(Butler, 1881) 斑苔藓夜蛾**

Gerbatha granitalis Butler, 1881, *Transactions of the Entomological Society of London*, 1881: 194. Type locality: Japan, Tokyo.

Synonymy: *Bryophila glaucula* Staudinger, 1892; *Metachrostis leprosa* Warren, 1909; *Metachrostis fraudatricula*: Leech, 1900, misident.

Distribution. China (Heilongjiang, Jilin, Liaoning, Hebei, Shandong, Jiangsu, Zhejiang, Hunan, Jiangxi, Fujian); Russia (Irkutsk reg., Transbaikalia, Siberia), Korea, Japan.

寄主：苔藓（陈一心，1999）；（Lichenophyta）（Matov & Kononenko, 2012）。

注：《中国动物志》（夜蛾科）中将本种归入"*Cryphia* Hübner, 1818 藓夜蛾属"，称为"斑藓夜蛾"，在此予以变更。

Bryophila* (*Bryoleuca*) *orthogramma **(Boursin, 1954) 纹苔藓夜蛾**

Cryphia orthogramma Boursin, 1954, *Zeitschrift der Wiener Entomologischen Gesellschaft*, 39: 81, pl. 5: 10. Type locality: Russia, East Siberia, Irkutsk.

Synonymy: *Cryphia orthogramma aksuensis* Boursin, 1954 (ssp.); *Cryphia orthogramma ammochroa* Boursin, 1954 (ssp.); *Cryphia orthogramma taishanensis* Boursin, 1954 (ssp.); *Cryphia cancellata* Berio, 1977.

Distribution. China (Heilongjiang, Jilin, Liaoning, Shandong); Russia (Ural, Altai, Transbaikalia, Siberia), Korea, Japan, India(North); Central Asia, Europe.

寄主：汇合网衣（*Lecidea confluens*）、（*Xanthoria janetina*）、（*Sticta parietina*）（Matov & Kononenko, 2012）。

注：《中国动物志》（夜蛾科）中将本种归入"*Cryphia* Hübner, 1818 藓夜蛾属"，称为"纹藓夜蛾"，在此予以变更。

Genus ***Bryomoia*** Staudinger, 1892 **黑绿藓夜蛾属**

Bryomoia Staudinger, 1892, *in* N.M. Romanoff (ed.), *Mémoires sur les Lépidoptères*, 6: 397. Type species: *Bryophila melachlora* Staudinger, 1892 [Russia, Primorye terr., Askold I.].

Synonymy: *Bryomoea* Hampson, 1908, emend..

The genus name *Bryomoia* was proposed by Staudinger (1892) as a subgenus of *Bryophila* Treitschke, 1825. Monotypic Chinese genus; 1 species is in the Three Provinces of Northeast China.

注：首次给出中文属名。

Bryomoia melachlora **(Staudinger, 1892) 黑绿藓夜蛾**

Bryophila (?) (*Bryomoia*) *melachlora* Staudinger, 1892, *in* N.M. Romanoff (ed.), *Mémoires sur les Lépidoptères*, 6: 397, pl. 5: 10. Type locality: Russia, Primorye terr., Askold I..

Distribution. China (Jilin, Liaoning, Taiwan); Russia (Siberia), Korea, Japan.

注：首次给出中文种名。

Genus *Athaumasta* Hampson, 1906 虚冬藓夜蛾属

Athaumasta Hampson, 1906, *Catalogue of the Lepidoptera Phalaenae in the British Museum*, 6: 380, repl. name for *Thaumasta* Staudinger, 1871. Type species: *Polia expressa* Lederer, 1855 [Russia, Altai Mts.].

Synonymy: *Thaumasta* Staudinger, 1871, preocc. (Gilst, 1848 [Crustacea]).

It contains about 20 species, while some species from West China and Mongolia are not yet described; 1 species is in the Three Provinces of Northeast China.

注：《中国动物志》（夜蛾科）中将本属归入冬夜蛾亚科，称为“虚冬夜蛾属”，根据属变动，其中文种名在此予以变更。

Athaumasta koreana Ronkay & Kononenko, 1998 朝虚冬藓夜蛾

Athaumasta koreana Ronkay & Kononenko, 1998, *Illustrated Catalogue of the Noctuidae in Korea*. Insects of Korea, 3: 201, 390, fig. 517. Type locality: North Korea, Gangwon, Mts. Kumgang-san, Manmulsang rocks, 750 m.

Synonymy: *Athaumasta siderigera*: Ronkay and Park, 1993: 55, nec Christoph, 1893, misident.

Distribution. China (Liaoning); Russia (West of the Primorye terr.), North Korea.

Genus *Stenoloba* Staudinger, 1892 兰纹夜蛾属

Stenoloba Staudinger, 1892, *Die Macrolepidopteren des Amurgebiets. I. Thiel. Rhopalocera, Sphinges, Bombyces, Noctuae. Mémoires sur les Lépidoptéres*, 6: 381. Type species: *Dichagyris jankowskii* Oberthür, 1884 [Russia, East Siberia].

Synonymy: *Neothripa* Hampson, 1894; *Conicochyta* Hampson, 1918.

The genus presently includes 75 species, arranged in 14 species groups; among them, 4 species are in the Three Provinces of Northeast China.

注：《中国动物志》（夜蛾科）中将本属归入“绮夜蛾亚科”，在此予以变更。

Stenoloba assimilis (Warren, 1909) 暗基兰纹夜蛾

Metachrostis assimilis Warren, 1909, *in* A.Seitz (ed.), *Die Gross-Schmetterlinge der Erde*, 3: 22, pl. 4: i. Type locality: Japan, Tsushima I.

Distribution. China (Jilin, Liaoning, Taiwan); Russia (Siberia), Korea, Japan.

寄主：（Lichenophyta）（Matov & Kononenko, 2012）。

注：首次给出中文种名。

Stenoloba jankowskii (Oberthür, 1884) 兰纹夜蛾

Dichagyris jankowskii Oberth ü r, 1884, *Études d'Entomologie*, 10: 28, pl. 3: 5. Type locality: Russia,

Primorye terr., Bezverkhovo [Sidemi].

Synonymy: *Edema nivilinea* Leech, 1888; *Stenoloba jancousci* Hampson, 1910, emend..

Distribution. China (Heilongjiang, Jilin, Liaoning, Zhejiang, Yunnan); Russia (Siberia), Korea, Japan.

Stenoloba oculata Draudt, 1950 灰兰纹夜蛾

Stenoloba oculata Draudt, 1950, *Mitteilungen der MünchnerEntomologischen Gesellschaft*, 40: 132, pl. 8: 20. Type locality: China, Hunan, Hoeng-shan; Zhejiang, Tianmushan[West Tien-mu-shan].

Distribution. China (Heilongjiang, Jilin, Liaoning, Jiangsu, Zhejiang, Hunan, Jiangxi), Korea, Japan.

寄主：（Lichenophyta）（Matov & Kononenko, 2012）。

Stenoloba clara (Leech, 1889) 细兰纹夜蛾

Selepa manleyi clara Leech, 1889, *Proceedings of the Zoological Society of London*, 1889: 479. Type locality: Japan, Yokohama.

Distribution. China (Jilin, Liaoning, Fujian), Korea, Japan.

Subfamily **Noctuinae** Latreille, 1809 夜蛾亚科

Noctuidae Latreille, 1809, *Genera Crustaceorum Insectorum* 4: 191, 224 (as Noctuaelites), incorrect original spelling which had to be emended to Noctuidae; official list of family-group names, opinion 450. Type genus: *Noctua* Linnaeus, 1758.

Synonymy: Noctuo-Bombycites Latreille, 1809; Noctuida Leach, [1815] 1830; Noctuides Leach, [1815] 1830; Noctuelidi Boisduval, [1828] 1829; Noctuelidae Duponchel, [1845] 1844; Noctuelites Guenée, 1852; Noctues Swinhoe, 1890.

The concept of this subfamily has been changed dramatical by moving part of genera traditionally placed by earlier authors to Amphipyrinae (sensu auctorum) and Cuculliinae (sensu auctorum) or Hadeninae (sensu Kitching & Rawlins, 1999 [1998]) to tribes Elaphriini, Prodenini, Caradrinini, Dipterygini, Actinotini, Phlogophorini, Apameini in subfamily Xyleninae (Fibiger & Lafontaine, 2005; Lafontaine & Fibiger, 2006) and subsequent downgrading the status of Xyleninae, Hadeninae and Noctuinae to tribes. The modern concept of Noctuinae was introduced by Lafontaine & Schmidt (2010) with a system of tribes within it. They defined tribes formerly placed to subfamily Xyleninae (Fibiger & Lafontaine, 2005; Lafontaine & Fibiger, 2006) and downgraded the status of the subfamilies Xyleninae (sensu Fibiger & Lafontaine, 2005), Hadeninae and Noctuidae to tribes.

In modern concept Noctuinae is a large subfamily of Noctuidae. It enumerates over 6100 species all over the world. The Noctuinae are distributed in all continents except the Antarctic region. The maximal diversity of the subfamily is observed in the forest zone, mountains, steppes and savannas of the temperate belt of Northern and Southern hemispheres. The subfamily divided into tribes Balsini, Pseudeustrotiini, Prodenini, Elaphriini,

Caradrinini, Dypterygiini, Actinotiini, Phlogophorini, Apameini, Xylenini, Hadenini and Noctuini. Trophical connections and habitat preferences are very variable within tribes. The herbaceous polyphagous species are dominating in most tribes. Most species of Prodenini, Elaphriini, Caradrinini are widely polyphagous, feeding of herbaceous plants; Actinotiini are olygophagous on Clusiaceae; Apameini – mainly olygophagous stem borers of monocotyledonous herbaceous plants. The tribe Xylenini includes many forest species, most of them are polyphagous, feeding on arboreous plants.

The subfamily Noctuinae represents in the Three Provinces of Northeast China by 14 tribes, 146 genera and 359 species.

References. Hampson, 1903 (Agrotinae, part), 1905 (Hadeninae, part), 1906 (Cucullianae, part), 1909 (Acronyctinae, part); Kozhantschikov, 1937 (Agrotinae); Zolotarenko, 1970 (Agrotinae); Chen, 1982, 1999 (Noctuidae: Hadeninae, Amphipyrinae, Noctuinae); Sugi, 1982 (Cuculliinae, part; Amphipyrinae, part; Hadeninae, Noctuinae); Holloway, 1989 (Noctuinae, Hadeninae, Amphipyrinae); Fibiger, 1990, 1993, 1997 (Noctuinae); Speidel *et al.*, 1996 (Noctuinae, Hadeninae, Amphipyrinae); Kononenko *et al.*, 1998 (Cuculliinae, part; Amphipyrinae, part; Hadeninae, part; Noctuinae); Kitching, Rawlins, 1999 [1998] (Hadeninae, Noctuinae); Ronkay *et al.*, 2001 (Hadeninae); Hacker *et al.*, 2002 (Hadeninae); Kononenko, 2003 (Cuculliinae, part; Amphipyrinae, Hadeninae, Noctuinae); Fibiger, Hacker, 2005 (Xyleninae, Hadeninae, Noctuinae); Fibiger, Lafontaine, 2005 (Xyleninae, Hadeninae, Noctuinae); Kononenko, 2005 (Xyleninae, Hadeninae, Noctuinae); Zilli *et al.*, 2005 (Noctuidae, Apameini); Lafontaine, Fibiger, 2006 (Xyleninae, Hadeninae, Noctuinae); Fibiger, Hacker, 2007 (Xyleninae); Kononenko, Han, 2007 (Cuculliinae, Xyleninae, Hadeninae, Noctuinae); Matov *et al.*, 2008 (Xyleninae, Hadeninae, Noctuinae); Lafontaine, Schmidt, 2010 (Noctuinae); Holloway, 2011 (Noctuinae); Fibiger *et al.*, 2011 (Xyleninae, Hadeninae, Noctuinae); Zahiri *et al.*, 2013b (Noctuinae); Kononenko, 2016 (Noctuinae).

Tribe **Pseudeustrotiini** Beck, 1996 清文夜蛾族

Pseudeustrotiini Beck, 1996, *Neue entomologische Nachrichten*, 36: 16. Type genus: *Pseudeustrotia* Warren, 1913.

Pseudeustrotia, the type genus of *Pseudeustrotiini* was traditionally associated by previous authors with the subfamily Eustrotiinae as close relatives of *Deltote* species. The tribe *Pseudeustrotiini* was proposed by Beck (1996) because the larva of *Pseudeustotia candidula* has a full complement of prolegs, two SV setae on A1 and SV1 on A9 is hair-like, all characteristics of the Noctuinae (in modern sense) and not of the Eustrotiinae. A position for the *Pseudeustrotia* in the "higher" trifids was confirmed by Speidel *et al.* (1996a) because of the raised nodular sclerite (epaulette) on the tympanal membrane and the presence of fully developed basal abdominal brushes and pockets in the male. Speidel *et al.* raised the group to subfamily status and treated it as one of several unresolved sister groups to the Cuculliinae *s. l.* and Noctuinae *s. l.* clade because of the absence of a corona on the male valva. Fibiger & Lafontaine (2005) restored the tribal status of Pseudeustrotinae and placed it to the subfamily Xyleninae; later its position was defined as a tribe of the Noctuinae (Lafontaine & Schmidt, 2010). The subfamily includes 2 genera, *Pseudeustrotia* and *Anterastria*, while some other genera (*Phyllophila*, presently in Eustrotiinae) might be associated with *Pseudeustrotiini.*

The tribe numbers 2 genera and 2 species are in the Three Provinces of Northeast China.

References. Chen, 1982, 1999 (Noctuidae); Sugi, 1982 (Amphipyrinae, part); Speidel *et al.*, 1996 (Pseudeustrotiinae); Kononenko *et al.*, 1998 (Amphipyrinae, part); Kitching, Rawlins, 1999 [1998] (Hadeninae); Kononenko, 2003 (Amphipyrinae, part); Fibiger, Hacker, 2005 (Xyleninae: Pseudeustrotiini); Fibiger, Lafontaine, 2005 (Xyleninae: Pseudeustrotiini); Kononenko, 2005 (Xyleninae: Pseudeustrotiini); Lafontaine, Fibiger, 2006 (Xyleninae: Pseudeustrotiini); Fibiger, Hacker, 2007 (Xyleninae: Pseudeustrotiini); Kononenko, Han, 2007 (Xyleninae: Pseudeustrotiini); Matov *et al.*, 2008 (Pseudeustrotiini); Lafontaine, Schmidt, 2010 (Noctuidae, Pseudeustrotiini); Fibiger *et al.*, 2011 (Xyleninae: Pseudeustrotiini); Kononenko, 2016 (Noctuinae, Pseudeustrotiini).

注：首次给出中文族名。

Genus ***Pseudeustrotia*** Warren, 1913 清文夜蛾属

Pseudeustrotia Warren, 1913, *in* A.Seitz (ed.), *Die Gross-Schmetterlinge der Erde*, 3: 288. Type species: *Noctua candidula* [Denis & Schiffermüller], 1775 [Austria].

The genus includes 6 species, distributed in the Palaearctic, Nearctic and Oriental regions; 1 species is in the Three Provinces of Northeast China.

注：首次给出中文属名。

Pseudeustrotia candidula **([Denis & Schiffermüller], 1775) 清文夜蛾**

Noctua candidula [Denis & Schifferm ü ller], 1775, *Ankündung eines Systematisches Werkes von den Schmetterlinge der Wiener Gegend. Wien*: 89. Type locality: Austria, Vienna.

Synonymy: *Noctua pusilla* Vieweg, 1790; *Eustrotia candidula leechiana* Bryk, 1948 (ssp.).

Distribution. China (Heilongjiang, Jilin, Xinjiang); Russia (European part, Siberia), Mongolia, Korea, Japan; Asia Minor, Europe.

寄主：荞麦（*Fagopyrum esculentum*）、萹蓄 (*Polygonum aviculare*)、拳参（*Polygonum bistorta*）、卷茎蓼（*Polygonum convolvulus*）、酸模（*Rumex acetosa*）、小酸模（*Rumex acetosella*）、长叶酸模（*Rumex longifolius*）、湿草地东北酸模（*Rumex tyrsiflorus*）、蓝沼草属（*Molinia*）、梯牧草属（*Phleum*）、黑三棱属（*Sparganium*）植物（Matov & Kononenko, 2012）。

注：《中国动物志》（夜蛾科）中将本种归入“*Eustrotia* Hübner, [1821] 文夜蛾属”，在此予以变更。

Genus ***Anterastria*** Sugi, 1982 黑俚夜蛾属

Anterastria Sugi, 1982, *Moths of Japan*, 1: 818, 2: 373, 464. Type species: *Erastria atrata* Butler, 1881 [Japan].

The genus contains 3 species, distributed in the Chinese subregion of the Palaearctic and Nearctic regions; 1 species is in the Three Provinces of Northeast China.

注：首次给出中文属名。

***Anterastria atrata* (Butler, 1881) 黑俚夜蛾**

Erastria atrata Butler, 1881, *Transactions of the Entomological Society of London*, 1881: 188. Type locality: Japan, Tokyo, Yokohama.

Synonymy: *Erastria sidemiata* Oberthür, 1884.

Distribution. China (Heilongjiang, Jilin, Liaoning, Fujian, Sichuan); Russia (Siberia), Korea, Japan.

寄主：香薷属（*Elsholtzia*）、薄荷属（*Mentha*）、紫苏属（*Perilla*）植物（Matov & Kononenko, 2012）。

注：《中国动物志》（夜蛾科）中将本种归入"*Lithacodia* Hübner, 1818 俚夜蛾属"，在此予以变更。

Tribe **Prodeniini** Forbes, 1954 灰翅夜蛾族

Prodeniini Forbes, 1954, *Memoirs of the Cornell University agriculture Experimental Station*, 329: 168. Type genus: *Prodenia* Guenée, 1852.

The tribe Prodeniini includes 3 genera with many species and most diverse in tropical area over the World (distributed world wide *Spodoptera* and African *Ariathisa* Walker, 1865, and *Neolaphygma* Hampson, 1908). 1 genus and 4 species are in the Three Provinces of Northeast China.

References. Chen, 1982, 1999 (Noctuidae); Sugi, 1982 (Amphipyrinae, part); Speidel *et al.*, 1996 (Amphipyrinae, part); Kononenko *et al.*, 1998 (Amphipyrinae, part); Kitching, Rawlins, 1999 [1998] (Hadeninae); Pogue, 2002 (Amphipyrinae, part); Kononenko, 2003 (Amphipyrinae, part); Fibiger, Hacker, 2005 (Xyleninae: Prodeniini); Fibiger, Lafontaine, 2005 (Xyleninae: Prodeniini); Kononenko, 2005 (Xyleninae: Prodeniini); Lafontaine, Fibiger, 2006 (Xyleninae: Prodeniini); Fibiger, Hacker, 2007 (Xyleninae: Prodeniini); Kononenko, Han, 2007 (Xyleninae: Prodeniini); Matov *et al.*, 2008 (Xyleninae: Prodeniini); Lafontaine, Schmidt, 2010 (Noctuidae, Prodeniini); Fibiger *et al.*, 2011 (Xyleninae: Prodeniini).

注：首次给出中文族名。

Genus ***Spodoptera*** Guenée, 1852 灰翅夜蛾属

Spodoptera Guenée, 1852, *in* Boisduval & Guenée, *Histoire Naturelle des Insectes. Species Général des Lépidoptères*, 5: 153. Type specics: *Hadena mauritia* Boisduval, 1833 [Mauritius].

Synonymy: *Laphygma* Guenée, 1852; *Prodenia* Guenée, 1852; *Calogramma* Guenée, 1852; *Ariathisa* Walker, 1865; *Eulaphygma* Butler, 1890; *Rusidrina* Staudinger, 1892; *Spodosoma*: Anonymous, 1966, missp..

The genus includes more than 40 species, distributed worldwide, mainly in the arid and semi-arid zones of the tropical and subtropical areas. 9 species are in the Palaearctic region; 4 species are in the Three Provinces of

Northeast China. The genus *Spodoptera* on global level has been revised by Pogue (2002).

Spodoptera exigua **(Hübner, [1808]) 甜菜夜蛾**

Noctua exigua Hübner, [1808], *Sammlung Europaischer Schmetterlinge. Eulen*, 4: pl. 78: 362 *Noctua*. Type locality: Europe.

Synonymy: *Noctua fulgens* Geyer, [1832]; *Caradrina pygmaea* Rambur, 1834; *Caradrina junceti* Zeller, 1847; *Laphygma caradrinoides* Walker, 1856; *Caradrina flavimaculata* Harvey, 1876; *Caradrina sebghana* Austaut, 1880; *Caradrina venosa* Butler, 1880; *Laphygma exigua antipodea* Warren, 1914 (ssp.); *Douzdrina protector* de Laever, 1985.

Distribution. China (except Northwest region) ; Russia (European part, Ural, Siberia), Korea, Japan, Nepal, India, Thailand, Vietnam, Pakistan, Mongolia; Asia Minor, Middle East, Central Asia, Europe.

寄主：甜菜、棉、马铃薯、豆类、番茄、多种蔬菜（陈一心，1999）；欧洲凤尾蕨（*Pteris cretica*）、黄细心（*Boerhavia diffusa*）、马齿苋（*Portulaca oleracea*）、康乃馨（*Dianthus caryophyllus*）、缕丝花（*Gypsophila elegans*）、圆锥石头花（*Gypsophila paniculata*）、长芒苋（*Amaranthus palmeri*）、反枝苋（*Amaranthus retroflexus*）、刺苋（*Amaranthus spinosus*）、苋（*Amaranthus tricolor*）、沿海甜菜（*Beta maritima*）、甜菜（*Beta vulgaris*）、藜（*Chenopodium album*）、俄罗斯蓟（*Salsola iberica*）、菠菜（*Spinacia oleracea*）、春蓼（*Persicaria maculosa*）、萹蓄(*Polygonum aviculare*)、酸模叶蓼（*Polygonum lapathifolium*）、皱叶酸模（*Rumex crispus*）、日本金缕梅（*Hamamelis japonica*）、兔眼蓝莓（*Vaccinium ashei*）、茶（*Camellia sinensis*）、四倍体西瓜（*Citrullus vulgaris*）、甜瓜（*Cucumis melo*）、瓠果（*Cucumis sativus*）、西葫芦（*Cucurbita pepo*）、异株栝楼（*Trichosanthes dioica*）、拟南芥（*Arabidopsis thaliana*）、芸苔（*Brassica campestris*）、青菜（*Brassica chinensis*）、黑芥（*Brassica nigra*）、甘蓝（*Brassica oleracea*）、芜青（*Brassica rapa*）、洋基芥（*Brassica tournefortii*）、疣果匙荠（*Bunias orientalis*）、芝麻菜（*Eruca vesicaria*）、豆瓣菜（*Nasturtium officinale*）、萝卜（*Raphanus sativus*）、咖啡黄葵（*Abelmoschus esculentus*）、草棉（*Gossypium herbaceum*）、印度棉（*Gossypium indicum*）、大麻槿（*Hibiscus cannabinus*）、木槿（*Hibiscus syriacus*）、圆叶锦葵（*Malva rotundifolia*）、冬葵（*Malva verticillata*）、蓖麻（*Ricinus communis*）、榅桲（*Cydonia oblonga*）、（Malus domestica）、苹果（Malus pumila）、黑橄榄树（*Bucida buceras*）、花生（*Arachis hypogaea*）、黄芪（*Astragalus crotalariae*）、树锦鸡儿（*Caragana arborescens*）、菽麻（*Crotalaria juncea*）、印度黄檀（*Dalbergia sissoo*）、（*Faba bona*）、大豆（*Glycine max*）、家山黧豆（*Lathyrus sativus*）、兵豆（*Lens culinaris*）、羽扇豆（*Lupinus arizonicus*）、紫苜蓿（*Medicago sativa*）、印度草木樨（*Melilotus indica*）、黄香草木樨（*Melilotus officinalis*）、芒柄花（*Ononis arvensis*）、（*Phaseolus lanatus*）、菜豆（*Phaseolus vulgaris*）、豌豆（*Pisum sativum*）、刺田菁（*Sesbania bispinosa*）、埃及车轴草（*Trifolium alexandrinum*）、黑吉豆（*Vigna mungo*）、豇豆（*Vigna unguiculata*）、日本七叶树（*Aesculus turbinata*）、葡萄柚（*Citrus paradisi*）、甜橙（*Citrus sinensis*）、亚麻（*Linum usitatissimum*）、天竺葵（*Pelargonium hortorum*）、菱叶葡萄（*Cissus rhombifolia*）、葡萄（*Vitis vinifera*）、东北土当归（*Aralia continentalis*）、鹅掌藤

（*Schefflera arboricola*）、旱芹（*Apium graveolens*）、芫荽（*Coriandrum sativum*）、野胡萝卜（*Daucus carota*）、欧防风（*Pastinaca sativa*）、洋香菜（*Petroselinum sativum*）、南非栀子（*Gardenia thunbergia*）、草原龙胆（*Eustoma grandiflorum*）、欧洲夹竹桃（*Nerium oleander*）、辣椒（*Capsicum annuum*）、五彩椒（*Capsicum frutescens*）、白夜丁香（*Cestrum diurnum*）、枸杞（*Lycium chinense*）、番茄（*Lycopersicon esculentum*）、光烟草（*Nicotiana glauca*）、黄花烟草（*Nicotiana rustica*）、烟草（*Nicotiana tabacum*）、碧冬茄（*Petunia hybrida*）、茄（*Solanum melongena*）、马铃薯（*Solanum tuberosum*）、昼颜（*Calystegia sepium*）、田旋花（*Convolvulus arvensis*）、甘薯（*Ipomoea batatas*）、全缘叶牵牛花（*Ipomoea hederacea* var. *integriuscula*）、芝麻（*Sesamum indicum*）、地黄（*Rehmannia glutinosa*）、琴叶毛蕊花（*Verbascum chinense*）、长叶车前（*Plantago lanceolata*）、鸡骨柴（*Elsholtzia fruticosa*）、紫苏（*Perilla frutescens*）、黄芩（*Scutellaria baicalensis*）、魁蒿（*Artemisia princeps*）、红花（*Carthamus tinctorius*）、杭菊（*Chrysanthemum morifolium*）、（*Geraea canescens*）、向日葵（*Helianthus annuus*）、莴苣（*Lactuca sativa*）、欧洲千里光（*Senecio vulgaris*）、早生一枝黄花（*Solidago gigantea*）、唐菖蒲（*Gladiolus gandavensis*）、香葱（*Allium ascalonicum*）、洋葱（*Allium cepa*）、葱（*Allium fistulosum*）、韭菜（*Allium tuberosum*）、（*Allium odorum*）、石刁柏（*Asparagus officinalis*）、姜（*Zingiber officinale*）、燕麦（*Avena sativa*）、狗牙根（*Cynodon dactylon*）、穇子（*Eleusine coracana*）、稻（*Oryza sativa*）、象草（*Pennisetum purpureum*）、高粱（*Sorghum bicolor*）、玉米（*Zea mays*）、蜀葵（*Alcea rosea*）、铁角蕨属（*Asplenium*）、贝壳杉属（*Agathis*）、南天竹属（*Nandina*）、针晶粟草属（*Gisekia*）、滨藜属（*Atriplex*）、补血草属（*Limonium*）、栎属（*Quercus*）、醉蝶花属（*Cleome*）、碎米荠属（*Cardamine*）、黄麻属（*Corchorus*）、苘麻属（*Abutilon*）、榆属（*Ulmus*）、蔷薇属（*Rosa*）、木蓝属（*Indigofera*）、驴食草属（*Onobrychis*）、老鹳草属（*Geranium*）、春黄菊属（*Anthemis*）、菊苣属（*Cichorium*）、大丁草属（*Gerbera*）、蒲公英属（*Taraxacum*）、竹芋属（*Maranta*）、结缕草属（*Zoysia*）、花叶万年青属（*Dieffenbachia*）植物（Matov & Kononenko, 2012）。

***Spodoptera litura* (Fabricius, 1775) 斜纹夜蛾**

Noctua litura Fabricius, 1775, *Entomologiae Systematica emendata et Aucta. Secundum Classes, Ordines, Genera, Species Adjectis Synonimis, Locis, Observationibus, Descriptionibus*: 601. Type locality: East India.

Synonymy: *Noctua histrionica* Fabricius, 1775; *Noctua elata* Fabricius, 1781; *Prodenia ciligera* Guenée, 1852; *Prodenia tasmanica* Guenée, 1852; *Prodenia glaucistriga* Walker, 1856; *Prodenia subterminalis* Walker, 1856; *Prodenia declinata* Walker, 1857; *Mamestra alparsa* Walker, 1862; *Prodenia evanescens* Butler, 1884.

Distribution. China (Heilongjiang, Jilin, Liaoning, Shandong, Jiangsu, Zhejiang, Hunan, Fujian, Guangdong, Hainan, Guizhou, Yunnan); Russia (Siberia), (subtropical and tropical Asia, from Pakistan throughout India, South China, Indonesia to Philippines and New Guinea.)

寄主：甘薯、棉、芋、荷、向日葵、烟草、芝麻、玉米、高粱、瓜类、豆类及多种蔬菜等（陈一心，1999）；侧柏（*Platycladus orientalis*）、黄金侧柏（*Thuja orientalis*）、番荔枝（*Annona squamosa*）、莲（*Nelumbo nucifera*）、蓟罂粟（*Argemone mexicana*）、罂粟（*Papaver somniferum*）、黄

细心（*Boerhavia diffusa*）、落葵（*Basella alba*）、康乃馨（*Dianthus caryophyllus*）、凹头苋（*Amaranthus blitum*）、苋（*Amaranthus tricolor*）、皱果苋（*Amaranthus viridis*）、青葙（*Celosia argentea*）、穗状鸡冠花（*Celosia plumosa*）、甜菜（*Beta vulgaris*）、藜（*Chenopodium album*）、（*Chenopodium macrospermum*）、菠菜（*Spinacia oleracea*）、木麻黄（*Casuarina equisetifolia*）、山地木麻黄（*Casuarina junghuhniana*）、茶（*Camellia sinensis*）、山柿（*Diospyros montana*）、龙珠果（*Passiflora foetida*）、番木瓜（*Carica papaya*）、冬瓜（*Benincasa hispida*）、西瓜（*Citrullus lanatus*）、红瓜（*Coccinia grandis*）、甜瓜（*Cucumis melo*）、黄瓜（*Cucumis sativus*）、南瓜（*Cucurbita moschata*）、棱角丝瓜（*Luffa acutangula*）、丝瓜（*Luffa cylindrica*）、苦瓜（*Momordica charantia*）、云南木鳖（*Momordica dioica*）、佛手瓜（*Sechium edule*）、芥蓝（*Brassica alboglabra*）、芸苔（*Brassica campestris*）、擘蓝（*Brassica caulorapa*）、青菜（*Brassica chinensis*）、大叶芥菜（*Brassica juncea*）、黑芥（*Brassica nigra*）、甘蓝（*Brassica oleracea*）、芜青（*Brassica rapa*）、萝卜（*Raphanus sativus*）、翼籽辣木（*Moringa pterygosperma*）、（*Shorea macroptera*）、可可（*Theobroma cacao*）、黄麻（*Corchorus capsularis*）、长蒴黄麻（*Corchorus olitorius*）、咖啡黄葵（*Abelmoschus esculentus*）、蜀葵（*Althaea rosea*）、草棉（*Gossypium herbaceum*）、印度棉（*Gossypium indicum*）、朱槿（*Hibiscus rosasinensis*）、山黄麻（*Trema orientalis*）、榴莲蜜（*Artocarpus integer*）、无花果（*Ficus carica*）、菩提树（*Ficus religiosa*）、桑（*Morus alba*）、黑桑（*Morus nigra*）、三叶橡胶（*Hevea brasiliensis*）、（*Macaranga bancana*）、木薯（*Manihot esculenta*）、蓖麻（*Ricinus communis*）、森林苹果（*Malus sylvestris*）、白桃（*Persica vulgaris*）、欧洲李（*Prunus domestica*）、大叶桉（*Eucalyptus robusta*）、细叶桉（*Eucalyptus tereticornis*）、玫瑰苹果（*Eugenia malaccensis*）、红胶木（*Lophostemon confertus*）、番石榴（*Psidium guajava*）、榄仁树（*Terminalia alata*）、毛榄仁（*Terminalia tomentosa*）、台湾相思（*Acacia confusa*）、阿拉伯金合欢（*Acacia nilotica*）、花生（*Arachis hypogaea*）、木豆（*Cajanus cajan*）、鹰嘴豆（*Cicer arietinum*）、蝴蝶花豆（*Clitorea ternatea*）、瓜尔豆（*Cyamopsis tetragonoloba*）、毛鱼藤（*Derris elliptica*）、大豆（*Glycine max*）、野大豆（*Glycine soja*）、扁豆（*Lablab purpureus*）、家山黧豆（*Lathyrus sativus*）、紫苜蓿（*Medicago sativa*）、菜豆（*Phaseolus vulgaris*）、豌豆（*Pisum sativum*）、决明子（*Senna obtusifolia*）、刺田菁（*Sesbania bispinosa*）、大花田菁（*Sesbania grandiflora*）、埃及车轴草（*Trifolium alexandrinum*）、胡卢巴（*Trigonella foenum-graecum*）、黑吉豆（*Vigna mungo*）、绿豆（*Vigna radiata*）、班巴拉花生（*Vigna subterranea*）、豇豆（*Vigna unguiculata*）、龙眼（*Dimocarpus longan*）、荔枝（*Litchi chinensis*）、柠檬（*Citrus limon*）、柚（*Citrus maxima*）、温州蜜柑（*Citrus unshiu*）、山小橘（*Glycosmis pentaphylla*）、杧果（*Mangifera indica*）、（*Mangifera pentandra*）、亚麻（*Linum usitatissimum*）、丁香罗勒（*Geranium gratissimum*）、凤仙花（*Impatiens balsamina*）、葡萄（*Vitis vinifera*）、美丽帽柱木（*Mitragyna speciosa*）、山石榴（*Randia dumetorum*）、（*Randia uliginosa*）、旱芹（*Apium graveolens*）、野胡萝卜（*Daucus carota*）、茴香（*Foeniculum vulgare*）、刺黄果（*Carissa carandas*）、欧丁香（*Syringa vulgaris*）、辣椒（*Capsicum annuum*）、五彩椒（*Capsicum frutescens*）、夜香树（*Cestrum nocturnum*）、（*Duboisea leichhardtii*）、澳洲毒茄（*Duboisea myoporoides*）、枸杞（*Lycium chinense*）、番茄（*Lycopersicon esculentum*）、假酸浆（*Nicandra physalodes*）、光烟草（*Nicotiana glauca*）、烟草（*Nicotiana tabacum*）、刺天茄（*Solanum indicum*）、茄（*Solanum melongena*）、龙葵

（*Solanum nigrum*）、马铃薯（*Solanum tuberosum*）、蕹菜（*Ipomoea aquatica*）、红薯（*Ipomoea batatas*）、厚藤（*Ipomoea pes-caprae*）、圆叶牵牛（*Ipomoea purpurea*）、马氏破布木（*Cordia macleodii*）、（*Angelonia salicariaefolia*）、长叶蝴蝶草（*Torenia asiatica*）、芝麻（*Sesamum indicum*）、三消草（*Ruellia tuberosa*）、五色梅（*Lantana camara*）、柚木（*Tectona grandi*）、绿薄荷（*Mentha arvensis*）、紫苏（*Perilla frutescens*）、翠菊（*Callistephus chinensis*）、红花（*Carthamus tinctorius*）、茼蒿（*Chrysanthemum coronarium*）、杭菊（*Chrysanthemum morifolium*）、菊花（*Dendranthema morifolium*）、丝棉草（*Gnaphalium luteoalbum*）、小葵子（*Guizotia abyssinica*）、向日葵（*Helianthus annuus*）、莴苣（*Lactuca sativa*）、野莴苣（*Lactuca serriola*）、百日菊（*Zinnia violacea*）、凤眼莲（*Eichhornia crassipes*）、小果野蕉（*Musa acuminata*）、大蕉（*Musa paradisiaca*）、蕉麻（*Musa textilis*）、（*Allium ampeloprasum*）、香葱（*Allium ascalonicum*）、洋葱（*Allium cepa*）、葱（*Allium fistulosum*）、韭菜（*Allium tuberosum*）、窄叶泽泻（*Alisma canaliculatum*）、欧洲慈姑（*Sagittaria sagittifolia*）、唐菖蒲（*Gladiolus gandavensis*）、参薯（*Dioscorea alata*）、五叶薯蓣（*Dioscorea pentaphylla*）、饭包草（*Commelina benghalensis*）、姜花（*Hedychium coronarium*）、姜（*Zingiber officinale*）、竹芋（*Maranta arundinacea*）、高粱（*Andropogon sorghum*）、巴拉草（*Brachiaria mutica*）、稻（*Oryza sativa*）、甘蔗（*Saccharum officinarum*）、小麦（*Triticum aestivum*）、玉米（*Zea mays*）、油棕（*Elaeis guineensis*）、五彩芋（*Caladium bicolor*）、野芋（*Colocasia antiquorum*）、芋（*Colocasia esculenta*）、大薸（*Pistia stratiotes*）、马蹄犁头尖（*Typhonium trilobatum*）、无根藤属（*Cassytha*）、胡椒属（*Piper*）、千日红属（*Gomphrena*）、蓼属（*Polygonum*）、堇菜属（*Viola*）、杨属（*Populus*）、秋海棠属（*Begonia*）、醉蝶花属（*Cleome*）、刺蒴麻属（*Triumfetta*）、锦葵属（*Malva*）、黄花棯属（*Sida*）、艾麻属（*Laportea*）、大戟属（*Euphorbia*）、草莓属（*Fragaria*）、苹果（*Malus pumila*）、梨属（*Pyrus*）、蔷薇属（*Rosa*）、刀豆属（*Canavalia*）、猪屎豆属（*Crotalaria*）、镰扁豆属（*Dolichos*）、木蓝属（*Indigofera*）、太平洋栗（*Inocarpus*）、金合欢属（*Mimosa*）、黧豆属（*Mucuna*）、天竺葵属（*Pelargonium*）、髯管花属（*Geniostoma*）、咖啡属（*Coffea*）、巴戟天属（*Morinda*）、墨苜蓿属（*Richardia*）、素馨属（*Jasminum*）、曼陀罗属（*Datura*）、碧冬茄属（*Petunia*）、金鱼草属（*Antirrhinum*）、母草属（*Lindernia*）、车前属（*Plantago*）、马刺花属（*Plectranthus*）、鞘蕊花属（*Solenostemon*）、藿香蓟属（*Ageratum*）、艾纳香属（*Blumea*）、金盏花属（*Calendula*）、苏利南野菊属（*Clibadium*）、大丽花属（*Dahlia*）、金腰箭属（*Synedrella*）、黄花蔺属（*Limnocharis*）、吊兰属（*Chlorophytum*）、亚马逊石蒜属（*Eucharis*）、天门冬属（*Asparagus*）、美人蕉属（*Canna*）、山姜属（*Alpinia*）、细穗草属（*Lepturus*）、黍属（*Panicum*）、高粱属（*Sorghum*）、芻雷草属（*Thuarea*）、海芋属（*Alocasia*）植物（Matov & Kononenko, 2012）。

***Spodoptera depravata* (Butler, 1879) 淡剑灰翅夜蛾**

Agrotis depravata Butler, 1879, *Annals and Magazine of Natural History*, 5 (4): 360. Type locality: Japan, Yokohama.

Synonymy: *Rusidrina rasdolnia* Staudinger, 1892; *Luperina inutilis* Alphéraky, 1897; *Euxoa kaolina* Bryk, 1948.

Distribution. China (Liaoning, Jiangsu, Zhejiang, Hubei, Hunan, Fujian); Russia (Siberia), Korea, Japan.

寄主：结缕草、栗（陈一心，1999）；高粱（*Sorghum bicolor*）、日本结缕草（*Zoysia japonica*）（Matov & Kononenko, 2012）。

***Spodoptera apertura* (Walker, 1856) 敞灰翅夜蛾**

Prodenia apertura Walker, 1856, *List of the Specimens of Lepidopterous Insects in the Collection of the British Museum*, 32: 654. Type locality: Sri Lanka.

Synonymy: *Prodenia synstictis* Hampson, 1896; *Prodenia leucophlebia* Hampson, 1902.

Distribution. China (Jilin, Liaoning, Hubei, Hunan, Zhejiang, Guangdong); South Korea, Sri Lanka, Indonesia (Borneo, Sulawesi, New Guinea), India, Australia, Madagascar; Central Africa.

寄主：茅膏菜属植物（陈一心，1999）。

Tribe **Elaphriini** Beck, 1996 孤夜蛾族

Elaphriinae Beck, 1996, *Neue entomologische Nachrichten*, 36: 16, 54, repl. name for Erastriinae Herrich-Schäffer, 1845. Type genus: *Elaphria* Hübner, 1818.

Synonymy: Galgulina Beck, 1996.

Small sized moths, characterized by structure of larva (presence of prominent subbasal hump) and genitalia (in the male genitalia costal margin of valva with unsclerotized band in costa, behind costal process; in the female genitalia the posterior part of the 7th segment has a prominent, heavily sclerotized antevaginal crest and ventrally a wrinkled or setose narrow, heavily sclerotized band). The tribe is poorly studied. It includes 10 genera with total 208 species, known mainly from North and South America; many species are still not yet described. European taxa of Elaphriini were reviewed by Fibiger & Hacker (2007). The only 3 genera of Elaphriini, *Elaphria* with 4 species, *Galgula* with 1 species and *Strotihypera* (Kononenko, Han, 2011; Han, Kononenko, 2015) with 3 species are known from the Old World. The tribe Elaphriini numbers 2 genera and 2 species are in the Three Provinces of Northeast China.

References. Chen, 1982, 1999 (Noctuidae: Amphipyrinae, part); Sugi, 1982 (Amphipyrinae, part); Speidel *et al.*, 1996 (Amphipyrinae, part); Kononenko *et al.*, 1998 (Amphipyrinae, part); Kitching, Rawlins, 1999 [1998] (Hadeninae); Kononenko, 2003 (Amphipyrinae, part); Fibiger, Hacker, 2005 (Xyleninae: Elaphriini); Fibiger, Lafontaine, 2005 (Xyleninae: Elaphriini); Kononenko, 2005 (Xyleninae: Elaphriini); Lafontaine, Fibiger, 2006 (Xyleninae: Elaphriini); Fibiger, Hacker, 2007 (Xyleninae: Elaphriini); Kononenko, Han, 2007 (Xyleninae: Elaphriini); Matov *et al.*, 2008 (Xyleninae: Elaphriini); Lafontaine, Schmidt, 2010 (Noctuidae, Elaphriini); Pogue, Sullivan, 2011 (Xyleninae: Elaphriini); Kononenko, Han, 2011 (Xyleninae: Elaphriini); Fibiger *et al.*, 2011 (Xyleninae: Elaphriini); Kononenko, 2016 (Xyleninae: Elaphriini).

注：首次给出中文族名。

Genus *Elaphria* Hübner, 1818 孤夜蛾属

Elaphria Hübner, 1818, *Zuträge zur Sammlung Exotischer Schmetterlinge*, 1: 16. Type species: *Elaphria grata* Hübner, 1818 [USA, Florida].

Synonymy: *Agrotis* Hübner, 1808, suppr. (ICZN, 1966 Op. 789); *Hapalotis* Hübner, [1821] 1816; *Monodes* Guenée, 1852; *Carbona* Schaus, 1906; *Psilomonodes* Warren, 1911; *Mesostrota* Möschler, 1890, missp.; *Agrotis* auct.; *Erastria* auct..

Elaphria is a large genus in the New World, predominantly Neotropical region, genus with 127 species; 14 of them are known from North America, 1 from the Palaearctic region, 1 from China, three from India and 1 from Africa (Pogue, Bolling, 2003). 1 species is in the Three Provinces of Northeast China.

Elaphria venustula (Hübner, 1790) 美纹孤夜蛾

Phalaena Noctua venustula Hübner, 1790, *Beiträge zur Geschichte der Schmetterlinge*, 2 (3): 78, pl. 3 (4): 7. Type locality: Germany, Augsburg.

Synonymy: *Pyralis hybnerana* Fabricius, 1794.

Distribution. China (Heilongjiang, Jilin, Xinjiang); Russia (European part, Siberia), Korea, Japan, Mongolia, Kazakhstan; Middle East, Europe.

寄主：葡萄、委陵菜（陈一心，1999）；帚石楠（*Calluna vulgaris*）、羽衣草（*Alchemilla vulgaris*）、蕨麻（*Potentilla anserina*）、洋委陵菜（*Potentilla erecta*）、匍匐委陵菜（*Potentilla reptans*）、（*Potentilla tormentilla*）、秦氏悬钩子（*Rubus caesius*）、树莓（*Rubus fruticosus*）、覆盆子（*Rubus idaeus*）、金雀儿（*Cytisus scoparius*）、金雀花（*Genista tinctoria*）、驴食草（*Onobrychis viciifolia*）、金丝桃属（*Hypericum*）、草莓属（*Fragaria*）、蔷薇属（*Rosa*）、月见草属（*Oenothera*）、山梅花属（*Philadelphus*）、紫菀属（*Aster*）、莴苣属（*Lactuca*）、蓝沼草属（*Molinia*）植物（Matov & Kononenko, 2012）。

Genus *Strotihypera* Kononenko & Han, 2011 窄弧夜蛾属

Strotihypera Kononenko & Han, 2011, *Zootaxa*, 2011. Type species: *Erastria flavipuncta* Leech, 1889 [Japan].

Synonymy: *Hyperstrotia* auct..

The genus includes 3 species, distributed in the Eastern Palaearctic (Sino-Himalayan and Chinese subregions) and north of the Oriental regions; 1 species is in the Three Provinces of Northeast China.

注：首次给出中文属名。

Strotihypera flavipuncta (Leech, 1889) 窄弧夜蛾

Erastria flavipuncta Leech, 1889, *Proceedings of the Zoological Society of London*, 1889: 524, pl. 52: 3.

Type locality: Japan, Yokohama.

Distribution. China (Jilin); Russia (Siberia), Korea, Japan.

寄主：栎属（*Quercus*）植物（Matov & Kononenko, 2012）。

注：首次给出中文种名。

Tribe **Caradrinini** Boisduval, 1840 逸夜蛾族

Caradrinides Boisduval, 1840, *Genera et Index Methodicus Europaeorum Lepidopterorum*, 1840: 136. Type genus: *Caradrina* Ochsenheimer, 1816.

Synonymy: Caradrinidi Stephens, 1850.

Small to medium sized species with grey-yellowish or pale brownish ground colour and inconspicuous wing pattern and in most cases with milky-white hindwing. The tribe Caradrinini arranged in 2 subtribes Caradrinina and Athetiina. The tribe Caradrinini in Eurasia has been extensively revised by Hacker (2004). Most species of the tribe Caradrinini occur in the arid and semi-arid areas of the subtropical and southern temperate regions of Central and West Palaearctic. The larvae are polyphagous and might damage a multitude of various garden vegetables. Many species are migratory. Some species-groups are hard to determine correctly and require careful genitalia preparations. The tribe Caradrinini is represented by 5 genera and 18 species are in the Three Provinces of Northeast China.

References. Chen, 1982, 1999 (Noctuidae: Amphipyrinae, part); Sugi, 1982 (Amphipyrinae, part); Speidel *et al.*, 1996 (Amphipyrinae, part); Kononenko, 1997b (Amphipyrinae); Kononenko *et al.*, 1998 (Amphipyrinae, part); Kitching, Rawlins, 1999 [1998] (Hadeninae); Kononenko, 2003 (Amphipyrinae, part); Hacker, 2004 (Amphipyrinae: Caradrinini); Fibiger, Hacker, 2005 (Xyleninae: Caradrinini); Fibiger, Lafontaine, 2005 (Xyleninae: Caradrinini); Kononenko, 2005 (Xyleninae: Caradrinini); Lafontaine, Fibiger, 2006 (Xyleninae: Caradrinini); Fibiger, Hacker, 2007 (Xyleninae: Caradrinini); Kononenko, Han, 2007 (Xyleninae: Caradrinini); Matov *et al.*, 2008 (Xyleninae: Caradrinini); Lafontaine, Schmidt, 2010 (Noctuinae: Caradrinini); Han, Kononenko, 2011 (Xyleninae: Caradrinini); Fibiger *et al.*, 2011 (Xyleninae: Caradrinini); Kononenko, 2016 (Xyleninae: Caradrinini: Caradrinina, Athetiina).

注：首次给出中文族名。

Subtribe **Caradrinina** Boisduval, 1840 逸夜蛾亚族

Caradrinides Boisduval, 1840, *Genera et Index Methodicus Europaeorum Lepidopterorum*, 1840: 136. Type genus: *Caradrina* Ochsenheimer, 1816.

Synonymy: Caradrinidi Stephens, 1850.

注：首次给出中文亚族名。

Genus ***Caradrina*** Ochsenheimer, 1816 逸夜蛾属

Caradrina Ochsenheimer, 1816, *Die Schmetterlinge von Europa*, 4: 80. Type species: *Phalaena morpheus*

Hufnagel, 1766 [Germany].

Synonymy: *Charadrina*: Agassiz [1847] 1846, emend.; *Amphidrina* Staudinger, 1891; *Platyperigea* Smith, 1894 (subgen.); *Eremodrina* Boursin, 1937 (subgen.); *Paradrina* Boursin, 1937 (subgen.); *Boursinidrina* Hacker, 2004 (subgen.); *Kalchbergiana* Hacker, 2004 (subgen.); *Lewantrina* Hacker, 2004 (subgen.); *Weigertrina* Hacker, 2004 (subgen.).

The genus includes 155 species, divided into 8 subgenera (*Caradrina* s. str.; *Platyperigea* Smith, 1894; *Boursinidrina* Hacker, 2004; *Kalchbergiana* Hacker, 2004; *Eremodrina* Boursin, 1937; *Lewantrina* Hacker, 2004 (subgen.); *Weigeritrina* Hacker, 2004 and *Paradrina* Boursin, 1937). 2 species are in the Three Provinces of Northeast China.

Subgenus ***Platyperigea*** Smith, 1894 **阔逸夜蛾亚属**

Platyperigea Smith, 1894, *Transactions of the American Entomological Society 21*: 60. Type species: *Platyperigea comma* Smith, 1894 [USA].

注：首次给出中文亚属名。

Caradrina* (*Platyperigea*) *montana **Bremer, 1861 暗灰逸夜蛾**

Caradrina montana Bremer, 1861, *Bulletin de la Classe Physico-Mathématique de l'Académie Impériale des Sciences de St.-Pétersbourg*, 1861: 49, pl. 4: 8. Type locality: Russia, Transbaikalia, [Yablonovyi Range] "Apfel-Gebirge".

Synonymy: *Caradrina montana menetriesi* Kretschmar, 1863 (ssp.); *Caradrina montana extima* Walker, 1865 (ssp.); *Caradrina cubicularis cinerascens* Tengström, 1869 (var.); *Caradrina montana rougemonti* Spuler, 1906 (ssp.); *Athetis* (*Caradrina*) *grisea apathetica* Püngeler, 1914 (var.); *Athetis fuscicornis sachalinensis* Matsumura, 1925 (ssp.); *Athetis melancholica* Draudt, 1934; *Athetis* (*Hymenodrina*) *grisea kaolina* Bryk, 1948; *Platyperigea montana weigerti* Hacker, 1993 (ssp.); *Caradrina albina* auct..

Distribution. China (Heilongjiang, Jilin, Liaoning, Xizang); Russia (European part, Siberia), Mongolia, Korea; Central Asia, Europe, North America.

寄主：萹蓄 (*Polygonum aviculare*)、酸模（*Rumex acetosa*）、小酸模（*Rumex acetosella*）、直穗酸模（*Rumex longifolius*）、黄花柳（*Salix caprea*）、覆盆子（*Rubus idaeus*）、毛蕊花（*Verbascum thapsus*）、山柳菊（*Hieracium umbellatum*）、曲尾藓属（*Dicranum*）、（*Alfalfa*）、车前属（*Plantago*）、苦苣菜属（*Sonchus*）、蒲公英属（*Taraxacum*）植物（Matov & Kononenko, 2012）。

Subgenus ***Eremodrina*** Boursin, 1937 **邻逸夜蛾亚属**

Eremodrina Boursin, 1937, *Entomologische Rundschau*, 54: 72. Type species: *Caradrina vicina* Staudinger, 1870 [South Russia].

Caradrina (Eremodrina) morosa Lederer, 1853 暮逸夜蛾

Caradrina morosa Lederer, 1853, *Verhandlungen des Zoologisch-Botanischen vereins in Wien*, 1853: pl. 5: 3. Type locality: East Kazakhstan, West Altai, Ust-Kamengorsk, Ust-Buchtarminsk.

Synonymy: *Elaphris* [sic!] *morosa permorosa* Bryk, 1948, partim; *Caradrina morosa explanata* Hacker, 2004 (ssp.).

Distribution. China (Heilongjiang, Jilin, Xinjiang, Hebei); Russia (Altai, Transbaikalia, Siberia), Mongolia, Korea, Kazakhstan.

寄主：树锦鸡儿（*Caragana arborescens*）、车轴草属（*Trifolium*）、苦苣菜属（*Sonchus*）、蒲公英属（*Taraxacum*）植物（Matov & Kononenko, 2012）。

Genus ***Hoplodrina*** Boursin, 1937 **筱夜蛾属**

Hoplodrina Boursin, 1937, *Entomologische Rundschau* 54: 367. Type species: *Phalaena* (*Noctua*) *alsines* Brahm, 1791 [Germany].

Synonymy: *Resperdina* Beck, 1996.

The genus includes 19 species, distributed in the Mediterranean, Central Asian and Chinese subregions of the Palaearctic and Oriental regions. 18 species are in the Palaearctic region; 2 species are in the Three Provinces of Northeast China.

Hoplodrina octogenaria (Goeze, 1781) 北筱夜蛾

Phalaena (*Noctua*) *octogenaria* Goeze, 1781, *Entomologische Beyträge zu des Ritter Linne Zwölften Ausgebe Natursystems,* 3 (3): 227. Type locality: [not stated].

Synonymy: *Phalaena* (*Noctua*) *alsines* Brahm, 1791; *Hoplodrina sordida* Haworth, 1809; *Caradrina sericea* Speyer, 1867; *Caradrina* (*Athetis*) *alsinides* Constanti, 1922; *Caradrina alsines amurensis* Staudinger, 1892 (spp.); *Hoplodrina melendizi* Agenjo, 1941.

Distribution. China (Heilongjiang, Jilin, Liaoning, Nei Mongol, Xinjiang); Russia (European part, Siberia), Mongolia, Korea, Japan; Middle East, Central Asia, Asia Minor, Europe.

寄主：繁缕、酸模、堇菜等属（陈一心，1999）；繁缕（*Stellaria media*）、荞麦（*Fagopyrum esculentum*）、欧洲报春（*Primula vulgaris*）、里文堇菜（*Viola riviniana*）、东陵山柳（*Salix phylicifolia*）、异株荨麻（*Urtica dioica*）、紫叶景天（*Sedum telephium*）、覆盆子（*Rubus idaeus*）、金雀儿（*Cytisus scoparius*）、百脉根（*Lotus corniculatus*）、羊角芹（*Aegopodium podagraria*）、野胡萝卜（*Daucus carota*）、毛地黄（*Digitalis purpurea*）、曲芒发草（*Deschampsia flexuosa*）、草地早熟禾（*Poa pratensis*）、毛蕊花属（*Ranunculus*）、（*Alsine*）、蝇子草属（*Silene*）、拟漆姑属（*Spergularia*）、蓼属（*Polygonum*）、酸模属（*Rumex*）、苜蓿属（*Medicago*）、驴食草属（*Onobrychis*）、亚麻属（*Linum*）、拉拉藤属（*Galium*）、旋花属（*Convolvulus*）、毛蕊花属（*Verbascum*）、车前属（*Plantago*）、野芝麻属

（*Lamium*）、风铃草属（*Campanula*）、牛蒡属（*Arctium*）、山柳菊属（*Hieracium*）、苦苣菜属（*Sonchus*）、蒲公英属（*Taraxacum*）植物（Matov & Kononenko, 2012）。

注：《中国动物志》（夜蛾科）中引用本种同物异名“*Phalaena* (*Noctua*) *alsines* Brahm, 1791”，在此予以变更。

Hoplodrina euryptera **Boursin, 1937 斑筱夜蛾**

Hoplodrina euryptera Boursin, 1937, *Entomologische Rundschau*, 54 (34): 429. Type locality: China, Sichuan [Setchuan], Giufu-shan.

Synonymy: *Hoplodrina paromoea marginepicta* Bryk, 1948 (ssp.); *Hoplodrina implacata* auct., nec Wileman & West, 1929.

Distribution. China (Heilongjiang, Jilin, Liaoning, Hunan, Taiwan); Russia (Siberia), Korea, Japan.

注：《中国动物志》（夜蛾科）中引用本种同物异名“*Hoplodrina implacata* Wileman & West, 1929”，在此予以变更。

Genus ***Stygiodrina*** Boursin, 1937 污夜蛾属

Stygiodrina Boursin, 1937, *Entomologische Rundschau*, 54 (34): 366. Type species: *Caradrina maurella* Staudinger, 1888. Type locality: Russia, Amur reg., Blagoveschensk; “Ussuri” [Primorye terr.]; Suifun [Razdolnoe]; Sidemi [Bezverkhovo].

Monotypic East Palaearctic genus; 1 species is in the Three Provinces of Northeast China.

注：首次给出中文属名。

Stygiodrina maurella **(Staudinger, 1888) 污夜蛾**

Caradrina maurella Staudinger, 1888, *Entomologische Zeitung. Entomologischen Vereine zu Stettin*, 49: 255. Type locality: Russia, Amur reg., Blagoveschensk; “Ussuri” [Primorye terr.]; Suifun [Razdolnoe]; Sidemi [Bezverkhovo].

Synonymy: *Caradrina eutapaishana* Berio, 1977.

Distribution. China (Heilongjiang, Jilin, Liaoning); Russia (Tuva, Transbaikalia, Siberia), Korea.

注：首次给出中文种名。

Genus ***Chilodes*** Herrich-Schäffer, 1849 陲夜蛾属

Chilodes Herrich-Schäffer, 1849, *Systematische Bearbeitung der Schmetterlinge von Europa*, 2: 226. Type species: *Noctua ulvae* Hübner, 1809 [Europe].

Synonymy: *Hypostilbia* Hampson, 1908.

The genus includes 4 species, all in the Palaearctic region; 1 species is in the Three Provinces of Northeast China.

***Chilodes distracta* (Eversmann, 1848) 斑陲夜蛾**

Caradrina distracta Eversmann, 1848, *Bulletin de la Société Impériale des Naturalistes de Moscou*, 21 (3): 217. Type locality: Russia [Transbaikalia], Kiachta.

Synonymy: *Proxenus* (*Hydrilla*) *cinerea* Alphéraky, 1889; *Hypostilbia megastigma* Püngeler, 1907; *Erastria umbrivaga* Krulikowsky, 1907.

Distribution. China (Heilongjiang); Russia (European part, Siberia), Mongolia; Asia Minor, Central Asia, Europe.

寄主：芦苇属（*Phragmites*）植物（Matov & Kononenko, 2012）。

Subtribe **Athetiina** Lafontaine & Fibiger, 2005 委夜蛾亚族

Athetiina Lafontaine & Fibiger, 2005, *Esperiana*, 11: 11. Type genus: *Athetis* Hübner, [1821].

Synonymy: Athetiina Lafontaine & Schmidt, 2010.

This subtribe is established by Lafontaine & Fibiger (2005) for a large worldwide group of genera of which the largest genus is *Athetis*; although "*Athetis*" probably includes a number of unrecognized genera and is in need of revision. As many as 250 species can be associated with the Athetina. The most obvious synapomorphy is the form of the uncus, which is vestigial or absent in all species. The subtribe includes Nearctic genus *Anorthodes* Smith and the genus *Athetis* with subgenera *Athetis* s. str., *Proxenus* and *Hydrillula*. The subtribe is represented in the Three Provinces of Northeast China by 1 genus and 12 species.

注：首次给出中文亚族名。

Genus ***Athetis*** Hübner, [1821] 1816 委夜蛾属

Athetis Hübner, [1821] 1816, *Verzeichniss bekannter Schmettlinge* [sic!]: 209. Type species: *Noctua dasychira* Hübner, [1817] (= *Noctua furvula* Hübner, [1808]) [Europe].

Synonymy: *Proxenus* Herrich-Schäffer, 1845 (subgen.); *Hydrilla* Guenée, 1852, preocc. (Boisduval, 1840 [Lepidoptera, Noctuidae]); *Elydna* Walker, 1858; *Nebrissa* Walker, 1862; *Curgia* Butler, [1863], preocc.; *Radinacra* Butler, 1878; *Dadica* Moore, 1881; *Radinogoes* Butler, 1886; *Leucocosmia* Butler, 1886; *Floccifera* Hampson, 1894; *Stygiathetis* Hampson, 1908; *Strepselydna* Warren, 1911; *Hydrillula* Tams, 1938 (subgen.), repl. name; *Tectorea* Berio, 1955.

The genus is distributed worldwide with centres of diversity in Africa (subgenus *Hydrillula*), East and Southeast Asia, and Australia (subgenera *Athetis* and *Proxenus*). The genus contains over 200 species worldwide, however, taking into account the high diversity and uniform external morphology of the genus in further revision it is expected that many undescribed taxa may be discovered. About 90 species are known from the Palaearctic region, among them 3 subgenera 12 species are in the Three Provinces of Northeast China.

Subgenus *Athetis* Hübner, [1821] 1816 委夜蛾亚属

Athetis Hübner, [1821] 1816, *Verzeichniss bekannter Schmettlinge* [sic!]: 209. Type species: *Noctua dasychira* Hübner, [1817] (=*Noctua furvula* Hübner, [1808]) [Europe].

Synonymy: *Radinacra* Butler, 1878.

Athetis (*Athetis*) *gluteosa* (Treitschke, 1835) 后委夜蛾

Caradrina gluteosa Treitschke, 1835, *Die Schmetterlinge von Europa*, 10 (2): 80. Type locality: "Shrimia".

Synonymy: *Hydrilla uliginosa* Boisduval, 1840; *Caradrina grisescens* Poujade, 1887; *Hydrilla kitti* Rebel, 1913; *Athetis lapidosa* Berio, 1977.

Distribution. China (Heilongjiang, Jilin, Liaoning, Qinghai, Sichuan, Xizang); Russia (Siberia), Mongolia, Korea, Japan; Asia Minor, Central Asia, Europe.

寄主：低矮草本植物（陈一心，1999）；海边香豌豆（*Lathyrus maritimus*）、马蹄豆（*Hippocrepis comosa*）、蓼属（*Polygonum*）、锦葵属（*Malva*）、车前属（*Plantago*）、野芝麻属（*Lamium*）、莴苣属（*Lactuca*）、苦苣菜属（*Sonchus*）、蒲公英属（*Taraxacum*）植物（Matov & Kononenko, 2012）。

Athetis (*Athetis*) *cinerascens* (Motschulsky, 1860) 灰委夜蛾

Cosmia cinerascens Motschulsky, 1860, *Etudes Entomologiques*, 9: 34. Type locality: Japan.

Synonymy: *Radinacra palpalis* Butler, 1878.

Distribution. China (Heilongjiang, Jilin, Liaoning, Hubei); Korea, Japan.

Athetis (*Athetis*) *furvula* (Hübner, [1808]) 委夜蛾

Noctua furvula Hübner, [1808], *Sammlung Europaischer Schmetterlinge. Eulen*, 2: pl. 83: 390. Type locality: Europe.

Synonymy: *Noctua dasychira* Hübner [1817]; *Caradrina lenta* Treitschke, 1825; *Caradrina tristis* Bremer, 1861; *Caradrina lenta lentina* Staudinger, 1888 (ssp.); *Athetis furvula tristifascia* Bryk, 1948 (ssp.).

Distribution. China (Heilongjiang, Jilin, Liaoning, Nei Mongol, Xinjiang, Hebei, Xizang); Russia (Siberia), Mongolia, Korea, Japan; Central Asia, Europe.

寄主：虎杖（*Polygonum cuspidatum*）、酸模（*Rumex acetosa*）、车前属（*Plantago*）、蒲公英属（*Taraxacum*）植物（Matov & Kononenko, 2012）。

Athetis (*Athetis*) *funesta* (Staudinger, 1888) 阴委夜蛾

Caradrina funesta Staudinger, 1888, *Entomologische Zeitung. Entomologischen Vereine zu Stettin*, 49: 255. Type locality: Russia, [Primorye terr.] Askold I..

Synonymy: *Acosmetia lugubris* Graeser, 1888 [1889]; *Glottula squalida* Leech, 1899.

Distribution. China (Heilongjiang, Jilin, Liaoning, Shanxi); Russia (Tuva, Transbaikalia, Siberia), Mongolia, Korea, Japan.

Subgenus ***Hydrillula*** Tams, 1938 **干委夜蛾亚属**

Hydrillula Tams, 1938, *Entomologist*, 71: 123. Type species: *Noctua pallustris* Hübner, [1808] Europe.

Synonymy: *Nebrissa* Walker, 1862; *Tectorea* Berio, 1955.

注：首次给出中文亚属名。

Athetis (*Hydrillula*) *lapidea* Wileman, 1911 石委夜蛾

Athetis lapidea Wileman, 1911, *Transactions of the Entomological Society of London*, 1911: 218. Type locality: Japan, Yamato, Yoshino.

Synonymy: *Arenostola suzukii* Matsumura, 1926.

Distribution. China (Heilongjiang, Jilin, Liaoning, Sichuan); Russia (Siberia), Korea, Japan.

Athetis (*Hydrillula*) *pallustris* (Hübner, 1808) 干委夜蛾

Noctua pallustris Hübner, 1808, *Sammlung Europaischer Schmetterlinge. Eulen*, 2: pl. 79: 367. Type locality: Europe.

Synonymy: *Caradrina exilis* Eversmann, 1842; *Caradrina lutea* Freyer, 1845; *Hydrilla aboleta* Guenée, 1852; *Caradrina pallustris melanochroa* Staudinger, 1892 (var.); *Petilampa sayana* Hampson, 1909; *Hydrilla raebeli* Dannehl, 1925; *Athetis palustris* auct., nec Hübner, 1808; *Athetis luteola* auct., nec Freyer, 1845.

Distribution. China (Heilongjiang, Jilin); Russia (European part, Siberia), Mongolia, Kazakhstan; Asia Minor, Europe.

寄主：车前属植物（陈一心，1999）；两栖蓼（*Polygonum amphibium*）、萹蓄 (*Polygonum aviculare*)、繁缕（*Stellaria media*）、旋果蚊子草（*Filipendula ulmaria*）、秦氏悬钩子（*Rubus caesius*）、长叶车前（*Plantago lanceolata*）、大车前（*Plantago major*）、酸模属（*Rumex*）、野芝麻属（*Lamium*）、风铃草属（*Campanula*）、莴苣属（*Lactuca*）、蒲公英属（*Taraxacum*）植物（Matov & Kononenko, 2012）。

Subgenus ***Proxenus*** Herrich-Schäffer, 1850 **原委夜蛾亚属**

Proxenus Herrich-Schäffer, 1850, *Systematische Bearbeitung der Schmetterlinge von Europa*, 2: 190, 240. Type species: *Caradrina hospes* Freyer, [1831] 1833 [Italy, Sicily].

Synonymy: *Elydna* Walker, 1858; *Dadica* Moore, 1881; *Leucocosmia* Butler, 1886; *Radinogoes* Butler, 1886; *Strepselydna* Warren, 1911.

注：首次给出中文亚属名。

***Athetis* (*Proxenus*) *lepigone* (Möschler, 1860) 二点委夜蛾**

Hydrilla lepigone Möschler, 1860, *Wiener Entomologische Monatschrift*, 4: 273. Type locality: South Russia, Sarepta.

Synonymy: *Senta nigrosignata* Graeser, 1888 [1889]; *Caradrina tristis lugens* Staudinger, 1892 (var.); *Hypostilbia banghaasi* Wagner, 1913; *Arenostola sounkeana* Matsumura, 1927; *Radinogoes exquisita* Bryk, 1942.

Distribution. China (Heilongjiang, Jilin, Liaoning, Hebei); Russia (European part, Siberia), Mongolia, Korea, Japan, Kazakhstan; Europe.

寄主：茶（*Camellia sinensis*）、苔景天（*Sedum acre*）、海滨山黧豆（*Lathyrus japonicus*）、海边香豌豆（*Lathyrus maritimus*）、玉米（*Zea mays*）、酸模属（*Rumex*）、繁缕属（*Stellaria*）、委陵菜属（*Potentilla*）、车轴草属（*Trifolium*）、当归属（*Angelica*）、车前属（*Plantago*）、风毛菊属（*Leontodon*）、苦苣菜属（*Sonchus*）、蒲公英属（*Taraxacum*）、雀稗属（*Paspalum*）、狗尾草属（*Setaria*）植物（Matov & Kononenko, 2012）。

***Athetis* (*Proxenus*) *correpta* (Püngeler, 1907) 暗黑委夜蛾**

Senta correpta Püngeler, 1907, *Deutsche Entomologische Zeitschrift Iris*, 19: 221, pl. 8: 2. Type locality: China, Thianshan, Urumtschi.

Synonymy: *Radinogoes subargentea* Caradja, 1926; *Radinogoes jezoensis* Matsumura, 1931; *Proxenus pacificus* Bryk, 1942.

Distribution. China (Heilongjiang, Jilin, Liaoning); Russia (European part, Siberia), Mongolia, Korea, Japan.

注：首次给出中文种名。

***Athetis* (*Proxenus*) *dissimilis* (Hampson, 1909) 黑委夜蛾**

Proxenus dissimilis Hampson, 1909, *Catalogue of the Lepidoptera Phalaenae in the British Museum*, 8: 431, pl. 133: 15. Type locality: Japan.

Distribution. China (Heilongjiang, Jilin, Liaoning, Taiwan); South Korea, Japan, India, Philippines, Indonesia (Java, Borneo).

注：首次给出中文种名。

***Athetis* (*Proxenus*) *albisignata* (Oberthür, 1879) 白斑委夜蛾**

Caradrina albisignata Oberthür, 1879, *Diagnoses d'Espéces nouvelles de Lépidoptères de l'ile Askold I.*: 14. Type locality: Russia, Primorye terr., Askold I..

Synonymy: *Caradrina caeca* Oberthür, 1879; *Elydna alignata faja* Bryk, 1942 (ssp.); *Dadica* [sic!] *differentiata* Bryk, 1948.

Distribution. China (Heilongjiang, Jilin, Liaoning, Shaanxi); Russia (Siberia), Korea, Japan.

寄主：杭菊（*Chrysanthemum morifolium*）、酸模属（*Rumex*）、栗属（*Castanea*）、菜豆属（*Phaseolus*）、蒿属（*Artemisia*）、蒲公英属（*Taraxacum*）植物（Matov & Kononenko, 2012）。

Athetis (Proxenus) pallidipennis Sugi, 1982 **亚白斑委夜蛾**

Athetis pallidipennis Sugi, 1982, *Moth of Japan*, 1: 767, 2: pl. 188: 42, 43. Type locality: Japan, Honshu, Gumma Pref., [coll. Sugi, NIAES, Tsukuba].

Distribution. China (Heilongjiang, Jilin, Liaoning); Russia (Siberia), Korea, Japan.

注：首次给出中文种名。

***Athetis* (*Proxenus*) *lineosa* (Moore, 1881) 线委夜蛾**

Dadica lineosa Moore, 1881, *Proceedings of the Zoological Society of London,* 1881: 349. Type locality: India, Punjab.

Distribution. China (Heilongjiang, Jilin, Liaoning, Hebei, Henan, Zhejiang, Hubei, Hunan, Fujian, Hainan, Guangxi, Sichuan, Yunnan); Russia (Siberia), Korea, Japan, Nepal, India, Thailand, Vietnam, Myanmar.

寄主；日本天剑（*Calystegia japonica*）、魁蒿（*Artemisia princeps*）、翠菊（*Callistephus chinensis*）、台湾蒲公英（*Taraxacum platycarpum*）（Matov & Kononenko, 2012）。

Tribe **Dypterygiini** Forbes, 1954 **翅夜蛾族**

Dypterygiini Forbes, 1954, *Memoirs of the Cornell University agriculture Experimental Station*, 329: 168 (as Dipterygiini, an incorrect original spelling, as formed from an unjustified emend. corrected in Dypterygiini [art. 35. 4. 1 ICZN]). Type genus: *Dipterygia* Agassiz, 1846 (emend. of *Dypterygia* Stephens, 1829).

Synonymy: Maniidae Herrich-Schäffer, [1851] 1845; Dypterygiini Forbes, 1954, incorrect original spelling; Polyphaenina Beck, 1996; Thalpophilina Beck, 1996; Tracheina Beck, 1996.

Medium to large sized and species with broad wings; ground colour usually dark-brown, in some genera with green or metallic-green elements of wing pattern; in some genera hindwing yellow with distinct terminal band. Externally Dipterygiini are highly heterogeneous, but the male and female genitalia show many monophyletic character states (Fibiger & Lafontaine, 2005). In the male genitalia pleural sclerite straight or evenly curved along posterior margin of tegumen; valva in most cases without a strong subapical constriction; digitus absent or fused into wall of valva; transtilla sclerotized; juxta triangular, tapered dorsally. In the female genitalia ovipositor quadrangular with posterior margin often slightly concave; ductus bursae heavily sclerotised; ventral wall of ostium slightly extended posteriorly. The European Dypterygiini have been revised by Fibiger & Hacker (2007) with listing of Palaearctic genera. Here we follow this revision, with the exception of the genus *Phoebophilus*, which is transferred to the tribe Xylenini.

The diversity of the tribe is centred in the tropics with isolated species and genera extending into the Holarctic region. This tribe is represented by 9 genera and 13 species are in the Three Provinces of Northeast

China, inhabiting mainly nemoral broad-leaved forest zone.

References. Chen, 1982, 1999 (Noctuidae: Amphipyrinae, part); Sugi, 1982 (Amphipyrinae, part); Speidel *et al.*, 1996 (Amphipyrinae, part); Kononenko *et al.*, 1998 (Amphipyrinae, part); Kononenko, 2003 (Amphipyrinae, part); Kitching, Rawlins, 1999 [1998] (Hadeninae); Fibiger, Lafontaine, 2005 (Xyleninae: Caradrinini); Fibiger, Hacker, 2005 (Xyleninae: Dypterygiini); Kononenko, 2005, 2016 (Xyleninae: Dypterygiini); Lafontaine, Fibiger, 2006 (Xyleninae: Dypterygiini); Kononenko, Han, 2007 (Xyleninae: Dypterygiini); Fibiger, Hacker, 2007 (Xyleninae: Dypterygiini); Matov *et al.*, 2008 (Xyleninae: Dypterygiini); Lafontaine, Schmidt, 2010 (Noctuidae, Dypterygiini); Fibiger *et al.*, 2011 (Xyleninae: Dypterygiini).

注：首次给出中文族名。

Genus ***Dypterygia*** Stephens, 1829 **翅夜蛾属**

Dypterygia Stephens, 1829, *Illustrations of British Entomology* (Haustellata), 2:167. Type species: *Phalaena scabriuscula* Linnaeus, 1758 [Europe].

Synonymy: *Dipterygia*: Agassiz, 1846, emend..

The genus includes 13 species, distributed in different zoogeographical regions, the affinity some of them to *Dypterigia* requires revision; 3 species are known from the Palaearctic region; 1 species is in the Three Provinces of Northeast China.

***Dypterygia caliginosa* (Walker, 1858) 暗翅夜蛾**

Hadena caliginosa Walker, 1858, *List of the Specimens of Lepidopterous Insects in the Collection of the British Museum*, 15: 1729. Type locality: North China.

Distribution. China (Heilongjiang, Jilin, Liaoning, Hebei, Shaanxi, Zhejiang, Hubei, Hunan, Fujian, Hainan, Guizhou, Yunnan); Russia (Siberia), Korea, Japan.

寄主：蓼属（*Polygonum*）、酸模属（*Rumex*）植物（Matov & Kononenko, 2012）。

Genus ***Trachea*** Ochsenheimer, 1816 **陌夜蛾属**

Trachea Ochsenheimer, 1816, *Die Schmetterlinge von Europa*, 4: 75. Type species: *Phalaena atriplicis* Linnaeus, 1758 [Europe].

Synonymy: *Achatia* Hübner, [1806], suppr. (ICZN, 1926 Op. 97; 1954 Op. 278); *Achatis* Billberg, 1820; *Heraema* Staudinger, 1892; *Trachaea* Schaus, 1923, missp.; *Chandata* auct..

The genus includes 12-15 species, distributed in the Chinese subregion, Palaearctic and Oriental regions, besides this around 30 species of uncertain systematics referred for the Neotropical region in "*Trachea*" (Poole, 1987). About 10 species are known from the Palaearctic region, among them 5 species are in the Three Provinces of Northeast China.

Trachea atriplicis (Linnaeus, 1758) 陌夜蛾

Phalaena atriplicis Linnaeus, 1758; *Systema Naturae* (Edn. 10) 1: 517. Type locality: Europe.

Synonymy: *Hadena gnoma* Butler, 1878; *Trachea similis* Staudinger, 1892; *Trachea atriplicis gnosis* Bryk, 1948 (f.); *Trachea atriplicis tristina* Bryk, 1948 (ssp.).

Distribution. China (Heilongjiang, Jilin, Liaoning, Hunan, Fujian, Jiangxi); Russia (Siberia), Korea, Japan, Iran; Asia Minor, Central Asia, Europe.

寄主：（*Atriplex hortense*）、（*Atriplex patulum*）、鞑靼滨藜（*Atriplex tatarica*）、甜菜（*Beta vulgaris*）、藜（*Chenopodium album*）、（*Chenopodium bonus-henricus*）、菊叶香藜（*Chenopodium vulvaria*）、荞麦（*Fagopyrum esculentum*）、长鬃蓼（*Persicaria blumei*）、春蓼（*Persicaria maculosa*）、高山蓼（*Polygonum alpinum*）、萹蓄 (*Polygonum aviculare*)、拳参（*Polygonum bistorta*）、卷茎蓼（*Polygonum convolvulus*）、酸模叶蓼（*Polygonum lapathifolium*）、卵叶酸模（*Rumex confertus*）、簇花酸模（*Rumex conglomeratus*）、羊蹄（*Rumex japonicum*）、直穗酸模（*Rumex longifolius*）、豌豆（*Pisum sativum*）、玉米（*Zea mays*）、大黄属（*Rheum*）、荨麻属（*Urtica*）、旋花属（*Convolvulus*）、车前属（*Plantago*）、野芝麻属（*Lamium*）、蒲公英属（*Taraxacum*）植物（Matov & Kononenko, 2012）。

Trachea punkikonis Matsumura, 1927 明陌夜蛾

Trachea punkikonis Matsumura, 1927, *Insecta Matsumurana*, 3: 116. Type locality: Taiwan, Punkiko.

Synonymy: *Hadena lucia* Butler, 1878, nec Felder et Rogenhofer, 1874; *Trachea punkikonis lucilla* Sugi, 1982 (ssp.); *Trachea lucilla* auct., nec *lucia* Butler, 1878, nec Sugi, 1982.

Distribution. China (Heilongjiang, Jilin, Taiwan); Russia (Siberia), Korea, Japan.

注：首次给出中文种名。

Trachea melanospila Kollar, 1844 黑环陌夜蛾

Trachea melanospila Kollar, 1844, *in* Hügel, *Kaschmir und des Reich der Sieh*, 4 (2): 480. Type locality: “Himalaya Massuri”.

Synonymy: *Hadena kosakka* Oberthür, 1880.

Distribution. China (Heilongjiang, Jilin, Hubei, Hainan, Sichuan, Yunnan); Russia (Siberia), Korea, Japan, Nepal, India(North), Sri Lanka, Vietnam, Thailand.

Trachea tokiensis (Butler, 1884) 纷陌夜蛾

Hadena tokiensis Butler, 1884, *Études d’Entomologie*, 10: 17, pl. 2: 6. Type locality: Japan.

Synonymy: *Hadena tociensis*: Hampson, 1908, emend..

Distribution. China (Jilin, Liaoning, Yunnan); Russia (Siberia), Korea, Japan.

Trachea mandschurica (Graeser, 1889) 贺陌夜蛾

Pachnobia mandschurica Graeser, 1889, *Berliner entomologische Zeitschrift*, 33: 256. Type locality: Russia, Amurland, Raddefka, Blagoweschtschensk[Amur reg., Radde, Blagoveschensk].

Distribution. China (Heilongjiang, Jilin, Liaoning, Nei Mongol, Hubei); Russia (Siberia), Korea, Japan.

注：《中国动物志》（夜蛾科）中将本种归入"*Heraema* Staudinger, 1892 贺夜蛾属"，在此予以变更。

Genus ***Mormo*** Ochsenheimer, 1816 莫夜蛾属

Mormo Ochsenheimer, 1816, *Die Schmetterlinge von Europa*, 4: 70. Type species: *Phalaena maura* Linnaeus, 1758 [Mauritania].

Synonymy: *Lemur* Hübner, [1806], suppr. (ICZN, 1926 Op. 97; 1954 Op. 278); *Lemures* Ochsenheimer, 1816; *Lemur* Hübner, 1822, preocc. (Linnaeus, 1758 [Mammalia]); *Mania* Treitschke, 1825, repl. name; *Lemuris*: Curtis, 1831, missp.; *Acyra* Gilst, 1848, repl. name; *Arcyra*: Neave, 1939, missp..

The genus includes 7 species, distributed in the Mediterranean and Chinese subregions of Palaearctic and Oriental regions. 5 species are known in the Palaearctic region; 1 species is in the Three Provinces of Northeast China.

Mormo muscivirens Butler, 1878 黑带莫夜蛾

Mormo muscivirens Butler, 1878, *Annals and Magazine of Natural History*, (5)1: 289. Type locality: Japan, Yokohama.

Distribution. China (Heilongjiang, Jilin, Liaoning, Hunan); Russia (Siberia), Korea, Japan.

Genus ***Orthogonia*** Felder & Felder, 1862 胖夜蛾属

Orthogonia Felder & Felder, 1862, *Wiener Entomologische Monatschrift*, 6: 38. Type species: *Orthogonia sera* Felder & Felder, 1862 [China], by monotypy.

Synonymy: *Orthogonia* Hampson, 1908.

The genus includes 9 species, mainly distributed in the Chinese subregion of the Palaearctic region; 1 species is in the Three Provinces of Northeast China.

Orthogonia sera Felder & Felder, 1862 胖夜蛾

Orthogonia sera Felder & Felder, 1862, *Wiener Entomologische Monatschrift*, 6: 38. Type locality: China, Zhejiang[Che-Kiang], Ningbo[Ning-po].

Synonymy: *Orthogonia crispina* Butler, 1878.

Distribution. China (Heilongjiang, Jilin, Liaoning, Zhejiang, Jiangxi, Sichuan, Yunnan); Russia (Siberia),

Korea, Japan.

寄主：兰屿木耳菜（*Reynoutria elliptica*）、直穗酸模（*Rumex longifolius*）、水蛇麻（*Fatoua villosa*）、鼠尾草（*Salvia officinalis*）（Matov & Kononenko, 2012）。

Genus *Olivenebula* Kishida & Yoshimoto, 1977 霉裙剑夜蛾属

Olivenebula Kishida & Yoshimoto, 1977, *Tyo to Ga*, 28: 144. Type species: *Polyphaenis oberthueri* Staudinger, 1892 [Russia, East].

The genus includes 4 species, distributed in the Chinese and Mediterranean subregions of the Palaearctic and Oriental regions, in the Palaearctic 2 species; 1 species is in the Three Provinces of Northeast China.

注：首次给出中文属名。

Olivenebula oberthueri (Staudinger, 1892) 霉裙剑夜蛾

Polyphaenis oberthueri Staudinger, 1892, *in* N.M. Romanoff (ed.), *Mémoires sur les Lépidoptères*, 6: 454. Type locality: Russia, Primorye terr., Askold I..

Distribution. China (Heilongjiang, Jilin, Liaoning, Xinjiang, Shaanxi, Henan, Hubei, Fujian, Sichuan, Yunnan, Taiwan); Russia (Siberia), Korea, Japan.

注：《中国动物志》（夜蛾科）中将本种归入"*Polyphaenis* Boisduval, 1840 裙剑夜蛾属"，在此予以变更。

Genus *Triphaenopsis* Butler, 1878 带夜蛾属

Triphaenopsis Butler, 1878, *Annals and Magazine of Natural History*, (5)1: 163. Type species: *Triphaenopsis lucilla* Butler, 1878 [Japan, Yokohama].

The genus includes 9 species, distributed in the Chinese subregion of the Palaearctic and partially Oriental regions; 1 species is in the Three Provinces of Northeast China.

Triphaenopsis lucilla Butler, 1878 明带夜蛾

Triphaenopsis lucilla Butler, 1878, *Annals and Magazine of Natural History*, (5)1: 163. Type locality: Japan, Yokohama.

Distribution. China (Heilongjiang, Jilin, Fujian, Taiwan); Russia (Siberia), Korea, Japan.

寄主：紫竹（*Phyllostachys nigra*）、（*Phyllostachys heterocila*）、库页赤竹（*Sasa kurilensis*）（Matov & Kononenko, 2012）。

Genus *Xenotrachea* Sugi, 1958 路夜蛾属

Xenotrachea Sugi, 1958, *Tinea*, 4: 184. Type species: *Hadena albidisca* Moore, 1867 [India, Bangladesh].

The genus includes 14 species, rather uniform externally, distributed in the Chinese subregion of the

Palaearctic and in mountains of the Oriental regions, 6 species known from the Palaearctic region; 1 species is in the Three Provinces of Northeast China.

Xenotrachea niphonica **Kishida & Yoshimoto, 1979 北路夜蛾**

Xenotrachea niphonica Kishida & Yoshimoto, 1979, *Tinea*, 10: 268, fig. 1, 10. Type locality: Japan, Hokkaido, Junsai-numa.

Synonymy: *Xenotrachea albidisca*: sen. Sugi, 1958, nec Wileman, 1911, misidentification.

Distribution. China (Jilin, Liaoning, Guangxi, Hainan, Sichuan, Yunnan); Russia (Siberia), Korea, Japan.

注：首次给出中文种名。

Genus ***Karana*** Moore, 1882 驳夜蛾属

Karana Moore, 1882, *in* Hewitson & Moore, *Description of new Indian lepidopterous insects from the collection of the late Mr. W.S. Atkinson. Heterocera*, 2: 106. Type species: *Karana decorata* Moore, 1882 [India].

The genus includes 8 species, distributed in the Chinese subregion of the Palaearctic and Oriental regions; 1 species is in the Three Provinces of Northeast China.

Karana laetevirens **(Oberthür, 1884) 暗驳夜蛾**

Diphthera laetevirens Oberthür, 1884, *Études d'Entomologie*, 10: 17, pl. 2: 6. Type locality: Russia, [Bezverkhovo] Sidemi.

Distribution. China (Jilin); Russia (Siberia), Korea, Japan.

注：首次给出中文种名。

Genus ***Dipterygina*** Sugi, 1954 寅夜蛾属

Dipterygina Sugi, 1954, *Tinea*, 1: 25. Type species: *Dipterygia japonica* Leech, 1889 [Japan].

The genus includes more than 10 species (including some undescribed ones), distributed mainly in the Oriental region; 1 species is in the Three Provinces of Northeast China.

Dipterygina cupreotincta **Sugi, 1954 铜寅夜蛾**

Dipterygina cupreotincta Sugi, 1954, *Tinea* 1: 26, pl. 2: 9. Type locality: Japan, Tokyo, Takoo Mt.

Distribution. China (Jilin, Liaoning, Fujian, Hainan, Taiwan); South Korea, Japan.

Tribe **Actinotiini** Beck, 1996 炫夜蛾族

Actinotiina Beck, 1996, *Neue entomologische Nachrichten*, 36: 17. Type genus: *Actinotia* Hübner, [1821].

Established as subtribe of Ipimorphini.

Medium and small sized species with bright reddish-brown or olive-brown coloration and reduced wing pattern formed by longitudinal streak and prominent W-mark in subterminal field. The list of the Palaearctic Actinotiini presented by Fibiger & Hacker (2007). The genus *Hyppa* excluded here from Actinotiini and transferred to the tribe Xylenini. The larvae feed on Clusiaceae, often on *Hypericium.* The tribe Actinotiini represented by 1 genus and 2 species in the Three Provinces of Northeast China.

References. Chen, 1982, 1999 (Noctuidae: Amphipyrinae, part); Sugi, 1982 (Amphipyrinae, part); Speidel *et al.*, 1996 (Amphipyrinae, part); Kononenko *et al.*, 1998 (Amphipyrinae, part); Kitching, Rawlins, 1999 [1998] (Hadeninae); Kononenko, 2003 (Amphipyrinae, part); Fibiger, Lafontaine, 2005 (Xyleninae: Actinotiini); Fibiger, Hacker, 2005 (Xyleninae: Actinotiini); Lafontaine, Fibiger, 2006 (Xyleninae: Actinotiini); Kononenko, Han, 2007 (Xyleninae: Actinotiini); Kononenko, 2005 (Xyleninae: Actinotiini); Fibiger, Hacker, 2007 (Xyleninae: Actinotiini); Matov *et al.*, 2008 (Xyleninae: Actinotiini); Lafontaine, Schmidt, 2010 (Noctuidae, Actinotiini); Fibiger *et al.*, 2011 (Xyleninae: Actinotiini); Kononenko, 2016 (Xyleninae: Actinotiini).

注：首次给出中文族名。

Genus ***Actinotia*** Hübner, [1821] 炫夜蛾属

Actinotia Hübner, [1821], *Verzeichniss bekannter Schmettlinge* [sic!]: 244. Type species: *Phalaena perspicillaris* Linnaeus, 1761 [Europe] (=*polyodon* Clerck, 1759).

Synonymy: *Radinotia* Beck, 1996; *Actinotis*: Shulze & Kükenthal, 1926, missp.; *Actinotis* auct., missp..

The genus includes 7 species, mainly distributed in the Chinese subregion of Palaearctic and Oriental regions, 5 species known from the Palaearctic; 2 species are in the Three Provinces of Northeast China.

Actinotia polyodon **(Clerck, 1759) 炫夜蛾**

Phalaena polyodon Clerck, 1759, *Icones Insectorum Rariorum cum Nominibus*, 1: pl. 2: 2. Type locality: [not stated] [Europe].

Synonymy: *Phalaena perspicillaris* Linnaeus, 1761.

Distribution. China (Liaoning, Xinjiang, Xizang); Russia (European part, Siberia), Mongolia, Korea, Japan, Nepal; Central Asia, Asia Minor, Europe.

寄主：连翘（陈一心，1999）；毛金丝桃（*Hypericum hirsutum*）、斑点金丝桃（*Hypericum maculatum*）、黑点叶金丝桃（*Hypericum perforatum*）、黄芪属（*Astragalus*）、槭树属（*Acer*）属植物（Matov & Kononenko 2012）。

Actinotia intermediata **(Bremer, 1861) 间纹炫夜蛾**

Cloantha intermediata Bremer, 1861, *Bulletin de l'Academie Impériale des Sciences de St.-Petersbroug*, 4: 489. Type locality: Russia, [Primorye terr., Khanka lake] "Kengka-See".

Synonymy: *Cloantha intermedia*: Bremer, 1864, emend..

Distribution. China (Heilongjiang, Jilin, Liaoning, Shaanxi, Hubei, Hunan, Zhejiang, Fujian, Hainan, Sichuan, Yunnan); Russia (Siberia), Korea, Japan, Nepal. India, Pakistan, Vietnam, Thailand.

寄主：（*Hypericum orectus*）、省沽油（*Staphylea bumalda*）（Matov & Kononenko, 2012）。

Tribe **Phlogophorini** Hampson, 1918 衫夜蛾族

Phlogophorinae Hampson, 1918, *Novitates zoologicae 25*: 190, 383. Type genus: *Phlogophora* Treitschke, 1825.

Synonymy: Phlogophorini Forbes, 1954.

One of the main characteristic of the tribe stated by Fibiger & Lafontaine (2005) is peculiar longitudinal folds in the forewings that give the wings a crumpled look when the moth is at rest. This character is typical for *Euplexia*, *Conservula* and *Phlogophora*, but not observed in some other genera of the Phlogophorini. The European fauna of Phlogophorini has been revised by Fibiger & Hacker (2007). The larvae are polyphagous, feeding on herbaceous plants. The tribe occurs worldwide but most diversely in temperate, subtropical and tropical regions. The tribe Phlogophorini includes 3 genera and 4 species in the Three Provinces of Northeast China, most species occur in zone of nemoral forest.

References. Chen, 1982, 1999 (Noctuidae: Amphipyrinae, part); Sugi, 1982 (Amphipyrinae, part); Speidel *et al.*, 1996 (Amphipyrinae, part); Kononenko, 1996 (Amphipyrinae, part); Kononenko *et al.*, 1998 (Amphipyrinae, part); Kitching, Rawlins, 1999 [1998] (Hadeninae); Kononenko, 2003 (Amphipyrinae, part); Fibiger, Lafontaine, 2005 (Xyleninae: Phlogophorini); Fibiger, Hacker, 2005 (Xyleninae: Phlogophorini); Kononenko, 2005 (Xyleninae: Phlogophorini); Lafontaine, Fibiger, 2006 (Xyleninae: Phlogophorini); Kononenko, Han, 2007 (Xyleninae: Phlogophorini); Kononenko, 2005, 2010 (Xyleninae: Phlogophorini); Fibiger, Hacker, 2007 (Xyleninae: Phlogophorini); Matov *et al.*, 2008 (Xyleninae: Phlogophorini); Lafontaine, Schmidt, 2010 (Noctuidae, Phlogophorini); Fibiger *et al.*, 2011 (Xyleninae: Phlogophorini).

Genus ***Phlogophora*** Treitschke, 1825 衫夜蛾属

Phlogophora Treitschke, 1825, *Die Schmetterlinge von Europa*, 5 (1): 369. Type species: *Phalaena meticulosa* Linnaeus, 1758 [Europe].

Synonymy: *Solenoptera* Duponchel, [1845] 1844, preocc. (Audinet – Serville, 1832 [Coleoptera]); *Brotolomia* Lederer, 1857, repl. name; *Habryntis* Lederer, 1857; *Racoptera* Scott, 1858; *Chutapha* Moore, 1882; *Mesolomia* Smith, 1893; *Madeuplexia* Viette, 1960; *Trigonophora* auct..

The genus includes around 60 species (Palaearctic, Nearctic, Oriental and Ethiopian regions), 17 species are in Palaearctic, among them 2 species are in the Three Provinces of Northeast China.

Phlogophora illustrata **(Graeser, 1888 [1889]) 北衫夜蛾**

Euplexia illustrata Graeser 1888 [1889], *Berliner entomologische Zeitschrift*, 32, II: 329. Type locality:

Russia, [Khabarovsk] “Chabarofka”, Vladivostok.

Distribution. China (Heilongjiang, Jilin, Liaoning); Russia (Siberia), Korea, Japan.

寄主：蕨属（*Pteridium*）植物（Matov & Kononenko, 2012）。

注：首次给出中文种名。

Phlogophora beatrix Butler, 1878 福衫夜蛾

Phlogophora beatrix Butler, 1878, *Annals and Magazine of Natural History*, (5)1: 193. Type locality: Japan, Hakodate.

Synonymy: *Phlogophora pallens* Oberthür, 1879.

Distribution. China (Heilongjiang, Jilin, Liaoning, Hunan, Xizang); Russia (Siberia), Korea, Japan.

Genus *Euplexia* Stephens, 1829 锦夜蛾属

Euplexia Stephens, 1829 [June], *The Nomenclature of British Insectes*: 41. Type species: *Phalaena lucipara* Linnaeus, 1758 [Sweden].

Synonymy: *Berrhaea* Walker, 1858; *Epa* Bethune-Backer, 1906.

The genus includes 64 species (Palaearctic, Nearctic, Oriental and Ethiopian regions); 9 species are in Palaearctic, among them 1 species is in the Three Provinces of Northeast China.

Euplexia lucipara (Linnaeus, 1758) 锦夜蛾

Phalaena lucipara Linnaeus, 1758, *Systema Naturae* (Edn. 10) 1: 518. Type locality: Europe.

Synonymy: *Phalaena dubia* Hufnagel, 1766; *Bombyx flavomacula* Fabricius, 1787; *Euplexia leonhardi* Rebel, 1909; *Euplexia lucipara brittannica* Turner, 1943 (ssp.).

Distribution. China (Heilongjiang, Jilin, Liaoning, Henan, Hubei, Sichuan, Yunnan); Russia (European part, Siberia), Korea, Japan; Asia Minor, Central Asia, Europe.

寄主：楞木、大豆、毛茛属、女贞属（陈一心，1999）；欧洲蕨（*Pteridium aquilinum*）、蹄盖蕨（*Athyrium filix-femina*）、刺叶鳞毛蕨（*Dryopteris carthusiana*）、欧洲鳞毛蕨（*Dryopteris filix-mas*）、（*Dryopteris spunulosa*）、叶升麻（*Actaea spicata*）、翠雀（*Delphinium grandiflorum*）、欧洲桤木（*Alnus glutinosa*）、胡榛（*Corylus avellana*）、黑果越橘（*Vaccinium myrtillus*）、毛黄连花（*Lysimachia vulgaris*）、里文堇菜（*Viola riviniana*）、耳柳（*Salix aurita*）、西葫芦（*Cucurbita pepo*）、啤酒花（*Humulus lupulus*）、异株荨麻（*Urtica dioica*）、多年生山靛（*Mercurialis perennis*）、正醋栗（*Ribes grossularia*）、黑穗状醋栗（*Ribes nigra*）、红茶藨子（*Ribes rubrum*）、鹅莓（*Ribes uva-crispa*）、秦氏悬钩子（*Rubus caesius*）、树莓（*Rubus fruticosus*）、覆盆子（*Rubus idaeus*）、珍珠梅（*Sorbaria sorbifolia*）、绣线菊（*Spiraea salicifolia*）、柳兰（*Chamaenerion angustifolium*）、沼生柳叶菜（*Epilobium palustre*）、牧地山黧豆（*Lathyrus pratensis*）、草木樨（*Melilotus officinalis*）、豌豆（*Pisum sativum*）、红车轴草（*Trifolium pratense*）、白车轴草（*Trifolium repens*）、广布野豌豆（*Vicia cracca*）、野豌豆（*Vicia sepium*）、欧洲红瑞木（*Cornus sanguinea*）、蔓生盘叶忍冬（*Lonicera caprifolium*）、欧洲绣球荚蒾（*Viburnum opulus*）、西洋接骨木（*Sambucus nigra*）、莳萝（*Anethum graveolens*）、林当归（*Angelica sylvestris*）、牛防风（*Heracleum sphondylium*）、颠茄草（*Atropa belladonna*）、番茄（*Lycopersicon esculentum*）、龙葵（*Solanum nigrum*）、马铃薯（*Solanum tuberosum*）、昼颜（*Calystegia sepium*）、毛地黄（*Digitalis purpurea*）、大车前（*Plantago major*）、匍筋骨草（*Ajuga reptans*）、宝盖草（*Lamium amplexicaule*）、紫花野芝麻（*Lamium maculatum*）、棕矢车菊（*Centaurea jacea*）、莴苣（*Lactuca sativa*）、多毛鹰齿菊（*Leontodon hispidus*）、母菊（*Matricaria recutita*）、吊钟千里光（*Senecio fuchsii*）、款冬花（*Tussilago farfara*）、复叶耳蕨属（*Aspidium*）、乌头属（*Aconitum*）、耧斗菜属（*Aquilegia*）、铁线莲属（*Clematis*）、毛蕊花属（*Ranunculus*）、白屈菜属（*Chelidonium*）、滨藜属（*Atriplex*）、藜属（*Chenopodium*）、蓼属（*Polygonum*）、酸模（*Rumex acetosa*）、栎属（*Quercus*）、桦木属（*Betula*）、金丝桃属（*Hypericum*）、大蒜芥属（*Sisymbrium*）、羽衣草属（*Alchemilla*）、委陵菜属（*Potentilla*）、羽扇豆属（*Lupinus*）、菜豆属（*Phaseolus*）、天竺葵属（*Pelargonium*）、凤仙花属（*Impatiens*）、山梅花属（*Philadelphus*）、梣属（*Fraxinus*）、女贞属（*Ligustrum*）、旋花属（*Convolvulus*）、薄荷属（*Mentha*）、水苏属（*Stachys*）、牛蒡属（*Arctium*）、金盏花属（*Calendula*）、蜂斗菜属（*Petasites*）、一枝黄花属（*Solidago*）、蒲公英属（*Taraxacum*）、水芋属（*Calla*）植物（Matov & Kononenko, 2012）。

Genus ***Auchmis*** Hübner, [1821] 1816 杰夜蛾属

Auchmis Hübner, [1821] 1816, *Verzeichniss bekannter Schmettlinge* [sic!]: 243. Type species: *Phalaena petroriza* Borkhausen, 1792 [Europe].

Synonymy: *Euscotia* Butler, 1889 (subgen.); *Trichoriza* Hampson, 1905; *Stenostigma* Warren, 1910.

The genus subdivided into subgenera *Auchmis* and *Euscotia*; it includes 17 species distributed in Central Asian and the Chinese subregions of the Palaearctic and Oriental regions, 10 species are known from the Palaearctic; 3 species are in the Three Provinces of Northeast China.

Subgenus ***Euscotia*** Butler, 1889 暗杰夜蛾亚属

Euscotia Butler, 1889, *Illustrations of Typical Specimens of Lepidoptera Heterocera in the Collection of the British Museum*, 7: 47. Type species: *Rhizogramma inextricata* Moore, 1881 [India].

The subgenus includes 2 species, distributed in the Chinese subregions of the Palaearctic and Oriental regions; only 1 species is in the Three Provinces of Northeast China.

注：首次给出中文亚属名。

Auchmis* (*Euscotia*) *saga **(Butler, 1878)** 沙杰夜蛾

Lithophane saga Butler, 1878, *Annals and Magazine of Natural History*, (5)1: 198. Type locality: Japan, Yokohama.

Synonymy: *Rhizogramma aurilegula* Oberthür, 1880.

Distribution. China (Heilongjiang, Jilin, Liaoning, Henan, Hubei, Sichuan, Yunnan, Taiwan); Russia (Siberia), Korea, Japan.

寄主：黄芦木（*Berberis amurensis*）（Matov & Kononenko, 2012）。

Tribe **Apameini** Guenée, 1841 秀夜蛾族

Apameini Guenée, 1841, *Annales de la Société Entomologique de France*, 10: 237 (as Apamidi, an original missp.; emended in Apameini). Type genus: *Apamea* Ochsenheimer, 1816.

Synonymy: Nonagridi Guenree, 1837; Gortynidae Duponchel, [1845] 1844; Xylophasides Guenée, 1854; Apameinae Tutt, 1896; Apameini Forbes, 1954, preocc.; Septidini Forbes, 1954; Calamiina Beck, 1996; Luperinina Beck, 1996; Mesapameina Beck, 1996; Eremobiina Beck, 1999.

The tribe Apameini comprises approximately 1000 species worldwide, with the highest diversity in the northern temperate and subtropical areas (although a few species are known from the tropics and the arctic). 3 subtribes Oxytripiina, Apameina and Sesamiina are in the Palaearctic region; the majority of genera belongs to the subtribe Apameina. The European Apameini have been revised by Zilli *et al.* (2005), the Eurasian and Holarctic *Apamea* generic group revised by Zilli *et al.* (2009) and by Mikkola *et al.* (2009) correspondently.

The most conspicuous key character of the tribe Apameini is flattened dorso-ventrally the heavily sclerotized ovipositor lobes in females; between these lobes there is a double rod-like structure; the pleural sclerites in the male genitalia are double, shaped like a double helix.

The larvae of all 3 subtribes, are closely associated mainly with monocotyledonous hosts plants, in the genus *Hydraecia* Guenée the larvae are polyphagous. In many genera larvae are stem borers, in the genus *Apamea* larvae of some species are underground inhabitances.

The tribe Apameini represented in the Three Provinces of Northeast China by 2 subtribes, 35 genera and 52 species.

References. Chen, 1982, 1999 (Noctuidae: Amphipyrinae, part); Sugi, 1982 (Amphipyrinae, part); Kononenko, 1985, 1995, 2001 (Amphipyrinae, part); Speidel *et al.*, 1996 (Amphipyrinae, part); Kononenko *et al.*, 1998 (Amphipyrinae, part); Kitching, Rawlins, 1999 [1998] (Hadeninae); Kononenko, 1986, 2003 (Amphipyrinae, part); Kononenko, 2005, 2006 (Xyleninae: Apameini); Fibiger, Lafontaine, 2005 (Xyleninae: Apameini); Fibiger, Hacker, 2005 (Xyleninae: Apameini); Zilli *et al.*, 2005 (Xyleninae: Apameini); Kononenko, 2005 (Xyleninae: Apameini), 2016 (Xyleninae: Apameini: Oxytripina, Apameina, Sesamiina); Lafontaine, Fibiger, 2006 (Xyleninae: Apameini); Kononenko, Han, 2007 (Xyleninae: Apameini); Fibiger, Hacker, 2007 (Xyleninae: Apameini); Matov *et al.*, 2008 (Xyleninae: Apameini); Zilli *et al.*, 2009 (Xyleninae: Apameini); Mikkola *et al.*, 2009 (Xyleninae: Apameini); Lafontaine, Schmidt, 2010 (Noctuidae, Apameini); Kononenko, Matov, 2012 (Xyleninae: Apameini); Fibiger *et al.*, 2011 (Xyleninae: Apameini).

注：首次给出中文族名。

Subtribe **Oxytripiina** Gozmany, 1970 蚀夜蛾亚族

Oxytripiinae Gozmany, 1970, *in* Szekessy, *Fauna Hungariae*, 102: 103 (as Oxytripiinae Gozmany, 1970, an original misspelling corrected in Oxytripiinae). Type genus: *Oxytrypia* Staudinger, 1871 (=*Oxitripia* Staudinger, 1871, misspelling).

The subtribe contains 1 genus widely distributed in the Palaearctic with 2 species; 1 species is in the Three Provinces of Northeast China.

注：首次给出中文亚族名。

Genus ***Oxytripia*** Staudinger, 1871 蚀夜蛾属

Oxytripia Staudinger, 1871, *in* Staudinger & Wocke, *Catalog der Lepidopteren des Europaeischen Faunengebiets*: 98. Type species: *Phalaena orbiculosa* Esper, 1799 [Hungary].

Synonymy: *Oxytripa*: Staudinger, 1871:423, missp..

The Palaearctic genus includes 2 species, Eurasian *O. orbiculosa* (Esper, 1799) and West Asian *O. stephania* (Sutton, 1964); 1 species is in the Three Provinces of Northeast China.

Oxytripia orbiculosa **(Esper, 1799)** 蚀夜蛾

Phalaena orbiculosa Esper, 1799, *Die Schmetterlinge in Abbildungen nach der Natur mit Beschreibungen*, 3 (3): 93, pl. 93: fig. 8. Type locality: Hungary, Szeged.

Synonymy: *Oxytripia orbiculosa ussurica* Schawerda, 1923 (var.); *Oxytripia orbiculosa noctivolans* Pinker, 1980 (ssp.).

Distribution. China (Heilongjiang, Jilin, Nei Mongol, Qinghai, Xizang, Xinjiang); Russia (Southeast European part, Ural, Siberia), Mongolia, Korea, Japan; South Europe.

寄主：鸢尾科植物（陈一心，1999）；沙生鸢尾（*Iris arenaria*）、短旗鸢尾（*Iris pumila*）（Matov & Kononenko, 2012）。

Subtribe **Apameina** Guenée, 1851 秀夜蛾亚族

Apameini Guenée, 1841, *Annales de la Société Entomologique de France* 10: 237 (as Apamidi, an original missp.; emended in Apameini). Type genus: *Apamea* Ochsenheimer, 1816.

Synonymy: Nonagridi Guenree, 1837; Gortynidae Duponchel, [1845] 1844; Xylophasides Guenée, 1854; Apameinae Tutt, 1896; Apameini Forbes, 1954, preocc.; Septidini Forbes, 1954; Calamiina Beck, 1996; Luperinina Beck, 1996; Mesapameina Beck, 1996; Eremobiina Beck, 1999.

The subtribe represented in the Three Provinces of Northeast China by 34 genera and 51 species.

Genus *Sidemia* Staudinger, 1892 袭夜蛾属

Sidemia Staudinger, 1892, *in* N.M. Romanoff (ed.), *Mémoires sur les Lépidoptères*, 6: 459. Type species: *Sidemia snelleni* Staudinger, 1892, repl. name; a junior synonym of *Noctua speciosa* Bremer, 1861 nec *Noctua speciosa* Hübner, 1808 [Russia, Primorye terr.].

The genus includes 8 species, distributed in the Palaearctic and Ethiopian regions; 4 species are in the Palaearctic region, among them 2 species are in the Three Provinces of Northeast China.

Sidemia spilogramma **(Rambur, 1871) 克袭夜蛾**

Valeria spilogramma Rambur, 1871, *Annales de la Société Entomologique de France*, 1 (5): 321. Type locality: Russia.

Synonymy: *Hadena christophi* Alphéraky, 1888.

Distribution. China (Heilongjiang, Jilin, Nei Mongol, Hebei, Shandong, Jiangsu, Hunan); Russia (Southeast European part, North Caucasus, Siberia), Mongolia, Korea, Japan, Balkans, Romania.

Sidemia bremeri **(Erschoff, 1870) 袭夜蛾**

Agrotis bremeri Erschoff, 1870, *Horae Societatis Entomologicae Rossicae*, 4: 152, 164. Type locality: Russia, "Siberia", an objective repl. name.

Synonymy: *Noctua speciosa* Bremer, 1861, nec Hübner, 1808; *Sidemia snelleni* Staudinger, 1892.

Distribution. China (Heilongjiang, Jilin, Shaanxi); Russia (Siberia), Mongolia, Korea, Japan.

Genus *Bambusiphila* Sugi, 1958 竹笋禾夜蛾属

Bambusiphila Sugi, 1958, *Tinea*, 4:184. Type species: *Polydesma vulgaris* Butler, 1886 [Japan, Tokyo].

Monotypic Sino-Himalayan genus; 1 species is in the Three Provinces of Northeast China.

注：首次给出中文属名。

Bambusiphila vulgaris **(Butler, 1886) 竹笋禾夜蛾**

Polydesma vulgaris Butler, 1886, *Transactions of the Entomological Society of London*, 1886: 135. Type locality: Japan, Tokyo.

Synonymy: *Atrachea vulgaris distincta* Warren, 1911 (f.).

Distribution. China (Liaoning, Jiangsu, Hubei, Hunan, Fujian, Jiangxi, Yunnan); South Korea, Japan.

寄主：竹笋（陈一心，1999）。

注：《中国动物志》（夜蛾科）中将本种归入到"*Oligia* Hübner, [1821] 1816 禾夜蛾属"，在此予以变更。

Genus *Staurophora* Reichenbach & Leipzig, 1817 干纹夜蛾属

Staurophora Reichenbach & Leipzig, 1817, *Jenaische Allgemeine Literatur-Zeitung*, l: 786, repl. name. Type species: *Phalaena celsia* Linnaeus, 1758 [Europe].

Synonymy: *Psylla* Germar, 1811, preocc. (Geoffroy, 1762 [Hemiptera]); *Diacopa* Hübner, [1820] 1816; *Celsia* Stephens, 1829 [June], nom. nud.; *Celsia* Stephens, 1829 [August]; *Calotaenia* Stephens, 1829, repl. name; *Jaspidia* Boisduval, 1840, preocc. Hübner, [1822] 1816 [Lepidoptera, Noctuidae]); *Callitaenia*: Agassiz, 1847, emend.; *Jaspidea*: Staudinger, 1861, missp..

The genus includes 1 species distributed across Eurasia.

Staurophora celsia **(Linnaeus, 1758) 干纹夜蛾**

Phalaena celsia Linnaeus, 1758, *Systema Naturae* (Edn. 10) 1: 507. Type locality: Europe.

Synonymy: *Celsia viridis* Stephens, 1829; *Calotaenia immaculata* Staudinger, 1871; *Calotaenia tenuis* Warren, 1910.

Distribution. China (Heilongjiang, Jilin, Nei Mongol, Xinjiang, Hebei, Shandong); Russia (European part, Siberia), Mongolia, Korea, Japan; Central Asia, Europe.

寄主：（*Agrostis capillaria*）、黄花茅（*Anthoxanthum odoratum*）、嘴唇花（*Calamagrostis elata*）、拂子茅（*Calamagrostis epigeios*）、鸭茅（*Dactylis glomerata*）、小穗发草（*Deschampsia caespitosa*）、葡萄冰草（*Elymus repens*）、那杜草（*Nardus stricta*）、银须草属（*Aira*）、看麦娘属（*Alopecurus*）、虉草属（*Phalaris*）、早熟禾属（*Poa*）植物（Matov & Kononenko, 2012）。

Genus *Helotropha* Lederer, 1857 蛮夜蛾属

Helotropha Lederer, 1857, *Die Noctuinen Europa's mit Zuziehung einiger bisher meist dazu gezahlter Arten des asiatischen Russland's, Kleinasien's, Syrien's und Labrador's*: 35, n8. Type species: *Noctua leucostigma* Hübner, 1808 [Europe].

Synonymy: *Celaena* auct., nec Stephens, 1829; *Gortyna* auct.; *Oligia* auct..

Holarctic boreal genus containing 4 species; 1 species distributed in the Palaearctic region, and in the Three Provinces of Northeast China.

Helotropha leucostigma **(Hübner, [1808]) 白纹蛮夜蛾**

Noctua leucostigma Hübner [1808], *Sammlung Europaischer Schmetterlinge. Eulen*, 2: pl. 80: 375. Type locality: Europe.

Synonymy: *Noctua fibrosa* Hübner, 1808; *Noctua lunina* Haworth, 1809; *Cerastis laevis* Butler, 1881; *Gortyna leucostigma kurilibia* Bryk, 1942 (ssp.); *Blepharidia yeterofuna* Bryk, 1942; *Celaena leucostigma scotica* Cockayne, 1955 (ssp.).

Distribution. China (Heilongjiang, Jilin, Liaoning, Nei Mongol, Xinjiang); Russia (European part, Siberia), Mongolia, Korea, Japan, Kazakhstan; Europe.

寄主：雅致柳叶菜（*Epilobium alpinum*）、柳叶菜（*Epilobium hirsutum*）、蒿柳（*Salix viminalis*）、沼生柳叶菜（*Epilobium palustre*）、向日葵（*Helianthus annuus*）、黄金鸢尾（*Iris flavissima*）、黄菖蒲（*Iris pseudacorus*）、紫苞鸢尾（*Iris ruthenica*）、膜苞鸢尾（*Iris scariosa*）、西伯利亚鸢尾（*Iris sibirica*）、尖叶苔草（*Carex acutiformis*）、（*Carex paludosa*）、狐狸薹草（*Carex vulpina*）、克拉莎（*Cladium mariscus*）、菖蒲（*Acorus calamus*）、拂子草（*Calamagrostis epigejos*）、沿沟草（*Catabrosa aquatica*）、鸭茅（*Dactylis glomerata*）、甜茅（*Glyceria aquatica*）、水甜茅（*Glyceria maxima*）、折甜茅（*Glyceria plicata*）、（*Glyceria spectabilis*）、酸沼草（*Molinia caerulea*）、芦苇（*Phragmites australis*）、玉米（*Zea mays*）、酸模属（*Rumex*）、当归属（*Angelica*）、黑三棱属（*Sparganium*）植物（Matov & Kononenko, 2012）。

注：《中国动物志》（夜蛾科）中引用同物异名“*Celaena leucostigma* (Hübner, [1808])”，在此予以变更。

Genus *Gortyna* Ochsenheimer, 1816 构夜蛾属

Gortyna Ochsenheimer, 1816, *Die Schmetterlinge von Europa*, 4: 82. Type species: *Noctua flavago* [Denis & Schiffermüller], 1775 [Austria].

Synonymy: *Ochria* Hübner, [1821] 1816; *Cnephozeta* Billberg, 1820, repl. name; *Hydroecia* auct.; *Xanthoecia* Hampson, 1908; *Nytorga* Beck, 1996.

The genus includes 18 species from the Palaearctic and Oriental regions; 11 species are distribed in the Palaearctic region; 2 species are in the Three Provinces of Northeast China.

Gortyna fortis (Butler, 1878) 健构夜蛾

Ochria fortis Butler, 1878, *Annals and Magazine of Natural History*, (5)1: 83. Type locality: Japan, Yokohama.

Synonymy: *Xanthoecia flavago koreago* Bryk, 1948 (ssp.).

Distribution. China (Heilongjiang, Jilin, Liaoning, Fujian); Russia (Siberia), Korea, Japan.

寄主：阳芋（*Solanum tuberosum*）、牛蒡（*Arctium lappa*）、乌头属（*Aconitum*）、唐松草属（*Thalictrum*）、蓼属（*Polygonum*）、酸模属（*Rumex*）、当归属（*Angelica*）、蒿属（*Artemisia*）、飞蓬属（*Erigeron*）、蜂斗菜属（*Petasites*）、千里光属（*Senecio*）植物（Matov & Kononenko, 2012）。

Gortyna basalipunctata Graeser, 1888 [1889] 基点构夜蛾

Gortyna basalipunctata Graeser, 1888 [1889], *Berliner entomologische Zeitschrift*, 32, II: 341. Type locality: Russia, [Khabarovsk] “Chabarofka”, [Primorie terr.] “Ussuri gebiete”.

Synonymy: *Gortyna intermixta* Swinhoe, 1891.

Distribution. China (Heilongjiang, Jilin, Liaoning, Shaanxi, Sichuan); Russia (Siberia), Korea, Japan,

India(North).

寄主：玉米（陈一心，1999）；大豆（*Glycine max*）、番茄（*Lycopersicon esculentum*）、马铃薯（*Solanum tuberosum*）、向日葵（*Helianthus annuus*）、玉米（*Zea mays*）、蓼属（*Polygonum*）、菜豆属（*Phaseolus*）、蒿属 (*Artemisia*)、蜂斗菜属（*Petasites*）植物（Matov & Kononenko, 2012）。

Genus ***Hydraecia*** Guenée, 1841 角剑夜蛾属

Hydraecia Guenée, 1841, *Annales de la Société Entomologique de France*, 10: 237. Type species: *Phalaena micacea* Esper, 1789 [Austria].

Synonymy: *Hydroecia*: Agassiz, 1846, emend.; *Hydrooecia*: Herrich-Schäffer, 1849, missp. *Hadroecia*: Schaus, 1894, missp..

The Holarctic genus includes 20 species, 10 species are in the Palaearctic, among them 3 species are in the Three Provinces of Northeast China.

Hydraecia ultima Holst, 1965 黑角剑夜蛾

Hydraecia ultima Holst, 1965, *Entomologiscke Meddelelser*, 34: 54, fig. 1a-d, g-j. Type locality: Denmark.

Synonymy: *Hydraecia micacea* auct..

Distribution. China (Heilongjiang); Russia (European part, Siberia), Mongolia, North Korea, Japan; Europe.

寄主：水生酸模（*Rumex hydrolapathum*）、直穗酸模（*Rumex longifolius*）、沼泽酸模（*Rumex palustris*）（Matov & Kononenko, 2012）。

注：首次给出中文种名。

Hydraecia mongoliensis Urbahn, 1967 蒙角剑夜蛾

Hydraecia mongoliensis Urbahn, 1967, *Reichenbachia*, 9: 250, fig. 1. Type locality: Mongolia, Central Aimak.

Distribution. China (Heilongjiang, Jilin, Nei Mongol); Russia (Siberia), Mongolia, Japan, Kazakhstan.

寄主：（*Allium victoralis*）（Matov & Kononenko, 2012）。

Hydraecia petasitis Doubleday, 1847 苏角剑夜蛾

Hydraecia petasitis Doubleday, 1847, *Zoologist*, 5: 1914. Type locality: Great Britain.

Synonymy: *Gortyna vindelica* Freyer, 1849; *Hydroecia petasitis amurensis* Staudinger, 1892 (var.); *Hydroecia micacea viola* Bryk, 1948 (ssp.).

Distribution. China (Heilongjiang, Jilin, Liaoning, Shaanxi); Russia (European part, Siberia), Mongolia, Korea, Japan, Kazakhstan; Europe.

寄主：向日葵（*Helianthus annuus*）、（*Andromeda polyfolia*）、白色蜂斗菜（*Petasites albus*）、紫

蜂斗菜（*Petasites hybridus*）、蜂斗菜（*Petasites japonicus*）、（*Petasites officinalis*）、（*Petasites spurius*）、菖蒲（*Acorus calamus*）、蓼属（*Polygonum*）、蒿属（*Artemisia*）、千里光属（*Senecio*）、鸢尾属（*Iris*）植物（Matov & Kononenko, 2012）。

Genus *Amphipoea* Billberg, 1820 **奂夜蛾属**

Amphipoea Billberg 1820, *Enumeratio insectorum in museo G.J. Billberg*: 87. Type species: *Phalaena nictitans* Linnaeus, 1767 [Europe] (=*Phalaena oculea* Linnaeus, 1761).

Synonymy: *Apamea* auct.; *Hydroecia* auct..

Holarctic genus includes 25 species, 16 species are in the Palaearctic region, among them 5 species are in the Three Provinces of Northeast China.

Amphipoea fucosa **(Freyer, 1830) 麦奂夜蛾**

Noctua fucosa Freyer, 1830, *Beiträge zur Geschichte Europaischer Schmetterlinge mit Abbildungen, etc.*, 3: 152, pl. 141: 2. Type locality: [not stated].

Synonymy: *Hydroecia nictitans paludis* Tutt, 1888 (r.); *Hydroecia fucosa pallescens* Staudinger, 1900 (var.).

Distribution. China (Heilongjiang, Jilin, Liaoning, Nei Mongol, Xinjiang, Shanxi, Qinghai, Hebei, Henan, Hubei, Hunan, Yunnan); Russia (European part, Siberia), Mongolia, Korea, Japan; Central Asia, Europe.

寄主：小麦、大麦、玉米（陈一心，1999）；野燕麦（*Avena fatua*）、燕麦（*Avena sativa*）、（*Bolboschoenus maritimus*）、海滨三棱草（*Bromis erectus*）、（*Bromis inermis*）、鸭茅（*Dactylis glomerata*）、发草（*Deschampsia caespitosa*）、匍匐冰草（*Elymus repens*）、偃麦草（*Elytrigia repens*）、大羊茅（*Festuca gigantea*）、羊茅（*Festuca ovina*）、草甸羊茅（*Festuca pratensis*）、栽培二棱大麦（*Hordeum distichon*）、鼠大麦（*Hordeum murinum*）、大麦（*Hordeum vulgare*）、丝带草（*Phalaris arundinacea*）、梯牧草（*Phleum pratense*）、早熟禾（*Poa annua*）、草地早熟禾（*Poa pratensis*）、黑麦（*Secale cereale*）、小麦（*Triticum aestivum*）、二粒麦（*Triticum dicoccum*）、玉米（*Zea mays*）、鸢尾属（*Iris*）、冰草属（*Agropyron*）、拂子茅属（*Calamagrostis*）、黑麦草属（*Lolium*）植物（Matov & Kononenko, 2012）。

Amphipoea lucens **(Freyer, 1845) 鲁奂夜蛾**

Apamea lucens Freyer, 1845, *Neuere Beiträge zur Schmetterlingskunde mit Abbildungen nach der Natur*, 5: 143, pl. 468: 3, 4. Type locality: [not stated].

Distribution. China (Heilongjiang, Jilin); Russia (European part, Siberia), Mongolia, Korea, Japan; Central Asia, Europe.

寄主：白毛羊胡子草（*Eriophorum vaginatum*）、燕麦（*Avena sativa*）、发草（*Deschampsia caepitosa*）、匍匐冰草（*Elymus repens*）、栽培二棱大麦（*Hordeum distichon*）、大麦（*Hordeum vulgare*）、酸沼草（*Molinia caerulea*）、日本麦氏草（*Molinia japonica*）、糙隐子草（*Cleistognes squarrosa*）、早熟禾（*Poa annua*）、黑麦（*Secale cereale*）、鸭茅（*Dactylis glomerata*）、拂子茅属（*Calamagrostis*）、小麦属（*Triticum*）植物（Matov & Kononenko, 2012）。

注：首次给出中文种名。

***Amphipoea burrowsi* (Chapman, 1912) 波奂夜蛾**

Hydraecia burrowsi Chapman, 1912, *Entomologist's Record and Journal of Variation*, 24: 109. Type locality: Russia, Vladivostok.

Distribution. China (Heilongjiang, Jilin); Russia (Siberia to Kamchatka and Magadan), Korea, Japan.

***Amphipoea ussuriensis* (Petersen, 1914) 北奂夜蛾**

Hydraecia ussuriensis Petersen, 1914, *Horae Societatis Entomologicae Rossicae*, 41 (4): 14. Type locality: Japan, Hakodate; Russia, [Primorye terr.] "Ussuri".

Distribution. China (Heilongjiang, Jilin, Liaoning); Russia (East Siberia), Korea, Japan.

***Amphipoea asiatica* (Burrows, 1912) 亚奂夜蛾**

Hydroecia asiatica Burrows, 1912, *Transactions of the Entomological Society of London*, 1912: 747. Type locality: Kyrgyzstan, Syrt Naryn.

Synonymy: *Apamea lucens motojondensis* Bryk, 1948 (ssp.); *Apamea asiatica magna* Heydemann, 1941 (ssp.).

Distribution. China (Heilongjiang, Jilin, Xinjiang, Shaanxi, Shanxi, Sichuan, Yunnan); Russia (Siberia), Mongolia, Korea, Japan; Central Asia.

Genus *Rhizedra* Warren, 1911 内夜蛾属

Rhizedra Warren, 1911, *in* A.Seitz (ed.), *Die Gross-Schmetterlinge der Erde*, 3: 234. Type species: *Noctua lutosa* Hübner, [1803] [Europe].

Monotypic Palaearctic genus, introduced to North America; 1 species is in the Three Provinces of Northeast China.

***Rhizedra lutosa* (Hübner, [1803]) 内夜蛾**

Noctua lutosa Hübner, [1803], *Sammlung Europaischer Schmetterlinge. Eulen*, 2: pl. 48: 232 [Europe].

Synonymy: *Noctua crassicornis* Haworth, 1809; *Nonagria pilicornis* Haworth, 1812; *Nonagria vectis* Curtis, 1833; *Noctua bathygera* Boie, 1835; *Rhizedra lutosa griseata* Warren, 1911 (ssp.); *Rhizedra pallidipennis* Warren, 1911.

Distribution. China (Heilongjiang, Jilin, Nei Mongol, Xinjiang); Russia (European part, Siberia), Mongolia,

Korea, Japan; Central Asia, Asia Minor, Europe.

寄主：芦苇（陈一心，1999）；菖蒲（*Acorus calamus*）、丝带草（*Phalaris arundinacea*）、芦苇（*Phragmites australis*）（Matov & Kononenko, 2012）。

Genus *Sedina* Urbahn, 1933 苔夜蛾属

Sedina Urbahn, 1933, *Stettiner Entomologische Zeitung*, 94: 150. Type species: *Arsilonche buettneri* Hering, 1858 [Poland].

Monotypic Palaearctic genus; 1 species is in the Three Provinces of Northeast China.

注：首次给出中文属名。

Sedina buettneri (Hering, 1858) 苔夜蛾

Arsilonche buettneri Hering, 1858, *Entomologische Zeitung. Entomologischen Vereine zu Stettin*, 19: 442, pl. 3. Type locality: Poland, Grabov.

Synonymy: *Nonagria bloomeri* Heidemann, 1859; *Simyra moltrechti* Bang-Haas, 1927; *Archanara pumilana* Bryk, 1942; *buttneri* auct. missp..

Distribution. China (Jilin); Russia (European part, Ural, Transbaikalia, Siberia), Mongolia, Korea, Japan, Kazakhstan; Europe.

寄主：尖叶苔草（*Carex acutiformis*）、甜茅（*Glyceria aquatica*）、（*Glyceria maritima*）、水甜茅（*Glyceria maxima*）（Matov & Kononenko, 2012）。

注：首次给出中文种名。

Genus *Nonagria* Ochsenheimer, 1816 杆夜蛾属

Nonagria Ochsenheimer, 1816, *Die Schmetterlinge von Europa*, 4: 82. Type species: *Noctua typhae* Thunberg, 1784 [Sweden].

Synonymy: *Enteriona* Sodoffsky, 1837; *Phragmatiphila* Hampson, 1908; *Conicophoria* Matsumura, 1929.

The genus includes 13 species, distributed in the Palearctic, Oriental and Ethiopian regions. Among them 1 species occurs in the Three Provinces of Northeast China.

Nonagria puengeleri (Schawerda, 1923) 香蒲杆夜蛾

Sidemia puengeleri Schawerda, 1923, *Verhandlungen des Zoologisch-Botanischen vereins in Wien*, 73: Sber, 89. Type locality: Russia, [Primorye terr., Ussuriisk] Siberia, Nikolsk, “Ussuri”.

Synonymy: *Sidemia albipuncta* Bang-Haas, 1927; *Archanara puengeleri pringlei* Wiltshire, 1958 (ssp.).

Distribution. China (Heilongjiang); Russia (Siberia), Korea, Japan.

寄主：香蒲（*Typha latifolia*）（陈一心，1999；Matov & Kononenko, 2012）。

Genus *Longaletedes* Beck, 1999 蛀亮夜蛾属

Longaletedes Beck, 1999, *Atalanta*, 22: 214. Type species: *Leucania elymi* Treitschke, 1825 [Germany, Prussia].

The genus included 17 species known from the Holarctic and Oriental regions. 1 species distributed in the Palaearctic region, and the Three Provinces of Northeast China.

注：首次给出中文属名。

Longaletedes elymi (Treitschke, 1825) 蛀亮夜蛾

Leucania elymi Treitschke, 1825, *Die Schmetterlinge von Europa*, 5 (2): 294. Type locality: Germany, Prussia.

Synonymy: *Tapinostola procera saturatior* Staudinger, 1889 (var.); *Tapinostola elymi procera* Staudinger, 1889 (r.); *Tapinostola elymi askoldensis* Turner, 1929 (var.); *Arenostola elymi punctivena* Draudt, 1950 (ssp.); *Oligia obsolescens* Berio, 1977.

Distribution. China (Heilongjiang, Xinjiang); Russia (European part, Siberia), Mongolia, Korea, Japan; Central Asia, Europe.

寄主：沙丘野麦（陈一心，1999）。

注：《中国动物志》（夜蛾科）中将本种归入"*Photedes* Lederer, 1858 亮夜蛾属"，在此予以变更。

Genus *Archanara* Walker, 1866 锹额夜蛾属

Archanara Walker, 1866, *List of the specimens of lepidopterous insects in the collection of the British Museum*, 35: 1737. Type species: *Archanara nonogriella* Walker, 1866 [China].

Synonymy: *Nonagria* auct..

The genus includes 19 species, distributed in the Holarctic and Oriental regions. 7 species are known in the Palaearctic region; 1 species is in the Three Provinces of Northeast China.

Archanara neurica (Hübner, [1808]) 黑纹锹额夜蛾

Noctua neurica Hübner, [1808], *Sammlung Europaischer Schmetterlinge. Eulen*, 2: pl. 82: 381. Type locality: Europe.

Synonymy: *Nonagria edelsteni* Tutt, 1908; *Archanara neurica affinis* Rotschild, 1920; *Nonagria phragmiticola* Staudinger, 1892; *Archanara resoluta* Viidalepp & Remm, 1982, nec Hampson, 1910, misident.

Distribution. China (Heilongjiang, Jilin, Liaoning, Hebei); Russia (European part, Siberia), Mongolia, Korea.

寄主：芦苇（陈一心，1999）；丝带草（*Phalaris arundinacea*）、芦苇（*Phragmites australis*）（Matov & Kononenko, 2012）。

Genus *Hypocoena* Hampson, 1908 脉夜蛾属

Hypocoena Hampson, 1908, *Catalogue of the Lepidoptera Phalaenae in the British Museum*, 7: 19. Type species: *Leucania rufostrigata* Packard, 1867 [Canada].

Holarctic genus includes 8 species; only *H. stigmatica* presented in the Palaearctic region, and the Three Provinces of Noetheast China.

注：首次给出中文属名。

Hypocoena stigmatica **(Eversmann, 1855) 暗纹脉夜蛾**

Leucania stigmatica Eversmann, 1855, *Bulletin de la Société Impériale des Naturalistes de Moscou*, 1855: 191. Type locality: East Siberia, Irkutsk.

Synonymy: *Tapinostola lagunica* Graeser, 1889; *Hypocoena stigmatica dispersa* Wolff, 1970 (ssp.).

Distribution. China (Heilongjiang); Russia (Siberia), Mongolia, Korea, Kazakhstan, Iceland, Denmark(Faroe Isl.); Central Asia, North America.

寄主：禾本科（Poaceae）植物（Matov & Kononenko, 2012）。

注：《中国动物志》（夜蛾科）中将本种归入"*Photedes* Lederer, 1858 亮夜蛾属"，在此予以变更。

Genus *Minigrapta* Kononenko & Matov, 2012 小黑夜蛾属

Minigrapta Kononenko & Matov, 2012, *Zootaxa*, 3590: 17 (2012). Type species: *Xanthograpta basinigra* Sugi, 1982 [Japan].

The genus contains 2 species, distributed in the Chinese subregion of the Palaearctic region; 1 species is in the Three Provinces of Northeast China.

注：首次给出中文属名。

Minigrapta basinigra **(Sugi, 1982) 小黑夜蛾**

Xanthograpta basinigra Sugi, 1982, *Moths of Japan*, 1: 822, 2: 384, pl. 198: 8. Type locality: Japan, Hokkaido, Tokachi, Taiki.

Distribution. China (Jilin, Liaoning); Russia (Siberia), Korea, Japan.

注：首次给出中文种名。

Genus *Ogilia* Kononenko, 2016 钩肾夜蛾属

Ogilia Kononenko, 2016, *Proceedings of the Witt Museum, Munich* 5: 150. Type species: *Oligia leuconephra* Hampson, 1908 [Russia, vic. Khabarovsk "Ussuri", Kazakevichi].

Monotypic East Palaearctic genus; 1 species is in the Three Provinces of Northeast China.

注：首次给出中文属名。

***Ogilia leuconephra* (Hampson, 1908) 钩肾夜蛾**

Oligia leuconephra Hampson, 1908, *Catalogue of the Lepidoptera Phalaenae in the British Museum*, 7: 389, pl. 118: 2. Type locality: [Russia, vic. Khabarovsk] "Ussuri", Kazakevichi.

Synonymy: *Oligia fasciuncula albiluna* Kozhanchikov, 1929 (ssp.).

Distribution. China (Jilin, Liaoning); Russia (Ural, Altai, Transbaikalia, Siberia), Mongolia, Korea, Japan.

注：首次给出中文种名。

Genus *Ctenostola* Sugi, 1982 清纹夜蛾属

Ctenostola Sugi, 1982, *Moths of Japan*, 1: 753, 2: 367, 463. Type species: *Archanara sparganoides* Bang-Haas, 1927.

Monotypic Chinese genus; 1 species is in the Three Provinces of Northeast China.

注：首次给出中文属名。

***Ctenostola sparganoides* (Bang-Haas, 1927) 清纹夜蛾**

Archanara sparganoides Bang-Haas, 1927, *Horae Macrolepidopterologiae Regionis Palaearcticae*, 1: 87, pl. 10: 38. Type locality: Russia, Primorye terr., Partizansk ["Sutschansk"].

Distribution. China (Jilin); Russia (Siberia), Korea, Japan.

寄主；苔属（*Carex*）植物（Matov & Kononenko, 2012）。

注：首次给出中文种名。

Genus *Rotoa* Strand, 1942 辐夜蛾属

Rotoa Strand, 1942, *Folia Zoologica et Hydrobiologica*, 11: 390, repl. name. Type species: *Tapinostola distincta* Bang-Haas, 1912 [Russia, Primorye terr.].

Synonymy: *Rosenia* Schawerda, 1922, preocc. Waagen & Wentzel, 1866 [Protozoa]; *Pseudosideridis* Viidalepp, 1971.

Monotypic Chinese genus. 1 species is in the Three Provinces of Northeast China.

注：首次给出中文属名。

***Rotoa distincta* (Bang-Haas, 1912) 辐夜蛾**

Tapinostola distincta Bang-Haas, 1912, *Deutsche Entomologische Zeitschrift Iris*, 26: 154, pl. 6. Type locality: Russia, [Primorye terr., Ussuriisk] "Nikolsk".

Synonymy: *Pseudosideridis albolineata* Viidalepp, 1971.

Distribution. China (Heilongjiang, Nei Mongol); Russia (Siberia).

注：首次给出中文种名。

Genus ***Globia*** Fibiger, Zilli, Ronkay & Goldstein, 2009 球额夜蛾属

Globia Fibiger, Zilli, Ronkay & Goldstein, 2009, *Noctuidae Europaeae*, 8: 109. Type species: *Phalaen* (*Noctua*) *algae* Esper, 1789 [Europe].

Synonymy: *Capsula* Fibiger, Zilli, Ronkay & Goldstein, 2005, preocc. (Shumacher, 1817 [Mollusca]).

Holarctic genus, and containing 7 species; among them 2 species are in the Three Provinces of Northeast China.

注：首次给出中文属名。

Globia sparganii **(Esper, 1790) 黑三棱球额夜蛾**

Noctua sparganii Esper, 1790, *Die Schmetterlinge in Abbildungen nach der Natur mit Beschreibungen*, 4 (2): 481, pl. 147: 2, 3. Type locality: Europe.

Synonymy: *Nonagria sparganii strigosa* Staudinger, 1892 (var.); *Archanara sparganii algaeoides* Wiltshire, 1947 (ssp.).

Distribution. China (Heilongjiang, Jilin); Russia (European part, Siberia), Mongolia, Korea, Japan; Central Asia, Middle East, Asia Minor, Europe.

寄主：黄菖蒲（*Iris pseudacorus*）、湖滨藨草（*Scirpus lacustris*）、水葱（*Scirpus tabernaemontani*）、芦苇（*Phragmites australis*）、直立黑三棱（*Sparganium erectum*）、黑三棱（*Sparganium ramosum*）、小黑三棱（*Sparganium simplex*）、水烛（*Typha angustifolia*）、香蒲（*Typha latifolia*）（Matov & Kononenko, 2012）。

注：《中国动物志》（夜蛾科）中将本种归入"*Archanara* Walker, 1866 锹额夜蛾属"，在此予以变更。

Globia aerata **(Butler, 1878) 条球额夜蛾**

Leucania aerata Butler, 1878, *Annals and Magazine of Natural History*, (5)1: 80. Type locality: Japan, Hakodate.

Distribution. China (Heilongjiang, Jilin, Liaoning, Shandong); Russia (Siberia), Mongolia, Korea, Japan.

注：《中国动物志》（夜蛾科）中将本种归入"*Archanara* Walker, 1866 锹额夜蛾属"，在此予以变更。

Genus ***Pabulatrix*** Sugi, 1982 伴白夜蛾属

Pabulatrix Sugi, 1982, *Moths of Japan*, 1: 741. Type species: *Noctua pabulatricula* Brahm, 1791 [Europe].

Synonymy: *Eremobina* auct., nec McDunnough, 1937.

The Holarctic genus includes 6 species; only 1 species is in the Palaearctic region, and the Three Provinces of Northeast China.

注：首次给出中文属名。

Pabulatrix pabulatricula (Brahm, 1791) 伴白夜蛾

Noctua pabulatricula Brahm, 1791, *Insekten Kalender für Sammler und Oeconomen*, 2: 170. Type locality: Europe.

Synonymy: *Phalaena* (*Noctua*) *connexa* Borkhausen, 1792; *Noctua elota* Hübner, [1813]; *Hadena fraudulenta* Staudinger, 1888.

Distribution. China (Jilin, Liaoning); Russia (European part, Siberia, Kamchatka, Magadan), Mongolia, Korea, Japan, Kazakhstan; Asia Minor, Europe.

寄主：毛茅草（*Holcus mollis*）、酸沼草（*Molinia caerulea*）、虉草（*Phalaris arundinacea*）、雀麦属（*Bromus*）、发草属（*Deschampsia*）植物（Matov & Kononenko, 2012）。

注：首次给出中文种名。

Genus *Apamea* Ochsenheimer, 1816 秀夜蛾属

Apamea Ochsenheimer, 1816, *Die Schmetterlinge von Europa*, 4: 75. Type species: *Noctua basilinea* [Denis & Schiffermüller], 1775 [Austria] (=*Phalaena sordens* Hufnagel, 1766).

Synonymy: *Xylena* Hübner, [1806], suppr. (ICZN, 1926 Op. 9); *Abromias* Billberg, 1820; *Septis* Hübner, [1821] 1816; *Xylophasia* Stephens, 1829; *Hama* Stephens, 1829; *Crymodes* Guenée, 1841; *Agrostobia* Boie, 1835; *Syma* Stephens, 1850; *Dimya* Moore, 1882, preocc. (Lesson, 1827 [Aves]); *Eurabila* Butler, 1889; *Eleemosia* Prout, 1901; *Protagrotis* Hampson, 1903; *Agroperina* Hampson, 1908; *Trichoplexia* Hampson, 1808; *Heteromma* Warren, 1911, preocc. (Menge, 1856 [Arachnida]); *Heterommiola* Strand, 1912, repl. name; *Furvabromias* Beck, [1992] 1991; *Apaconjunctdonta* Beck, [1992] 1991; *Alexapamea* Varga, Ronkay & Ronkay, 2009 (subgen.); *Hirudapamea* Varga, Ronkay & Ronkay, 2009 (subgen.); *Mixapamea* Varga, Ronkay & Ronkay, 2009 (subgen.); *Digitapamea* Varga, Ronkay & Ronkay, 2009 (subgen.); *Vovapamea* Varga, Ronkay & Ronkay, 2009 (subgen.); *Cernauncus* Varga, Ronkay & Ronkay, 2009 (subgen.); *Hadena* auct.; *Trachea* auct.; *Parastichtis* auct..

The genus *Apamea* is mainly distributed in the Northern Hemisphere, with over 60 species currently recognized from the Holarctic region and about 90 in the Palaearctic region; a few species are also recorded from the Oriental and African regions (Zilli *et al.*, 2009). The genus is most diverse in open steppe and mountain habitat in the southern part of the temperate zone. Most species are associated with open grasslands dominated by Poaceae, the host plant for the larvae of the great majority of *Apamea* species. The larvae are mainly olygophagous, many species are stem-borers or live in ruts of Poaceae grasses. The genus is represented in the Three Provinces of Northeast China by 3 subgenera and 12 species.

Subgenus *Vovapamea* Zilli, Varga, Ronkay & Ronkay, 2009 沃秀夜蛾亚属

Vovapamea Zilli, Varga, Ronkay & Ronkay, 2009, *Witt Catalogue*, Vol. 3: 40. Type species: *Apamea brunnescens* Kononenko, 1985 [Russia, South Primorye terr.].

Monotypic Chinese subgenus.

注：首次给出中文亚属名。

***Apamea* (*Vovapamea*) *brunnescens* Kononenko, 1985 棕秀夜蛾**

Apamea brunnescens Kononenko, 1985, *Tinea*, 11 (28): 221, fig. 1, 2. Type locality: Russia, South Primorye terr., Kedrovaya Pad nature reservation.

Distribution. China (Heilongjiang, Jilin); Russia (Siberia), Korea.

注：首次给出中文种名。

Subgenus ***Digitapamea*** Zilli, Varga, Ronkay & Ronkay, 2009 指秀夜蛾亚属

Digitapamea Zilli, Varga, Ronkay & Ronkay, 2009, *Witt Catalogue*, Vol. 3: 40. Type species: *Mesapamea concinnata* Heinicke, 1959 [China, Wa-ssu-kow].

Sino-Pacific subgenus includes 5 species; among them, 1 species is in the Three Provinces of Northeast China.

注：首次给出中文亚属名。

***Apamea* (*Digitapamea*) *hampsoni* Sugi, 1963 汉秀夜蛾**

Apamea hampsoni Sugi, 1963, *Tinea*, 6: 10, pl. 4: 1-4. Type locality: Japan, Shizouka Pref., Odaru [coll. S. Sugi, Tokyo].

Distribution. China (Heilongjiang, Jilin); Russia (Siberia); South Korea, Japan.

Subgenus ***Apamea*** Ochsenheimer, 1816 秀夜蛾亚属

Apamea Ochsenheimer, 1816, *Die Schmetterlinge von Europa*, 4: 75. Type species: *Noctua basilinea* [Denis & Schiffermüller], 1775 [Austria] (=*Phalaena sordens* Hufnagel, 1766).

The subgenus shares most characters described for the genus; 10 species are in the Three Provinces of Northeast China.

***Apamea* (*Apamea*) *remissa* (Hübner, 1808) 迴秀夜蛾**

Noctua remissa Hübner, 1808, *Sammlung Europaischer Schmetterlinge. Eulen*, 2: pl. 90: 423. Type locality: Europe.

Synonymy: *Noctua obscura* Haworth, 1809; *Noctua gemina* Hübner, [1813]; *Hadena gemina submissa* Treitschke, 1825 (var.); *Polia w-latinum divitis* Bryk, 1942.

Distribution. China (Heilongjiang, Jilin, Xinjiang); Russia (European part, Siberia, Kamchatka), Mongolia,

Korea, Japan, Kazakhstan; Asia Minor, Europe, North America.

寄主：拂子茅属（陈一心，1999）；拂子茅（*Calamagrostis epigeios*）、（*Calamagrostis villosa*）、（*Juncus acutiflorus*）、鸭茅（*Dactylis glomerata*）、小穗发草（*Deschampsia caespitosa*）、曲芒发草（*Deschampsia flexuosa*）、偃麦草（*Elytrigia repens*）、高羊茅（*Festuca arundinacea*）、大羊茅（*Festuca gigantea*）、紫羊茅（*Festuca rubra*）、短柄草（*Brachypodium sylvaticum*）、绒毛草（*Holcus lanatus*）、芦苇（*Phragmites australis*）、早熟禾（*Poa annua*）、黑麦（*Secale cereale*）、报春花属（*Primula*）、蒲公英属（*Taraxacum*）、唐菖蒲属（*Gladiolus*）、剪股颖属（*Agrostis*）、小麦属（*Triticum*）植物（Matov & Kononenko, 2012）。

***Apamea* (*Apamea*) *aquila* Donzel, 1837 毁秀夜蛾**

Apamea aquila Donzel, 1837, *Annales de la Société Entomologique de France*, 6: 476, pl. 18: 6. Type locality: France, Basses Alpes.

Synonymy: *Hadena funerea* Heinemann, 1859; *Hadena hepatica discrepans* Staudinger, 1892 (var.); *Parastichtis funerea oriens* Warren, 1911 (ssp.); *Parastichtis funerea brunneata* Warren, 1911 (ssp.); *Parastichtis funerea geminea* Bryk, 1948 (ssp.); *Perigea affinis* Draudt, 1950; *Apamea striata*: Chang, 1991, nec Haruta; *Apamea aquila substriata* Hreblay & Ronkay, 1997 (ssp.).

Distribution. China (Heilongjiang, Jilin, Liaoning, Hubei, Taiwan); Russia (North Caucasus, Siberia), Mongolia, Korea, Japan; Asia Minor, Europe.

寄主：酸沼草（*Molinia caerulea*）、日本麦氏草（*Molinia japonica*）（Matov & Kononenko, 2012）。

***Apamea* (*Apamea*) *striata* Haruta & Sugi, 1958 纹秀夜蛾**

Apamea striata Haruta & Sugi, 1958, *Tinea*, 4: 271, fig. 5. Type locality: Japan, Gumma Pref., Kazawa-Spa.

Distribution. China (Heilongjiang, Jilin, Liaoning, Yunnan); Russia (Siberia), Korea, Japan.

***Apamea* (*Apamea*) *crenata* (Hufnagel, 1766) 旋秀夜蛾**

Phalaena crenata Hufnagel, 1766, *Berlinisches Magazine*, 3 (4): 402. Type locality: Germany, Berlin.

Synonymy: *Noctua rurea* Fabricius, 1775; *Phalaena* (*Noctua*) *luculenta* Esper, [1788]; *Phalaena* (*Noctua*) *alopecurus* Esper, [1790]; *Noctua combusta* Hübner, [1808]; *Noctua hirticornus* Haworth, 1812; *Xylina borealis* Curtis, 1829; *Hadena rurea corsa* Schawerda, 1928 (var.).

Distribution. China (Heilongjiang, Jilin, Liaoning, Xinjiang, Qinghai, Sichuan); Russia (European part, Siberia, Kamchatka), Mongolia, Korea, Japan; Central Asia, Asia Minor, Europe.

寄主：报春、酸模属（陈一心，1999）；旋果蚊子草（*Filipendula ulmaria*）、树莓（*Rubus fruticosus*）、吊钟千里光（*Senecio fuchsii*）、毛果一枝黄花（*Solidago virgaurea*）、（*Juncus acutiflorus*）、（*Luzula luzuloides*）、（*Carex verna*）、匍匐冰草（*Agropyron repens*）、黄花茅（*Anthoxanthum odoratum*）、燕麦草（*Arrhenatherum elatius*）、无芒雀麦（*Bromus inermis*）、野青茅

（*Calamagrostis arundinacea*）、拂子茅（*Calamagrostis epigeios*）、鸭茅（*Dactylis glomerata*）、小穗发草（*Deschampsia caespitosa*）、曲芒发草（*Deschampsia flexuosa*）、巨羊茅（*Festuca altissima*）、高羊茅（*Festuca arundinacea*）、黑麦草（*Lolium perenne*）、酸沼草（*Molinia caerulea*）、虉草（*Phalaris arundinacea*）、梯牧草（*Phleum pratense*）、扁鞘早熟禾（*Poa chaixii*）、黑麦（*Secale cereale*）、玉米（*Zea mays*）、酸模属（*Rumex*）、报春花属（*Primula*）、莴苣属（*Lactuca*）、剪股颖属（*Agrostis*）、小麦属（*Triticum*）植物（Matov & Kononenko, 2012）。

***Apamea* (*Apamea*) *anceps* ([Denis & Schiffermüller], 1775) 污秀夜蛾**

Phalaena (*Noctua*) *anceps* [Denis & Schiffermüller], 1775, *Ankündung eines Systematisches Werkes von den Schmetterlinge der Wiener Gegend. Wien*: 81. Type locality: Austria, Vienna.

Synonymy: *Noctua sordida* Borkhausen, 1792; *Noctua contigua* Haworth, 1809; *Apamea infesta* Ochsenheimer, 1816; *Caradrina renardii* Boisduval, 1829; *Hadena engelhartii* Durloo, 1889; *Luceria pyxina* Bang-Haas, 1910; *Enargia siegeli* Berio, 1985; *Apamea pyxina sassanidica* Hacker, 1990 (ssp.); *Apamea anceps mazeli* Lutran, 1993 (ssp.); *Apamea anceps maritima* Sugi, 1994 (ssp.).

Distribution. China (Heilongjiang, Jilin, Shaanxi); Russia (European part, Siberia), Mongolia, Japan; Central Asia, Asia Minor, Middle East, Europe, North Africa.

寄主：小麦、杂草（陈一心，1999）；匍匐冰草（*Agropyron repens*）、燕麦（*Avena sativa*）、鸭茅（*Dactylis glomerata*）、偃麦草（*Elytrigia repens*）、二行大麦（*Hordeum distichum*）、大麦（*Hordeum vulgare*）、沙丘野麦（*Elymus arenarius*）、早熟禾（*Poa annua*）、黑麦（*Secale cereale*）、硬粒小麦（*Triticum durum*）、小麦（*Triticum aestivum*）、玉米（*Zea mays*）、短柄草属（*Brachypodium*）植物（Matov & Kononenko, 2012）。

***Apamea* (*Apamea*) *sordens* (Hufnagel, 1766) 秀夜蛾**

Phalaena sordens Hufnagel, 1766, *Berlinisches Magazine*, 3: 306. Type locality: Germany, Berlin.

Synonymy: *Noctua basilinea* [Denis & Schiffermüller], 1775; *Phalaena* (*Noctua*) *marmorosa* Esper, 1788; *Noctua nebulosa* Vieweg, 1789; *Apamea finitima* Guenée, 1852; *Hadena cinefacta* Graeser, 1888 [1889]; *Hadena basilinea grisescens* Staudinger, 1889 (var.); *Hadena basilinea basistriga* Staudinger, 1892 (var.); *Hadena cerivana* Smith, 1900; *Hadena basilinea alcara* Dannehl, 1933 (ssp.).

Distribution. China (Heilongjiang, Jilin, Nei Mongol, Xinjiang, Qinghai, Gansu, Sichuan, Yunnan); Russia (European part, Ural, Siberia, Kamchatka, Magadan), Mongolia, Korea, Japan, Kazakhstan; Central Asia, Asia Minor, Caucasus, Transcaucasia, Europe.

寄主：小麦、蒲公英（陈一心，1999）；金雀儿（*Cytisus scoparius*）、（*Agrostis spicaventi*）、西伯利亚剪股颖（*Agrostis stolonifera*）、燕麦草（*Arrhenatherum elatius*）、燕麦（*Avena sativa*）、直立雀麦（*Bromus erectus*）、无芒雀麦（*Bromus inermis*）、拂子茅（*Calamagrostis epigeios*）、鸭茅（*Dactylis glomerata*）、小穗发草（*Deschampsia caespitosa*）、沙丘野麦（*Elymus arenarius*）、偃麦草（*Elytrigia repens*）、高羊茅（*Festuca arundinacea*）、草甸羊茅（*Festuca pratensis*）、紫羊茅（*Festuca rubra*）、二行大麦（*Hordeum distichum*）、大麦（*Hordeum vulgare*）、丝带草（*Phalaris arundinacea*）、梯牧草（*Phleum pratense*）、芦苇（*Phragmites australis*）、黑麦（*Secale cereale*）、小麦（*Triticum aestivum*）、偃麦草（*Elytrigia repens*）、玉米（*Zea mays*）、蒲公英属（*Taraxacum*）、薹草属（*Carex*）、藨草属（*Scirpus*）植物（Matov & Kononenko, 2012）。

Apamea (_Apamea_) _leucodon_ (Eversmann, 1837) 白脉秀夜蛾

Hadena leucodon Eversmann, 1837, *Bulletin de la Société Impériale des Naturalistes de Moscou*, 1837: 37. Type locality: Russia, Saratschik.

Synonymy: *Hadena bidens* Eversmann, 1856; *Hadena leucodon anatolica* Rebel, 1913 (ssp.).

Distribution. China (Heilongjiang, Nei Mongol); Russia (Southeast European part, Siberia), Mongolia; Central Asia, Asia Minor, Middle East, Europe.

寄主：看麦娘属（*Alopecurus*）、雀麦属（*Bromus*）、大麦属（*Hordeum*）植物（Matov & Kononenko, 2012）。

注：首次给出中文种名。

Apamea (_Apamea_) _oblonga_ (Haworth, 1809) 修秀夜蛾

Noctua oblonga Haworth, 1809, *Lepidoptera Britannica sistens Digestionem novam Insectorum Lepidopterorum quae in Magna Britannia Reperiuntur, Larvarum pabulo, Temporeque Pascendi; Expansione Alarum; Mensibusque Volandi; Synonymis Atque Locis Observationibusque Variis*, 2: 188. Type locality: England, Wisbeach.

Synonymy: *Noctua lunulina* Haworth, 1809; *Noctua abjecta* Hübner, [1809-1813]; *Luperina fribolus* Boisduval, [1837] 1832; *Hadena abjecta variegata* Staudinger, 1871 (var.).

Distribution. China (Heilongjiang, Jilin, Xinjiang, Ningxia); Russia (European part, Siberia), Mongolia, Korea, Japan, Pakistan, India; Central Asia, Asia Minor, Europe.

寄主：早熟禾属（陈一心，1999）；牛尾草（*Festuca elatior*）、羊茅（*Festuca ovina*）、紫羊茅（*Festuca rubra*）、芦苇（*Phragmites australis*）、早熟禾（*Poa annua*）、（*Poa borreri*）、（*Poa distans*）、（*Poa maritima*）、林地早熟禾（*Poa nemoralis*）、草地早熟禾（*Poa pratensis*）、拂子茅属（*Calamagrostis*）、发草属（*Deschampsia*）、披碱草属（*Elymus*）、偃麦草属（*Elytrigia*）、大麦属（*Hordeum*）、碱茅属（*Puccinellia*）、黑麦属（*Secale*）、小麦属（*Triticum*）植物（Matov & Kononenko, 2012）。

Apamea (_Apamea_) _lateritia_ (Hufnagel, 1766) 荒秀夜蛾

Phalaena lateritia Hufnagel, 1766, *Berlinisches Magazine*, 3: 306. Type locality: Germany, Berlin.

Synonymy: *Noctua molochina* Hübner, [1803]; *Xylina unicolora* Zetterstedt, 1840; *Hadena lateritia borealis*

Strand, 1903 (ab.); *Hadena lateritia decolor* Stertz, 1915 (var.); *Hadena lateritia soldana* Noack, 1925 (ssp.); *Hadena lateritia alpium* Dannehl, 1937 (f.); *Agroperina lateritia kunashirina* Bryk, 1942 (ssp.); *Hadena lateritia expallescens* Staudinger, 1882 (ssp.); *Apamea lateritia cinerascens* Zilli, Varga, Ronkay & Ronkay, 2009 (ssp.); *Apamea lateritia obfuscata* Hreblay & Ronkay, 1998 (ssp.).

Distribution. China (Heilongjiang, Jilin, Nei Mongol, Xinjiang, Qinghai, Guangxi, Xizang); Russia (European part, Siberia, Kamchatka), Mongolia, Korea, Japan; Central Asia, Asia Minor, Europe.

寄主：拂子茅（*Calamagrostis epigeios*）、小穗发草（*Deschampsia caespitosa*）、曲芒发草（*Deschampsia flexuosa*）、牛尾草（*Festuca elatior*）、羊茅（*Festuca ovina*）、那杜草（*Nardus stricta*）、梯牧草（*Phleum pratense*）、玉米（*Zea mays*）、看麦娘属（*Alopecurus*）、雀麦属（*Bromus*）、鸭茅属（*Dactylis*）、大麦属（*Hordeum*）、芦苇属（*Phragmites*）、早熟禾属（*Poa*）、小麦属（*Triticum*）植物（Matov & Kononenko, 2012）。

***Apamea (Apamea) veterina* (Lederer, 1853) 负秀夜蛾**

Hadena veterina Lederer, 1853, *Verhandlungen des Zoologisch-Botanischen vereins in Wien*, 3: 370, pl. 2: 4. Type locality: East Kazakhstan, West Altai [Ust-Kamengorsk, Ust-Buchtarminsk].

Synonymy: *Hadena haelsseni* Graeser, 1889 [1890]; *Hadena veterina mandschurica* Staudinger, 1892 (var.); *Parastichtis jezoensis* Matsumura, 1926; *Parastichtis veterina coreina* Bryk, 1948 (ssp.).

Distribution. China (Heilongjiang, Jilin, Liaoning, Nei Mongol, Xinjiang, Hebei, Shanxi); Russia (Altai, Siberia), Korea, Japan, India, Kazakhstan, Pakistan.

Genus *Loscopia* Beck, [1992] 1991 赭秀夜蛾属

Loscopia Beck, [1992] 1991, *Atalanta*, 22: 212. Type species: *Phalaena* (*Noctua*) *scolopacina* Esper, 1788 [Germany].

Synonymy: *Sinapamea* Rakosy, 1996.

The genus includes 3 species; among them 1 species is in the Three Provinces of Northeast China.

注：首次给出中文属名。

***Loscopia scolopacina* (Esper, 1788) 赭秀夜蛾**

Phalaena (*Noctua*) *scolopacina* Esper, 1788, *Die Schmetterlinge in Abbildungen nach der Natur mit Beschreibungen*, 4 (1): 394, pl. 130: 1. Type locality: Germany, Leipzig.

Synonymy: *Noctua abbreviata* Haworth, 1809; *Hadena hammoniasis* Sauber, 1899; *Parastichtis scolopacina subbrunnea* Warren, 1911 (f.); *Apamea scolopacina subbrunnea* Inoue & Sugi, 1958 (ssp.).

Distribution. China (Heilongjiang, Hebei); Russia (European part, Ural), Mongolia, Korea, Japan, India, Kazakhstan; Asia Minor, Europe.

寄主：地杨梅属（陈一心，1999）；五福花（*Adoxa moschatellina*）、疏毛地杨梅（*Luzula pilosa*）、

尾穗苔草（*Carex spicata*）、麻根薹草（*Carex sylvatica*）、沼泽荸荠（*Eleocharis palustris*）、（*Scirpus silvatica*）、（*Scirpus palustris*）、（*Trichophorum cespitosum*）、大看麦娘（*Alopecurus pratensis*）、短柄草（*Brachypodium sylvaticum*）、凌风草（*Briza media*）、假苇拂子茅（*Calamagrostis pseudophragmites*）、鸭茅（*Dactylis glomerata*）、（*Dactylis polygama*）、犬草（*Elymus caninus*）、偃麦草（*Elytrigia repens*）、蔗甜茅（*Glyceria notata*）、（*Melica uniflora*）、粟草（*Milium effusum*）、丝带草（*Phalaris arundinacea*）、林地早熟禾（*Poa nemoralis*）、普通早熟禾（*Poa trivialis*）、桦木属（*Betula*）、灯心草属（*Juncus*）、地杨梅属（*Luzula*）、冰草属（*Agropyron*）、银须草属（*Aira*）、发草属（*Deschampsia*）植物（Matov & Kononenko, 2012）。

Genus ***Lateroligia*** Zilli, Fibiger & Ronkay, 2005 褐斑秀夜蛾属

Lateroligia Zilli, Fibiger & Ronkay, 2005, *Noctuidae Europaeae*, 8: 157. Type species: *Noctua ophiogramma* Esper, 1794 [Europe].

Palaearctic monotypic genus. 1 genus is in the Three Provinces of Northeast China.

注：首次给出中文属名。

Lateroligia ophiogramma **(Esper, 1794) 褐斑秀夜蛾**

Noctua ophiogramma Esper, 1794, *Die Schmetterlinge in Abbildungen nach der Natur mit Beschreibungen*, 4 (1): pl. 182: 2. Type locality: not stated, Europe.

Synonymy: *Noctua biloba* Haworth, 1809.

Distribution. China (Heilongjiang, Jilin, Liaoning, Xinjiang); Russia (European part, Siberia), Mongolia, North Korea, Japan, India; Central Asia, Asia Minor, Europe, North America.

寄主：黄菖蒲（*Iris pseudacorus*）、紫苞鸢尾（*Iris ruthenica*）、山鸢尾（*Iris setosa*）、假苇拂子茅（*Calamagrostis pseudophragmites*）、甜茅（*Glyceria aquatica*）、浮甜茅（*Glyceria fluitans*）、（*Glyceria maritima*）、水甜茅（*Glyceria maxima*）、（*Glyceria spectabilis*）、丝带草（*Phalaris arundinacea*）、金丝雀绶草（*Phalaris canariensis*）、芦苇（*Phragmites australis*）、水甜茅(*Poa aquatica*)、芦竹属（*Arundo*）植物（Matov & Kononenko, 2012）。

Genus ***Leucapamea*** Sugi, 1982 亚秀夜蛾属

Leucapamea Sugi, 1982, *Moths of Japan*, 1: 741, 742, 2: 364, 462. Type species: *Apamea kawadai* Sugi, 1955 [Japan].

The genus includes 6 species, distributed in the Chinese subregion of the Palaearctic and Oriental regions. 4 species are known from the Palaearctic region; among them 1 species is in the Three Provinces of Northeast China.

注：首次给出中文属名。

Leucapamea askoldis (Oberthür, 1880) 亚秀夜蛾

Hadena askoldis Oberthür, 1880, *Études d'Entomologie*, 5: 72, pl. 3: 13. Type locality: Russia, South Primorye terr., Askold I..

Synonymy: *Apamea nivalis* Butler, 1881; *Trachea yoshinoensis* Wileman, 1911; *Metachrostis niphandothauma* Bryk, 1948; *Trachea ascoldis*: Hampson, 1908, emend..

Distribution. China (Heilongjiang, Jilin, Liaoning, Xinjiang, Hubei, Fujian, Sichuan); Russia (Altai, Siberia), Korea, Japan.

寄主：禾本科（Poaceae）植物（Matov & Kononenko, 2012）。

Genus *Anapamea* Sugi, 1982 齿秀夜蛾属

Anapamea Sugi, 1982, *Moths of Japan*, 1: 745, 2: 365, 462. Type species: *Mamestra cuneata* Leech, 1889 [Japan].

The genus includes 5 species, distributed in the Chinese subregion of the Palaearctic and partly Oriental regions; 3 species are in the Three Provinces of Northeast China.

注：首次给出中文属名。

Anapamea incerta (Staudinger, 1892) 尖齿秀夜蛾

Mesotrosta incerta Staudinger, 1892, *in* N.M. Romanoff (ed.), *Mémoires sur les Lépidoptères*, 6: 572. Type locality: Russia, Amur reg., Radde ["Raddefka"].

Synonymy: *Apamea minor* Sugi, 1963.

Distribution. China (Jilin, Liaoning); Russia (Siberia), Korea, Japan.

注：首次给出中文种名。

Anapamea apameoides (Draudt, 1950) 笋齿秀夜蛾

Oligia apameoides Draudt, 1950, *Mitteilungen der MünchnerEntomologischen Gesellschaft*, 40: 99, pl. 7: 3. Type locality: China, Hunan, Hoeng-shan[=Hengshan].

Distribution. China (Liaoning, Henan, Zhejiang, Hunan); South Korea, Japan.

寄主：竹笋及多种杂草（陈一心，1999）。

注：曾用中文种名"笋秀夜蛾"。

Anapamea cuneatoides Poole, 1989 齿秀夜蛾

Anapamea cuneatoides Poole, 1989, *Lepidopterorum Catalogus* (*New Series*), 118 (1): 78, repl. name.

Synonymy: *Mamestra cuneata* Leech, 1889.

Distribution. China (Liaoning, Henan, Fujian); South Korea, Japan.

寄主：竹笋（陈一心，1999）。

Genus *Atrachea* Warren, 1911 纬夜蛾属

Atrachea Warren, 1911, *in* A.Seitz (ed.), *Die Gross-Schmetterlinge der Erde*, 3: 175. Type species: *Spaelotis nitens* Butler, 1878 [Japan].

Synonymy: *Trichosternum* Draudt, 1950; *Zutragum* Nye, 1975; *Hadena* auct.; *Trachea* auct.; *Apamea* auct..

The genus contains 12 species distributed in the Chinese subregion of the Palaearctic region; among them, 4 species are in the Three Provinces of Northeast China.

Atrachea nitens **(Butler, 1878) 纬夜蛾**

Spaelotis nitens Butler, 1878, *Annals and Magazine of Natural History*, (5)1: 164. Type locality: Japan, Yokohama.

Synonymy: *Naenia muscosa* Butler, 1878.

Distribution. China (Heilongjiang, Jilin, Liaoning, Zhejiang, Hunan); Russia (Siberia, Kuril Isl.), Korea, Japan.

寄主：蓝茎冰草（*Agropyron tsukushiense*）（Matov & Kononenko, 2012）。

Atrachea jankowskii **(Oberthür, 1879) 北纬夜蛾**

Hadena jankowskii Oberthür, 1879, *Diagnoses d'Espéces nouvelles de Lépidoptères de l'ile Askold*: 14. Type locality: Russia, Primorye terr., Askold I..

Synonymy: *Hadena jancousci*: Hampson, 1908, emend..

Distribution. China (Heilongjiang); Russia (Siberia), Korea, Japan.

寄主：禾本科（Poaceae）植物（Matov & Kononenko, 2012）。

注：首次给出中文种名。

Atrachea alpherakyi **Kononenko, 1986 绿纬夜蛾**

Atrachea alpherakyi Kononenko, 1986, *Sistematica ecologia nasekomyh Dal'nego Vostoka*: 41, fig. 5, Type locality: RFE, Primorye terr., Askold Isl. Repl. name for *Calotaenia jankowskii* Alphéraky, 1897.

Synonymy: *Calotaenia jankowskii* Alphéraky, 1897, nec Oberthür, 1879.

Distribution. China (Heilongjiang, Jilin); Russia (Siberia), Korea.

Atrachea japonica **(Leech, 1889) 日纬夜蛾**

Euplexia japonica Leech, 1889, *Proceedings of the Zoological Society of London*, 1889: 505, pl. 51: 4. Type locality: Japan.

Synonymy: *Hadena succincta* Graeser, 1890.

Distribution. China (Heilongjiang); Russia (Siberia), Korea, Japan.

Genus *Mesoligia* Boursin, 1965 草禾夜蛾属

Mesoligia Boursin, 1965, *Bulletin mensuel de la Société linneenne de Lyon*, 34: 185. Type species: *Noctua furuncula* [Denis & Schiffermüller], 1775 [Austria].

Synonymy: *Oligia* auct., *Miana* auct..

The genus includes 5 species, distributed in the Palaearctic and Ethiopian regions; 3 species are in the Palaearctic, among them 1 species is in the Three Provinces of Northeast China.

注：首次给出中文属名。

Mesoligia furuncula **([Denis & Schiffermüller], 1775) 草禾夜蛾**

Noctua furuncula [Denis & Schiffermüller], 1775, *Ankündung eines Systematisches Werkes von den Schmetterlinge der Wiener Gegend. Wien*: 89. Type locality: Austria, Vienna.

Synonymy: *Noctua spoliatricula* [Denis & Schiffermüller], 1775; *Noctua bicoloria* Villers, 1789; *Noctua vinctiuncula* Hübner, 1800-1803; *Noctua rufuncula* Haworth, 1809; *Noctua humeralis* Haworth, 1809; *Noctua terminalis* Haworth, 1809; *Hadena* (*Miana*) *furuncula insulicola* Staudinger, 1871 (var.); *Oligia furuncula reisseri* Schawerda, 1932 (r.); *Oligia furuncula semicretacea* Alphéraky, 1897 (var.); *Hadena pallidior*: Staudinger, 1882 nec Denis & Schiffermüller, 1775.

Distribution. China (Heilongjiang, Qinghai); Russia (European part, Siberia), Korea, Japan; Central Asia, Asia Minor, Middle East, Europe, North Africa.

寄主：羊茅、发草等属植物（陈一心，1999）；钝叶酸模（*Rumex obtusifolius*）、杂种车轴草（*Trifolium hybridum*）、（*Luzula luzuloides*）、普通剪股颖（*Agrostis canina*）、燕麦草（*Arrhenatherum elatius*）、无芒雀麦（*Bromus inermis*）、拂子茅（*Calamagrostis epigeios*）、鸭茅（*Dactylis glomerata*）、小穗发草（*Deschampsia caespitosa*）、沙丘野麦（*Elymus arenarius*）、偃麦草（*Elytrigia repens*）、高羊茅（*Festuca arundinacea*）、羊茅（*Festuca ovina*）、草甸羊茅（*Festuca pratensis*）、紫羊茅（*Festuca rubra*）、水甜茅（*Glyceria maxima*）、黑麦草（*Lolium perenne*）、草地早熟禾（*Poa pratensis*）、银须草属（*Aira*）植物（Matov & Kononenko, 2012）。

注：《中国动物志》（夜蛾科）中将本种归入"*Oligia* Hübner, [1821] 1816 禾夜蛾属"，在此予以变更。

Genus *Litoligia* Beck, 1999 斑禾夜蛾属

Litoligia Beck, 1999, *Herbipoliana*, 5/1: 574. Type species: *Noctua literosa* Haworth, 1809 [England].

The genus includes 2 Palaearctic species; among them 1 species is in the Three Provinces of Northeast China.

注：首次给出中文属名。

Litoligia fodinae (Oberthür, 1880) **斗斑禾夜蛾**

Miana fodinae Oberthür, 1880, *Études d'Entomologie*, 5: 73, pl. 3: 12. Type locality: Russia, Primorye terr., Askold I..

Synonymy: *Gerbatha subfasciata* Butler, 1881; *Miana elbergi* Viidalepp, 1964; *Parastichis fraudatricula*: Matsumura, 1925, nec Hübner, [1803].

Distribution. China (Heilongjiang, Jiangxi); Russia (Siberia), Korea, Japan.

注：《中国动物志》（夜蛾科）中将本种归入"*Oligia* Hübner, [1821] 1816 禾夜蛾属"，在此予以变更。

Genus *Chandata* Moore, 1882 仿陌夜蛾属

Chandata Moore, 1882, *in* Hewitson & Moore, *Description of new Indian lepidopterous insects from the collection of the late Mr. W. S.Atkinson. Heterocera*, 2: 113. Type species: *Chandata partita* Moore, 1882 [India, Darjiling].

The genus includes 8 species, distributed in the Chinese subregion of the Palaearctic and mainly Oriental regions; 1 species is in the Three Provinces of Northeast China.

注：首次给出中文属名。

***Chandata bella* (Butler, 1881) 美仿陌夜蛾**

Lamprosticta bella Butler, 1881, *Transactions of the Entomological Society of London*, 1881: 183. Type locality: Japan, Tokyo.

Synonymy: *Euplexia graeseri* Staudinger, 1892.

Distribution. China (Heilongjiang, Jilin, Taiwan); Russia (Siberia), Korea, Japan.

寄主：（*Carex amplifolia*）、（*Carex blepharicarpa*）、（*Carex distata*）（Matov & Kononenko, 2012）。

注：《中国动物志》（夜蛾科）中将本种归入"*Trachea* Ochsenheimer, 1816 陌夜蛾属"，在此予以变更。

Genus *Mesapamea* Heinicke, 1959 中秀夜蛾属

Mesapamea Heinicke, 1959, *Deutsche Entomologische Zeitschrift* (*N.F.*), 6: 103. Type species: *Apamea moderata* Eversmann, 1843 [Russia, Urals].

Synonymy: *Lampetia* Boie, 1937, preocc. (Meigen, 1880 [Diptera]); *Hadena* auct.; *Parastichtis* auct.; *Luperina* auct..

The genus includes 26 species, distributed in the Palaearctic and Nearctic regions, with 16 species are in the Palaearctic region; 1 species is in the Three Provinces of Northeast China.

注：首次给出中文属名。

***Mesapamea concinnata* Heinicke, 1959 美中秀夜蛾**

Mesapamea concinnata Heinicke, 1959, *Deutsche Entomologische Zeitschrift, N.F.*, 6 (1-3): 109, pl. 11: 5. Type locality: Japan, Asamayama.

Synonymy: *Mesapamea secalis* auct., nec Linnaeus, 1758.

Distribution. China (Heilongjiang, Jilin); Russia (Siberia), Korea, Japan.

注：首次给出中文种名。

Genus *Resapamea* Varga & Ronkay, 1992 衍陆夜蛾属

Resapamea Varga & Ronkay, 1992, *Acta Zoologica Hungarica*, 38: 114. Type species: *Hadena hedeni* Graeser, 1889 [Russia, Amur reg.].

Holarctic genus containing 8 species, 5 of them distributed in the Palaearctic region. 1 species is in the Three Provinces of Northeast China.

注：首次给出中文属名。

***Resapamea hedeni* (Graeser, 1889) 衍陆夜蛾**

Hadena hedeni Graeser, 1889, *Berliner entomologische Zeitschrift*, 32: 333. Type locality: Russia, Amur reg., Radde, Pokrovka, Blagoveschensk, "Raddefka, Pokrofka, Blagowestchensk".

Synonymy: *Cosmia vulpecula* Eversmann, 1852, unused senior synonym; *Hadena subaquila* Graeser, 1892; *Hadena radicosa* Graeser, 1892; *Hadena subornata* Staudinger, 1896; *Hadena terrago* Alphéraky, 1897; *Hadena terrago umbrata* Herz, 1903 (ab.); *Luperina lacunosa* W. Kozhantshikov, 1925; *Luperina eversmanni* Kozhanchikov, 1936; *Palluperina rubrina* Bryk, 1942; *Luperina takanensis* Marumo, 1932; *Mesapamea vulpecula vargai* Hacker, 1988 (ssp.); *Resapamea vulpecula rhodochrea* Ronkay & Varga, 1992 (ssp.).

Distribution. China (Heilongjiang, Nei Mongol); Russia (European part, Siberia, Kamchatka), Mongolia, Korea, Japan; Central Asia, Asia Minor, Europe.

寄主：大看麦娘（*Alopecurus pratensis*）、（*Bromis inermis*）、鸭茅（*Dactylis glomerata*）、偃麦草（*Elytrigia repens*）、大麦（*Hordeum vulgare*）、梯牧草（*Phleum pratense*）（Matov & Kononenko, 2012）。

注：《中国动物志》（夜蛾科）中将本种归入"*Luperina* Boisduval, 1829 陆夜蛾属"，在此予以变更。

Genus *Xylomoia* Staudinger, 1892 纹夜蛾属

Xylomoia Staudinger, 1892, *in* N.M. Romanoff (ed.), *Mémoires sur les Lépidoptères*, 6: 451. Type species: *Xylomoia separata* Staudinger, 1892, [Russia, Primorye terr., Partizansk].

Synonymy: *Xylomoea* Hampson, 1909, emend..

A Holarctic genus, includes 7 species, 5 of them are distributed in the Palaearctic region; among them 2 species are in the Three Provinces of Northeast China.

Xylomoia graminea (Graeser, [1889] 1888) 纵纹夜蛾

Nonagria graminea Graeser, [1889] 1888, *Berliner entomologische Zeitschrift*, 32, II: 343. Type locality: Russia, "Chabarofka" [Khabarovsk].

Synonymy: *Xylomoia separata* Staudinger, 1892.

Distribution. China (Heilongjiang, Jilin); Russia (European part, Siberia), Korea, Japan; Europe.

寄主：芦苇（*Phragmites australis*）（Matov & Kononenko, 2012）。

Xylomoia fusei Sugi, 1976 高岭纹夜蛾

Xylomoia fusei Sugi, 1976, *Tinea*, 10: 63, fig. 1. Type locality: Japan, Gumma Pref., Itakura.

Distribution. China (Heilongjiang, Liaoning); Russia (Siberia), Korea, Japan; Europe.

寄主：芦苇（*Phragmites australis*）（Matov & Kononenko, 2012）。

Subtribe **Sesamiina** Fibiger & Goldstein, 2005 蛀茎夜蛾亚族

Sesamiina Fibiger & Goldstein, 2005, *in* Lafontaine & Fibiger 2005, *Esperiana*, 11: 12. Type genus: *Sesamia* Guenée, 1852.

The subtribe comprises over 200 species and 10 genera, primarily distributed in the tropical and subtropical areas of the Old World. 3 genera ang 4 species are in the Three Provinces of Northeast China.

Genus ***Virgo*** Staudinger, 1892 条夜蛾属

Virgo Staudinger, 1892, *in* N.M. Romanoff (ed.), *Mémoires sur les Lépidoptères*, 6: 467. Type species: *Nonagria amoena* Staudinger, 1888 [Russia, Primorye terr., Partizansk] (=*Nephelodes datanidia* Butler, 1885).

The genus includes 3 species from the Chinese subregion of the Palaearctic region; 1 species is in the Three Provinces of Northeast China.

Virgo datanidia (Butler, 1885) 条夜蛾

Nephelodes datanidia Butler, 1885, *Cistula Entomologica*, 3: 132. Type locality: Japan.

Synonymy: *Nonagria amoena* Staudinger, 1888; *Virgo datanidia parthenon* Bryk, 1948 (f.).

Distribution. China (Heilongjiang, Jilin, Shaanxi, Zhejiang, Hunan); Russia (Siberia), Korea, Japan.

Genus ***Sesamia*** Guenée, 1852 蛀茎夜蛾属

Sesamia Guenée, 1852, *in* Boisduval & Guenée, *Histoire Naturelle des Insectes. Species Général des Lépidoptères*, 5: 95. Type species: *Cossus nonagrioides* Lefebvre, 1827 [Italy].

Synonymy: *Microsemyra* Butler, 1883; *Busseola* Thurau, 1904; *Calamistis* Hampson, 1908; *Hydrostola* Warren,

1912; *Hydrostyloides* Strand, 1920; *Semasia* Matsumura, 1928, missp.; *Conicophora* Matsumura, 1929.

The genus includes 54 species, distributed in the Palaearctic, Oriental and Ethiopian regions, with great diversity in the North Africa and tropical Asia areas. 6 species are in the Palaearctic, among them 2 species are in the Three Provinces of Northeast China.

***Sesamia turpis* (Butler, 1879) 污蛀茎夜蛾**

Nonagria turpis Butler, 1879, *Annals and Magazine of Natural History*, (5)4: 259. Type locality: Japan, Yokohama.

Synonymy: *Sideridis sinuosella* Bryk, 1948.

Distribution. China (Heilongjiang, Jilin, Hubei, Taiwan); Russia (Siberia), Korea, Japan.

注：《中国动物志》（夜蛾科）中将本种归入"*Nonagria* Ochsenheimer, 1816 杆夜蛾属"，在此予以变更。

***Sesamia confusa* (Sugi, 1982) 混蛀茎夜蛾**

Epipsammia confusa Sugi, 1982, *Moths of Japan*, 1: 752, pl. 185: 16-18. Type locality: Japan, Tsushima I., Mitsushima-cho, Gaya.

Distribution. China (Jilin); Russia (Siberia), Korea, Japan.

寄主：中国芒（*Miscanthus sinensis*）（Matov & Kononenko, 2012）。

Genus ***Acrapex*** Hampson, 1894 亚茎夜蛾属

Acrapex Hampson, 1894, *The Fauna of British India, including Ceylon and Burma*, 2: 286. Type species: *Leucania prisca* Walker, 1866 [Ceylon].

Synonymy: *Acutipenna* Hampson, 1894.

The genus includes about 86 species, distributed in the Palaearctic, Oriental and Ethiopian regions, with great diversity in South Africa and tropical Asia areas; 1 species is in the Three Provinces of Northeast China.

注：首次给出中文属名。

***Acrapex azumai* Sugi, 1970 小亚茎夜蛾**

Acrapex azumai Sugi, 1970, *Tinea*, 8: 221, pl. 71: 31. Type locality: Japan, Okinawa, Koza.

Distribution. China (Jilin); South Korea, Japan.

注：首次给出中文种名。

Tribe **Xylenini** Guenée, 1837 木冬夜蛾族

Xylenidae Guenée, 1837, *Annales de la Société Entomologique de France*, 6: 227 (as Xylinides). Type genus: *Xylena* Ochsenheimer, 1816.

Synonymy: Xylinides Guenée, 1837, incorrect original spelling; Xylinidi Guenée, 1837, incorrect original spelling; Calocampini Grote, 1890; Xyleninae Heslop, 1960, nomen nudum; Xylenini Franclemont & Todd, 1983, nomen nudum; Xylocampidi Tutt, 1896.

The Xyleninae has been restored as a subfamily of Noctuidae by Fibiger & Lafontaine (2005) in order to define a monophyletic unit within the large polyphyletic assemblage Hadeninae *sensu lato* by Kitching and Rawlins (1998). In that interpretation it included 14 tribes, most of which are presently defined as tribes of Noctuinae (Lafontaine & Schmidt, 2010; Fibiger *et al.*, 2011; Zahiri *et al.*, 2012). Later, the subfamily status of Xyleninae was downgraded by Lafontaine & Schmidt (2010) to the tribe Xylenini of the subfamily Noctuinae. Subsequently this point of view was accepted for the Checklist of European Noctuoidea (Fibiger *et al.*, 2011). In current treatment, the tribe includes in Palaearctic fauna 3 subtribes: Cosmiina, Xyleninae, Antitypina and Pseudohadenina (Fibiger *et al.*, 2011). Xylenini is one of the largest tribes of Noctuidae. Most species are associated with the arboreal zone of the Holarctic region and mountains in the north of the Oriental region. The external and genital morphology as well as peculiarities of biology are very diverse within the tribe and characterized in diagnosis of the subtribes and genera.

The tribe Xylenini is represented in the Three Provinces of Northeast China with three subtribes, 33 genera and 77 species.

References. Hampson, 1906 (Cucullianae, part; Acronictinae, part); Chen, 1982, 1999 (Noctuidae: Cuculliinae, part; Amphipyrinae, part); Sugi, 1982 (Cuculliinae, part; Amphipyrinae, part); Kononenko, 1990; (Cuculliinae, part; Amphipyrinae, part); Speidel *et al.*, 1996 (Cuculliinae, part; Amphipyrinae, part); Kononenko *et al.*, 1998 (Cuculliinae, part; Amphipyrinae, part); Kitching, Rawlins, 1999 [1998] (Hadeninae, part); Ronkay *et al.*, 2001 (Hadeninae); Kononenko, 1989 (Amphipyrinae, part), 2000 (Cuculliinae, part); Kononenko, 1976, 1977, 1978, 1988, 1982, 1984, 2003e (Cuculliinae, part; Amphipyrinae, part); Kononenko, 2005, 2006 (Xyleninae: Xylenini); Fibiger, Lafontaine, 2005 (Xyleninae: Xylenini); Fibiger, Hacker, 2005 (Xyleninae: Xylenini); Kononenko, 2005 (Xyleninae: Xylenini), 2016 (Xyleninae: Xylenina, Cosmiina, Pseudohadenina, Antitypina); Lafontaine, Fibiger, 2006 (Xyleninae: Xylenini); Kononenko, Han, 2007 (Xyleninae: Xylenini); Fibiger, Hacker, 2007 (Xyleninae: Xylenini, part); Matov *et al.*, 2008 (Xyleninae: Xylenini); Lafontaine, Schmidt, 2010 (Noctuidae, Xyleninae); Babics *et al.*, 2012 (Xyleninae: Xylenini); Fibiger *et al.*, 2011 (Xyleninae: Xylenini); Ronkay *et al.*, 2017 (Xyleninae, Xylenini).

Subtribe **Xylenina** Guenée, 1837 木冬夜蛾亚族

Xylenidae Guenée, 1837, *Annales de la Société Entomologique de France* 6: 227 (as Xylinides). Type genus: *Xylena* Ochsenheimer, 1816.

Synonymy: Xanthidi Stephens, 1850; Lithophanini Forbes & Franclemont, 1954; Conistrina Beck, 1996; Agrocholina Beck, 1996.

The subtribe Xylenina is most diverse in mountains between the Palaearctic and Oriental regions. Many species are associated with mid or high forested habitats and are especially diverse in the deciduous broad-leaf forest zone. In the Three Provinces of Northeast China, the subtribe is represented by 20 genera and 42 species, concentrated mainly in the broad-leaf forest.

Genus *Hyppa* Duponchel, 1844 回夜蛾属

Hyppa Duponchel, 1844, *Catalogue Mèthodique des Lépidoptères d'Europe Distribuls en Families, Tribus et Genres, etc.*, (2): 161. Type species: *Phalaena rectilinea* Esper, 1788 [Germany].

Holarctic boreal genus containing 5 species, one species known from the Palaearctic region, include the Three Provinces of Northeast China.

Hyppa rectilinea (Esper, 1788) 回夜蛾

Phalaena rectilinea Esper, 1788, *Die Schmetterlinge in Abbildungen nach der Natur mit Beschreibungen*, 4 (2) Abt. l: pl. 127 fig. i; 1796. Type locality: Germany, Schwaben.

Synonymy: *Meganephria albopicta* Matsumura, 1925.

Distribution. China (Heilongjiang, Jilin, Xinjiang); Russia (European part, Siberia, Kamchatka), Mongolia, Korea, Japan, Kazakhstan; Asia Minor, Europe.

寄主：柳、悬钩子属植物（陈一心，1999）；欧洲蕨（*Pteridium aquilinum*）、金冠鳞毛蕨（*Aspidium filix-mas*）、欧洲鳞毛蕨（*Dryopteris filix-mas*）、萹蓄 (*Polygonum aviculare*)、垂枝桦（*Betula pendula*）、熊果（*Arctostaphylos uvaursi*）、帚石楠（*Calluna vulgaris*）、黑果越橘（*Vaccinium myrtillus*）、笃斯越橘（*Vaccinium uliginosum*）、越橘（*Vaccinium vitis-idaea*）、耳柳（*Salix aurita*）、黄花柳（*Salix caprea*）、秦氏悬钩子（*Rubus caesius*）、覆盆子（*Rubus idaeus*）、树莓（*Rubus fruticosus*）、北欧花楸（*Sorbus aucuparia*）、柳兰（*Chamaenerion angustifolium*）、峨参（*Anthriscus sylvestris*）、大车前（*Plantago major*）、耧斗菜属（*Aquilegia*）、堇菜属（*Viola*）、苹果属（*Malus*）、柳叶菜属（*Epilobium*）、梣属（*Fraxinus*）、野芝麻属（*Lamium*）、苦苣菜属（*Sonchus*）植物（Matov & Kononenko, 2012）。

Genus *Brachylomia* Hampson, 1906 柳冬夜蛾属

Brachylomia Hampson, 1906, *Catalogue of the Lepidoptera Phalaenae in the British Museum*, 6: x, 220. Type species: *Cleoceris populi* Strecker, 1898 [USA].

Synonymy: *Iteophaga* Boursin, 1965, *Bombycia* auct..

Holarctic boreal genus containing 12 species; 3 species known in the Palaearctic, among them 1 species is in the Three Provinces of Northeast China.

Brachylomia viminalis (Fabricius, 1777) 维柳冬夜蛾

Noctua viminalis Fabricius, 1777, *Genera Insectorum erorumque characteres naturales adjecta mantissa specierum nuper detectarum*: 284. Type locality: Germany.

Synonymy: *Phalaena* (*Noctua*) *saliceti* Borkhausen, 1792; *Phalaena* (*Bombyx*) *stricta* Esper, [1789]; *Noctua scripta* Hübner, [1800-1813]; *Bombycia viminalis asiatica* Bang-Haas, 1907 (var.); *Bombycia viminalis obscura* Staudinger, 1871 (ssp.).

Distribution. China (Heilongjiang, Gansu, Xinjiang); Russia (European part, Siberia, Kamchatka), Mongolia, Japan; Asia Minor, Central Asia, Europe.

寄主：白柳（*Salix alba*）、灌木柳（*Salix arbuscula*）、黄华柳（*Salix atrocinerea*）、耳柳（*Salix aurita*）、黄花柳（*Salix caprea*）、灰柳（*Salix cinerea*）、爆竹柳（*Salix fragilis*）、无毛柳（*Salix glabra*）、五蕊柳（*Salix pentandra*）、东陵山柳（*Salix phylicifolia*）、沼柳（*Salix rosmarinifolia*）、蒿柳（*Salix viminalis*）、锐刺山楂（*Crataegus oxyacantha*）、杨属（*Populus*）、蔷薇科（*Prunus*）、椴树属（*Tilia*）植物（Matov & Kononenko, 2012）。

Genus ***Parastichtis*** Hübner, [1821] 1816 **伍夜蛾属**

Parastichtis Hübner, [1821] 1816, *Verzeichniss bekannter Schmettlinge* [sic!]: 213. Type species: *Noctua suspecta* Hübner, [1817] [Europe].

Synonymy: *Parastictus* Agassiz, 1846, missp.; *Dyschorista* Lederer, 1857; *Taeniosea* Grote, 1874; *Amathes* auct..

Holarctic boreal genus includes 2 species, both distributed in the Palaearctic region. Only 1 species is in the Three Provinces of Northeast China.

Parastichtis suspecta **(Hübner, [1817]) 伍夜蛾**

Noctua suspecta Hübner, [1817], *Sammlung Europaischer Schmetterlinge. Eulen*, 2: pl. 138: 633. Type locality: Europe.

Synonymy: *Caradrina iners* Germar, 1817; *Noctua coneger* Hübner, [1817]; *Caradrina iners* Treitschke, 1825 nec Germar, 1817; *Bryophila discivaria* Walker, 1856; *Taeniosea gentilis* Grote, 1874; *Taeniosea perbellis* Grote, 1874; *Amathes iners tibetica* Draudt, 1950 (f.); *Parastichtis suspecta erythema* Henriot, 1960 (ssp.); *Oligia karafutonis* Matsumura, 1925.

Distribution. China (Jilin, Qinghai, Xizang); Russia (European part, Siberia, Kamchatka), Mongolia, Korea, Japan, Kazakhstan; Asia Minor, Europe, North America.

寄主：钻天杨（*Populus nigra*）、欧洲山杨（*Populus tremula*）、黄花柳（*Salix caprea*）、绢柳（*Salix lapponum*）、五蕊柳（*Salix pentandra*）、东陵山柳（*Salix phylicifolia*）、大黄柳（*Salix raddeana*）、覆盆子（*Rubus idaeus*）、欧洲花楸（*Sorbus aucuparia*）、栎属（*Quercus*）、桤木属（*Alnus*）、桦木属（*Betula*）、车前属（*Plantago*）、蒲公英属（*Taraxacum*）植物（Matov & Kononenko, 2012）。

Genus *Tiliacea* Tutt, 1896 尖冬夜蛾属

Tiliacea Tutt, 1896, *Entomologist's Record and Journal of Variation*, 8: 17. Type species: *Phalaena citrago* Linnaeus, 1758 [Europe].

Synonymy: *Aurxanthia* Beck, [1992] 1991; *Helladica* Beck, 1999.

Palaearctic genus containing 16 species; among them 2 species are in the Three Provinces of Northeast China.

Tiliacea japonago (Wileman & West, 1929) 日美尖冬夜蛾

Cosmia japonago Wileman & West, 1929, *Novitates Zoologicae*, 35: 2. Type locality: Japan, Hokkaido, Oshima Pref., Junsai Numa.

Distribution. China (Heilongjiang, Jilin, Liaoning); Russia (Siberia), Korea, Japan.

注：《中国动物志》（夜蛾科）中将本种归入“*Xanthia* Ochsenheimer, 1816 美冬夜蛾属”，在此予以变更。

Tiliacea auragides (Draudt, 1950) 棕尖冬夜蛾

Cosmia auragides Draudt, 1950, *Mitteilungen der Münchner Entomologischen Gesellschaft*, 40: 82, pl. 6: 2. Type locality: China, Tianmushan[West Tien-mu-shan].

Synonymy: *Cirrhia tigrina* Kononenko, 1978.

Distribution. China (Jilin, Liaoning); Russia (Siberia), Korea, Japan.

注：首次给出中文种名。

Genus *Xanthia* Ochsenheimer, 1816 美冬夜蛾属

Xanthia Ochsenheimer, 1816, *Die Schmetterlinge von Europa*, 4: 82. Type species: *Noctua flavago* Fabricius, 1787 [Germany], a junior primary homonym of *Noctua flavago* [Denis & Schiffermüller], 1775, junior synonym of *Phalaena* (*Noctua*) *lutea* Ström, 1783. A subjective repl. name is *Phalaena togata* Esper, 1788.

Synonymy: *Xantha* Billberg, 1820, emend.; *Citria* Hübner, [1821] 1816; *Mellinia* Hübner, [1821] 1816; *Euthemonia* Gilst, 1848 preocc. (Stephens, 1828 [Lepidoptera, Arctiidae]); *Spudaea* Snellen, 1867 (subgen.).

Holarctic genus containing 2 species; among them 1 species is in the Three Provinces of Northeast China.

Xanthia togata (Esper, 1788) 黄紫美冬夜蛾

Phalaena Noctua togata Esper, 1788, *Die Schmetterlinge in Abbildungen nach der Natur mit Beschreibungen*, 4 (1): pl. 134: 1, [1791] : 336. Type locality: Germany, Leipzig.

Synonymy: *Phalaena lutea* Ström, 1783, nec Stoll in Cramer, 1781; *Phalaena* (*Noctua*) *flavago* Fabricius,

1787; *Phalaena rubago* Donovan, 1801; *Noctua silago* Hübner, 1803; *Cosmia lutea postlutea* Bryk, 1942 (ssp.).

Distribution. China (Heilongjiang, Jilin, Liaoning, Xinjiang); Russia (European part, Siberia, Kamchatka), Mongolia, Korea, Japan; Central Asia, Asia Minor, Europe.

寄主：黄华柳（陈一心，1999）；黑果越橘（*Vaccinium myrtillus*）、笃斯越橘（*Vaccinium uliginosum*）、欧洲山杨（*Populus tremula*）、黄华柳（*Salix atrocinerea*）、耳柳（*Salix aurita*）、黄花柳（*Salix caprea*）、灰柳（*Salix cinerea*）、（*Salix myrsinifolia*）、东陵山柳（*Salix phylicifolia*）、光榆（*Ulmus glabra*）、异株荨麻（*Urtica dioica*）、稠李（*Padus racemosa*）、覆盆子（*Rubus idaeus*）、欧洲白蜡（*Fraxinus excelsior*）、酸模属（*Rumex*）、桦木属（*Betula*）、李属（*Prunus*）、车轴草属（*Trifolium*）、拉拉藤属（*Galium*）、毛蕊花属（*Verbascum*）、车前属（*Plantago*）、牛蒡属（*Arctium*）、蒿属（*Artemisia*）、飞廉属（*Carduus*）、蒲公英属（*Taraxacum*）植物（Matov & Kononenko, 2012）。

Genus ***Cirrhia*** Hübner, [1821] 1816 **点美冬夜蛾属**

Cirrhia Hübner, [1821] 1816, *Verzeichniss bekannter Schmettlinge* [sic!]: 234. Type species: *Phalaena fulvago* Clerck, sensu Linnaeus, 1761 (=*Phalaena icteritia* Hufnagel, 1766) [Germany].

Synonymy: *Mellinia* Hübner, 1821.

The genus includes 11 species, distributed in the Palaearctic, that in south of the forest zone from Europe through Middle East and Central Asia to Chinese and Sino-Himalayan subregions. 3 species are in the Three Provinces of Northeast China.

注：首次给出中文种名。

***Cirrhia icteritia* (Hufnagel, 1766) 淡黄点美冬夜蛾**

Phalaena icteritia Hufnagel, 1766, *Berlinisches Magazine*, 3: 296. Type locality: Germany, Berlin.

Synonymy: *Noctua cerago* [Denis et Schiffermüller], 1775; *Phalaena* (*Noctua*) *flavescens* Esper, 1788; *Cosmia fulvago korecosmia* Bryk, 1948 (ssp.); *Cirrhia fulvago* auct., nec Clerk, 1759, nec Linnaeus, 1761.

Distribution. China (Heilongjiang, Jilin, Qinghai); Russia (European part, Siberia, Kamchatka), Mongolia, Korea, Japan; Central Asia, Asia Minor, Europe.

寄主：山新杨（*Populus bolleana*）、毛桦（*Betula pubescens*）、黑果越橘（*Vaccinium myrtillus*）、笃斯越橘（*Vaccinium uliginosum*）、钻天杨（*Populus nigra*）、新疆杨（*Populus pyramidalis*）、欧洲山杨（*Populus tremula*）、白柳（*Salix alba*）、黄华柳（*Salix atrocinerea*）、（*Salix australior*）、黄花柳（*Salix caprea*）、灰柳（*Salix cinerea*）、爆竹柳（*Salix fragilis*）、（*Salix gilgina*）、尖果密穗柳（*Salix oxycarpa*）、东陵山柳（*Salix phylicifolia*）、龙江柳（*Salix sachalinensis*）、细穗柳（*Salix tenuijulis*）、三蕊柳（*Salix triandra*）、光榆（*Ulmus glabra*）、暗红委陵菜（*Potentilla atrosanguinea*）、酸模属（*Rumex*）、栎属（*Quercus*）、椴树属（*Tilia*）、梨属（*Pyrus*）、车前属（*Plantago*）、山柳菊属（*Hieracium*）、风毛菊属（*Leontodon*）、蒲公英属（*Taraxacum*）植物（Matov & Kononenko, 2012）。

注：《中国动物志》（夜蛾科）中将本种归入“*Xanthia* Ochsenheimer, 1816 美冬夜蛾属”，在此予以变更。

***Cirrhia tunicata* (Graeser, [1890] 1889) 齿点美冬夜蛾**

Xanthia tunicata Graeser, [1890]1889, *Berliner entomologische Zeitschrift*, 33, II: 259. Type locality: Russia, [Primorye terr.] "Ussuri".

Synonymy: *Cosmia siphuncula* Hampson, 1906; *Cosmia fulvescens* Draudt, 1934.

Distribution. China (Heilongjiang, Jilin, Liaoning, Nei Mongol, Hebei); Russia (Siberia), Mongolia, Korea, Japan, Kazakhstan.

注：《中国动物志》（夜蛾科）中将本种归入"*Xanthia* Ochsenheimer, 1816 美冬夜蛾属"，在此予以变更。

***Cirrhia fasciata* (Kononenko, 1978) 暗点美冬夜蛾**

Cirrhia fasciata Kononenko, 1978, *Entomologisckoe Obozrenie*, 57 (4): 898, fig. 10, 11. Type locality: Russia, Primorye terr., Ussuriisky Nature Reserve.

Distribution. China (Heilongjiang, Jilin); Russia (Siberia), Korea.

注：首次给出中文种名。

Genus ***Vulpechola*** L. Ronkay, G. Ronkay, Gyulai & Varga, 2017 志田冬夜蛾属

Vulpechola L. Ronkay, G. Ronkay, Gyulai & Varga, 2017, *The Witt catalogue. A taxonomic atlas of the Eurasian and North African Noctuoidea* 9: 73. Type species: *Xanthia vulpecula* Lederer, 1853 [East Kazakhstan, West Altai].

Sino-Siberian monotypic genus. 1 species is in the Three Provinces of Northeast China.

***Vulpechola vulpecula* (Lederer, 1853) 狐志田冬夜蛾**

Xanthia vulpecula Lederer, 1853, *Verhandlungen des Zoologisch-Botanischen vereins in Wien*, 3: 374, pl. 3: 5. Type locality: East Kazakhstan, West Altai [Ust-Kamengorsk, Ust-Buchtarminsk].

Synonymy: *Xanthia vulpina* Eversmann, 1855.

Distribution. China (Heilongjiang, Jilin, Xinjiang); Russia (Siberia), Mongolia, Korea, Japan.

寄主：黑果越橘（*Vaccinium myrtillus*）、异株荨麻（*Urtica dioica*）、覆盆子（*Rubus idaeus*）、水曲柳（*Fraxinus mandshurica*）、杨属（*Populus*）、柳属（*Salix*）、车前属（*Plantago*）植物（Matov & Kononenko, 2012）。

注：《中国动物志》（夜蛾科）中将本种归入"*Episema* Ochsenheimer, 1816 志冬夜蛾属"，在此予以变更。

Genus ***Hyalobole*** Warren, 1911 明冬夜蛾属

Hyalobole Warren, 1911, *Novitates Zoologicae*, 18: 141. Type species: *Hyalobole orthosioides* Warren, 1911 [India, Darjiling].

The genus contains 14 species, distributed mainly in Sino-Himalayan and Sino-Pacific subregions of the Palaearctic and mountains of the Oriental region. The genus divided on 4 subgenera: *Hyalobole* Warren, 1911; *Anstineobole* Ronkay, Ronkay, Gyulai & Varga, 2017; *Tzoobole* L. Ronkay, G. Ronkay, Gyulai & Varga, 2017; *Fuobole* L. Ronkay, G. Ronkay, Gyulai & Varga, 2017. 1 species is in the Three Provinces of Northeast China.

注：首次给出中文属名。

Subgenus ***Anstineobole*** Ronkay, Ronkay, Gyulai & Varga, 2017 **白明冬夜蛾亚属**

Anstineobole Ronkay, Ronkay, Gyulai & Varga, 2017, *The Witt catalogue. A taxonomic atlas of the Eurasian and North African Noctuoidea* 9: 112. Type species: *Agrochola albimacula* Kononenko, 1978 [Russia, South Primorye terr.].

注：首次给出中文亚属名。

Hyalobole **(*Anstineobole*)** ***albimacula*** **(Kononenko, 1978) 白明冬夜蛾**

Agrochola albimacula Kononenko, 1978, *Entomologisckoe Obozrenie*, 57 (4): 894, fig. 6, 7. Type locality: Russia, Primorye terr., Kedrovaya Pad nature reserve.

Distribution. China (Jilin); Russia (Siberia), Korea.

注：首次给出中文种名。

Genus ***Incertobole*** L. Ronkay, G. Ronkay, Gyulai & Varga, 2017 **短角明冬夜蛾属**

Incertobole L. Ronkay, G. Ronkay, Gyulai & Varga, 2017, *The Witt catalogue. A taxonomic atlas of the Eurasian and North African Noctuoidea* 9: 104. Type species: *Dasycampa evelina* Butler, 1879 [Japan].

Sino-Pacific monotypic genus. 1 species is in the Three Provinces of Northeast China.

注：首次给出中文亚属名。

Incertobole evelina **(Butler, 1879) 短角明冬夜蛾**

Dasycampa evelina Butler, 1879, *Annals and Magazine of Natural History*, (5)4: 363. Type locality: Japan, Yokohama.

Synonymy: *Orrhodia canicostata* Graeser, 1888 [1889]; *Orrhodia ciliata* Staudinger, 1892.

Distribution. China (Heilongjiang, Jilin); Russia (Siberia), Korea, Japan.

寄主：槲树（*Quercus dentata*）、蒙古栎（*Quercus mongolica*）（Matov & Kononenko, 2012）。

注：首次给出中文种名。

Genus ***Telorta*** Warren, 1910 **遥冬夜蛾属**

Telorta Warren, 1910, *in* A.Seitz (ed.), *Die Gross-Schmetterlinge der Erde*, 3: 156. Type species: *Mesogona divergens* Butler, 1879 [Japan].

The genus includes 10 species distributed in the Chinese subregion of the Palaearctic and Oriental regions. 1 species in the Three Provinces of Northeast China.

Telorta divergens **(Butler, 1879) 遥冬夜蛾**

Mesogona divergens Butler, 1879, *Annals and Magazine of Natural History*, (5)4: 364. Type locality: Japan, Yokohama.

Synonymy: *Xanthia coriacea* Graeser, 1888 [1889].

Distribution. China (Heilongjiang, Jilin, Liaoning, Zhejiang, Hunan, Xizang); Russia (Siberia), Korea, Japan.

Genus ***Etlorta*** Kononenko, 2016 厄冬夜蛾属

Etlorta Kononenko, 2016, *Proceedings of the Witt Museum, Munich* 5: 181. Type species: *Gortyna edentata* Leech, 1889 [Japan].

Chinese monotypic genus. 1 species is in the Three Provinces of Northeast China.

注：首次给出中文属名。

Etlorta edentata **(Leech, 1889) 厄冬夜蛾**

Gortyna edentata Leech, 1889, *Proceedings of the Zoological Society of London*, 1889: 485, pl. 51: 9. Type locality: Japan, Yokohama.

Synonymy: *Cosmia trapezoides*: Staudinger, 1882, part.

Distribution. China (Jilin); Russia (Siberia), Korea, Japan.

寄主：大叶栎（*Quercus crispula*）、青冈 (*Quercus glauca*)、李属（*Prunus*）植物（Matov & Kononenko, 2012）。

注：首次给出中文种名。

Genus ***Conistra*** Hübner, [1821] 1816 峦冬夜蛾属

Conistra Hübner, [1821] 1816, *Verzeichniss bekannter Schmettlinge* [sic!]: 229. Type species: *Noctua veronicae* Hübner, 1813 [Europe].

Synonymy: *Heteromorpha* Failla-Tedaldi, 1890, preocc. (Hübner, 1822 [Lepidoptera, Noctuidae]); *Orrhodiella* Spuler, 1907 (subgen.), repl. name; *Glaee*: Hübner, [1806], suppr. (ICZN, 1926 Op. 97; 1954 Op. 278); *Orrhodia* Hübner, [1821] 1816; *Gloia* Hübner, [1822]; *Glaea*: Stephens, 1829, emend.; *Glaea*: Curtis, 1829, missp.; *Gloea* Agassiz, 1846, emend.; *Dasycampa* Guenée, 1837 (subgen.); *Omalosoma* Walker, 1857, unavail.; *Peperida* Hreblay, 1993 (subgen.).

The Old World temperate genus includes about 40 species, mainly distributed in the Mediterranean, north of Ethiopian, Central Asian, Chinese subregions of the Palaearctic and mountains of the Oriental regions; 30 species are known from the Palaearctic region; among them, 9 species of 2 genera in the Three Provinces of Northeast China.

Subgenus *Conistra* Hübner, [1821] 1816 峦冬夜蛾亚属

Conistra Hübner, [1821] 1816, *Verzeichniss bekannter Schmettlinge*: 229. Type species: *Noctua veronicae* Hübner, 1813 [Europe].

***Conistra* (*Conistra*) *grisescens* Draudt, 1950 苏峦冬夜蛾**

Conistra ardescens grisescens Draudt, 1950, *Mitteilungen der Münchner Entomologischen Gesellschaft*, 40: 81. Type locality: China, Jiangsu.

Distribution. China (Heilongjiang, Jilin, Jiangsu); Russia (Siberia), Korea, Japan.

寄主：蒙古栎、地榆、蔷薇、日本樱花、灯台树（陈一心，1999）；麻栎（*Quercus acutissima*）、枹栎（*Quercus serrata*）、千金榆（*Carpinus cordata*）、野蔷薇（*Rosa multiflora*）、苹果（Malus pumila）、椴树属（*Tilia*）、李属（*Prunus*）、槭树属（*Acer*）植物（Matov & Kononenko, 2012）。

注：首次给出中文种名。

***Conistra* (*Conistra*) *ardescens* (Butler, 1879) 日峦冬夜蛾**

Dasycampa ardescens Butler, 1879, *Annals and Magazine of Natural History*, (5)4: 364. Type locality: Japan, Yokohama.

Distribution. China (Heilongjiang, Liaoning); Russia (Siberia), Korea, Japan.

寄主：蒙古栎、日本樱花、蔷薇、灯台树（陈一心，1999）；麻栎（*Quercus acutissima*）、青冈(*Quercus glauca*)、枹栎（*Quercus serrata*）、椴树属（*Tilia*）、榆属（*Ulmus*）、李属（*Prunus*）、槭树属（*Acer*）植物（Matov & Kononenko, 2012）。

注：首次给出中文种名。

***Conistra* (*Conistra*) *fletcheri* Sugi, 1958 绯纤峦冬夜蛾**

Conistra fletcheri Sugi, 1858, *Tinea*, 4: 211, pl. 31: 5, 6. Type locality: Japan, Shizuoka Pref., Umegashima.

Distribution. China (Heilongjiang, Liaoning); Russia (Siberia), Korea, Japan.

注：首次给出中文种名。

Subgenus *Dasycampa* Guenée, 1837 锈峦冬夜蛾亚属

Dasycampa Guenée, 1837, *Annales de la Société Entomologique de France*, 6: 224. Type species: *Noctua rubiginea* [Denis & Schiffermüller], 1775 [Austria].

Synonymy: *Omalosoma* Walker, 1857, unavailable name.

Conistra **(*Dasycampa*)** ***castaneofasciata*** **(Motschulsky, [1861] 1860) 褐锈峦冬夜蛾**

Oporina castaneofasciata Motschulsky, [1861] 1860, *Études d'Entomologie*, 9: 34. Type locality: Japan.

Synonymy: *Conistra fornax* Butler, 1878; *Conistra castaneofasciata punctillum* Draudt, 1950 (ssp.).

Distribution. China (Heilongjiang, Jilin, Liaoning, Shaanxi, Yunnan); Russia (Siberia), Korea, Japan.

寄主：麻栎、蒙古栎（陈一心，1999）；鹅耳枥属（*Carpinus*）、椴树属（*Tilia*）、山楂属（*Crataegus*）、槭树属（*Acer*）植物（Matov & Kononenko, 2012）。

注：首次给出中文种名。

Conistra **(*Dasycampa*)** ***filipjevi*** **Kononenko, 1978 北锈峦冬夜蛾**

Conistra filipjevi Kononenko, 1978, *Entomologisckoe Obozrenie*, 57 (4): 891, fig. 1, 3. Type locality: Russia, Primorye terr., Yakovlevka.

Distribution. China (Heilongjiang, Liaoning); Russia (Siberia), Korea, Japan.

注：首次给出中文种名。

Conistra **(*Dasycampa*)** ***albipuncta*** **(Leech, 1889) 白锈峦冬夜蛾**

Cerastis albipuncta Leech, 1889, *Proceedings of the Zoological Society of London*, 1889: 514, pl. 51: 10. Type locality: Japan, Yokohama.

Synonymy: *Conistra ardescens unimacula* Warren, 1910 (ab.); *Conistra unimacula* Sugi, 1958.

Distribution. China (Heilongjiang, Liaoning); Russia (Siberia), Korea, Japan.

寄主：白桦、鹅耳枥、麻栎、栎属、槭树科等植物（陈一心，1999）；麻栎（*Quercus acutissima*）、蒙古栎（*Quercusmongolica*）、千金榆（*Carpinus cordata*）、苹果（Malus pumila）、桦木属（*Betula*）、李属（*Prunus*）、槭树属（*Acer*）植物（Matov & Kononenko, 2012）。

注：首次给出中文种名。

Genus ***Jodia*** Hübner, 1818 亮冬夜蛾属

Jodia Hübner, 1818, *Zuträge zur Sammlung Exotischer Schmetterlinge*, l: 15. Type species: *Noctua croceago* [Denis & Schiffermüller], 1775 [Austria].

Synonymy: *Xanthia* Boisduval, 1828, nec Ochsenheimer, 1816 (Lepidoptera, Noctuidae); *Lampetia* Curtis, 1829 nec Meigen, 1800 (Diptera); *Xantholeuca* Stephens, 1831; *Hoporina* Blanchard, 1840; *Hosporina*: Duponchel, 1845, missp.; *Oporina*: Agassiz, 1846, emend..

The genus contains 2 species, distributed in the Mediterranean and Chinese subregions of the Palaearctic region. 1 species is in the Three Provinces of Northeast China.

Jodia sericea (Butler, 1878) 丝亮冬夜蛾

Hoporina sericea Butler, 1878, *Annals and Magazine of Natural History*, (5)1: 168. Type locality: Japan, Yokohama.

Distribution. China (Heilongjiang, Liaoning); Russia (Siberia), Korea, Japan.

寄主：麻栎（*Quercus acutissima*）、槲树（*Quercus dentata*）、蒙古栎（*Quercus mongolica*）（Matov & Kononenko, 2012）。

Genus *Teratoglaea* Sugi, 1958 曲翼冬夜蛾属

Teratoglaea Sugi, 1958, *Tinea*, 4: 208. Type species: *Teratoglaea pacifca* Sugi, 1958 [Japan, Tokyo, Mt. Takao].

Sino-Pacific genus includes 2 species; 1 species occurs in the Three Provinces of Northeast China.

Teratoglaea pacifica Sugi, 1958 曲翼冬夜蛾

Teratoglaea pacifica Sugi, 1958, *Tinea*, 4: 209, pl. 31: 20. Type locality: Japan, Tokyo, Mt. Takao.

Distribution. China (Heilongjiang, Jilin); Russia (Siberia), Korea, Japan.

寄主：冷杉（*Abies fabri*）、红松（*Pinus koraiensis*）（Matov & Kononenko, 2012）。

Genus *Lithophane* Hübner, [1821] 1816 石冬夜蛾属

Lithophane Hübner, [1821] 1816, *Verzeichniss bekannter Schmettlinge* [sic!]: 242. Type species: *Noctua petrifcata* [Denis & Schiffermüller], 1775 [Austria] (=*Phalaena hepatica* Clerck, 1759).

Synonymy: *Graptolitha* Hübner, [1821] 1816; *Rhizolitha* Curtis, 1829; *Prolitha* Berio, 1980 (subgen.); *Ornitopia* Beck, 1996; *Dubiphane* Beck, 1996; *Epilitha* Beck, 1996; *Xylina* auct..

The Holarctic genus includes over 75 species, distributed in the temperate Palaearctic, the mountains of the Oriental, the temperate and boreal Nearctic regions. The Palaearctic has 20 species, among them 7 species are in the Three Provinces of Northeast China.

Lithophane ustulata (Butler, 1878) 焦石冬夜蛾

Agrotis ustulata Butler, 1878, *Annals and Magazine of Natural History*, 1 (5): 162. Type locality: Japan, Yokohama.

Distribution. China (Heilongjiang, Jilin); Russia (Siberia), Korea, Japan.

寄主：槲树（*Quercus dentata*）（Matov & Kononenko, 2012）。

Lithophane pruinosa **(Butler, 1878) 霜石冬夜蛾**

Xylina pruinosa Butler, 1878, *Annals and Magazine of Natural History*, 1 (5): 197. Type locality: Japan, Yokohama.

Synonymy: *Xylina brachyptera* Staudinger, 1892; *Xylina ornithopus japonica* Neuburger, 1903 (var.); *Lithophane ornithopus* auct., nec Hufnagel, 1766.

Distribution. China (Heilongjiang, Jilin); Russia (Siberia), Korea, Japan.

寄主：蒙古栎（*Quercus mongolica*）（Matov & Kononenko, 2012）。

Lithophane plumbealis **(Matsumura, 1926) 灰石冬夜蛾**

Parastichtis plumbealis Matsumura, 1926, *Insecta Matsumurana*, 1: 56, pl. 1: 14. Type locality: Japan, Hokkaido, Sapporo.

Distribution. China (Heilongjiang); Russia (Siberia), Korea, Japan.

寄主：椴树属（*Tilia*）植物（Matov & Kononenko, 2012）。

Lithophane venusta **(Leech, 1889) 亮石冬夜蛾**

Lamprosticta venusta Leech, 1889, *Proceedings of the Zoological Society of London*, 1889: 504, pl. 51: 5. Type locality: Japan, Yokohama.

Synonymy: *Lithophane venusta yazakii* Yoshimoto, 1988 (ssp.); *Lithophane venusta fibigeri* Hreblay & Ronkay, 1998 (ssp.).

Distribution. China (Heilongjiang, Jilin, Taiwan); Russia (Siberia), Korea, Japan, Nepal.

寄主：槲树（*Quercus dentata*）、青冈（*Quercus glauca*）、蒙古栎（*Quercus mongolica*）（Matov & Kononenko, 2012）。

注：首次给出中文种名。本种具有 3 个亚种：*L. v. venusta* 分布于中国东北地区、俄罗斯远东地区、朝鲜、韩国、日本；*L. v. yazakii* 分布于中国台湾；*L. v. fibigeri* 分布于尼泊尔。

Lithophane socia **(Hufnagel, 1766) 石冬夜蛾**

Phalaena socia Hufnagel, 1766, *Berlinisches Magazine*, 3: 418. Type locality: Germany, Berlin.

Synonymy: *Phalaena hepatica* Clerk, 1759; *Noctua petrificata* [Denis et Schiffermüller], 1775; *Noctua petrificosa*: Hübner, [1803]; *Noctua petrolignea*: Hübner, [1821]; *Xylophasia incognita* Butler, 1885.

Distribution. China (Heilongjiang, Jilin, Liaoning); Russia (European part, Siberia), Mongolia, Korea, Japan, Nepal, Kazakhstan; Europe.

寄主：椤、椴、栎等属（陈一心，1999）；繁缕（*Stellaria media*）、欧洲桤木（*Alnus glutinosa*）、灰桤木（*Alnus incana*）、毛桦（*Betula pubescens*）、桦叶鹅耳枥（*Carpinus betulus*）、黑果越橘（*Vaccinium myrtillus*）、越橘（*Vaccinium vitis-idaea*）、毛黄连花（*Lysimachia vulgaris*）、耳柳（*Salix aurita*）、黄花柳（*Salix caprea*）、（*Salix myrsinifolia*）、东陵山柳（*Salix phylicifolia*）、细叶沼柳（*Salix rosmarinifolia*）、欧洲小叶椴（*Tilia cordata*）、糠椴（*Tilia manshurica*）、春榆（*Ulmus japonica*）、（*Malus domestica*）、稠李（*Padus racemosa*）、欧洲李（*Prunus domestica*）、黑刺李（*Prunus spinosa*）、覆盆子（*Rubus idaeus*）、北欧花楸（*Sorbus aucuparia*）、欧银花（*Lonicera xylosteum*）、欧洲绣球荚蒾（*Viburnum opulus*）、欧鼠李（*Frangula alnus*）、欧丁香（*Syringa vulgaris*）、樟味藜属（*Camphorosma*）、盐节木属（*Halocnemum*）、猪毛菜属（*Salsola*）、酸模属（*Rumex*）、栎属（*Quercus*）、山楂属（*Crataegus*）、梨属（*Pyrus*）、蔷薇属（*Rosa*）、槭树属（*Acer*）、梣属（*Fraxinus*）、女贞属（*Ligustrum*）、车前属（*Plantago*）、蒿属（*Artemisia*）植物（Matov & Kononenko, 2012）。

Lithophane rosinae (Püngeler, 1906) 松香石冬夜蛾

Xylina rosinae Püngeler, 1906, *Deutsche Entomologische Zeitschrift Iris*, 19: 95. Type locality: "Amur".

Distribution. China (Jilin); Russia (Siberia), Korea, Japan.

寄主：麻栎（*Quercus acutissima*）、柿（*Diospyros kaki*）、朴树（*Celtis sinensis*）、鹅耳枥属（*Carpinus*）、柳属（*Salix*）、椴树属（*Tilia*）植物（Matov & Kononenko, 2012）。

注：首次给出中文种名。

Lithophane consocia (Borkhausen, 1792) 暗石冬夜蛾

Phalaena Noctua consocia Borkhausen, 1792, *Naturgeschichte der Europaischen Schmetterlinge nach Systematischer Ordnung*, 4: 341. Type locality: Germany, Frankfurt.

Synonymy: *Xylina ingrica* Herrich-Schäffer, 1850; *Xylina cinerosa* Guenée, 1852; *Agriopis jezoensis* Matsumura, 1931; *Xylina ingrica grisea* Graeser, 1888[1889] (var.).

Distribution. China (Heilongjiang, Jilin, Shaanxi); Russia (European part, Siberia, Kamchatka), Korea, Japan, Kazakhstan; Europe.

寄主：肢桤木（陈一心，1999）；矮桤木（*Alnus fruticosa*）、欧洲桤木（*Alnus glutinosa*）、（*Alnus incana*）、胡榛（*Corylus avellana*）、黄花柳（*Salix caprea*）、欧洲花楸（*Sorbus aucuparia*）、桦木属（*Betula*）植物（Matov & Kononenko, 2012）。

Genus *Litholomia* Grote, 1875 仿石冬夜蛾属

Litholomia Grote, 1875, *Canadian Entomologist*, 7: 206. Type species: *Scopelosoma napaea* Morrison, 1874, [USA, Massachusetts].

Holarctic genus includes 4 species; 2 species are known from Chinese subregion of the Palaearctic region. 1 species is in the Three Provinces of Northeast China.

***Litholomia pacifica* (Kononenko, 1978) 北仿石冬夜蛾**

Lithophane pacifica Kononenko, 1978, *Entomologisckoe Obozrenie*, 57 (4): 894, fig. 4, 5. Type locality: Russia, Primorye terr., Lazovsky distr., Lysaya Benevskaya Mt..

Distribution. China (Heilongjiang, Jilin); Russia (Siberia), North Korea.

Genus ***Xylena*** Ochsenheimer, 1816 木冬夜蛾属

Xylena Ochsenheimer, 1816, *Die Schmetterlinge von Europa*, 4: 85. Type species: *Phalaena exsoleta* Linnaeus, 1758 [Europe].

Synonymy: *Xylites* Reichenbach & Leipzig, 1817; *Xylaena* Hübner, [1822], missp.; *Xylina* Treitschke, 1826, emend.; *Calocampa* Stephens, 1829; *Xsylina* Frivaldsky, 1835, missp.; *Hylina* Freyer, 1840, missp.; *Callicampa* Agassiz [1847] 1846, emend.; *Calocampa* Stichel, 1908, missp.; *Monoxylena* Beck, 1996.

The genus is divided into subgenera *Xylena* s. str. and *Lithomoia*. Holarctic genus, contains 26 species, mainly distributed in the temperate zone of the Palaearctic, Nearctic and in mountains of the Oriental and Neotropical regions; 10 species are known in the Palaearctic region, among them 5 species of 2 subgenera are in the Three Provinces of Northeast China.

Subgenus ***Lithomoia*** Hübner, [1821] 1816 珂木冬夜蛾亚属

Lithomoia Hübner, [1821] 1816, *Verzeichniss bekannter Schmettlinge* [sic!]: 244. Type species: *Noctua solidaginis* Hübner, [1803] [Europe].

Synonymy: *Calocampa* auct., nec Stephens, 1829; *Chloantha* auct., nec Boisduval, Rambur & Graslin, 1836.

注：首次给出中文亚属名。

***Xylena* (*Lithomoia*) *solidaginis* (Hübner, [1803]) 珂木冬夜蛾**

Noctua solidaginis Hübner, [1803], *Sammlung Europaischer Schmetterlinge. Eulen*, 2: pl. 53: 256. Type locality: Europe.

Synonymy: *Calocampa solidaginis cinerascens* Staudinger, 1871 (var.); *Colocampa* [sic!] *solidaginis rangnowi* Stichel, 1908 (var.).

Distribution. China (Heilongjiang, Xinjiang); Russia (European part, Siberia, Kamchatka), Korea, Japan, Kazakhstan; Europe.

寄主：乌饭树属、一枝黄花属植物（陈一心，1999）；西伯利亚落叶松（*Larix sibirica*）、西伯利亚红松（*Pinus sibirica*）、香杨梅（*Myrica gale*）、安多杜鹃（*Andromeda polifolia*）、熊果（*Arctostaphylos uvaursi*）、帚石楠（*Calluna vulgaris*）、杜香（*Ledum palustre*）、黑果越橘（*Vaccinium myrtillus*）、笃斯越橘（*Vaccinium uliginosum*）、越橘（*Vaccinium vitisidaea*）、耳柳（*Salix aurita*）、黄花柳（*Salix caprea*）、（*Salix hebbiana*）、东陵山柳（*Salix phylicifolia*）、正醋栗（*Ribes grossularia*）、鹅莓（*Ribes *uva-crispa*）、（*Amelanchier confusa*）、兴安悬钩子（*Rubus chamaemorus*）、覆盆子（*Rubus idaeus*）、北欧花楸（*Sorbus aucuparia*）、道格拉斯绣线菊（*Spiraea douglasi*）、宽叶绣线菊（*Spiraea latifolia*）、（*Spiraea pinnatifidum*）、药鼠李（*Rhamnus cathartica*）、小米草（*Euphrasia stricta*）、沼生马先蒿（*Pedicularis palustris*）、毛果一枝黄花（*Solidago virgaurea*）、桦木属（*Betula*）、欧石南属（*Erica*）、山楂属（*Crataegus*）、李属（*Prunus*）、忍冬属（*Lonicera*）植物（Matov & Kononenko, 2012）。

注：《中国动物志》（夜蛾科）中将本种归入"*Lithomoia* Hübner, [1821] 1816 珂冬夜蛾属"，在此予以变更。

Subgenus *Xylena* Ochsenheimer, 1816 木冬夜蛾亚属

Xylena Ochsenheimer, 1816, *Die Schmetterlinge von Europa*, 4: 85. Type species: *Phalaena exsoleta* Linnaeus, 1758 [Europe].

Xylena (*Xylena*) *exsoleta* (Linnaeus, 1758) 木冬夜蛾

Phalaena exsoleta Linnaeus, 1758, *Systema Naturae* (Edn. 10) 1: 515. Type locality: Europe. Lectotype designated by Mikkola & Honey, 1992 (RM, Stockholm).

Synonymy: *Calocampa fumosa* Butler, 1878; *Calocampa exoleta impudica* Staudinger, 1888 (var.); *Xylena exsoleta canaria* Pinker, 1968 (ssp.).

Distribution. China (Heilongjiang, Liaoning, Yunnan); Russia (European part, Siberia, Kamchatka), Mongolia, Korea, Japan; Middle East, Central Asia, Asia Minor, Europe, North Africa.

寄主：落叶松属（*Larix*）、驴蹄草（*Caltha palustris*）、剪秋罗（*Lychnis flos-cuculi*）、甜菜（*Beta vulgaris*）、李氏碱柴（*Salsola richteri*）、梨叶木蓼（*Atraphaxis pyrifolia*）、（*Calligonum griseum*）、（*Polygonum bucharica*）、酸模（*Rumex acetosa*）、直穗酸模（*Rumex longifolius*）、香杨梅（*Myrica gale*）、黑果越橘（*Vaccinium myrtillus*）、新疆杨（*Populus pyramidalis*）、（*Salix australior*）、细穗柳（*Salix tenuijulis*）、甘蓝（*Brassica oleracea*）、欧亚香花芥（*Hesperis matronalis*）、欧洲菘蓝（*Isatis tinctoria*）、（*Malcolmia turkestanica*）、菠菜（*Spinacia oleracea*）、啤酒花（*Humulus lupulus*）、柏大戟（*Euphorbia cyparissias*）、大景天（*Sedum maximum*）、紫叶景天（*Sedum telephium*）、正醋栗（*Ribes grossularia*）、黑茶藨子（*Ribes nigrum*）、（*Amygdalus bucharica*）、杏（*Armeniaca vulgaris*）、毛樱桃（*Cerasus tomentosa*）、（*Cotoneaster acutiuscula*）、（*Cotoneaster hissarica*）、（*Cotoneaster insignis*）、（*Cotoneaster multiflora*）、（*Cotoneaster nummularia*）、（*Cotoneaster nummularioides*）、（*Cotoneaster suavis*）、（Malus domestica）、（*Malus hissarica*）、稠李（*Padus racemosa*）、西洋梨（*Pyrus communis*）、秦氏悬钩子（*Rubus caesius*）、覆盆子（*Rubus idaeus*）、（*Sorbus aucuparia*）、柳兰（*Chamaenerion angustifolium*）、（*Cytisus albus*）、（*Faba bona*）、北方拉拉藤（*Hedysarum coronarium*）、香豌豆（*Lathyrus odorosus*）、牧地山黧豆（*Lathyrus pratensis*）、（*Onobrychis viciifolia*）、匍匐芒柄草（*Ononis repens*）、（*Thermopsis dolichocarpa*）、红车轴草（*Trifolium pratense*）、亚麻（*Linum usitatissimum*）、葡萄（*Vitis vinifera*）、林当归（*Angelica *silvestris*）、峨参（*Anthriscus silvestris*）、蓬子菜（*Galium verum*）、（*Apocynum scabrum*）、烟草（*Nicotiana tabacum*）、马铃薯（*Solanum tuberosum*）、（*Scrophularia canina*）、（*Phlomis bucharica*）、飞廉（*Carduus nutans*）、茼蒿（*Chrysanthemum coronarium*）、法国菊（*Chrysanthemum leucanthemum*）、（*Matricaria arvensis*）、（*Taraxacum dens-leonis*）、蒲公英（*Taraxacum vulgare*）、泽泻（*Alisma plantago-aquatica*）、（*Gladiolus italicus*）、黄菖蒲（*Iris pseudacorus*）、（*Ornithogalum ponticum*）、洋葱（*Allium cepa*）、飞燕草属（*Consolida*）、罂粟属（*Papaver*）、石竹属（*Dianthus*）、蝇子草属（*Silene*）、滨藜属（*Atriplex*）、藜属（*Chenopodium*）、栎属（*Quercus*）、桤木属（*Alnus*）、榛属（*Corylus*）、椴树属（*Tilia*）、大戟属（*Tithymalus*）、长生草属（*Sempervivum*）、山楂属（*Crataegus*）、李属（*Prunus*）、柳叶菜属（*Epilobium*）、染料木属（*Genista*）、苜蓿属（*Medicago*）、豌豆属（*Pisum*）、蚕豆属（*Vicia*）、柑橘属（*Citrus*）、老鹳草属（*Geranium*）、阿魏属（*Ferula*）、前胡属（*Peucedanum*）、天蓝绣球属（*Phlox*）、毛蕊花属（*Verbascum*）、水苏属（*Stachys*）、蓍属（*Achillea*）、春黄菊属（*Anthemis*）、蓟属（*Cirsium*）、刺头菊属（*Cousinia*）、莴苣属（*Lactuca*）、蜂斗菜属（*Petasites*）、千里光属（*Senecio*）、麻花头属（*Serratula*）、苦苣菜属（*Sonchus*）、三肋果属（*Tripleurospermum*）、独尾草属（*Eremurus*）、天门冬属（*Asparagus*）植物（Matov & Kononenko, 2012）。

***Xylena (Xylena) vetusta* (Hübner, 1809-1813) 老木冬夜蛾**

Noctua vetusta Hübner, 1809-1813, *Sammlung Europaischer Schmetterlinge. Eulen*, 2: pl. 97: 459. Type locality: Europe.

Distribution. China (Heilongjiang, Liaoning, Yunnan); Russia (European part, Siberia, Kamchatka), Mongolia, Korea, Japan; Middle East, Central Asia, Europe, North Africa.

寄主：飞廉、麦瓶草、芒柄花属等植物（陈一心，1999）；驴蹄草（*Caltha palustris*）、草甸毛茛（*Ranunculus acris*）、匍枝毛茛（*Ranunculus repens*）、甜菜（*Beta vulgaris*）、拳参（*Polygonum bistorta*）、水生酸模（*Rumex hydrolapathum*）、垂枝桦（*Betula pendula*）、香杨梅（*Myrica gale*）、帚石楠（*Calluna vulgaris*）、甘蓝（*Brassica oleracea*）、欧洲山杨（*Populus tremula*）、黄花柳（*Salix caprea*）、爆竹柳（*Salix fragilis*）、啤酒花（*Humulus lupulus*）、黑茶藨子（*Ribes nigrum*）、覆盆子（*Rubus idaeus*）、柳叶菜属（*Epilobium*）、兵豆（*Lens culinaris*）、百脉根（*Lotus corniculatus*）、紫苜蓿（*Medicago sativa*）、豌豆（*Pisum sativum*）、红车轴草（*Trifolium pratense*）、挪威枫（*Acer platanoides*）、葡萄（*Vitis vinifera*）、林当归（*Angelica silvestris*）、睡菜（*Menyanthes trifoliata*）、阳芋（*Solanum tuberosum*）、海紫菀（*Aster tripolium*）、（*Crepis paludosa*）、黄菖蒲（*Iris pseudacorus*）、胀囊苔草（*Carex vesicaria*）、沼泽荸荠（*Eleocharis palustris*）、大水烛（*Schoenoplectus lacustris*）、水葱（*Scirpus tabernaemontani*）、西伯利亚剪股颖（*Agrostis stolonifera*）、紫羊茅（*Festuca rubra*）、（*Glyceria maritima*）、水甜茅（*Glyceria maxima*）、芦苇（*Phragmites australis*）、木贼属（*Equisetum*）、落叶松属（*Larix*）、翠雀属（*Delphinium*）、飞燕草属（*Consolida*）、唐松草属（*Thalictrum*）、蝇子草属（*Silene*）、荞麦属（*Fagopyrum*）、越橘属（*Vaccinium*）、珍珠菜属（*Lysimachia*）、椴树属（*Tilia*）、大戟属（*Euphorbia*）、李属（*Prunus*）、芒柄花属（*Ononis*）、拉拉

藤属（*Galium*）、野芝麻属（*Lamium*）、飞廉属（*Carduus*）、矢车菊属（*Centaurea*）、蓟属（*Cirsium*）、蒲公英属（*Taraxacum*）、天门冬属（*Asparagus*）、克拉莎属（*Cladium*）、荸荠属（*Heleocharis*）植物（Matov & Kononenko, 2012）。

***Xylena* (*Xylena*) *confusa* Kononenko & Ronkay, 1998 远东木冬夜蛾**

Xylena confusa Kononenko & Ronkay, 1998, *Japan Heterocerists' Journal*, 197: 361, figs 1, 4, 7, 10. Type locality: Russia, Primorye terr., Jakovlevka; North Korea; East China.

Distribution. China (Liaoning); Russia (Siberia), Korea.

寄主：草莓属（*Fragaria*）、路边青属（*Geum*）、车轴草属（*Trifolium*）、车前属（*Plantago*）、一枝黄花属（*Solidago*）、蒲公英属（*Taraxacum*）植物（Matov & Kononenko, 2012）。

注：首次给出中文种名。

***Xylena* (*Xylena*) *formosa* (Butler, 1878) 台木冬夜蛾**

Calocampa formosa Butler, 1878, *Annals and Magazine of Natural History*, (5)1: 196. Type locality: Japan, Yokohama.

Distribution. China (Heilongjiang, Liaoning); Russia (Siberia), Korea.

寄主：生白芍（*Paeonia lactiflora*）、虎杖（*Polygonum cuspidatum*）、烟草（*Nicotiana tabacum*）、豌豆（*Pisum sativum*）、酸模属（*Rumex*）、路边青属（*Geum*）、车轴草属（*Trifolium*）、车前属（*Plantago*）、紫菀属（*Aster*）、菊属（*Chrysanthemum*）植物（Matov & Kononenko, 2012）。

注：首次给出中文种名。

Genus ***Orbona*** Hübner, [1821] 1816 奥峦冬夜蛾属

Orbona Hübner, [1821] 1816, *Verzeichniss bekannter Schmettlinge* (15): 232. Type species: *Cerastis serotina* Ochsenheimer, 1816 [Europe].

Synonymy: *Conistra* auct..

Monotypic genus in the temperate zone of the Palaearctic region.

***Orbona fragariae* (Vieweg, 1790) 奥峦冬夜蛾**

Noctua fragariae Vieweg, 1790, *Tabellarisches Verzeichniss der in der Churmark Brandenburg Einheimischen Schmetterlinge*, 2: 56, pl. 2: 4, 5. Type locality: Germany, Brandenburg.

Synonymy: *Noctua domiduca* Borkhausen, 1792; *Noctua* (*Bombyx*) *orbona* Rossi, 1794; *Orbona fragariae*: Esper, 1794, nec Vieweg, 1790; *Cerastis serotina* Ochsenheimer, 1816; *Orbona fragariae pallidior* Warren, 1909 (f.).

Distribution. China (Heilongjiang, Jilin, Liaoning); Russia (European part, Ural, Siberia), Korea, Japan; Europe.

寄主：野草莓（*Fragaria vesca*）、稠李（*Padus racemosa*）、黑刺李（*Prunus spinosa*）、（*Chenopodium bonus-henricus*）、秦氏悬钩子（*Rubus caesius*）、欧洲花楸（*Sorbus aucuparia*）、野豌豆（*Vicia sepium*）、欧洲千里光（*Senecio vulgaris*）、铁线莲属（*Clematis*）、酸模属（*Rumex*）、柳属（*Salix*）、拉拉藤属（*Galium*）、车前属（*Plantago*）、野芝麻属（*Lamium*）、蒲公英属（*Taraxacum*）、拂子茅属（*Calamagrostis*）、早熟禾属（*Poa*）植物（Matov & Kononenko, 2012）。

Genus ***Eupsilia*** Hübner, [1821] 1816 **犹冬夜蛾属**

Eupsilia Hübner, [1821] 1816, *Verzeichniss bekannter Schmettlinge*: 231. Type species: *Phalaena satellitia* Linnaeus, 1767 [Germany] (=*Phalaena fransversa* Hufnagel, 1766).

Synonymy: *Scopelosoma* Curtis, 1836; *Mecoptera* Guenée, 1837; *Dichagramma* Grote, 1864.

The Holarctic genus contains 30 species, distributed in the temperate zone of the Palaearctic (mainly in Chinese subregion), and Nearctic, the mountains of the Oriental regions. 8 species are known in the Palaearctic, among them 4 species are in the Three Provinces of Northeast China.

Eupsilia contracta **(Butler, 1878) 晰线犹冬夜蛾**

Mesogona contracta Butler, 1878, *Annals and Magazine of Natural History*, (5)1: 169. Type locality: Japan, Yokohama.

Distribution. China (Heilongjiang); Russia (Siberia), Korea, Japan; Europe.

寄主：蒙古栎（*Quercus mongolica*）、日本桤木（*Alnus japonica*）（Matov & Kononenko, 2012）。

注：首次给出中文种名。

Eupsilia transversa **(Hufnagel, 1766) 槲犹冬夜蛾**

Phalaena Noctua transversa Hufnagel, 1766, *Berlinisches Magazine*, 3: 418. Type locality: Germany, Berlin.

Synonymy: *Phalaena* (*Noctua*) *satellitia* Linnaeus, 1767.

Distribution. China (Heilongjiang, Jilin); Russia (European part, Siberia), Korea, Japan; Central Asia, Middle East, Asia Minor, Europe.

寄主：槲、山毛榉（陈一心，1999）；英国栎（*Quercus pedunculata*）、灰桤木（*Alnus incana*）、垂枝桦（*Betula pendula*）、桦叶鹅耳枥（*Carpinus betulus*）、胡榛（*Corylus avellana*）、胡桃（*Juglans regia*）、黑果越橘（*Vaccinium myrtillus*）、新疆杨（*Populus pyramidalis*）、欧洲山杨（*Populus tremula*）、白柳（*Salix alba*）、黄花柳（*Salix caprea*）、欧洲小叶椴（*Tilia cordata*）、欧洲大叶椴（*Tilia platyphylla*）、高加索朴（*Celtis caucasica*）、高精细榆树（*Ulmus campestris*）、（*Ulmus effusa*）、光榆（*Ulmus glabra*）、欧洲白榆（*Ulmus laevis*）、欧洲榆（*Ulmus minor*）、红茶藨子（*Ribes rubrum*）、欧洲甜樱桃（*Cerasus avium*）、（*Crataegus turkestanica*）、旋果蚊子草（*Filipendula ulmaria*）、（*Malus domestica*）、（*Malus hissarica*）、稠李（*Padus racemosa*）、紫叶稠李（*Padus virginiana*）、野生樱桃李（*Prunus divaricata*）、欧洲李（*Prunus domestica*）、黑刺李（*Prunus spinosa*）、西洋梨（*Pyrus communis*）、野玫瑰（*Rosa canina*）、（*Rosa cocanica*）、伞房蔷薇（*Rosa corymbifera*）、（*Rosa maracandica*）、树莓（*Rubus fruticosus*）、覆盆子（*Rubus idaeus*）、石生悬钩子（*Rubus saxatilis*）、欧洲花楸（*Sorbus aucuparia*）、欧洲七叶树（*Aesculus hippocastanum*）、紫苜蓿（*Medicago sativa*）、杦树（*Schleichera trijuga*）、栓皮槭（*Acer campestre*）、挪威枫（*Acer platanoides*）、（*Acer turkestanicum*）、杧果（*Mangifera indica*）、欧卫矛（*Euonymus europaeus*）、欧鼠李（*Frangula alnus*）、繁果忍冬（*Lonicera tatarica*）、欧银花（*Lonicera xylosteum*）、欧洲绣球荚蒾（*Viburnum opulus*）、欧洲白蜡（*Fraxinus excelsior*）、欧丁香（*Syringa vulgaris*）、小檗属（*Berberis*）、山毛榉属（*Fagus*）、唐棣属（*Amelanchier*）、栒子属（*Cotoneaster*）、绣线菊属（*Spiraea*）、蒲公英属（*Taraxacum*）植物（Matov & Kononenko, 2012）。

Eupsilia boursini Sugi, 1958 布犹冬夜蛾

Eupsilia boursini Sugi, 1958, *Tinea*, 4: 205, pl. 31: 17. Type locality: Japan, Nagano Pref., Shiga Hights.

Distribution. China (Jilin); Russia (Siberia), Korea, Japan.

寄主：华东椴（*Tilia japonica*）（Matov & Kononenko, 2012）。

注：首次给出中文种名。

Eupsilia kurenzovi Kononenko, 1976 齿犹冬夜蛾

Eupsilia kurenzovi Kononenko, 1976, *Proceedings of the Zoological Institute, Academy of Sciences of the USSR*, 67: 65, fig. 3. Type locality: Russia, Primorie terr., Kedrovay Pad nature reserve.

Distribution. China (Jilin); Russia (Siberia).

注：首次给出中文种名。

Genus *Pygopteryx* Staudinger, 1887 殿夜蛾属

Pygopteryx Staudinger, 1887, *in* N.M. Romanoff (ed.), *Mémoires sur les Lépidoptères*, 3: 230. Type species: *Pygopteryx suava* Staudinger, 1887 [Russia, Primorye terr.]. *Pygopteryx* was originally placed in the family Notodontidae.

Synonymy: *Prionoxanthia* Draudt, 1950.

The genus includes 2 species known from the Chinese subregion of the Palaearctic and Oriental regions. 1 species is in the Three Provinces of Northeast China.

Pygopteryx suava Staudinger, 1887 殿夜蛾

Pygopteryx suava Staudinger, 1887, *in* N.M. Romanoff (ed.), *Mémoires sur les Lépidoptères*, 3: 230, pl. [17] 13: 4. Type locality: Russia, [Primorye terr., Bezverkhovo] Sidemi, Vladivostok.

Synonymy: *Prionoxanthia cinnamomina* Draudt, 1950.

Distribution. China (Heilongjiang, Jilin, Hebei, Shandong); Russia (Siberia), Korea, Japan.

Genus *Antivaleria* Sugi, 1958 绿鹰冬夜蛾属

Antivaleria Sugi, 1958, *Tyo to Ga*, 30: 203. Type species: *Hadena viridimacula* Graeser, 1888 [1889] [Russia, Vladivostok].

Synonymy: *Valeria* auct., nec Stephens, 1829; *Valeriodes* auct., nec Warren, 1913.

The genus includes 3 species, distributed in the Chinese subregion of the Palaearctic and Oriental regions. 1 species is in the Three Provinces of Northeast China.

Antivaleria viridimacula (Graeser, 1888 [1889]) 绿鹰冬夜蛾

Hadena viridimacula Graeser, 1888[1889], *Berliner entomologische Zeitschrift*, 32, II: 331. Type locality: Russia, Vladivostok.

Synonymy: *Berrhea japonica* Leech, 1889; *Valeria viridimacula aurata* Bryk, 1942 (ssp.).

Distribution. China (Heilongjiang, Jilin, Liaoning, Zhejiang); Russia (Siberia), Korea, Japan.

寄主：羊蹄甲（陈一心，1999）；柳栎（*Quercus salicina*）、苹果属（*Malus*）、李属（*Prunus*）植物（Matov & Kononenko, 2012）。

Subtribe **Cosmiina** Guenée, 1852 兜夜蛾亚族

Cosmidae Guenée, 1852, *in* Boisduval & Guenée, *Histoire Naturelle des Insectes. Species Général des Lépidoptères*, 6: 1. Type genus: *Cosmia* Ochsenheimer, 1816.

Synonymy: Zenobianae Hampson, 1918; Ipimorphini Beck, 1989.

The Cosmiini were treated in rank of the tribe in the subfamily Xyleninae (Fibiger & Lafontaine, 2005), subsequently they were downgraded to the subtribe of Xylenini *sensu lato* (Lafontaine & Fibiger, 2006). Fibiger & Hacker (2007) listed the Palaearctic genera included in Cosmiina. The subtribe distributed worldwide, but most diverse in the temperate arboreal zone of the Holarctic and in the north area of the Oriental regions. The subtribe Cosmiina is represented in the Three Provinces of Northeast China by 9 genera and 28 species. Most genera are concentrated in the broad-leaved forest zone.

Genus *Enargia* Hübner, [1821] 1816 清夜蛾属

Enargia Hübner, [1821] 1816, *Verzeichniss bekannter Schmettlinge* [sic!]: 235. Type species: *Phalaena paleacea* Esper, 1788 [Germany].

Synonymy: *Euperia* Guenée, 1839; *Cosmia* auct.; *Dyschorista* auct.; *Sidemia* auct..

The Holarctic genus includes 12 species; 10 species are known in the Palaearctic, mainly from Mediterranean and Chinese subregions, among them 1 species is distribed in the Three Provinces of Northeast China.

Enargia paleacea (Esper, 1788) 清夜蛾

Phalaena paleacea Esper, 1788, *Die Schmetterlinge in Abbildungen nach der Natur mit Beschreibungen*, 4 (1): 323, pl. 122: 3, 4. Type locality: Germany, Leipzig.

Synonymy: *Noctua fulvago* Hübner, 1803; *Noctua angulago* Hübner, 1812.

Distribution. China (Heilongjiang, Jilin, Liaoning); Russia (European part, Siberia, Kamchatka), Mongolia, Korea, Japan, Kazakhstan; Europe.

寄主：桦、槲（陈一心，1999）；桦（*Betula alba*）、垂枝桦（*Betula pendula*）、白桦（*Betula platyphylla*）、（*Betula pubescens*）、北美杨树（*Populus balsamifera*）、钻天杨（*Populus nigra*）、欧洲山杨（*Populus tremula*）、桤木属（*Alnus*）、栎属（*Quercus*）植物（Matov & Kononenko, 2012）。

Genus *Ipimorpha* Hübner, [1821] 1816 逸色夜蛾属

Ipimorpha Hübner, [1821] 1816, *Verzeichniss bekannter Schmettlinge*: 238. Type species: *Noctua subtusa* [Denis & Schiffermüller], 1775 [Austria].

Synonymy: *Zenobia* Oken, 1815, suppr. (ICZN, 1956 Op. 417); *Plastenis* Boisduval, 1840; *Zenobia* Agassiz, 1846; *Iphimorpha* Möschler, 1886, missp.; *Retusia* Beck, 1996.

Holarctic boreal genus with 9 species; 4 species are known in the Palaearctic region, among them 3 species are in the Three Provinces of Northeast China.

Ipimorpha retusa (Linnaeus, 1761) 逸色夜蛾

Phalaena retusa Linnaeus, 1761, *Fauna Suecica* (Edn. 2): 321. Type locality: Sweden.

Synonymy: *Phalaena* (*Noctua*) *vetula* Hübner, 1788; *Phalaena* (*Tortrix*) *merianana* Lang, 1789; *Phalaena* (*Noctua*) *chrysoglossa* Beckwith, 1794; *Noctua gracilis* Haworth, 1809; *Cosmia curvata* Butler, 1886.

Distribution. China (Heilongjiang, Jilin, Liaoning, Zhejiang, Henan); Russia (European part, Siberia), Mongolia, Korea, Japan, Kazakhstan; Asia Minor, Europe.

寄主：杨、柳（陈一心，1999）；欧洲桤木（*Alnus glutinosa*）、（*Alnus incana*）、加杨（*Populus canadensis*）、钻天杨（*Populus nigra*）、欧洲山杨（*Populus tremula*）、白柳（*Salix alba*）、黄华柳（*Salix atrocinerea*）、耳柳（*Salix aurita*）、黄花柳（*Salix caprea*）、灰柳（*Salix cinerea*）、爆竹柳（*Salix fragilis*）、五蕊柳（*Salix pentandra*）、杞柳（*Salix purpurea*）、蒿柳（*Salix viminalis*）、马兜铃属（*Aristolochia*）、桦木属（*Betula*）植物（Matov & Kononenko, 2012）。

Ipimorpha subtusa ([Denis & Schiffermüller], 1775) 杨逸色夜蛾

Noctua subtusa [Denis & Schiffermüller], 1775, *Ankündung eines Systematisches Werkes von den Schmetterlinge der Wiener Gegend. Wien*: 88. Type locality: Austria, Vienna.

Distribution. China (Heilongjiang, Jilin, Liaoning, Guangdong); Russia (European part, Siberia), Mongolia,

Korea, Japan, Kazakhstan; Asia Minor, Europe.

寄主：加杨（*Populus canadensis*）、辽杨（*Populus maximowiczii*）、钻天杨（*Populus nigra*）、欧洲山杨（*Populus tremula*）、（*Populus petrowskiana*）、（*Populus virginiana*）、桤木属（*Alnus*）、柳属（*Salix*）、苹果属（*Malus*）、梨属（*Pyrus*）植物（Matov & Kononenko, 2012）。

Ipimorpha contusa **(Freyer, 1849) 点逸色夜蛾**

Noctua contusa Freyer, 1849, *Neuere Beiträge zur Schmetterlingskunde mit Abbildungen nach der Natur*, 6: pl. 534: 2, pl. 550. Type locality: Germany.

Synonymy: *Ipomorpha contusa pergrandis* Bryk, 1948 (ssp.).

Distribution. China (Liaoning); Russia (European part, Siberia), Mongolia, Korea, Japan, Kazakhstan; Asia Minor, Europe.

寄主：欧洲山杨（*Populus tremula*）、（*Populus petrowskiana*）、柳属（*Salix*）植物（Matov & Kononenko, 2012）。

注：首次给出中文种名。

Genus *Brachyxanthia* Butler, 1878 楔胸夜蛾属

Brachyxanthia Butler, 1878, *Annals and Magazine of Natural History*, (5)1: 169. Type species: *Brachyxanthia peculiaris* Butler, 1878 [Japan, Yokohama] (=*Xanthia zelotypa* Lederer, 1853).

Monotypic Palaearctic genus.

Brachyxanthia zelotypa **(Lederer, 1853) 楔胸夜蛾**

Xanthia zelotypa Lederer, 1853, *Verhandlungen des Zoologisch-Botanischen vereins in Wien*, 3: 373, pl. 3, fig. 4. Type locality: East Kazakhstan, West Altai [Ust-Kamengorsk, Ust-Buchtarminsk].

Synonymy: *Brachyxanthia peculiaris* Butler, 1878; *Brachyxanthia jezoensis* Matsumura, 1925.

Distribution. China (Heilongjiang, Jilin); Russia (Siberia), Mongolia, Korea, Japan.

寄主：亚欧唐松草（*Thalictrum minus*）、柳属（*Salix*）植物（Matov & Kononenko, 2012）。

Genus *Cosmia* Ochsenheimer, 1816 兜夜蛾属

Cosmia Ochsenheimer, 1816, *Die Schmetterlinge von Europa*, 4: 84.Type species: *Phalaena diffinis* Linnaeus, 1767 [Portugal, Lusitania].

Synonymy: *Cosmia* Hübner, [1806], suppr. (ICZN, 1926 Op. 97; 1954 Op. 278); *Calymnia* Hübner, [1823] 1816 (subgen.); *Eustegnia* Hübner, [1821] 1816; *Ulmia* Fibiger & Hacker, 2007 (subgen.); *Silva* Fibiger & Hacker, 2007 (subgen.); *Nemus* Fibiger & Hacker, 2007 (subgen.).

Holarctic genus, includes 22 species, distributed mainly in the South taiga and the zone of broad-leaved forests. 20 species are known from the Palaearctic with the centre of diversity in the Mediterranean and Chinese subregions, among them 10 species of 3 subgenera and 1 unassociated species are in the Three Provinces of Northeast China.

Subgenus *Ulmia* Fibiger & Hacker, 2007 联兜夜蛾亚属

Ulmia Fibiger & Hacker, 2007, *Noctuidae Europaeae*, Vol. 10: 156. Type species: *Phalaena* (*Noctua*) *affinis* Linnaeus, 1767 [Italy].

注：首次给出中文亚属名。

Cosmia (*Ulmia*) *affinis* (Linnaeus, 1767) 联兜夜蛾

Phalaena affinis Linnaeus, 1767, *Systema Naturae* (Edn. 12) 1: 848. Type locality: Italy.

Synonymy: *Noctua palliata* Fabricius, 1794; *Calymnia affinis magna* Warren, 1911 (ssp.).

Distribution. China (Heilongjiang, Jilin, Liaoning, Xizang); Russia (European part, Siberia), Korea, Japan; Asia Minor, Europe.

寄主：蒙古栎（*Quercus mongolica*）、银白杨（*Populus alba*）、糠椴（*Tilia manshurica*）、高加索朴（*Celtis caucasica*）、朴树（*Celtis sinensis*）、高精细榆树（*Ulmus campestris*）、（*Ulmus effusa*）、光榆（*Ulmus glabra*）、欧洲榆（*Ulmus minor*）、榔榆（*Ulmus parvifolia*）、英国榆（*Ulmus procera*）、黑刺李（*Prunus spinosa*）、桦木属（*Betula*）、柳属（*Salix*）、榉属（*Zelkova*）植物（Matov & Kononenko, 2012）。

Cosmia (*Ulmia*) *unicolor* (Staudinger, 1892) 一色兜夜蛾

Calymnia affinis unicolor Staudinger, 1892 (var.), *in* N.M. Romanoff (ed.), *Mémoires sur les Lépidoptères*, 6: 502. Type locality: Russia, Vladivostok, [Primorye terr.] Askold I., [Bezverkhovo] Sidemi, Amur reg., [Radde] Raddefka, Blagoveschensk.

Distribution. China (Heilongjiang, Jilin, Liaoning, Nei Mongol); Russia (Siberia), Korea, Japan.

寄主：紫椴（*Tilia amurensis*）、糠椴（*Tilia mandshurica*）、栎属（*Quercus*）、榛属（*Corylus*）、榆属（*Ulmus*）、槭树属（*Acer*）植物（Matov & Kononenko, 2012）。

Cosmia (*Ulmia*) *cara* (Butler, 1881) 肖果兜夜蛾

Dyrzela cara Butler, 1881, *Transactions of the Entomological Society of London*, 1881: 188. Type locality: Japan, Tokyo.

Synonymy: *Cosmia pembertoni* Holland, 1889; *Calymnia penicillata* Graeser, 1890.

Distribution. China (Heilongjiang, Jilin, Liaoning); Russia (Siberia), Korea, Japan.

寄主：裂叶榆（*Ulmus laciniata*）、榔榆（*Ulmus parvifolia*）、春榆（*Ulmus propinqua*）（Matov &

Kononenko, 2012）。

Cosmia (_Ulmia_) _restituta_ Walker, 1857 白斑兜夜蛾

Cosmia restituta Walker, 1857, *List of the Specimens of Lepidopterous Insects in the Collection of the British Museum*, 10: 490. Type locality: Nepal.

Synonymy: *Calymnia dieckmanni* Graeser, 1888 [1889]; *Calymnia picta* Staudinger, 1888.

Distribution. China (Heilongjiang, Jilin, Liaoning, Gansu); Russia (Siberia), Korea, Japan.

寄主：裂叶榆（*Ulmus laciniata*）、榔榆（*Ulmus parvifolia*）（Matov & Kononenko, 2012）。

Cosmia (_Ulmia_) _inconspicua_ (Draudt, 1950) 小斑兜夜蛾

Calymnia inconspicua Draudt, 1950, *Mitteilungen der Münchner Entomologischen Gesellschaft*, 40: 126, pl. 9: 15. Type locality: China, Shanxi, Mienshan.

Synonymy: *Cosmia apicimacula* Sugi, 1954.

Distribution. China (Jilin, Liaoning); Russia (Siberia), Korea, Japan.

寄主：榉树（*Zelkova serrata*）（Matov & Kononenko, 2012）。

注：首次给出中文种名。

Cosmia (_Ulmia_) _achatina_ Butler, 1879 玛瑙兜夜蛾

Cosmia achatina Butler, 1879, *Annals and Magazine of Natural History*, (5)4: 365. Type locality: Japan, Yokohama.

Distribution. China (Heilongjiang, Jilin, Liaoning, Nei Mongol, Taiwan); Korea, Japan, Nepal.

Subgenus *Calymnia* Hübner, [1821] 亮兜夜蛾亚属

Calymnia Hübner, [1821], *Verzeichniss bekannter Schmettlinge* (15): 235. Type species: *Phalaena trapezina* Linnaeus, 1758 [Europe].

注：首次给出中文亚属名。

Cosmia (_Calymnia_) _trapezina_ (Linnaeus, 1758) 亮兜夜蛾

Phalaena trapezina Linnaeus, 1758, *Systema Naturae* (Edn. 10) 1: 510. Type locality: Europe.

Synonymy: *Phalaena rhombica* Hufnagel, 1766; *Calymnia badiofasciata* Teich, 1883; *Calymnia obscura* Aurivillius, 1888; *Mesogona exigua* Butler, 1881; *Calymnia trapezina saturata* Staudinger, 1892 (var.).

Distribution. China (Heilongjiang, Jilin, Liaoning); Russia (European part, Ural, Siberia, Transbaikalia), Korea, Japan, Mongolia, Kazakhstan; Asia Minor, Middle East, Caucasus, Transcaucasia, Europe, North Africa.

寄主：榆（陈一心，1999）；垂枝欧洲山毛榉（*Fagus sylvatica*）、槲树（*Quercus dentata*）、蒙古

栎（*Quercus mongolica*）、无梗花栎（*Quercus petraea*）、夏栎（*Quercus robur*）、（*Quercus rubra*）、北美红栎（*Quercus rubra*）、（*Alnus incana*）、垂枝桦（*Betula pendula*）、（*Betula pubescens*）、胡榛（*Corylus avellana*）、胡桃（*Juglans regia*）、黑果越橘（*Vaccinium myrtillus*）、北美杨树（*Populus balsamifera*）、白柳（*Salix alba*）、耳柳（*Salix aurita*）、黄花柳（*Salix caprea*）、爆竹柳（*Salix fragilis*）、五蕊柳（*Salix pentandra*）、东陵山柳（*Salix phylicifolia*）、欧洲小叶椴（*Tilia cordata*）、欧洲大叶椴（*Tilia platyphylla*）、光榆（*Ulmus glabra*）、欧洲白榆（*Ulmus laevis*）、欧洲榆（*Ulmus minor*）、高山茶藨子（*Ribes alpinum*）、黑茶藨子（*Ribes nigrum*）、珍珠梅（*Sorbaria sorbifolia*）、欧洲花楸（*Sorbus aucuparia*）、（*Sorbus suecica*）、驱疝花楸（*Sorbus torminalis*）、金雀儿（*Cytisus scoparius*）、栓皮槭（*Acer campestre*）、挪威枫（*Acer platanoides*）、欧亚槭（*Acer pseudoplatanus*）、欧鼠李（*Frangula alnus*）、药鼠李（*Rhamnus cathartica*）、东方沙枣（*Elaeagnus angustifolia*）、沙棘（*Hippophae rhamnoides*）、山葡萄（*Vitis amurensis*）、花叶冬青（*Ilex aquifolium*）、欧洲红瑞木（*Cornus sanguinea*）、繁果忍冬（*Lonicera tatarica*）、西洋接骨木（*Sambucus nigra*）、欧洲白蜡（*Fraxinus excelsior*）、水曲柳（*Fraxinus mandshurica*）、林生玄参（*Scrophularia nodosa*）、落叶松属（*Larix*）、大黄属（*Rheum*）、小檗属（*Berberis*）、鹅耳枥属（*Carpinus*）、猕猴桃属（*Actinidia*）、绣线菊属（*Spiraea*）植物（Matov & Kononenko, 2012）。

Subgenus *Nemus* Fibiger & Hacker, 2007 果兜夜蛾亚属

Nemus Fibiger & Hacker, 2007, *Noctuidae Europaea*, Vol. 10: 156. Type species: *Noctua pyralina* [Denis & Schiffermüller], 1775 [Austria, Vienna].

Cosmia (*Nemus*) *pyralina* ([Denis & Schiffermüller], 1775) 果兜夜蛾

Noctua pyralina ([Denis & Schiffermüller], 1775), *Ankündung eines Systematisches Werkes von den Schmetterlinge der Wiener Gegend. Wien*: 88. Type locality: Austria, Vienna.

Synonymy: *Phalaena* (*Noctua*) *corusca* Esper, 1788; *Calymnia pyralina dannehli* Hartig, 1924 (f.); *Calymnia pyralina roesleri* Dannehl, 1926 (var.).

Distribution. China (Heilongjiang, Jilin, Liaoning); Russia (European part, Siberia), Korea, Japan, Kazakhstan; Asia Minor, Europe.

寄主：苹果、梨（陈一心，1999）；胡桃楸（*Juglans mandshurica*）、无毛柳（*Salix glabra*）、紫椴（*Tilia amurensis*）、糠椴（*Tilia mandshurica*）、高精细榆树（*Ulmus campestris*）、光榆（*Ulmus glabra*）、欧洲白榆（*Ulmus laevis*）、榔榆（*Ulmus parvifolia*）、英国榆（*Ulmus procera*）、穗状唐棣（*Amelanchier spicata*）、（*Malus domestica*）、毛山荆子（*Malus mandshurica*）、苹果（Malus pumila）、三叶海棠（*Malus sieboldii*）、（*Malus sylvestris*）、亚洲稠李（*Padus asiatica*）、稠李（*Padus racemosa*）、欧洲李（*Prunus domestica*）、（*Prunus institia*）、黑刺李（*Prunus spinosa*）、栎属（*Quercus*）、桦木属（*Betula*）、杨属（*Populus*）、锦葵属（*Malva*）、山楂属（*Crataegus*）、梨属（*Pyrus*）、胡枝子属（*Lespedeza*）植物（Matov & Kononenko, 2012）。

Cosmia (_Nemus_) _moderata_ (Staudinger, 1888) 凡兜夜蛾

Calymnia moderata Staudinger, 1888, *Entomologische Zeitung. Entomologischen Vereine zu Stettin*, 49: 257. Type locality: Russia, Primorye terr., [Razdolnaya river] Suifun, [Khabarovsk] "Chabarowka".

Synonymy: *Calymnia grandifica* Graeser, 1888 [1889].

Distribution. China (Heilongjiang, Jilin, Liaoning, Henan, Yunnan); Russia (Siberia), Korea, Japan.

寄主：蒙古栎（*Quercus mongolica*）、核桃楸（*Juglans manshurica*）、紫椴（*Tilia amurensis*）（Matov & Kononenko, 2012）。

Cosmia (_Nemus_) _trapezinula_ (Filipjev, 1927) 净兜夜蛾

Calymnia trapezinula Filipjev, 1927, *Annuaire du Musée Zoologique de l'Académie des Sciences d'URSS*, 28: 252. Type locality: Russia, [Primorye terr.] Tigrovyi.

Synonymy: *Calymnia eugeniae* Kardakoff, 1928; *Calymnia spurcopyga* Alphèraky, 1895.

Distribution. China (Heilongjiang); Russia (Siberia), Korea, Japan.

寄主：（*Corylus heterophila*）（Matov & Kononenko, 2012）。

Subgenus *incertae sedis* 隶属的亚属未明确

Cosmia camptostigma (Ménétriès, 1859) 曲纹兜夜蛾

Heliothis camptostigma Ménétriès, 1859, *Bulletin de la Classe Physico-Mathématique de l'Académie Impériale des Sciences de St.-Pétersbourg*, 17: 219. Type locality: Russia, [south of Khabarovsk terr., Primorye terr.] "l'Amur", "l'Oussjuri".

Synonymy: *Cosmia distincta* Butler, 1878.

Distribution. China (Heilongjiang, Jilin, Liaoning, Nei Mongol); Russia (Siberia), Korea, Japan.

寄主：麻栎（*Quercus acutissima*）、青冈 (*Quercus glauca*)、蒙古栎（*Quercus mongolica*）、枹栎（*Quercus serrata*）、萝卜（*Raphanus sativus*）、紫椴（*Tilia amurensis*）、豌豆（*Pisum sativum*）、胡枝子属（*Lespedeza*）、榆属（*Ulmus*）植物（Matov & Kononenko, 2012）。

Genus ***Dimorphicosmia*** Sugi, 1982 异灿夜蛾属

Dimorphicosmia Sugi, 1982, *Moths of Japan*, 1: 775; 2: 372, 463. Type species: *Acontia variegata* Oberthür, 1879 [Russia, East].

Monotypic Sino- Pacific genus. 1 species is in the Three Provinces of Northeast China.

注：首次给出中文属名。

***Dimorphicosmia variegata* (Oberthür, 1879) 异灿夜蛾**

Acontia variegata Oberthür, 1879, *Diagnoses d'Espéces nouvelles de Lépidoptères de l'ile Askold I.*: 15. Type locality: Russia, [Primorye terr.] Askold I..

Synonymy: *Acontia flavomaculata* Oberthür, 1879.

Distribution. China (Heilongjiang, Jilin, Liaoning); Russia (Siberia), Korea, Japan.

寄主：紫椴（*Tilia amurensis*）、糠椴（*Tilia mandshurica*）、蒙古栎（*Quercus mongolica*）（Matov & Kononenko, 2012）。

Genus ***Xanthocosmia*** Sugi, 1982 番夜蛾属

Xanthocosmia Sugi, 1982, *Moths of Japan*, 1: 775; 2: 372, 463. Type species: *Eugraphia jankowskii* Oberthür, 1884 [Russia, East], repl. name for *Warrenia* Viidalepp, 1971.

Synonymy: *Warrenia* Viidalepp, 1971, preocc. (Prout, 1910 [Lepidoptera, Geometridae]).

Monotypic Sino-Pacific species.

***Xanthocosmia jankowskii* (Oberthür, 1884) 番夜蛾**

Eugraphia jankowskii Oberthür, 1884, *Études d'Entomologie*, 10: 25, pl. 3: 8. Type locality: Russia, [Primorye terr. Bezverkhovo] "Sidemi".

Distribution. China (Heilongjiang, Jilin); Russia (Siberia), Korea, Japan.

寄主：日本常绿橡树（*Quercus acuta*）、柳栎（*Quercus salicina*）、紫椴（*Tilia amurensis*）、华东椴（*Tilia japonica*）、糠椴（*Tilia mandshurica*）（Matov & Kononenko, 2012）。

Genus ***Chasminodes*** Hampson, 1908 白夜蛾属

Chasminodes Hampson, 1908, *Catalogue of the Lepidoptera Phalaenae in the British Museum*, 7: 4. Type species: *Acontia albonitens* Bremer, 1861 [Russia, Primorye terr., "der 'Ussuri' -Mündung und dem Noor"].

The genus contains 17 species, distributed mainly in the Chinese subregion of the Palaearctic and partially in the Oriental regions. 8 species are in the Three Provinces of Northeast China.

***Chasminodes albonitens* (Bremer, 1861) 白夜蛾**

Acontia albonitens Bremer, 1861, *Bulletin de l'Academie Impériale des Sciences de St.-Petersbroug*, 3: 490. Type locality: Russia, Primorye terr., "der 'Ussuri'-Mündung und dem Noor".

Synonymy: *Chasminodes harutai* Sugi, 1955.

Distribution. China (Heilongjiang, Jilin, Liaoning, Shaanxi, Shanxi, Hebei, Jiangsu, Zhejiang, Hunan); Russia (Siberia), Korea, Japan.

寄主：紫椴（*Tilia amurensis*）、华东椴（*Tilia japonica*）（Matov & Kononenko, 2012）。

Chasminodes bremeri Sugi & Kononenko, 1981 布白夜蛾

Chasminodes bremeri Sugi & Kononenko, 1981, *Tinea*, 11: 50. Type locality: Russia, [Primorye terr.] "Ussuri".

Distribution. China (Heilongjiang, Jilin, Liaoning); Russia (Siberia), Korea, Japan.

寄主：糠椴（*Tilia manshurica*）、马氏椴（*Tilia maximowicziana*）（Matov & Kononenko, 2012）。

Chasminodes sugii Kononenko, 1981 杉白夜蛾

Chasminodes sugii Kononenko, 1981, *Tinea*, 11: 50. Type locality: Russia, [Primorye terr., Bezverkhovo] Sidemi.

Synonymy: *Chasminodes albonitens* Sugi, 1955, nec Bremer, 1861, partim.

Distribution. China (Heilongjiang, Jilin, Liaoning); Russia (Siberia), Korea, Japan.

寄主：蒙古栎（*Quercus mongolica*）（Park *et al.*, 2006）；紫椴（*Tilia amurensis*）、华东椴（*Tilia japonica*）（Matov & Kononenko, 2012）。

Chasminodes aino Sugi, 1956 日白夜蛾

Chasminodes aino Sugi, 1956, *Tinea*, 3: 144. Type locality: Japan, Hokkaido, Shibecha.

Synonymy: *Chasminodes albonitens* Sugi, 1955, nec Bremer, 1861, partim.

Distribution. China (Jilin); Russia (Siberia), Korea, Japan.

寄主：糠椴（*Tilia manshurica*）、马氏椴（*Tilia maximowicziana*）（Matov & Kononenko, 2012）。

Chasminodes pseudalbonitens Sugi, 1955 条白夜蛾

Chasminodes pseudalbonitens Sugi, 1955, *Kontyu*, 23: 79, pl. 14: 6. Type locality: Japan, Gumma Pref., Kumanotaira.

Distribution. China (Jilin); Russia (Siberia), Korea, Japan.

寄主：椴树属（*Tilia*）植物（Matov & Kononenko, 2012）。

Chasminodes ussurica Kononenko, 1982 乌苏里白夜蛾

Chasminodes ussurica Kononenko, 1982, *Entomologisckoe Obozrenie*, 51 (3): 595, fig. 19, 23. Type locality: Russia, South Primorye terr., Pyazanovka.

Distribution. China (Jilin); Russia (Amur reg., Khabarovsk and Primorye terr.), Korea.

Chasminodes cilia (Staudinger, 1888) 点线白夜蛾

Leocyma cilia Staudinger, 1888, *Entomologische Zeitung. Entomologischen Vereine zu Stettin*, 49: 263. Type locality: Russia, Vladivostok.

Distribution. China (Jilin); Russia (Siberia), Korea, Japan.

寄主：椴树属（*Tilia*）植物（Matov & Kononenko, 2012）。

Chasminodes atrata (Butler, 1884) 黑白夜蛾

Chasmina atrata Butler, 1884, *Annals and Magazine of Natural History*, (5)13: 274. Type locality: Japan, Honshu [Yesso].

Synonymy: *Leocyma borussica* Staudinger, 1888.

Distribution. China (Jilin, Liaoning); Russia (Siberia), Korea, Japan.

寄主：椴树属（*Tilia*）植物（Matov & Kononenko, 2012）。

Genus *Gyrospilara* Kononenko, 1989 黑脉邪夜蛾属

Gyrospilara Kononenko, 1989, *Japan Heterocerist's Journal*, 152: 28. Type species: *Argyrospila formosa* Graeser, [1889] 1888 [Russia, Khabarovsk terr. "Nikolaevsk"] (=*Oria extraordinaria* Draudt, 1950 [China]).

Monotypic Chinese genus. 1 species is in the Three Provinces of Northeast China.

注：首次给出中文属名。

Gyrospilara formosa (Graeser, [1889] 1888) 黑脉邪夜蛾

Argyrospila formosa Graeser, [1889] 1888, *Berliner entomologische Zeitschrift*, 32, II: 345. Type locality: Russia, Khabarovsk terr., Nikolaevsk.

Synonymy: *Oria extraordinaria* Draudt, 1950.

Distribution. China (Heilongjiang, Jilin, Shaanxi, Yunnan, Xizang, Taiwan); Russia (Siberia), Korea.

注：《中国动物志》（夜蛾科）中将本种归入"*Argyrospila* Herrich-Schäffer, 1851 邪夜蛾属"，在此予以变更。

Genus *Antha* Staudinger, 1892 斜额夜蛾属

Antha Staudinger, 1892, *in* N.M. Romanoff (ed.), *Mémoires sur les Lépidoptères*, 6: 448. Type species: *Antha pretiosa* Staudinger, 1892 [Russia, Primorye terr.] (=*Leptina grata* Butler, 1881).

Monotypic Chinese species; 1 species is in the Three Provinces of Northeast China.

Antha grata (Butler, 1881) 斜额夜蛾

Leptina grata Butler, 1881, *Transactions of the Entomological Society of London*, 1881: 172. Type locality:

Japan, Tokyo.

Synonymy: *Antha pretiosa* Staudinger, 1892.

Distribution. China (Heilongjiang, Jilin, Fujian, Guizhou, Yunnan); Russia (Siberia), Korea, Japan.

Subtribe **Antitypina** Forbes & Franclemont, 1954 亚模夜蛾亚族

Antitypini Forbes& Franclemont, 1954, *in* Forbes, 1954, *Memoirs of the Cornell University agriculture Experimental Station*, 329: 159. Type genus: *Antitype* Hübner, [1821].

Synonymy: Dichoniina Beck, 1996; Dryobotina Beck, 1996; Polymixina Beck, 1996.

注：首次给出中文亚族名。

Genus ***Dryobotodes*** Warren, 1910 卓夜蛾属

Dryobotodes Warren, 1910, *in* A.Seitz (ed.), *Die Gross-Schmetterlinge der Erde*, 3: 133. Type species: *Noctua protea* [Denis & Schiffermüller], 1775 [Austria] (=*Noctua eremita* Fabricius, 1775).

Synonymy: *Dichonioxia* Berio, 1980 (subgen.), repl. name; *Dichoniopsis* Berio, 1980, preocc. (Warren, 1913 [Lepidoptera, Noctuidae]); *Monobotodes* Beck, [1992] 1991; *Robortodes* Beck, [1992] 1991.

The genus includes 16 species, mainly distributed in the Mediterranean and partly in Chinese subregions of the Palaearctic region, some species are known from the Oriental region. 12 species are known from the Palaearctic; 1 species is in the Three Provinces of Northeast China.

Subgenus ***Dryobotodes*** Warren, 1910 卓夜蛾亚属

Dryobotodes Warren, 1910, *in* A.Seitz (ed.), *Die Gross-Schmetterlinge der Erde*, 3: 133. Type species: *Noctua protea* [Denis & Schiffermüller], 1775 [Austria] (=*Noctua eremita* Fabricius, 1775).

***Dryobotodes* (*Dryobotodes*) *pryeri* (Leech, 1900) 天目卓夜蛾**

Eurois pryeri Leech, 1900, *Transactions of the Entomological Society of London*, 1900: 92. Type locality: Japan.

Synonymy: *Polia aino* Matsumura, 1926; *Eumichtis praetermissa* Draudt, 1950.

Distribution. China (Heilongjiang, Jilin, Shaanxi, Zhejiang); Russia (Siberia), Korea, Japan.

寄主：蒙古栎（*Quercus mongolica*）、垂桤木（*Alnus pendula*）（Park *et al.*, 2006; Matov & Kononenko, 2012）。

注：《中国动物志》（夜蛾科）中将本种归入“*Blepharita* Hampson, 1907 毛眼夜蛾属”，在此予以变更。

Genus ***Polymixis*** Hübner, [1820] 1816 展冬夜蛾属

Polymixis Hübner, [1820] 1816, *Verzeichniss bekannter Schmettlinge* [sic!]: 205. Type species: *Phalaena polymita* Linnaeus, 1761 [Sweden].

Synonymy: *Polia* Hübner, [1806], rejected name; suppr. (ICZN, 1926 Op. 97; 1954 Op. 278); *Eumichtis* Hübner, [1821] 1816 (subgen.); *Epunda* Duponchel, 1845; *Eumichthis*: Agassiz, 1846, emend.; *Pseudopolia* Turati, 1924; *Myxinia* Berio, 1985 (subgen.); *Parabrachionycha* Hacker, 1990 (subgen.); *Eremophysa* Boursin, 1958 (subgen.); *Brandticola* Hacker & Ronkay, 1993 (subgen.); *Bousinixis* Hacker & Ronkay, 1993 (subgen.); *Bischoffia* Hacker & Ronkay, 1993 (subgen.); *Simplitype* Berio, 1980 (subgen.); *Serpmixis* Beck, [1992] 1991 (subgen.); *Xanthomixis* Beck, 1996.

The genus includes about 85 species, mainly distributed in the Mediterranean, Central Asian and Sino-Himalayan subregions of the Palaearctic region; several species of uncertain position referred in the literature from the Neotropical region require revision; 67 species are known in the Palaearctic region, among them 1 species of 1 subgenus in the Three Provinces of Northeast China.

Subgenus ***Myxinia*** Berio, 1985 粘冬夜蛾亚属

Myxinia Berio, 1985, *Fauna d'Italia*, *Lepidoptera*, *Noctuidae*: 867. Type species: *Noctua rufocincta* Geyer, 1828 [Italy].

The subgenus includes 12 species, distributed mainly in the Mediterranean and Central Asian subregions of the Palaearctic region; 1 species is in the Three Provinces of Northeast China.

Polymixis* (*Myxinia*) *mandschurica **Boursin, 1970** 东北展冬夜蛾

Polymixis mandschurica Boursin, 1970, *Entomops*, 18: 51, fig. 64, 65. Type locality: [North East China] Manchuria, Erzendianzy.

Distribution. China (Heilongjiang); Russia (Southwest of Primorye terr.), North Korea.

注：首次给出中文种名。

Genus ***Blepharita*** Hampson, 1907 毛眼夜蛾属

Blepharita Hampson, 1907, *Annals and Magazine of Natural History*, (7) 19: 251. Type species: *Hadena amica* Treitschke, 1825 [Russia].

Monotypic Palaearctic genus. 1 species is in the Three Provinces of Northeast China.

Blepharita amica **(Treitschke, 1825)** 毛眼夜蛾

Hadena amica Treitschke, 1825, *Die Schmetterlinge von Europa*, 5 (1): 332. Type locality: "Podevin".

Synonymy: *Blepharita amica ussuriensis* Sheljuzhko, 1919 (ssp.).

Distribution. China (Heilongjiang, Jilin, Liaoning); Russia (European part, Siberia), Korea, Japan, Europe.

寄主：乌头属、稠李（陈一心，1999）；繁缕（*Stellaria media*）、（*Aconitum septentrionale*）、欧洲甜樱桃（*Cerasus avium*）、稠李（*Padus racemosa*）、羽扇豆（*Lupinus polyphyllus*）、（*Heracleum sphondylium*）、欧防风（*Pastinaca sativa*）、（*Lactuca muralis*）、莴苣（*Lactuca sativa*）、鹿根（*Rhaponticum carthamoides*）、蒲公英（*Taraxacum vulgare*）、李属（*Prunus*）、忍冬属（*Lonicera*）、牛蒡属（*Arctium*）、蒿属（*Artemisia*）、禾本科（Poaceae）植物（Matov & Kononenko, 2012）。

注：指名亚种 *Blepharita amica amica* 分布于欧洲至俄罗斯西伯利亚南部；乌苏里亚种 *B. amica ussuriensis* 分布在我国东北地区、俄罗斯东西伯利亚南部、朝鲜、韩国和日本。

Genus *Mniotype* Franclemont, 1941 拟毛眼夜蛾属

Mniotype Franclemont, 1941, *Entomological News*, 52: 201. Type species: *Hadena ducta* Grote, 1878 [USA].

Synonymy: *Ablephica* Berio, 1985; *Blepharamia* Berio, 1980; *Pseudomniotype* Beck, [1992] 1991; *Eremobastis* Pérez-Lopez & Morente-Benítez, 1996; *Hadena* auct.; *Crino* auct.; *Blepharita* auct..

The Holarctic genus contains about 80 species, distributed in the temperate Palaearctic, Nearctic and mountains of the Oriental regions; about 40 species are known from the Palaearctic region, among them 4 species are in the Three Provinces of Northeast China.

Mniotype satura ([Denis & Schiffermüller], 1775) 樱拟毛眼夜蛾

Noctua satura [Denis & Schiffermüller], 1775, *Ankündung eines Systematisches Werkes von den Schmetterlinge der Wiener Gegend. Wien*: 83. Type locality: Austria, Vienna.

Synonymy: *Phalaena porphyrea* Esper, 1789; *Blepharita amicissima* Bryk, 1948.

Distribution. China (Heilongjiang, Jilin, Xinjiang); Russia (European part, Siberia), Korea, Japan; Europe.

寄主：葎草、忍冬属（陈一心，1999）；葎草（*Humulus scandens*）、忍冬属（*Lonicera*）植物（Park *et al.*, 2006）；叶升麻（*Actaea spicata*）、萹蓄 (*Polygonum aviculare*)、食用大黄（*Rheum rhaponticum*）、黑果越橘（*Vaccinium myrtillus*）、葎草（*Humulus scandens*）、（*Malus domestica*）、稠李（*Padus racemosa*）、刺李（*Prunus spinosa*）、秦氏悬钩子（*Rubus caesius*）、树莓（*Rubus fruticosus*）、覆盆子（*Rubus idaeus*）、紫叶景天（*Sedum telephium*）、牧地山黧豆（*Lathyrus pratensis*）、红车轴草（*Trifolium pratense*）、（*Lonicera bungeana*）、刚毛忍冬（*Lonicera hispida*）、小叶忍冬（*Lonicera microphilla*）、黑果忍冬（*Lonicera nigra*）、繁果忍冬（*Lonicera tatarica*）、欧银花（*Lonicera xylosteum*）、颠茄草（*Atropa belladonna*）、大车前（*Plantago major*）、刺儿菜（*Cirsium arvense*）、毛果一枝黄花（*Solidago virgaurea*）、苣荬菜（*Sonchus arvensis*）、（*Listera ovata*）、凤尾蕨属（*Pteris*）、耧斗菜属（*Aquilegia*）、翠雀属（*Delphinium*）、帚石楠属（*Calluna*）、欧石南属（*Erica*）、柳属（*Salix*）、椴树属（*Tilia*）、樱属（*Cerasus*）、山楂属（*Crataegus*）、蔷薇属（*Rosa*）、羽扇豆属（*Lupinus*）、蚕豆属（*Vicia*）、槭树属（*Acer*）、拉拉藤属（*Galium*）、女贞属（*Ligustrum*）、玄参属（*Scrophularia*）、山柳菊属（*Hieracium*）、蒲公英属（*Taraxacum*）、火烧兰属（*Epipactis*）、禾本科（Poaceae）植物（Matov & Kononenko, 2012）。

注：《中国动物志》（夜蛾科）中将本种归入“*Blepharita* Hampson, 1907 毛眼夜蛾属”，在此予以变更。

Mniotype melanodonta (Hampson, 1906) 紫褐拟毛眼夜蛾

Eumichtis melanodonta Hampson, 1906, *Catalogue of the Lepidoptera Phalaenae in the British Museum*, 6: 327, pl. 104: 6. Type locality: Japan, Oiwake.

Distribution. China (Heilongjiang); Russia (Siberia), Korea, Japan.

寄主：酸果蔓属（*Oxycoccus*）、越橘属（*Vaccinium*）植物（Park *et al.*, 2006; Matov & Kononenko, 2012）。

注：《中国动物志》（夜蛾科）中将本种归入“*Blepharita* Hampson, 1907 毛眼夜蛾属”，在此予以变更。

Mniotype adusta (Esper, 1790) 焦拟毛眼夜蛾

Phalaena (*Noctua*) *adusta* Esper, 1790, *Die Schmetterlinge in Abbildungen nach der Natur mit Beschreibungen*, 4 (2): 483, pl. 149: 1, 2. Type locality: Germany.

Synonymy: *Noctua duplex* Haworth, 1809; *Noctua valida* Hübner, [1813]; *Noctua vultarina* Freyer, 1832; *Hadena vultarina* Freyer, 1832; *Hadena chardinyi* Duponchel, [1838] 1836; *Noctua chardinyi* Duponchel, 1838; *Noctua picticollis* Zetterstedt, [1839] 1840; *Hadena pavida* Boisduval, 1840; *Noctua pavida* Boisduval, 1840; *Hadena baltica* O. Hering, 1846; *Mamestra vicina* Alphéraky, 1882 (preocc.); *Hadena adusta grisescens* Standfuss, 1893 (var.) (preocc.); *Hadena adusta moesta* Staudinger, 1897 (var.); *Crino aterrima* Cinstantini, 1916; *Crino aterrina* Constantini, 1916; *Hadena adusta carpathica* Kaucki, 1922 (var.); *Crino adusta lappona* Rangnow, 1935 (ssp.); *Mniotype adusta poltavskyi* Volynkin, Matov & Behounek, 2014 (ssp.).

Distribution. China (Heilongjiang, Xinjiang, Qinghai, Xizang); Russia (European part, Siberia), Korea, Japan, Kazakhstan; Asia Minor, Europe.

寄主：柳、杨梅、桤木属植物（陈一心，1999）；葡萄叶铁线莲（*Clematis vitalba*）、（*Silene inflata*）、狗筋麦瓶草（*Silene vulgaris*）、萹蓄 (*Polygonum aviculare*)、小酸模（*Rumex acetosella*）、欧洲桤木（*Alnus glutinosa*）、香杨梅（*Myrica gale*）、帚石楠（*Calluna vulgaris*）、黑果越橘（*Vaccinium myrtillus*）、笃斯越橘（*Vaccinium uliginosum*）、耳柳（*Salix aurita*）、黄花柳（*Salix caprea*）、紫叶景天（*Sedum telephium*）、单子山楂（*Crataegus monogyna*）、锐刺山楂（*Crataegus oxyacantha*）、覆盆子（*Rubus idaeus*）、（*Galium mollugo*）、（*Gentiana pannonica*）、中亚苦蒿（*Artemisia absinthium*）、荒野蒿（*Artemisia campestris*）、北艾（*Artemisia vulgaris*）、大矢车菊（*Centaurea scabiosa*）、山柳菊（*Hieracium umbellatum*）、欧洲千里光（*Senecio vulgaris*）、毛果一枝黄花（*Solidago virgaurea*）、苣荬菜（*Sonchus arvensis*）、菊蒿（*Tanacetum vulgare*）、罂粟属（*Papaver*）、石竹属（*Dianthus*）、滨藜属（*Atriplex*）、欧石南属（*Erica*）、木樨草属（*Reseda*）、蜀葵属（*Althaea*）、梨属（*Pyrus*）、花楸属（*Sorbus*）、柳叶菜属（*Epilobium*）、亚麻属（*Linum*）、蓝盆花属（*Scabiosa*）、车前属（*Plantago*）、野芝麻属（*Lamium*）、牛至属（*Origanum*）、蓍属（*Achillea*）、牛蒡属（*Arctium*）、莴苣属（*Lactuca*）、风毛菊属（*Leontodon*）、蒲公英属（*Taraxacum*）、伞形科（Apiaceae）植物（Matov & Kononenko, 2012）。

注：《中国动物志》（夜蛾科）中将本种归入“*Blepharita* Hampson, 1907 毛眼夜蛾属”，在此予以变更。

***Mniotype bathensis* (Lutzau, 1900) 长白拟毛眼夜蛾**

Crino bathensis Lutzau, 1900, *Entomologische Zeitschrift, Frankfurt am Main*, 14: 162. Type locality: Germany.

Synonymy: *Hadena adusta moesta* Staudinger, 1897 (var.); *Crino adusta urupino* Bryk, 1942 (ssp.); *Polia urupolia* Bryk, 1942; *Blepharita hoenei* Sugi, 1959.

Distribution. China (Heilongjiang, Jilin, Liaoning); Russia (European part, Siberia), Korea, Japan; Europe.

寄主：舟形乌头（*Aconitum napellus*）、甜菜（*Beta vulgaris*）、（*Alnus incana*）、黑果越橘（*Vaccinium myrtillus*）、紫叶景天（*Sedum telephium*）、三花龙胆（*Gentiana triflora*）、荒野蒿（*Artemisia campestris*）、刺儿菜（*Cirsium arvense*）、毛果一枝黄花（*Solidago virgaurea*）、苣荬菜（*Sonchus arvensis*）、桦木属（*Betula*）、帚石楠属（*Calluna*）、悬钩子属（*Rubus*）、拉拉藤属（*Galium*）、烟草属（*Nicotiana*）、毛地黄属（*Digitalis*）、车前属（*Plantago*）、多榔菊属（*Doronicum*）、山柳菊属（*Hieracium*）、蒲公英属（*Taraxacum*）植物（Matov & Kononenko, 2012）。

Tribe **Orthosiini** Guenée, 1837 梦尼夜蛾族

Orthosides Guenée, 1837, *Annales de la Société Entomologique de France*, 6: 224. Type genus: *Orthosia* Ochsenheimer, 1816.

Synonymy: Taeniocampinae Grote, 1882; Orthosiina Beck, 1996; Perigraphina Beck, 1996.

The tribe was placed in the subfamily Hadeninae (sensu auctorum) (see below). The tribe Orthosiini represented by species with early spring imagines activity. The moths fly in the Three Provinces of Northeast China from mid March to early May. The larvae of most species are mainly dendrophilous polyphags feeding on various deciduoud trees, often on *Populus* and *Salix*. The tribe represented in the Three Provinces of Northeast China 8 genera and 22 species, inhabiting mainly lowland and montane forested habitats.

References. Hampson, 1905 (Hadeninae, part); Sugi, 1955, 1982 (Hadeninae, part); Chen, 1982, 1999 (Noctuidae); Kononenko, 1989b, 1990 (Hadeninae, part); Kononenko, Mikkola, 1989 (Hadeninae, part); Kononenko, Ronkay, 1995 (Hadeninae, part); Kitching, Rawlins, 1998 (Hadeninae, part); Kononenko, 1990 (Hadeninae, part); Hreblay, 1996 (Hadeninae, part); Hacker *et al.*, 2002 (Hadeninae: Orthosiini); Kononenko, 2003 (Hadeninae, part); Kononenko, 2005, 2010 (Hadeninae: Orthosiini); Fibiger, Lafontaine, 2005 (Hadeninae: Orthosiini); Fibiger, Hacker, 2005 (Hadeninae: Orthosiini); Lafontaine, Fibiger, 2006 (Hadeninae: Orthosiini); Kononenko, Han, 2007 (Hadeninae: Orthosiini); Ronkay *et al.*, 2010 (Hadeninae, Orthosiini); Matov *et al.*, 2008 (Hadeninae: Orthosiini); Lafontaine, Schmidt, 2010 (Noctuinae: Orthosiini); Fibiger *et al.*, 2011 (Hadeninae: Orthosiini).

Genus ***Panolis*** Hübner, [1821] 1816 小眼夜蛾属

Panolis Hübner, [1821] 1816, *Verzeichniss bekannter Schmettlinge* [sic!]: 214. Type species: *Noctua flammea* [Denis & Schiffermüller], 1775 [Austria].

Synonymy: *Ilarus* Boisduval, 1828.

The genus includes about 5 species, distributed in the Palaearctic and Oriental regions, among them 1 species is in the Three Provinces of Northeast China.

***Panolis japonica* Draudt, 1935 松小眼夜蛾**

Panolis flammea japonica Draudt, 1935, *in* A.Seitz (ed.), *Die Gross-Schmetterlinge der Erde*, 3, suppl.: 199, pl. 22: h. Type locality: Japan, Kobe.

Synonymy: *Panolis flammea sutshana* Draudt, 1935 (f.); *Panolis flammea* auct., nec Denis et Schiffermüller, 1775.

Distribution. China (Heilongjiang, Jilin); Russia (Siberia), Korea, Japan.

寄主：欧洲赤松、樟子松（陈一心，1999）；赤松（*Pinus densiflora*）、红松（*Pinus koraiensis*）（Matov & Kononenko, 2012）。

Genus ***Clavipalpula*** Staudinger, 1892 **克夜蛾属**

Clavipalpula Staudinger, 1892, *in* N.M. Romanoff (ed.), *Mémoires sur les Lépidoptères*, 6: 499. Type species: *Taeniocampa aurariae* Oberthür, 1880 [Russia, Primorye terr.].

The genus contains 2 species, 1 species is in the Three Provinces of Northeast China.

***Clavipalpula aurariae* (Oberthür, 1880) 克夜蛾**

Taeniocampa aurariae Oberth ü r, 1880, *Études d'Entomologie*, 5: 76, pl. 3: 6. Type locality: Russia, [Primorye terr.] Askold I..

Synonymy: *Perigrapha pfennigschmidti* Höne, 1917.

Distribution. China (Heilongjiang, Jilin, Liaoning); Russia (Siberia), Korea, Japan.

寄主：麻栎（*Quercus acutissima*）、槲树（*Quercus dentata*）、青冈 (*Quercus glauca*)、枹栎（*Quercus serrata*）（Park *et al.*, 2006; Matov & Kononenko, 2012）。

Genus ***Xylopolia*** Sugi, 1982 **丽栖夜蛾属**

Xylopolia Sugi, 1982, *Moths of Japan*, 1: 707, 708; 2: 356, 462. Type species: *Lamprosticta bella* Butler, 1881 [Japan].

The genus contains 5 species, distributed in the Chinese subregion of the Palaearctic region. 1 species is in the Three Provinces of Northeast China.

***Xylopolia bella* (Butler, 1881) 丽栖夜蛾**

Lamprosticta bella Butler, 1881, *Transactions of the Entomological Society of London*, 1881: 175. Type locality: Japan, Yokohama.

Synonymy: *Polia abikonis* Matsumura, 1926; *Egira bella amamiensis* Kishida & Yoshimoto, 1979 (ssp.); *Xylopolia bella koreana* Kononenko & Ronkay, 1995 (ssp.).

Distribution. China (Heilongjiang, Taiwan); South Korea, Japan.

注：《中国动物志》（夜蛾科）中将本种归入“*Egira* Duponchel, [1845] 1844 栖夜蛾属”，在此予以变更。

Genus ***Orthosia*** Ochsenheimer, 1816 梦尼夜蛾属

Orthosia Ochsenheimer, 1816, *Die Schmetterlinge von Europa*, 4: 79. Type species: *Noctua instabilis* [Denis & Schiffermüller], 1775 [Austria].

Synonymy: *Orthoa Billberg*, 1820; *Graphiphora* Hübner, [1806], rejected name (ICZN Op. 97).; *Monima* Hübner, [1821] 1816 (subgen.); *Taeniocampa* Guenée, 1839; *Cuphanoa* Hübner, [1821] 1816; *Microrthosia* Berio, 1980; *Poporthosia* Beck, 1996; *Cororthosia* Berio, 1980 (subgen.); *Semiophora* Stephens, 1829 (subgen.); *Graphiphora* Hübner, [1806], suppr. (ICZN, 1926 Op. 97); *Ancata* Căpuse, 1958; *Anacta* Kristensen, 1966; *Cyphonoa* Agassiz, [1874], emend.; *Euchoristea* Warren, 1910 (subgen.); *Erythrotis* Bryk, 1948 (subgen.); *Euchorista* Poole, 1989, missp.; *Cororthosia* Berio 1980 (subgen.).

The genus contains about more than 80 species, distributed in the temperate Holarctic and the mountains of the Oriental regions; in the palaearctic about 55 species. Among them, 11 species of 5 subgenera are in the Three Provinces of Northeast China.

Subgenus ***Orthosia*** Ochsenheimer, 1816 梦尼夜蛾亚属

Orthosia **(*Orthosia*)** ***incerta*** **(Hufnagel, 1766) 梦尼夜蛾**

Phalaena incerta Hufnagel, 1766, *Berlinisches Magazine*, 3: 298. Type locality: Germany, Berlin.

Synonymy: *Noctua instabilis* [Denis et Schiffermüller], 1775; *Phalaena* (*Noctua*) *colinita* Esper, 1790; *Phalaena* (*Noctua*) *trigutta* Esper, 1790; *Phalaena* (*Noctua*) *contracta* Esper, 1790; *Bombyx subsetaceus* Haworth, 1803; *Bombyx angustus* Haworth, 1803; *Bombyx fuscatus* Haworth, 1803; *Bombyx nebulosus* Haworth, 1803; *Taeniocampa pallida* Lampa, 1885; *Monima incerta subcarnea* Warren, 1909 (f.); *Monima incerta olivacea* Warren, 1909 (f.).

Distribution. China (Heilongjiang, Jilin, Liaoning, Xinjiang); Russia (European part, Ural, Siberia), Mongolia, Korea, Japan, Turkey, Morocco; Central Asia, Middle East, Transcaucasia, Caucasus, Europe.

寄主：柳、杨、栎、山楂、榆等（陈一心，1999）；山楂树（*Crataegus pinnatifida*）、苹果（*Malus pumila*）、桃树（*Prunus persica*）、红车轴草（*Trifolium pratense*）、杨属（*Populus*）、柳属（*Salix*）、桦木属（*Betula*）、栎属（*Quercus*）、榆属（*Ulmus*）、椴树属（*Tilia*）植物（Park *et al.*, 2006）；七瓣莲（*Trollius europaea*）、异株蝇子草(*Silene dioica*)、拳参（*Polygonum bistorta*）、夏栎（*Quercus robur*）、灰桤木（*Alnus incana*）、垂枝桦（*Betula pendula*）、（*Betula pubescens*）、桦叶鹅耳枥

（*Carpinus betulus*）、胡榛（*Corylus avellana*）、帚石楠（*Calluna vulgaris*）、黑果越橘（*Vaccinium myrtillus*）、笃斯越橘（*Vaccinium uliginosum*）、越橘（*Vaccinium vitis-idaea*）、银白杨（*Populus alba*）、北美杨树（*Populus balsamifera*）、加杨（*Populus canadensis*）、钻天杨（*Populus nigra*）、欧洲山杨（*Populus tremula*）、白柳（*Salix alba*）、耳柳（*Salix aurita*）、黄花柳（*Salix caprea*）、五蕊柳（*Salix pentandra*）、东陵山柳（*Salix phylicifolia*）、蒿柳（*Salix viminalis*）、欧洲小叶椴（*Tilia cordata*）、啤酒花（*Humulus lupulus*）、高精细榆树（*Ulmus campestris*）、高山茶藨子（*Ribes alpinum*）、黑茶藨子（*Ribes nigrum*）、沼委陵菜（*Comarum palustre*）、长柄矮生栒子（*Cotoneaster dammeri*）、单子山楂（*Crataegus monogyna*）、锐刺山楂（*Crataegus oxyacantha*）、山楂（*Crataegus pinnatifida*）、（*Malus domestica*）、苹果（*Malus pumila*）、稠李（*Padus racemosa*）、白桃（*Persica vulgaris*）、欧洲李（*Prunus domestica*）、黑刺李（*Prunus spinosa*）、野玫瑰（*Rosa canina*）、覆盆子（*Rubus idaeus*）、欧洲花楸（*Sorbus aucuparia*）、千屈菜（*Lythrum salicaria*）、柳兰（*Epilobium angustifolium*）、红车轴草（*Trifolium pratense*）、欧鼠李（*Frangula alnus*）、（*Lonicera capfirolium*）、繁果忍冬（*Lonicera tatarica*）、欧银花（*Lonicera xylosteum*）、欧洲绣球荚蒾（*Viburnum opulus*）、棕矢车菊（*Centaurea jacea*）、加拿大一枝黄花（*Solidago canadensis*）、毛果一枝黄花（*Solidago virgaurea*）、鸭茅（*Dactylis glomerata*）、酸模属（*Rumex*）、山毛榉属（*Fagus*）、杜鹃花属（*Rhododendron*）、安息香属（*Styrax*）、杏属（*Armeniaca*）、腺肋花楸属（*Aronia*）、草莓属（*Fragaria*）、梨属（*Pyrus*）、槭树属（*Acer*）、孀草属（*Knautia*）、醉鱼草属（*Buddleja*）、牛蒡属（*Arctium*）、蒲公英属（*Taraxacum*）植物（Matov & Kononenko, 2012）。

***Orthosia* (*Orthosia*) *evanida* (Butler, 1879) 晰线梦尼夜蛾**

Taeniocampa evanida Butler, 1879, *Annals and Magazine of Natural History*, (5)4: 362. Type locality: Japan, Yokohama.

Distribution. China (Heilongjiang, Liaoning); Russia (Siberia), Korea, Japan.

寄主：麻栎（*Quercus acutissima*）、枹栎（*Quercus serrata*）、椴树属（*Tilia*）、榆属（*Ulmus*）植物（Matov & Kononenko, 2012）。

注：首次给出中文种名。

***Orthosia* (*Orthosia*) *lizetta* (Butler, 1878) 黑线梦尼夜蛾**

Taeniocampa lizetta Butler, 1878, *Annals and Magazine of Natural History*, (5)1: 167. Type locality: Japan, Yokohama.

Synonymy: *Monima japonica* Warren, 1910; *Monima constabilis* Wileman, 1911; *Monima nigrolinea* Matsumura, 1926.

Distribution. China (Heilongjiang, Liaoning); Russia (Siberia), Korea, Japan.

寄主：枹栎（*Quercus serrata*）、朴树（*Celtis sinensis*）、椴树属（*Tilia*）、榆属（*Ulmus*）、苹果属（*Malus*）植物（Matov & Kononenko, 2012）。

注：首次给出中文种名。

***Orthosia* (*Orthosia*) *ussuriana* Kononenko, 1988 北梦尼夜蛾**

Orthosia ussuriana Kononenko, 1988, *Annales Entomologici Fennici*, 54: 103-105. Type locality: Russia, Primorje, Gornotaezhnoe.

Distribution. China (Heilongjiang); Russia (Siberia).

注：首次给出中文种名。

***Orthosia* (*Orthosia*) *paromoea* (Hampson, 1905) 匀梦尼夜蛾**

Monima paromoea Hampson, 1905, *Catalogue of the Lepidoptera Phalaenae in the British Museum*, 5: 429, pl. 90: 26. Type locality: Japan, Yokohama.

Distribution. China (Heilongjiang); Russia (Siberia), Korea, Japan.

寄主：日本金缕梅（*Hamamelis japonica*）、麻栎（*Quercus acutissima*）、槲树（*Quercus dentata*）、青冈 (*Quercus glauca*)、柳栎（*Quercus salicina*）、李属（*Prunus*）植物（Matov & Kononenko, 2012）。

注：首次给出中文种名。

Subgenus *Cororthosia* Berio 1980 单梦尼夜蛾亚属

Cororthosia Berio 1980, *Annali del Museo Civico di Storia Naturale Giacomo Doria*, 83: 8, 17. Type species: *Noctua gracilis* [Denis & Schiffermüller], 1775 [Austria].

注：首次给出中文亚属名。

***Orthosia* (*Cororthosia*) *ella* (Butler, 1878) 雾单梦尼夜蛾**

Taeniocampa ella Butler, 1878, *Annals and Magazine of Natural History*, (5)1: 167. Type locality: Japan, Yokohama.

Distribution. China (Heilongjiang); Russia (Transbaikalia, Siberia), Korea, Japan.

寄主：白桃（*Persica vulgaris*）、大豆（*Glycine max*）、海滨山黧豆（*Lathyrus japonicus*）、魁蒿（*Artemisia princeps*）、柳属（*Salix*）、苹果属（*Malus*）、梨属（*Pyrus*）、悬钩子属（*Rubus*）、蒲公英属（*Taraxacum*）植物（Matov & Kononenko, 2012）。

注：首次给出中文种名。

Subgenus *Euchoristea* Warren, 1910 灵梦尼夜蛾亚属

Euchoristea Warren, 1910, *in* Seitz, *Grosschmett. Erde*, 3: 93. Type species: *Apamea limbata* Butler, 1879 [Japan].

注：首次给出中文亚属名。

Orthosia (_Euchoristea_) _limbata_ (Butler, 1879) 灵梦尼夜蛾

Apamea limbata Butler, 1879, *Annals and Magazine of Natural History*, (5)4:360. Type locality: Japan, Yokohama.

Synonymy: *Orthosia limbata himalaya* Hreblay & Ronkay, 1998 (ssp.).

Distribution. China (Heilongjiang, Jilin, Jiangsu, Hubei, Taiwan); Korea (Central), Japan.

寄主：麻栎（*Quercus acutissima*）、枹栎（*Quercus serrata*）、李属（*Prunus*）植物（Park *et al.*, 2006）。

Subgenus *Erythrotis* Bryk, 1949 思梦尼夜蛾亚属

Erythrotis Bryk, 1949, *Arkiv för Zoologi*, 41 (A) 1: 78. Type species: *Erythrotis cedermarki* Bryk, 1949 [North Korea].

注：首次给出中文亚属名。

Orthosia (_Erythrotis_) _cedermarki_ (Bryk, 1948) 思梦尼夜蛾

Erythrotis cedermarki Bryk, 1948, *Arkiv för Zoologi*, 41(A) 1: 78. Type locality: North Korea, near Kyeongseong, Hamgyeong bugdo, Shuotsu.

Distribution. China (Heilongjiang); Russia (Siberia), Korea, Japan.

Orthosia (_Erythrotis_) _carnipennis_ (Butler, 1878) 联梦尼夜蛾

Taeniocampa carnipennis Butler, 1878, *Annals and Magazine of Natural History*, (5)1: 167. Type locality: Japan, Yokohama.

Distribution. China (Heilongjiang, Jilin); Russia (Siberia), Korea, Japan.

寄主：麻栎（*Quercus acutissima*）、槲树（*Quercus dentata*）、青冈 (*Quercus glauca*)、枹栎（*Quercus serrata*）、朴树（*Celtis sinensis*）、大麻（*Cannabis sativa*）、苹果（*Malus pumila*）、白桃（*Persica vulgaris*）、椴树属（*Tilia*）、榆属（*Ulmus*）、山楂属（*Crataegus*）、梨属（*Pyrus*）植物（Matov & Kononenko, 2012）。

Subgenus *Semiophora* Stephens, 1829 歌梦尼夜蛾亚属

Semiophora Stephens, 1829, *Illustrations of British Entomology*, 2: 138. Type species: *Phalaena gothica* Linnaeus, 1758 [Europe].

Synonymy: *Graphiphora* Hübner, [1806], rejected name; *Ancata* Căpuse, 1858.

注：首次给出中文亚属名。

***Orthosia* (*Semiophora*) *askoldensis* (Staudinger, 1892) 北歌梦尼夜蛾**

Taeniocampa gothica askoldensis Staudinger, 1892 (var.), *in* N.M. Romanoff (ed.), *Mémoires sur les Lépidoptères* 6: 496. Type locality: Russia, [Primorye terr.], Askold I., "Ussuri".

Synonymy: *Orthosia askoldensis turpika* Hreblay, 1991 (ssp.); *Orthosia gothica* auct., nec Linnaeus, 1758.

Distribution. China (Heilongjiang); Russia (Siberia), Korea, Japan.

寄主：蒙古栎（*Quercus mongolica*）、稠李（*Padus racemosa*）、山楂属（*Crataegus*）、梨属（*Pyrus*）、苹果属（*Malus*）植物（Matov & Kononenko, 2012）。

注：首次给出中文种名。

Subgenus *incertae sedis* 隶属的亚属未明确

***Orthosia odiosa* (Butler, 1878) 棕梦尼夜蛾**

Agrotis odiosa Butler, 1878, *Annals and Magazine of Natural History*, (5)1: 162. Type locality: Japan, Yokohama.

Distribution. China (Heilongjiang); Russia (Siberia), Korea, Japan.

寄主：青冈 (*Quercus glauca*)、柿（*Diospyros kaki*）、朴树（*Celtis sinensis*）、椴树属（*Tilia*）、苹果属（*Malus*）、山楂属（*Crataegus*）植物（Matov & Kononenko, 2012）。

注：首次给出中文种名。

***Orthosia coniortota* (Filipjev, 1927) 尘梦尼夜蛾**

Monima coniortota Filipjev, 1927, *Annuaire du Musée Zoologique de l'Académie des Sciences d'URSS*, 28: 251. Type locality: Russia, Primorye terr., Tigrovoi.

Distribution. China (Jilin); Russia (Siberia), Korea, Japan.

寄主：华东椴（*Tilia japonica*）、栎属（*Quercus*）、榆属（*Ulmus*）、山楂属（*Crataegus*）植物（Matov & Kononenko, 2012）。

Genus ***Anorthoa*** Berio, 1980 黄灰梦尼夜蛾属

Anorthoa Berio, 1980, *Annali del Museo Civico di Storia Naturale Giacomo Doria*, 83: 8, 17. Type species: *Noctua munda* [Denis & Schiffermüller], 1775 [Austria].

The genus contains 8 species, distributed in the Palaearctic and the mountains of the Oriental regions. Among them, 2 species are in the Three Provinces of Northeast China.

***Anorthoa munda* ([Denis & Schiffermüller], 1775) 黄灰梦尼夜蛾**

Noctua munda [Denis & Schiffermüller], 1775, *Ankündung eines Systematisches Werkes von den*

Schmetterlinge der Wiener Gegend. Wien: 76. Type locality: Austria, Vienna.

Synonymy: *Phalaena* (*Noctua*) *gemina* Borkhausen, 1792; *Bombyx geminatus* Haworth, 1803; *Bombyx bimaculatus* Haworth, 1803; *Perigrapha* (*Anortoa*) *munda plumbeata* Hreblay & Ronkay, 1998 (ssp.).

Distribution. China (Heilongjiang, Jilin, Nei Mongol); Russia (European part, Siberia), Mongolia, Korea, Japan; Central Asia, Middle East, Turkey, Europe, Morocco.

寄主：栎、榆、杨、李、柳等（陈一心，1999）；虎杖（*Polygonum cuspidatum*）、垂枝欧洲山毛榉（*Fagus sylvatica*）、麻栎（*Quercus acutissima*）、槲树（*Quercus dentata*）、枹栎（*Quercus serrata*）、欧洲桤木（*Alnus glutinosa*）、桦叶鹅耳枥（*Carpinus betulus*）、胡榛（*Corylus avellana*）、银白杨（*Populus alba*）、钻天杨（*Populus nigra*）、欧洲山杨（*Populus tremula*）、黄花柳（*Salix caprea*）、榆（*Ulmus pumila*）、啤酒花（*Humulus lupulus*）、葎草（*Humulus scandens*）、野草莓（*Fragaria vesca*）、稠李（*Padus racemosa*）、欧洲李（*Prunus domestica*）、黑樱桃木（*Prunus serotina*）、黑刺李（*Prunus spinosa*）、树莓（*Rubus fruticosus*）、覆盆子（*Rubus idaeus*）、（*Lonicera periclymenum*）、欧银花（*Lonicera xylosteum*）、欧洲白蜡（*Fraxinus excelsior*）、欧丁香（*Syringa vulgaris*）、桦木属（*Betula*）、椴树属（*Tilia*）、山楂属（*Crataegus*）、苹果属（*Malus*）、蔷薇属（*Rosa*）、合欢属（*Albizzia*）、蜜蜂花属（*Melissa*）、红门兰属（*Orchis*）植物（Matov & Kononenko, 2012）。

注：《中国动物志》（夜蛾科）中将本种归入“*Orthosia* Ochsenheimer, 1816 梦尼夜蛾属”，在此予以变更。

Anorthoa angustipennis **(Matsumura, 1926) 异黄灰梦尼夜蛾**

Monima angustipennis Matsumura, 1926, *Insecta Matsumurana*, 1: 12, pl. 1: 24. Type locality: Japan, Hokkaido.

Synonymy: *Monima fluvilinea* Matsumura, 1926.

Distribution. China (Heilongjiang); Russia (Siberia), Korea, Japan.

寄主：栎属（*Quercus*）、椴树属（*Tilia*）、榆属（*Ulmus*）、山楂属（*Crataegus*）、苹果属（*Malus*）植物（Matov & Kononenko, 2012）。

注：首次给出中文种名。

Genus ***Harutaeographa*** Yoshimoto, 1993 粗线夜蛾属

Harutaeographa Yoshimoto, 1993, *Moths of Nepal*, 2: 127. Type species: *Hadena fasciculata* Hampson, 1894 [North India, Sikkim].

The genus contains about 40 species, distributed in the temperate Palaearctic and the mountains of the Oriental regions. Among them, 1 species is in the Three Provinces of Northeast China.

注：首次给出中文属名。

Harutaeographa stenoptera **(Staudinger, 1892) 棕粗线夜蛾**

Taeniocampa stenoptera Staudinger, 1892, *in* N.M. Romanoff (ed.), *Mémoires sur les Lépidoptères*, 6: 497, pl. 9: 6. Type locality: Russia, [Primorye terr.] "Ussuri".

Distribution. China (Heilongjiang, Liaoning); Russia (Siberia), Korea, Japan.

注：首次给出中文种名。

Genus *Perigrapha* Lederer, 1857 连环夜蛾属

Perigrapha Lederer, 1857, *Die Noctuinen Europa's, mit Zuziehung einger bisher mesit dazu gezahlter Arten des Asiatischen Russland's, Kleinasien's, Syrien's, und Labrador's*: 38, 136. Type species: *Noctua i-cinctum* [Denis & Schiffermüller], 1775 [Austria].

Synonymy: *Harutaeographa* Yoshimoto, 1993 (subgen.); *Opacographa* Hreblay, 1996 (subgen.); *Rororthosia* Beck, 1999 (subgen.).

The genus contains 52 species, distributed in the temperate Palaearctic and in the mountains of the Oriental region. Among them 3 species are in the Three Provinces of Northeast China.

Perigrapha circumducta (Lederer, 1855) 围连环夜蛾

Orthosia circumducta Lederer, 1855, *Verhandlungen des Zoologisch-Botanischen vereins in Wien*, 5: 111, pl. 1: 9. Type locality: Kazakhstan, West Altai.

Synonymy: *Perigrapha circumducta pallescens* Draudt, 1934 (f.); *Perigrapha circumducta irkuta* Draudt, 1934 (f.).

Distribution. China (Jilin, Nei Mongol, Xinjiang, Shanxi); Russia (European part, Ural, Siberia), Mongolia, Korea, Japan; Central Asia, Europe.

寄主：（*Spiraea pinnatifidum*）、草莓属（*Fragaria*）、车前属（*Plantago*）、莴苣属（Lactuca）、蒲公英属（*Taraxacum*）植物（Matov & Kononenko, 2012）。

Perigrapha extincta Kononenko, 1989 狭缝连环夜蛾

Perigrapha extincta Kononenko, 1989, *Annales Entomologici Fennici*, 55: 79, figs 1, 3. Type locality: Russia, Primorye terr., Gornotaezhnoe.

Distribution. China (Heilongjiang); Russia (Siberia), Korea.

Perigrapha hoenei Püngeler, 1914 扁连环夜蛾

Perigrapha hoenei Püngeler, 1914, *Deutsche Entomologische Zeitschrift Iris*, 28: 40, pl. 3: 24. Type locality: Japan, Asamayama.

Synonymy: *Perigrapha sugitanii* Matsumura, 1926; *Perigramma* [sic!] *triangulifera* Warren, 1915; *Perigrapha hönei*: O. Bang-Haas, 1927, emend..

Distribution. China (Heilongjiang, Jilin, Liaoning); Russia (Siberia), Korea, Japan.

寄主：日本落叶松（*Larix leptolepis*）、麻栎（*Quercus acutissima*）、枹栎（*Quercus serrata*）、李属（*Prunus*）植物（Park *et al.*, 2006; Matov & Kononenko, 2012）。

Genus *Pseudopanolis* Inaba, 1927 伪小眼夜蛾属

Pseudopanolis Inaba, 1927, *Proceedings of the Imperial Academy, Tokyo*, 3: 449. Type species: *Pseudopanolis takao* Inaba, 1927 [Japan].

The genus contains 9 species, distributed in the North Africa, Eastern Palaearctic and mountains of the Oriental regions. Among them, 1 species is in the Three Provinces of Northeast China.

Pseudopanolis heterogyna (Bang–Haas, 1927) 长线伪小眼夜蛾

Aplecta (*Mamestra*) *heterogyna* Bang-Haas, 1927, *Horae Macrolepidopterologiae Regionis Palaearcticae*, 1: 85, pl. 10: 34. Type locality: Russia, [Primorye terr., Partizansk] "Sutschansk", Vladivostok.

Distribution. China (Heilongjiang, Jilin, Liaoning); Russia (Siberia), Korea, Japan.

寄主：红松（*Pinus koraiensis*）（Matov & Kononenko, 2012）。

Tribe **Hadenini** Guenée, 1837 盗夜蛾族

Hadenidi Guenée, 1837, *Annales de la Société Entomologique de France*, 6: 321; ibidem 7 (1838): 201. Type genus: *Hadena* Schrank, 1802.

Synonymy: Anartidae Heinemann, 1859; Polianae Hampson, 1902; Carpocapsidi Tutt, 1896; Mamestrinae Hampson, 1902; Clemathadina Beck, 1996; Conisaniina Beck, 1996; Coranartina Beck, 1996; Discestrina Beck, 1996; Pachetrina Beck, 1996; Poliina Beck, 1996; Hadina Beck, 1996.

This tribe in unit with tribes Orthosiini, Tholerini, Eriopygini and Leucaniini was traditionally treated by early authors as subfamily of the Hadeninae (sensu auctorum), however recently their status, was downgraded to tribes (Lafontaine & Schmidt, 2010) of the subfamily Noctuinae. The adults of Hadenini, Orthosiini, Tholerini, Eriopygini and Leucaniini have long-hairy eyes, which means that they have long hair-like setae between the many facets of the eyes. The larvae of most species are mainly polyphagous feeding on herbaceous Dicotyledont plants. Tribe Hadeninae is represented in the Three Provinces of Northeast China by 13 genera and 37 species occurring mainly in low and mid mountains forested and open habitats.

References. Hampson, 1906 (Hadeninae, part); McCabe, 1980 (Hadeninae, part); Chen, 1982, 1999 (Noctuidae: Hadeninae, part); Sugi, 1982 (Hadeninae, part); Kononenko, 1990 (Hadeninae, part), 2005, 2010 (Hadeninae: Hadenini); Varga, Ronkay, 1991 (Hadeninae, part); Kononenko, Spitzer, 1993 (Hadeninae, part); Kononenko *et al.*, 1998 (Hadeninae, part); Hacker, 1990, 1996, 1998 (Hadeninae, part); Behounek, 1993 (Hadeninae, part); Hacker *et al.*, 2002 (Hadeninae, Hadenini); Fibiger, Lafontaine, 2005 (Hadeninae: Hadenini); Fibiger, Hacker, 2005 (Hadeninae: Hadenini); Lafontaine, Fibiger, 2006 (Hadeninae: Hadenini); Matov *et al.*, 2008 (Hadeninae: Hadenini); Lafontaine, Schmidt, 2010 (Noctuinae: Hadenini); Fibiger *et al.*, 2011 (Hadeninae:

Hadenini).

Genus *Anarta* Ochsenheimer, 1816 窄眼夜蛾属

Anarta Ochsenheimer, 1816, *Die Schmetterlinge von Europa*, 4: 90. Type species: *Phalaena myrtilli* Linnaeus, 1761 [Sweden].

Synonymy: *Charellia* Sodoffsky, 1837; *Salacia* Boie, 1839, preocc. (Lamourox, 1816 [Coelenterata]); *Hadula* Staudinger, 1889 (subgen.); *Discestra* Hampson, 1905; *Aglossestra* Hampson, 1905; *Calocestra* Beck, 1961; *Cardiestra* Boursin, 1963; *Melanarta* Beck, 1991; *Calocestra* Beck, 1991 (subgen.); *Frifcestra* Beck, 2000; *Dianthcestra* Beck, 2000.

The genus contains 92 species, mainly distributed in the arid and montane areas, some species occurs in the subarctic. Among them, 2 species of 1 subgenus are in the Three Provinces of Northeast China.

Subgenus *Calocestra* Beck, 1991 歧夜蛾亚属

Calocestra Beck, 1991, *Atalanta, Würzburg*, 22 (2-4): 190. Type species: *Noctua pugnax* Hübner, [1824].

Anarta (*Calocestra*) *imperspicua* Hacker, 1998 北歧夜蛾

Anarta imperspicua Hacker, 1998, *Esperiana*, 6: 693, pl. VI: 109-111. Type locality: China, Xining.

Synonymy: *Hadula imperspicua mandschurica* Hacker, 1998 (ssp.).

Distribution. China (Liaoning, Qinghai); Russia (Altai, Transbaikalia); North Europe.

注：首次给出中文种名。

Anarta (*Calocestra*) *trifolii* (Hufnagel, 1766) 旋歧夜蛾

Phalaena trifolii Hufnagel, 1766, *Berlinisches Magazine*, 3: 398. Type locality: Germany, Berlin.

Synonymy: *Noctua chenopodii* [Denis et Schiffermüller], 1775; *Noctua verna* Esper, 1786; *Phalaena* (*Noctua*) *saucia* Esper, 1786; *Noctua infraina* Haworth, 1809; *Noctua contribulis* Duponchel, 1827; *Orthosia farkasii* Treitschke, 1835; *Hadena intermissa* Walker, 1857; *Hadena albifusa* Walker, 1857; *Apamea inquieta* Walker, 1857; *Apamea glaucovaria* Walker, 1860; *Mamestra trifolii major* Speter, 1875 (var.); *Mamestra canescens* Moore, 1878; *Scotogramma trifolii zermattensis* Draudt, 1943 (f.); *Cardepia taylori* Rothschild, 1913; *Scotogramma cinnamomina* Rothschild, 1913; *Scotogramma fruticosae* Dumont, 1925; *Noctua chenopodiata*: Villers, 1789.

Distribution. China (Heilongjiang, Jilin, Liaoning, Beijing, Xinjiang, Gansu, Ningxia, Qinghai, Hebei, Xizang); Russia (European part, Ural, Siberia, Kamchatka), Mongolia, Korea, Japan; Central Asia, Middle East, Europe, North America.

寄主：洋葱及多种草本植物（陈一心，1999）；肥皂草（*Saponaria officinalis*）、（*Silene inflata*）、

大爪草（*Spergula arvensis*）、凹头苋（*Amaranthus blitum*）、滨藜（*Atriplex halimus*）、（*Atriplex laciniatum*）、（*Atriplex laciniatum*）、（*Atriplex patulum*）、甜菜（*Beta vulgaris*）、藜（*Chenopodium album*）、（*Chenopodium bonus-henricus*）、红叶藜（*Chenopodium rubrum*）、菊叶香藜（*Chenopodium vulvaria*）、盐生草（*Halogeton glomeratus*）、（*Salsola fruticosa*）、俄罗斯蓟（*Salsola iberica*）、钾猪毛菜（*Salsola kali*）、菠菜（*Spinacia oleracea*）、萹蓄 (*Polygonum aviculare*)、波叶大黄（*Rheum undulatum*）、钝叶酸模（*Rumex obtusifolius*）、北美杨树（*Populus balsamifera*）、西瓜（*Citrullus vulgaris*）、甜瓜（*Cucumis melo*）、西葫芦（*Cucurbita pepo*）、岩生庭芥（*Alyssum saxatile*）、甘蓝（*Brassica oleracea*）、欧洲油菜（*Brassica napus*）、芜青（*Brassica rapa*）、密花独行菜（*Lepidium densiflorum*）、大蒜芥（*Sisymbrium altissimum*）、（*Ochradenus baccatus*）、大麻槿（*Hibiscus cannabinus*）、（*Malva nicaensis*）、大麻（*Cannabis sativa*）、大麻（*Cannabis sativa*）、蓖麻（*Ricinus communis*）、黑刺李（*Prunus spinosa*）、柳叶菜（*Epilobium hirsutum*）、花生（*Arachis hypogaea*）、金雀儿属（*Cytisus*）、大豆（*Glycine max*）、紫苜蓿（*Medicago sativa*）、白花草木樨（*Melilotus albus*）、黄香草木樨（*Melilotus officinalis*）、菜豆（*Phaseolus vulgaris*）、豌豆（*Pisum sativum*）、（*Prosopis farcta*）、红车轴草（*Trifolium pratense*）、广布野豌豆（*Vicia cracca*）、豇豆（*Vigna sinensis*）、亚麻（*Linum usitatissimum*）、骆驼蓬（*Peganum harmala*）、旱芹（*Apium graveolens*）、野胡萝卜（*Daucus carota*）、欧防风（*Pastinaca sativa*）、欧芹（*Petroselinum crispum*）、洋香菜（*Petroselinum sativum*）、番茄（*Lycopersicon esculentum*）、阳芋（*Solanum tuberosum*）、田旋花（*Convolvulus arvensis*）、毛蕊花（*Verbascum thapsus*）、（*Artemisia sieberi*）、野艾蒿（*Artemisia vulgaris*）、向日葵（*Helianthus annuus*）、莴苣（*Lactuca sativa*）、苦苣菜（*Sonchus oleraceus*）、孔雀草（*Tagetes patulus*）、百日菊（*Zinnia elegans*）、洋葱（*Allium cepa*）、蒜（*Allium sativum*）、石刁柏（*Asparagus officinalis*）、高粱（*Sorghum bicolor*）、玉米（*Zea mays*）、蒲公英属（*Taraxacum*）、马齿苋属（*Portulaca*）、石竹属（*Dianthus*）、棉属（*Gossypium*）、榆属（*Ulmus*）、苹果属（*Malus*）、老鹳草属（*Geranium*）、前胡属（*Peucedanum*）、烟草属（*Nicotiana*）、柳穿鱼属（*Linaria*）、车前草属（*Plantago*）、鼠尾草属（*Salvia*）、牛蒡属（*Arctium*）、紫菀属（*Aster*）、金盏花属（*Calendula*）、鸦葱属（*Scorzonera*）植物（Matov & Kononenko, 2012）。

Genus *Polia* Ochsenheimer, 1816 灰夜蛾属

Polia Ochsenheimer, 1816, *Die Schmetterlinge von Europa*, 4: 73. Type species: *Phalaena nebulosa* Hufnagel, 1766 [Germany].

Synonymy: *Chera* Hübner, [1821] 1816; *Polia* Boisduval, 1828, preocc. (Ochsenheimer, 1816 [Lepidoptera, Noctuidae]); *Aplecta* Guenée, 1852; *Anartodes* Culot, 1915; *Bombipolia* Beck, 1996; *Ripolia* Beck, 1996; *Antipolia* Beck, 1996.

The genus contains 24 species, mainly distributed in temperate zone, the mountains of the Palaearctic and Nearctic regions; 16 species are known from the Palaearctic. Among them, 6 species are in the Three Provinces of Northeast China.

***Polia bombycina* (Hufnagel, 1766) 蒙灰夜蛾**

Noctua bombycina Hufnagel, 1766, *Berlinisches Magazine* 3: 410. Type locality: Germany, Berlin.

Synonymy: *Noctua advena* [Denis et Schiffermüller], 1775; *Noctua nitens* Haworth, 1809; *Alyssia grisea* Butler, 1878; *Mamestra advena adjuncta* Staudinger, 1888 (var.); *Mamestra tetrica* Graeser, 1888 [1889]; *Mamestra advena mongolica* Staudinger, 1896 (var.); *Parastichtis sordida sachalinensis* Matsumura, 1931 (var.); *Aplecta advenina* Bryk, 1948; *Aplecta mongolica koreagena* Bryk, 1948 (ssp.); *Aplecta mongolica chidisana* Bryk, 1948 (ssp.); *Polia bombycina psammochroa* Varga, 1974 (ssp.).

Distribution. China (Heilongjiang, Jilin, Liaoning, Nei Mongol, Xinjiang, Qinghai, Hebei, Shandong); Russia (European part, Ural, Siberia), Mongolia, Korea, Japan; Central Asia, Near East, Europe.

寄主：苦苣菜、蓼、耆属植物（陈一心，1999）；葡萄叶铁线莲（*Clematis vitalba*）、萹蓄（*Polygonum aviculare*）、灰桤木（*Alnus incana*）、毛桦（*Betula pubescens*）、（*Hypericum quadrangulum*）、帚石楠（*Calluna vulgaris*）、杜香（*Ledum palustre*）、黑果越橘（*Vaccinium myrtillus*）、笃斯越橘（*Vaccinium uliginosum*）、白柳（*Salix alba*）、黄华柳（*Salix atrocinerea*）、黄花柳（*Salix caprea*）、啤酒花（*Humulus lupulus*）、旋果蚊子草（*Filipendula ulmaria*）、稠李（*Padus racemosa*）、秦氏悬钩子（*Rubus caesius*）、覆盆子（*Rubus idaeus*）、北美树莓（*Rubus odoratus*）、欧洲花楸（*Sorbus aucuparia*）、金雀儿（*Cytisus scoparius*）、草木樨（*Melilotus officinalis*）、匍匐芒柄草（*Ononis repens*）、刺芒柄花（*Ononis spinosa*）、红车轴草（*Trifolium pratense*）、欧洲红瑞木（*Cornus sanguinea*）、欧银花（*Lonicera xylosteum*）、林当归（*Angelica silvestris*）、野胡萝卜（*Daucus carota*）、虎耳草茴芹（*Pimpinella saxifraga*）、蓬子菜（*Galium verum*）、山石蚕（*Teucrium montanum*）、红辣椒千叶蓍（*Achillea millefolium*）、中亚苦蒿（*Artemisia absinthium*）、荒野蒿（*Artemisia campestris*）、野艾蒿（*Artemisia vulgaris*）、棕矢车菊（*Centaurea jacea*）、艾菊（*Chrysanthemum vulgare*）、菊芋（*Helianthus tuberosus*）、山柳菊（*Hieracium umbellatum*）、毛果一枝黄花（*Solidago virgaurea*）、菊蒿（*Tanacetum vulgare*）、野青茅（*Calamagrostis arundinacea*）、落叶松属（*Larix*）、酸模属（*Rumex*）、鹅耳枥属（*Carpinus*）、虎耳草属（*Saxifraga*）、羽衣草属（*Alchemilla*）、委陵菜属（*Potentilla*）、蔷薇属（*Rosa*）、染料木属（*Genista*）、蚕豆属（*Vicia*）、槭树属（*Acer*）、接骨木属（*Sambucus*）、前胡属（*Peucedanum*）、车前属（*Plantago*）、野芝麻属（*Lamium*）、百里香属（*Thymus*）、风铃草属（*Campanula*）、莴苣属（*Lactuca*）、苦苣菜属（*Sonchus*）、蒲公英属（*Taraxacum*）植物（Matov & Kononenko, 2012）。

***Polia hepatica* (Clerck, 1759) 窄灰夜蛾**

Phalaena hepatica Clerck, 1759, *Icones Isl. Rar.*: pl. 8: 3. Type locality: Europe.

Synonymy: *Phalaena* (*Noctua*) *trimaculosa* Esper, [1788]; *Phalaena* (*Noctua*) *tincta* Brahm, 1791; *Noctua argentina* Haworth, 1809; *Mamestra tincta obscurata* Staudinger, 1897 (var.).

Distribution. China (Heilongjiang, Jilin, Liaoning); Russia (European part, Ural, Siberia), Mongolia, Korea, Turkey, Kazakhstan; Europe.

寄主：萹蓄 (*Polygonum aviculare*)、垂枝桦（*Betula pendula*）、（*Betula pubescens*）、桦叶鹅耳枥（*Carpinus betulus*）、香杨梅（*Myrica gale*）、帚石楠（*Calluna vulgaris*）、黑果越橘（*Vaccinium myrtillus*）、笃斯越橘（*Vaccinium uliginosum*）、欧洲山杨（*Populus tremula*）、耳柳（*Salix aurita*）、黄花柳（*Salix caprea*）、秦氏悬钩子（*Rubus caesius*）、树莓（*Rubus fruticosus*）、覆盆子（*Rubus idaeus*）、匍匐芒柄草（*Ononis repens*）、刺芒柄花（*Ononis spinosa*）、林地水苏（*Stachys sylvatica*）、欧银花（*Lonicera xylosteum*）、毛果一枝黄花（*Solidago virgaurea*）、酸模属（*Rumex*）、桤木属（*Alnus*）、山楂属（*Crataegus*）、李属（*Prunus*）、蔷薇属（*Rosa*）、染料木属（*Genista*）、前胡属（*Peucedanum*）、车前属（*Plantago*）、野芝麻属（*Lamium*）、蒲公英属（*Taraxacum*）植物（Matov & Kononenko, 2012）。

注：首次给出中文种名。

Polia mortua **(Staudinger, 1888) 冥灰夜蛾**

Mamestra mortua Staudinger, 1888, *Entomologische Zeitung. Entomologischen Vereine zu Stettin*, 49: 249. Type locality: Russia, Primorye terr., Askold I..

Synonymy: *Mamestra nigerrima* Warren, 1888; *Mamestra afra* Graeser, 1888 [1889]; *Hadena kala* Swinhoe, 1900; *Polia szetschwana* Draeseke, 1928; *Polia persicariae minorita* Bryk, 1948 (ssp.).

Distribution. China (Heilongjiang, Jilin, Nei Mongol, Gansu, Sichuan, Guizhou, Yunnan, Xizang); Russia (Siberia), Korea, Japan, India, Nepal.

Polia nebulosa **(Hufnagel, 1766) 灰夜蛾**

Phalaena nebulosa Hufnagel, 1766, *Berlinisches Magazine*, 3: 418. Type locality: Germany, Berlin.

Synonymy: *Phalaena* (*Noctua*) *bimaculosa* Esper, 1788; *Phalaena thapsi* Brahm, 1791; *Phalaena grandis* Donovan, 1801; *Noctua plebeja* Hübner, [1803]; *Aplecta calabrica* Warren, 1910; *Aplecta nebulosa askolda* Oberthür, 1880 (var.); *Mamestra nebulosa lama* Staudinger, 1896 (var.).

Distribution. China (Heilongjiang, Jilin, Xinjiang, Qinghai, Gansu, Shanxi); Russia (European part, Ural, Siberia), Mongolia, Korea, Kazakhstan, Turkey; Europe.

寄主：桦、柳、榆属植物（陈一心，1999）；拳参（*Polygonum bistorta*）、帚石楠（*Calluna vulgaris*）、黑果越橘（*Vaccinium myrtillus*）、黄花九轮草（*Primula veris*）、欧洲报春（*Primula vulgaris*）、欧洲山杨（*Populus tremula*）、白柳（*Salix alba*）、耳柳（*Salix aurita*）、黄花柳（*Salix caprea*）、芸苔（*Brassica campestris*）、甘蓝（*Brassica oleracea*）、啤酒花（*Humulus lupulus*）、异株荨麻（*Urtica dioica*）、榆（*Ulmus pumila*）、羽衣草（*Alchemilla vulgaris*）、欧洲甜樱桃（*Cerasus avium*）、单子山楂（*Crataegus monogyna*）、旋果蚊子草（*Filipendula ulmaria*）、欧洲李（*Prunus domestica*）、黑刺李（*Prunus spinosa*）、秦氏悬钩子（*Rubus caesius*）、树莓（*Rubus fruticosus*）、覆盆子（*Rubus idaeus*）、紫苜蓿（*Medicago sativa*）、欧亚槭（*Acer peudoplatanus*）、轮花忍冬（*Lonicera caprifolia*）、圆盾状忍冬（*Lonicera periclymenum*）、欧银花（*Lonicera xylosteum*）、欧洲绣球荚蒾（*Viburnum opulus*）、西洋接骨木（*Sambucus nigra*）、金叶接骨木（*Sambucus racemosa*）、北方拉拉藤

(*Galium boreale*)、蓬子菜(*Galium verum*)、番茄(*Lycopersicon esculentum*)、马铃薯(*Solanum tuberosum*)、毛蕊花(*Verbascum thapsus*)、短柄野芝麻(*Lamium album*)、林地水苏(*Stachys silvatica*)、刺儿菜(*Cirsium arvense*)、山柳菊(*Hieracium umbellatum*)、(*Senecio fuchsii*)、毛果一枝黄花(*Solidago virgaurea*)、溪荪(*Iris nertschinskia*)、葱(*Allium fistulosum*)、燕麦(*Avena sativa*)、薏苡(*Coix lachryma-jobi*)、稻(*Oryza sativa*)、高粱(*Sorghum bicolor*)、小麦(*Triticum aestivum*)、玉米(*Zea mays*)、铁线莲属(*Clematis*)、酸模属(*Rumex*)、栎属(*Quercus*)、桤木属(*Alnus*)、桦木属(*Betula*)、金丝桃属(*Hypericum*)、欧石南属(*Erica*)、茶藨子属(*Ribes*)、苹果属(*Malus*)、蔷薇属(*Rosa*)、露珠草属(*Circaea*)、芒柄花属(*Ononis*)、草木樨属(*Melilotus*)、车轴草属(*Trifolium*)、梣属(*Fraxinus*)、女贞属(*Ligustrum*)、丁香属(*Syringa*)、玄参属(*Scrophularia*)、车前草属(*Plantago*)、羽衣草属(*Alchemilla*)、蒿属(*Artemisia*)、蒲公英属(*Taraxacum*)植物(Matov & Kononenko, 2012)。

***Polia goliath* (Oberthür, 1880) 鹏灰夜蛾**

Dichonia goliath Oberthür, 1880, *Études d'Entomologie*, 5: 68, pl. 6: 7. Type locality: Russia, Primorye terr., Askold I..

Distribution. China (Heilongjiang, Jilin, Gansu, Shanxi, Henan, Hubei, Sichuan); Russia (Siberia), Korea, Japan.

寄主：报春花属(*Primula*)、蚊子草属(*Filipendula*)、悬钩子属(*Rubus*)植物(Matov & Kononenko, 2012)。

***Polia vespertilio* (Draudt, 1934) 浅灰夜蛾**

Aplecta vespertilio Draudt, 1934, *in* A.Seitz (ed.), *Die Gross-Schmetterlinge der Erde* 3, suppl.: 109, pl. 14: i. Type locality: Siberia, [West Buryatia] "Irkutsk reg. ", East Sayan, Tunkinski Range.

Distribution. China (Heilongjiang); Russia (Ural, Siberia, Sakhalin), Mongolia, North Korea.

寄主：西伯利亚落叶松(*Larix sibirica*)(Matov & Kononenko, 2012)。

注：首次给出中文种名。

***Polia malchani* (Draudt, 1934) 俄灰夜蛾**

Aplecta malchani Draudt, 1934, *in* Seitz A. (ed.): *Die Gross–Schmetteriinge der Erde. Die Gross–Schmetterlinge des Palaearktischen Faunengebiet* 3, *Supplement*: 108, pl. 14: i. Type locality: Russia, Southwest Transbaikalia, Malchan Range.

Distribution. China (Heilongjiang); Russia (Ural, Altai, Sayan and Baikal area, Transbaikalia, Amur reg., Khabarovsk terr., North Sakhalin), Mongolia(Northern), North Korea.

寄主：西伯利亚落叶松(*Larix sibirica*)(Matov & Kononenko, 2012)。

注：首次给出中文种名。

Genus ***Lacanobia*** Billberg, 1820 安夜蛾属

Lacanobia Billberg, 1820, *Enumeratio Insectorum in Museo G.J. Billberg*: 87. Type species: *Phalaena w-latinum* Hufnagel, 1766 [Germany, Berlin].

Synonymy: *Diataraxia* Hübner, [1821] 1816 (subgen.); *Peucephila* Hampson, 1909; *Dianobia* Behounek, 1993 (subgen.); *Alinobia* Beck, 1996; *Contranobia* Beck, 1999.

The genus contains 26 species, distributed in the temperate zone of Holarctic region; 22 species of them are known from the Palaearctic region. Among them, 8 species of 3 subgenus are in the Three Provinces of Northeast China.

Subgenus ***Lacanobia*** Billberg, 1820 安夜蛾亚属

***Lacanobia* (*Lacanobia*) *dentata* (Kononenko, 1981) 齿安夜蛾**

Mamestra (*Lacanobia*) *dentata* Kononenko, 1981, *Procedings of Zoological Institute of the Academy of Sciences of the USSR*, 92: 95, fig. 6, 8. Type locality: Russia, Primorye terr., Barabash-Levada.

Distribution. China (Heilongjiang, Jilin); Russia (Southwest of the Primorye terr.), North Korea.

注：首次给出中文种名。

Subgenus ***Dianobia*** Behounek, 1993 横安夜蛾亚属

Dianobia Behounek, 1993, *Esperiana*, 3: 47. Type species: *Phalaena thalassina* Hufnagel, 1766 [Germany].

注：首次给出中文亚属名。

***Lacanobia* (*Dianobia*) *contrastata* (Bryk, 1942) 海安夜蛾**

Polia contrastata Bryk, 1942, *Deutsche Entomologische Zeitschrift Iris*, 56: 41. Type locality: Russia, Kuriles, Kunashir I., Tomari bay.

Synonymy: *Phalaena thalassina* auct., nec Hufnagel, 1766.

Distribution. China (Heilongjiang, Jilin, Nei Mongol, Xinjiang); Russia (Transbaikalia, Siberia), Korea, Japan.

寄主：桦、忍冬属、蓼属等植物（陈一心，1999）；小檗属（*Berberis*）、藜属（*Chenopodium*）、蓼属（*Polygonum*）、悬钩子属（*Rubus*）、忍冬属（*Lonicera*）、车前属（*Plantago*）、金盏花属（*Calendula*）、蒲公英属（*Taraxacum*）植物（Matov & Kononenko, 2012）。

注：《中国动物志》（夜蛾科）中引用本种同物异名“*Lacanobia thalassina* (Hufnagel, 1766)”，在此予以变更。

***Lacanobia* (*Dianobia*) *mongolica* Behounek, 1992 蒙安夜蛾**

Lacanobia mongolica Behounek, 1992, *Esperiana*, 3: 49, pl. B: 14-16, gen. 10. Type locality: Mongolia.

Distribution. China (Jilin); Russia (Transbaikalia, Siberia), Mongolia.

***Lacanobia* (*Dianobia*) *contigua* ([Denis & Schiffermüller], 1775) 桦安夜蛾**

Noctua contigua [Denis & Schiffermüller], 1775, *Ankündung eines Systematisches Werkes von den Schmetterlinge der Wiener Gegend. Wien*: 82. Type locality: Austria, Vienna.

Synonymy: *Phalaena histrio* Goeze, 1781; *Phalaena* (*Noctua*) *ariae* Esper, 1786; *Noctua spartii* Brahm, 1791; *Noctua dives* Haworth, 1809; *Noctua dives pulchellina* Haworth, 1809 (var.); *Hadena subcontigua* Eversmann, 1852; *Polia contigua amurensis* Spuler, 1908 (var.); *Mamestra contigua decolor* Bang-Haas, 1912 (var.); *Polia contigua spuleri* Wnukowsky, 1929 (var.); *Polia contigua olivaecula* Bryk, 1949 (ssp.); *Mamestra contigua griseomontana* Hartig, 1970 (ssp.).

Distribution. China (Heilongjiang, Jilin, Liaoning, Nei Mongol, Xinjiang, Shandong); Russia (Ural, Siberia, Altai), Mongolia, Korea, Japan, Turkey; Central Asia, Europe.

寄主：桦、栎、一枝黄花属植物（陈一心，1999）；欧洲蕨（*Pteridium aquilinum*）、黄花耧斗菜（*Aquilegia chrysantha*）、欧亚蝇子草（*Silene nutans*）、菠菜（*Spinacia oleracea*）、两栖蓼（*Polygonum amphibium*）、萹蓄（*Polygonum aviculare*）、酸模叶蓼（*Polygonum lapathifolium*）、小酸模（*Rumex acetosella*）、灰桤木（*Alnus incana*）、桦（*Betula alba*）、垂枝桦（*Betula pendula*）、（*Betula pubescens*）、胡榛（*Corylus avellana*）、香杨梅（*Myrica gale*）、胡桃（*Juglans regia*）、帚石楠（*Calluna vulgaris*）、黑果越橘（*Vaccinium myrtillus*）、笃斯越橘（*Vaccinium uliginosum*）、越橘（*Vaccinium vitis-idaea*）、毛黄连花（*Lysimachia vulgaris*）、北美杨树（*Populus balsamifera*）、欧洲山杨（*Populus tremula*）、耳柳（*Salix aurita*）、黄花柳（*Salix caprea*）、五蕊柳（*Salix pentandra*）、东陵山柳（*Salix phylicifolia*）、甘蓝（*Brassica oleracea*）、异株荨麻（*Urtica dioica*）、胭脂红景天（*Sedum spurium*）、旋果蚊子草（*Filipendula ulmaria*）、稠李（*Padus racemosa*）、黑刺李（*Prunus spinosa*）、秦氏悬钩子（*Rubus caesius*）、树莓（*Rubus fruticosus*）、覆盆子（*Rubus idaeus*）、欧洲花楸（*Sorbus aucuparia*）、金雀花（*Genista tinctoria*）、（*Pisum odoratum*）、豌豆（*Pisum sativum*）、荆豆（*Ulex europaeus*）、莳萝（*Anethum graveolens*）、林当归（*Angelica silvestris*）、野胡萝卜（*Daucus carota*）、欧洲前胡（*Peucedanum officinale*）、沼地前胡（*Peucedanum palustre*）、欧银花（*Lonicera xylosteum*）、欧卫矛（*Euonymus europaeus*）、欧鼠李（*Frangula alnus*）、蓬子菜（*Galium verum*）、龙胆草（*Gentiana cruciata*）、（*Gentiana pannonica*）、欧洲女贞（*Ligustrum vulgare*）、欧洲柳穿鱼（*Linaria vulgaris*）、林生玄参（*Scrophularia nodosa*）、车前（*Plantago asiatica*）、短柄野芝麻（*Lamium album*）、圆叶风铃草（*Campanula rotundifolia*）、毛头牛蒡（*Arctium tomentosum*）、（*Alchemill vulgaris*）、北艾（*Artemisia vulgaris*）、金盏花（*Calendula officinalis*）、刺儿菜（*Cirsium arvense*）、山莴苣（*Lactuca sibirica*）、新疆千里（*Senecio jacobaea*）、毛果一枝黄花（*Solidago virgaurea*）、阿尔泰蒲公英（*Taraxacum altaicum*）、药用蒲公英（*Taraxacum officinale*）、翠雀属（*Delphinum*）、小檗属（*Berberis*）、滨藜属（*Atriplex*）、藜属（*Chenopodium*）、栎属（*Quercus*）、茶藨子属（*Ribes*）、蔷薇属（*Rosa*）、金雀儿属（*Cytisus*）、羽扇豆属（*Lupinus*）、独活属（*Heracleum*）、蓝盆花属（*Scabiosa*）、梣属（*Fraxinus*）、矢车菊属（*Centaurea*）、泽兰属（*Eupatorium*）、向日葵属（*Helianthus*）、旋覆花属（*Inula*）、疆菊属

（*Tanacetum*）、葱属（*Allium*）植物（Matov & Kononenko, 2012）。

Lacanobia* (*Dianobia*) *suasa **([Denis & Schiffermüller], 1775) 俗安夜蛾**

Noctua suasa [Denis & Schiffermüller], 1775, *Ankündung eines Systematisches Werkes von den Schmetterlinge der Wiener Gegend. Wien*: 83. Type locality: Austria, Vienna.

Synonymy: *Phalaena* (*Noctua*) *dissimilis* Knoch, 1781; *Phalaena* (*Noctua*) *leucographa* Esper, [1790]; *Noctua denscanis* Haworth, 1809; *Noctua permixta* Geyer, [1833]; *Mamestra aliena* Duponchel, 1836, nec Hübner, [1809]; *Mamestra confluens* Eversmann, 1844; *Hadena pavida* Boisduval, 1840; *Mamestra dissimilis extincta* Staudinger, 1892 (var.); *Mamestra dissimilis laeta* Reuter, 1893 (var.); *Mamestra dissimilis turanica* Spuler, 1908 (ssp.).

Distribution. China (Heilongjiang, Jilin, Liaoning, Xinjiang); Russia (European part, Ural, Siberia), Mongolia, Korea, Japan, Turkey; Central Asia, Europe.

寄主：酸模、藜、车前等属植物（陈一心，1999；（*Atriplex laciniatum*）、甜菜（*Beta vulgaris*）、藜（*Chenopodium album*）、荞麦（*Fagopyrum esculentum*）、萹蓄（*Polygonum aviculare*）、酸模（*Rumex acetosa*）、卵叶酸模（*Rumex confertus*）、钝叶酸模（*Rumex obtusifolius*）、灰桤木（*Alnus incana*）、胡榛（*Corylus avellana*）、毛黄连花（*Lysimachia vulgaris*）、北美杨树（*Populus balsamifera*）、钻天杨（*Populus nigra*）、黄花柳（*Salix caprea*）、西瓜（*Citrullus vulgaris*）、甜瓜（*Cucumis melo*）、瓠果（*Cucumis sativus*）、西葫芦（*Cucurbita pepo*）、芸苔（*Brassica campestris*）、甘蓝（*Brassica oleracea*）、桑（*Morus alba*）、黑茶藨子（*Ribes nigrum*）、杏（*Armeniaca vulgaris*）、欧洲甜樱桃（*Cerasus avium*）、（*Malus domestica*）、稠李（*Padus racemosa*）、欧洲李（*Prunus domestica*）、覆盆子（*Rubus idaeus*）、欧洲花楸（*Sorbus aucuparia*）、大豆（*Glycine max*）、紫苜蓿（*Medicago sativa*）、豌豆（*Pisum sativum*）、红车轴草（*Trifolium pratense*）、林当归（*Angelica silvestris*）、旱芹（*Apium graveolens*）、野胡萝卜（*Daucus carota*）、欧防风（*Pastinaca sativa*）、洋香菜（*Petroselinum sativum*）、番茄（*Lycopersicon esculentum*）、烟草（*Nicotiana tabacum*）、马铃薯（*Solanum tuberosum*）、田旋花（*Convolvulus arvensis*）、金盏花（*Calendula officinalis*）、刺儿菜（*Cirsium arvense*）、向日葵（*Helianthus annuus*）、莴苣（*Lactuca sativa*）、燕麦（*Avena sativa*）、黍（*Panicum miliaceum*）、芦苇（*Phragmites australis*）、玉米（*Zea mays*）、落叶松属（*Larix*）、罂粟属（*Papaver*）、桦木属（*Betula*）、仙客来属（*Cyclamen*）、报春花属（*Primula*）、堇菜属（*Viola*）、萝卜属（*Raphanus*）、椴树属（*Tilia*）、锦葵属（*Malva*）、棉属（*Gossypium*）、百脉根属（*Lotus*）、羽扇豆属（*Lupinus*）、草木樨属（*Melilotus*）、芒柄花属（*Ononis*）、亚麻属（*Linum*）、缬草属（*Valeriana*）、拉拉藤属（*Galium*）、车前属（*Plantago*）、鼠尾草属（*Salvia*）、山柳菊属（*Hieracium*）、千里光属（*Senecio*）、苦苣菜属（*Sonchus*）、蒲公英属（*Taraxacum*）、葱属（*Allium*）、梯牧草属（*Phleum*）、小麦属（*Triticum*）植物（Matov & Kononenko, 2012）。

Subgenus ***Diataraxia*** Hübner, [1821] 1816 **华安夜蛾亚属**

Diataraxia Hübner, [1821] 1816, *Verzeichniss bekannter Schmettlinge* [sic!]: 219. Type species: *Noctua splendens* Hübner, [1808] [Europe].

***Lacanobia* (*Diataraxia*) *oleracea* (Linnaeus, 1758) 草安夜蛾**

Phalaena oleracea Linnaeus, 1758, *Systema Naturae* (Edn. 10) 1: 517. Type locality: Europe.

Synonymy: *Phalaena* (*Noctua*) *spinacea* Borkhausen, 1792; *Peucephila essoni* Hampson, 1909; *Monima albolineata* Matsumura, 192; *Mamestra variegata* Austaut, 1885.

Distribution. China (Heilongjiang, Jilin, Nei Mongol, Xinjiang); Russia (European part, Ural, Siberia), Mongolia, Korea, Japan, Mongolia, Turkey; Central Asia, Europe.

寄主：甘蓝、荨麻、榆、酸模属等植物（陈一心，1999）；蹄盖蕨（*Athyrium filix-femina*）、葡萄叶铁线莲（*Clematis vitalba*）、罂粟（*Papaver somniferum*）、膀胱麦瓶草（*Silene cucubalus*）、海滨蝇子草（*Silene maritima*）、单花雪轮（*Silene uniflora*）、狗筋麦瓶草（*Silene vulgaris*）、繁缕（*Stellaria media*）、（*Atriplex laciniatum*）、（*Atriplex litoralis*）、草地滨藜（*Atriplex patula*）、甜菜（*Beta vulgaris*）、藜（*Chenopodium album*）、（*Chenopodium bonushenricus*）、菊叶香藜（*Chenopodium vulvaria*）、菠菜（*Spinacia oleracea*）、两楼蓼（*Persicaria amphibia*）、酸模叶蓼（*Persicaria lapathifolia*）、春蓼（*Persicaria maculosa*）、萹蓄（*Polygonum aviculare*）、珠芽蓼（*Polygonum viviparum*）、皱叶酸模（*Rumex crispus*）、钝叶酸模（*Rumex obtusifolius*）、垂枝桦（*Betula pendula*）、帚石楠（*Calluna vulgaris*）、黑果越橘（*Vaccinium myrtillus*）、毛黄连花（*Lysimachia vulgaris*）、红花多枝柽柳（*Tamarix gallica*）、黄华柳（*Salix atrocinerea*）、黄花柳（*Salix caprea*）、蒿柳（*Salix viminalis*）、白泻根（*Bryonia alba*）、甜瓜（*Cucumis melo*）、黄瓜（*Cucumis sativus*）、西葫芦（*Cucurbita pepo*）、辣根（*Armoracia rusticana*）、芸苔（*Brassica campestris*）、欧洲油菜（*Brassica napus*）、甘蓝（*Brassica oleracea*）、芜青（*Brassica rapa*）、海滨芥（*Cakile maritima*）、萝卜（*Raphanus sativus*）、啤酒花（*Humulus lupulus*）、异株荨麻（*Urtica dioica*）、高精细榆树（*Ulmus campestris*）、苔景天（*Sedum acre*）、紫叶景天（*Sedum telephium*）、欧洲龙芽草（*Agrimonia eupatoria*）、（*Malus domestica*）、黑刺李（*Prunus spinosa*）、覆盆子（*Rubus idaeus*）、千屈菜（*Lythrum salicaria*）、小冠花（*Coronilla varia*）、（*Faba bona*）、海滨山黧豆（*Lathyrus japonicus*）、香豌豆（*Lathyrus odoratus*）、牧地山黧豆（*Lathyrus pratensis*）、百脉根（*Lotus corniculatus*）、草木樨（*Melilotus officinalis*）、多花菜豆（*Phaseolus multiflorus*）、菜豆（*Phaseolus vulgaris*）、豌豆（*Pisum sativum*）、刺槐（*Robinia pseudoacacia*）、中间车轴草（*Trifolium medium*）、凤仙花（*Impatiens balsamina*）、旱金莲（*Tropaeolum majus*）、八仙花（*Hydrangea macrophylla*）、旱芹（*Apium graveolens*）、野胡萝卜（*Daucus carota*）、欧防风（*Pastinaca sativa*）、洋香菜（*Petroselinum sativum*）、西洋接骨木（*Sambucus nigra*）、颠茄草（*Atropa belladonna*）、辣椒（*Capsicum annuum*）、番茄（*Lycopersicon esculentum*）、烟草（*Nicotiana tabacum*）、马铃薯（*Solanum tuberosum*）、昼颜（*Calystegia sepium*）、（*Chaenorrhinum minus*）、长叶婆婆纳（*Veronica longifolia*）、大车前（*Plantago *major*）、黄鼬瓣花（*Galeopsis tetrahit*）、短柄野芝麻（*Lamium album*）、红辣椒千叶蓍（*Achillea millefolium*）、中亚苦蒿（*Artemisia absinthium*）、野艾蒿（*Artemisia vulgaris*）、金盏花（*Calendula officinalis*）、刺儿菜（*Cirsium arvense*）、大丽花（*Dahlia pinnata*）、向日葵（*Helianthus annuus*）、莴苣（*Lactuca sativa*）、吊钟千里光（*Senecio fuchsii*）、欧洲千里光（*Senecio vulgaris*）、毛果一枝黄花（*Solidago virgaurea*）、苣荬菜（*Sonchus arvensis*）、百日菊（*Zinnia elegans*）、黄菖蒲（*Iris pseudacorus*）、圣母百合（*Lilium candidum*）、洋葱（*Allium cepa*）、蒜（*Allium sativum*）、石刁柏（*Asparagus officinalis*）、杰拉德氏灯心草（*Juncus gerardi*）、疏毛地杨梅（*Luzula pilosa*）、单鳞苞荸荠（*Eleocharis uniglumis*）、大看麦娘（*Alopecurus pratensis*）、贫育雀麦（*Bromus sterilis*）、玉米（*Zea mays*）、翠雀属（*Delphinium*）、石竹属（*Dianthus*）、大黄属（*Rheum*）、仙客来属（*Cyclamen*）、报春花属（*Primula*）、堇菜属（*Viola*）、秋海棠属（*Begonia*）、独行菜属（*Lepidium*）、紫罗兰属（*Matthiola*）、花葵属（*Lavatera*）、茶藨子属（*Ribes*）、花楸属（*Sorbus*）、羽扇豆属（*Lupinus*）、芒柄花属（*Ononis*）、亚麻属（*Linum*）、天竺葵属（*Pelargonium*）、旋花属（*Convolvulus*）、天芥菜属（*Heliotropium*）、毛蕊花属（*Verbascum*）、薄荷属（*Mentha*）、牛蒡属（*Arctium*）、紫菀属（*Aster*）、菊属（*Chrysanthemum*）、疆菊属（*Tanacetum*）、蒲公英属（*Taraxacum*）、芦竹属（*Arundo*）、甜茅属（*Glyceria*）、虉草属（*Phalaris*）植物（Matov & Kononenko, 2012）。

***Lacanobia* (*Diataraxia*) *splendens* (Hübner, [1808]) 华安夜蛾**

Noctua splendens Hübner, [1808], *Sammlung Europaischer Schmetterlinge. Eulen*, 2 and 3: pl. 85, f. 400. Type locality: Europe.

Distribution. China (Heilongjiang, Jilin, Liaoning, Xinjiang); Russia (European part, Ural, Siberia), Mongolia, Korea, Japan, Turkey; Central Asia, Europe.

寄主：酸模、车前等属植物（陈一心，1999）；沼泽蕨（*Nephrodium thelypteris*）、（*Lastrea thelypteris*）、沼泽蕨（*Thelypteris palustris*）、（*Thelypteris thelypteroides*）、欧洲耧斗菜（*Aquilegia vulgaris*）、欧洲耧斗菜（*Aquilegia vulgaris*）、甜菜（*Beta vulgaris*）、毒芹（*Cicuta virosa*）、睡菜（*Menyanthes trifoliata*）、欧白英（*Solanum dulcamara*）、荒野蒿（*Artemisia campestris*）、野艾蒿（*Artemisia vulgaris*）、莴苣（*Lactuca sativa*）、野莴苣（*Lactuca serriola*）、紫福玉草（*Prenanthes purpurea*）、酸模属（*Rumex*）、珍珠菜属（*Lysimachia*）、车轴草属（*Trifolium*）、打碗花属（*Calystegia*）、旋花属（*Convolvulus*）、车前属（*Plantago*）、牛蒡属（*Arctium*）、山柳菊属（*Hieracium*）、苦苣菜属（*Sonchus*）、蒲公英属（*Taraxacum*）植物（Matov & Kononenko, 2012）。

***Lacanobia* (*Diataraxia*) *aliena* (Hübner, 1809) 异安夜蛾**

Noctua aliena Hübner, 1809, *Sammlung Europaischer Schmetterlinge. Eulen*, 2: pl 94: 411. Type locality: Europe.

Synonymy: *Mamestra aliena amurensis* Staudinger, 1901 (var.), nec Spuler, 1908.

Distribution. China (Heilongjiang, Xinjiang, Gansu); Russia (European part, Ural, Siberia), Mongolia, Korea,

Japan, Turkey; Central Asia, Europe.

寄主：翘摇属植物（陈一心，1999）；白屈菜（*Chelidonium majus*）、甘蓝（*Brassica oleracea*）、啤酒花（*Humulus lupulus*）、莓叶委陵菜（*Potentilla fragarioides*）、稠李（*Padus racemosa*）、黑刺李（*Prunus spinosa*）、欧洲花楸（*Sorbus aucuparia*）、西伯利亚花楸（*Sorbus sibirica*）、绒毛花（*Anthyllis vulneraria*）、南欧扫帚豆（*Cytisus austriacus*）、变黑金雀儿（*Cytisus nigricans*）、金雀儿（*Cytisus scoparius*）、马蹄豆（*Hippocrepis comosa*）、百脉根（*Lotus corniculatus*）、紫苜蓿（*Medicago sativa*）、（*Melilotus album*）、黄香草木樨（*Melilotus officinalis*）、匍匐芒柄草（*Ononis repens*）、刺芒柄花（*Ononis spinosa*）、刺槐（*Robinia pseudoacacia*）、延边车轴草（*Trifolium montanum*）、红车轴草（*Trifolium pratense*）、白车轴草（*Trifolium repens*）、麻菀（*Aster linosyris*）、毛果一枝黄花（*Solidago virgaurea*）、毛蕊花属（*Ranunculus*）、藜属（*Chenopodium*）、荨麻属（*Urtica*）、大戟属（*Euphorbia*）、金合欢属（*Acacia*）、小冠花属（*Coronilla*）、乌足豆属（*Ornithopus*）、梣属（*Fraxinus*）、茄属（*Solanum*）、旋花属（*Convolvulus*）、牛蒡属（*Arctium*）、蒿属（*Artemisia*）、山柳菊属（*Hieracium*）、蒲公英属（*Taraxacum*）植物（Matov & Kononenko, 2012）。

Genus *Melanchra* Hübner, [1820] 1816 乌夜蛾属

Melanchra Hübner, [1820] 1816, *Verzeichniss bekannter Schmettlinge* [sic!]: 207. Type species: *Phalaena persicariae* Linnaeus, 1761 [Sweden].

The genus contains 14 species, distributed in the Palaearctic, Oriental and Nearctic regions; 4 species occur in the Palaearctic region. Among them, 2 species are in the Three Provinces of Northeast China.

Melanchra persicariae (Linnaeus, 1761) 乌夜蛾

Phalaena persicariae Linnaeus, 1761, *Fauna Suecica* (Edn. 2): 319. Type locality: Sweden.

Synonymy: *Phalaena sambuci* Hufnagel, 1766; *Phalaena graphica* Geoffroy, 1785; *Phalaena* (*Noctua*) *accipitrina* Esper, 1788; *Mamestra persicariae unicolor* Staudinger, 1871 (ssp.); *Polia persicariae ochrorenis* Kardakoff, 1928 (ab.); *Polia persicariae japonibia* Bryk, 1942 (ssp.).

Distribution. China (Heilongjiang, Jilin, Nei Mongol, Hebei, Shanxi, Shandong, Henan, Sichuan, Yunnan); Russia (European part, Ural, Siberia), Korea, Japan.

寄主：多食性、低矮草本植物，柳、桦、楸等木本植物（陈一心，1999）；欧洲鳞毛蕨（*Dryopteris filix-mas*）、欧洲蕨（*Pteridium aquilinum*）、荚果蕨（*Matteucia struthiopteris*）、欧洲落叶松（*Larix decidua*）、（*Aconitum intermedium*）、根北方乌头（*Aconitum septentrionale*）、叶升麻（*Actaea spicata*）、秋牡丹（*Anemone japonica*）、欧洲耧斗菜（*Aquilegia vulgaris*）、葡萄叶铁线莲（*Clematis vitalba*）、甜菜（*Beta vulgaris*）、菠菜（*Spinacia oleracea*）、春蓼（*Persicaria maculosa*）、水蓼（*Polygonum hydropiper*）、高山蓼（*Polygonum alpinum*）、酸模（*Rumex acetosa*）、垂枝欧洲山毛榉（*Fagus sylvatica*）、冬青栎（*Quercus ilex*）、欧洲桤木（*Alnus glutinosa*）、灰桤木（*Alnus incana*）、桦（*Betula alba*）、垂枝桦（*Betula pendula*）、胡榛（*Corylus avellana*）、帚石楠（*Calluna vulgaris*）、

（*Vaccinium myrtiluus*）、笃斯越橘（*Vaccinium uliginosum*）、越橘（*Vaccinium vitis-idaea*）、毛黄连花（*Lysimachia vulgaris*）、黄花柳（*Salix caprea*）、灰柳（*Salix cinerea*）、东陵山柳（*Salix phylicifolia*）、蒿柳（*Salix viminalis*）、甘蓝（*Brassica oleracea*）、大麻（*Cannabis sativa*）、啤酒花（*Humulus lupulus*）、异株荨麻（*Urtica dioica*）、玉树（*Crassula arborescens*）、高山茶藨子（*Ribes alpinum*）、红茶藨子（*Ribes rubrum*）、鹅莓（*Ribes uva-crispa*）、欧洲龙芽草（*Agrimonia eupatoria*）、（*Malus domestica*）、稠李（*Padus racemosa*）、欧洲李（*Prunus domestica*）、黑刺李（*Prunus spinosa*）、西洋梨（*Pyrus communis*）、秦氏悬钩子（*Rubus caesius*）、树莓（*Rubus fruticosus*）、覆盆子（*Rubus idaeus*）、欧洲花楸（*Sorbus aucuparia*）、千屈菜（*Lythrum salicaria*）、柳兰（*Chamaenerion angustifolium*）、大豆（*Glycine max*）、狭叶羽扇豆（*Lupinus angustifolius*）、黄香草木樨（*Melilotus officinalis*）、多花菜豆（*Phaseolus multiflorus*）、菜豆（*Phaseolus vulgaris*）、豌豆（*Pisum sativum*）、红车轴草（*Trifolium pratense*）、广布野豌豆（*Vicia cracca*）、梣叶槭（*Acer negundo*）、欧亚槭（*Acer pseudoplatanus*）、亚麻（*Linum usitatissimum*）、水金凤（*Impatiens noli-tangere*）、小花凤仙花（*Impatiens parviflora*）、旱金莲（*Tropaeolum majus*）、欧鼠李（*Frangula alnus*）、葡萄（*Vitis vinifera*）、薄叶山梅花（*Philadelphus tenuifolius*）、羊角芹（*Aegopodium podagraria*）、林当归（*Angelica sylvestris*）、旱芹（*Apium graveolens*）、野胡萝卜（*Daucus carota*）、（*Heracleum sphondylium*）、欧防风（*Pastinaca sativa*）、洋香菜（*Petroselinum sativum*）、亮叶芹（*Silaum silaus*）、欧银花（*Lonicera xylosteum*）、西洋接骨木（*Sambucus nigra*）、金叶接骨木（*Sambucus racemosa*）、番茄（*Lycopersicon esculentum*）、烟草（*Nicotiana tabacum*）、阳芋（*Solanum tuberosum*）、欧洲柳穿鱼（*Linaria vulgaris*）、林生玄参（*Scrophularia nodosa*）、鼬瓣花（*Galeopsis ladanum*）、林地水苏（*Stachys sylvatica*）、匍匐风铃草（*Campanula rapunculoides*）、中亚苦蒿（*Artemisia absinthium*）、荒野蒿（*Artemisia campestris*）、野艾蒿（*Artemisia vulgaris*）、金盏花（*Calendula officinalis*）、大矢车菊（*Centaurea scabiosa*）、刺儿菜（*Cirsium arvense*）、卷心菜蓟（*Cirsium oleraceum*）、大丽花（*Dahlia pinnata*）、欧洲山柳菊（*Hieracium sabaudum*）、山柳菊（*Hieracium umbellatum*）、莴苣（*Lactuca sativa*）、（*Mycelis muralis*）、金光菊（*Rudbeckia laciniata*）、吊钟千里光（*Senecio fuchsii*）、毛果一枝黄花（*Solidago virgaurea*）、玉米（*Zea mays*）、松属（*Pinus*）、翠雀属（*Delphinium*）、滨藜属（*Atriplex*）、杨属（*Populus*）、独行菜属（*Lepidium*）、蜀葵属（*Althaea*）、榆属（*Ulmus*）、醋栗属（*Grossularia*）、樱属（*Cerasus*）、草莓属（*Fragaria*）、金雀儿属（*Cytisus*）、染料木属（*Genista*）、芒柄花属（*Ononis*）、天竺葵属（*Pelargonium*）、卫矛属（*Euonymus*）、沙棘属（*Hippophae*）、常春藤属（*Hedera*）、梣属（*Fraxinus*）、蓝盆花属（*Scabiosa*）、碧冬茄属（*Petunia*）、天蓝绣球属（*Phlox*）、梓属（*Catalpa*）、车前属（*Plantago*）、野芝麻属（*Lamium*）、蓍属（*Achillea*）、紫菀属（*Aster*）、菊属（*Chrysanthemum*）、向日葵属（*Helianthus*）、苦苣菜属（*Sonchus*）、万寿菊属（*Tagetes*）、蒲公英属（*Taraxacum*）、款冬属（*Tussilago*）、重扇属（*Tradescantia*）植物（Matov & Kononenko, 2012）。

Melanchra postalba Sugi, 1982 后乌夜蛾

Melanchra postalba Sugi, 1982, *Moths of Japan*, 1: 702, pl. 173: 11. Type locality: Japan, Honshu, Gumma

Pref., Yuo.

Distribution. China (Jilin); Russia (Transbaikalia, Siberia), North Korea, Japan.

注：首次给出中文种名。

Genus ***Hypobarathra*** Hampson, 1905 后甘夜蛾属

Hypobarathra Hampson, 1905, *Catalogue of the Lepidoptera Phalaenae in the British Museum*, 5: x, 221. Type species: *Xylina icterias* Eversmann, 1843 [Russia, Ural].

The East Palaearctic – Oriental genus, contains 3 species; among them, 1 species is in the Three Provinces of Northeast China.

***Hypobarathra icterias* (Eversmann, 1843) 后甘夜蛾**

Xylina icterias Eversmann, 1843, *Bulletin de la Société Impériale des Naturalistes de Moscou*, 16 (35): 548. Type locality: Russia, Ural.

Distribution. China (Heilongjiang, Jilin, Liaoning, Nei Mongol, Guansu, Hebei, Shanxi, Xizang); Russia (Ural, Siberia), Korea, Japan, Kazakshstan.

Genus ***Ceramica*** Guenée, 1852 蜡夜蛾属

Ceramica Guenée, 1852, *in* Boisduval & Guenée, *Histoire Naturelle des Insectes. Species Général des Lépidoptères*, 5: 343. Type species: *Ceramica exusta* Guenée, 1852 [North America].

The genus contains 2 species, distributed in the temperate zone of Holarctic region. Among them, 1 species is in the Three Provinces of Northeast China.

注：首次给出中文属名。

***Ceramica pisi* (Linnaeus, 1758) 白线蜡夜蛾**

Phalaena pisi Linnaeus, 1758, *Systema Naturae* (Edn. 10) 1: 517. Type locality: Europe.

Synonymy: *Mamestra pisi pallens* Staudinger, 1882 (var.); *Mamestra pisi rukavaarae* Hoffmann, 1893 (var.); *Polia saltdalensis* Strand, 1903; *Polia pisi nyiwonis* Matsumura, 1925 (ssp.); *Polia pisi pisella* Bryk, 1941 (ssp.); *Polia pisi pardoi* Agenjo, 1945 (ssp.).

Distribution. China (Heilongjiang, Xinjiang); Russia (European part, Ural, Siberia, Kamchatka).

寄主：蕨类、山萝卜（陈一心，1999）；欧洲蕨（*Pteridium aquilinum*）、桂皮紫萁（*Osmunda asiatica*）、欧洲落叶松（*Larix decidua*）、西伯利亚落叶松（*Larix sibirica*）、欧洲云杉（*Picea abies*）、舟形乌头（*Aconitum napellus*）、条叶银莲花（*Anemone trullifolia*）、草甸毛茛（*Ranunculus acris*）、异珠女娄菜（*Melandrium album*）、甜菜（*Beta vulgaris*）、荞麦（*Fagopyrum esculentum*）、春蓼（*Persicaria maculosa*）、酸模叶蓼（*Polygonum lapathifolium*）、珠芽蓼（*Polygonum viviparum*）、酸模（*Rumex acetosa*）、直穗酸模（*Rumex longifolius*）、夏栎（*Quercus robur*）、灰桤木（*Alnus incana*）、欧洲桤木（*Alnus glutinosa*）、垂枝桦（*Betula pendula*）、毛桦（*Betula pubescens*）、香杨梅（*Myrica gale*）、斑点金丝桃（*Hypericum maculatum*）、帚石楠（*Calluna vulgaris*）、黑果越橘（*Vaccinium myrtillus*）、笃斯越橘（*Vaccinium uliginosum*）、毛黄连花（*Lysimachia vulgaris*）、钻天杨（*Populus nigra*）、欧洲山杨（*Populus tremula*）、耳柳（*Salix aurita*）、黄花柳（*Salix caprea*）、灰柳（*Salix cinerea*）、（*Salix myrsinifolia*）、五蕊柳（*Salix pentandra*）、东陵山柳（*Salix phylicifolia*）、匍匐柳（*Salix repens*）、欧洲油菜（*Brassica napus*）、高山菥蓂（*Thlaspi alpestre*）、麝香锦葵（*Malva moschata*）、异株荨麻（*Urtica dioica*）、欧洲醋栗（*Grossularia reclinata*）、黑茶藨子（*Ribes nigrum*）、鹅莓（*Ribes uva-crispa*）、锐刺山楂（*Crataegus oxyacantha*）、旋果蚊子草（*Filipendula ulmaria*）、野草莓（*Fragaria vesca*）、（*Malus domestica*）、森林苹果（*Malus sylvestris*）、稠李（*Padus racemosa*）、蕨麻（*Potentilla anserina*）、欧洲李（*Prunus domestica*）、苹果（*Pyrus malus*）、玫瑰（*Rosa rugosa*）、兴安悬钩子（*Rubus chamaemorus*）、树莓（*Rubus fruticosus*）、覆盆子（*Rubus idaeus*）、腺地榆（*Sanguisorba officinalis*）、珍珠梅（*Sorbaria sorbifolia*）、欧洲花楸（*Sorbus aucuparia*）、（*Sorbus suecica*）、千屈菜（*Lythrum salicaria*）、树锦鸡儿（*Caragana arborescens*）、金雀儿（*Cytisus scoparius*）、蚕豆（*Faba bona*）、香豌豆（*Lathyrus odoratus*）、草地老鹳草（*Lathyrus pratense*）、兵豆（*Lens esculenta*）、百脉根（*Lotus corniculatus*）、大百脉根（*Lotus uliginosus*）、紫苜蓿（*Medicago sativa*）、草木樨（*Melilotus officinalis*）、多花菜豆（*Phaseolus multiflorus*）、菜豆（*Phaseolus vulgaris*）、豌豆（*Pisum sativum*）、杂种车轴草（*Trifolium hybridum*）、中间车轴草（Trifolium media）、红车轴草（*Trifolium pratense*）、广布野豌豆（*Vicia cracca*）、救荒野豌豆（*Vicia sativa*）、林生野豌豆（*Vicia sylvatica*）、梣叶槭（*Acer negundo*）、亚麻（*Linum usitatissimum*）、羊角芹（*Aegopodium podagraria*）、欧白芷根（*Angelica archangelica*）、林当归（*Angelica silvestris*）、旱芹（*Apium graveolens*）、野胡萝卜（*Daucus carota*）、欧防风（*Pastinaca sativa*）、欧银花（*Lonicera xylosteum*）、欧洲绣球荚蒾（*Viburnum opulus*）、欧洲山萝卜（*Knautia arvensis*）、美丽龙胆（*Gentiana formosa*）、蓝色龙胆（*Gentiana pneumonanthe*）、欧白英（*Solanum dulcamara*）、马铃薯（*Solanum tuberosum*）、（*Galium album*）、蓬子菜（*Galium verum*）、欧洲柳穿鱼（*Linaria vulgaris*）、沼生马先蒿（*Pedicularis palustris*）、长叶婆婆纳（*Veronica longifolia*）、块根糙苏（*Phlomis tuberosa*）、林石蚕（*Teucrium scorodonia*）、北疆风铃草（*Campanula glomerata*）、红辣椒千叶蓍（*Achillea millefolium*）、北艾（*Artemisia vulgaris*）、麻菀（*Aster linosyris*）、海紫苑（*Aster tripolium*）、金盏花（*Calendula officinalis*）、棕矢车菊（*Centaurea jacea*）、刺儿菜（*Cirsium arvense*）、大丽花（*Dahlia pinnata*）、滨菊（*Leucanthemum vulgare*）、吊钟千里光（*Senecio fuchsii*）、毛果一枝黄花（*Solidago virgaurea*）、苣荬菜（*Sonchus arvensis*）、菊蒿（*Tanacetum vulgare*）、款冬花（*Tussilago farfara*）、石刁柏（*Asparagus officinalis*）、（*Juncus acutiflorus*）、燕麦（*Avena sativa*）、拂子茅（*Calamagrostis epigeios*）、发草（*Deschampsia caespitosa*）、酸沼草（*Molinia caerulea*）、虉草（*Phalaris arundinacea*）、梯牧草（*Phleum pratense*）、芦苇（*Phragmites australis*）、玉米（*Zea mays*）、香蒲（*Typha latifolia*）、耳蕨属（*Polystichum*）、翠雀属（*Delphinium*）、报春花属（*Primula*）、萝卜属（*Raphanus*）、榆属（*Ulmus*）、大戟属（*Euphorbia*）、唐棣属（*Amelanchier*）、月见草属（*Oenothera*）、羽扇豆属（*Lupinus*）、驴食草属（*Onobrychis*）、天竺葵属（*Pelargonium*）、欧芹属（*Petroselinum*）、蓝盆花属（*Scabiosa*）、烟草属（*Nicotiana*）、鼻花属（*Rhinanthus*）、玄参属

（*Scrophularia*）、车前属（*Plantago*）、莴苣属（*Lactuca*）、蒲公英属（*Taraxacum*）、葱属（*Allium*）、小麦属（*Triticum*）植物（Matov & Kononenko, 2012）。

注：《中国动物志》（夜蛾科）中将本种归入"*Lacanobia* Billberg, 1820 安夜蛾属"，在此予以变更。

Genus *Papestra* Sukhharea, 1973 褐青夜蛾属

Papestra Sukhareva, 1973, *Entomologicheskoe Obozrenie* [*Entomological Review*], 52: 409. Type species: *Phalaena biren* Goeze, 1781[Europe].

The genus contains 5 species, distributed in the boreal zone of the Holarctic region. Among them, 1 species is in the Three Provinces of Northeast China.

注：首次给出中文属名。

Papestra biren (Goeze, 1781) 褐青夜蛾

Phalaena (*Noctua*) *biren* Goeze, 1781, *Entomologische Beiträge zu des Ritter Linne Zwdlften*, 3 (3): 206. Type locality: Europe.

Synonymy: *Noctua glauca* Kleeman, 1792; *Noctua lappo* Duponchel, 1827; *Noctua aperta* Geyer, [1832]; *Hadena quadriposita* Zetterstedt, 1840; *Hadena poliostigma* Hampson, 1894; *Mamestra glauca taunensis* Fuchs, 1899 (var.); *Mamestra glauca paupercula* Püngeler, 1902 (var.); *Polia glauca puengeleri* Draudt, 1934 (f.); *Polia frustrata* McDunnough, 1946.

Distribution. China (Heilongjiang); Russia (European part, Ural, Siberia, Kamchatka), Mongolia, Korea, Japan, Turkey; Central Asia, Europe, America.

寄主：西伯利亚落叶松（*Larix sibirica*）、新疆云杉（*Picea obovata*）、紫花高乌头（*Aconitum excelsum*）、舟形乌头（*Aconitum napellus*）、蔓乌头（*Aconitum volubile*）、萹蓄 (*Polygonum aviculare*)、垂枝桦（*Betula pendula*）、香杨梅（*Myrica gale*）、熊果（*Arctostaphylos uva-ursi*）、帚石楠（*Calluna vulgaris*）、枞枝欧石楠（*Erica cinerea*）、杜香（*Ledum palustre*）、黑果越橘（*Vaccinium myrtillus*）、笃斯越橘（*Vaccinium uliginosum*）、越橘（*Vaccinium vitis-idaea*）、白柳（*Salix alba*）、黄华柳（*Salix atrocinerea*）、耳柳（*Salix aurita*）、黄花柳（*Salix caprea*）、灰柳（*Salix cinerea*）、绢柳（*Salix lapponum*）、黑柳（*Salix nigricans*）、东陵山柳（*Salix phylicifolia*）、匍匐柳（*Salix repens*）、（*Daphne metzereum*）、黑茶藨子（*Ribes nigrum*）、旋果蚊子草（*Filipendula ulmaria*）、稠李（*Padus racemosa*）、刺蔷薇（*Rosa acicularis*）、覆盆子（*Rubus idaeus*）、北美树莓（*Rubus odoratus*）、欧洲花楸（*Sorbus aucuparia*）、柳叶龙胆（*Gentiana asclepiadea*）、黄龙胆（*Gentiana lutea*）、（*Gentiana pannonica*）、短柄野芝麻（*Lamium album*）、山柳菊（*Hieracium umbellatum*）、苦苣菜（*Sonchus oleraceus*）、药用蒲公英（*Taraxacum officinale*）、款冬花（*Tussilago farfara*）、桤木属（*Alnus*）、莴苣属（*Lactuca*）植物（Matov & Kononenko, 2012）。

注：首次给出中文种名。

Genus *Hyssia* Guenée, 1852 艺夜蛾属

Hyssia Guenée, 1852, *in* Boisduval & Guenée, *Histoire Naturelle des Insectes. Species Général des Lépidoptères*, 5: 345. Type species: *Orthosia cavernosa* Eversmann, 1842 [Russia, Urals and Kazan].

The genus contains 21 species, distributed in the temperate zone of the Holarctic region. Among them 1 species is in the Three Provinces of Northeast China.

Hyssia cavernosa **(Eversmann, 1842) 艺夜蛾**

Mamestra cavernosa Eversmann, 1842, *Bulletin de la Société Impériale des Naturalistes de Moscou*, 54: 547, pl. 5: 3. Type locality: Russia, Kasan; [West Kazakhstan], Uralsk.

Synonymy: *Hyssia cavernosa korebia* Bryk, 1948 (ssp.); *Hyssia cavernosa gozmanyi* Kovács, 1968 (ssp.); *Hyssia cavernosa kaszabi* Kovács, 1968 (ssp.).

Distribution. China (Heilongjiang, Jilin, Xinjiang); Russia (Southeast European part, Siberia), Mongolia, Korea, Japan; Central Asia, Middle East, Europe.

寄主：（*Aristolochia clematilis*）、黑果越橘（*Vaccinium myrtillus*）、黑刺李（*Prunus spinosa*）、蝇子草属（*Silene*）、大戟属（*Euphorbia*）、车前属（*Plantago*）、蒲公英属（*Taraxacum*）植物（Matov & Kononenko, 2012）。

Genus *Mamestra* Ochsenheimer, 1816 甘蓝夜蛾属

Mamestra Ochsenheimer, 1816, *Die Schmetterlinge von Europa*, 4: 76. Type species: *Phalaena brassicae* Linnaeus, 1758 [Europe].

Synonymy: *Barathra* Hübner, [1821] 1816; *Copimamestra* Grote, 1883; *Mamistra*: Sodoffsky, 1837, emend..

The genus contains 14 species, distributed in the temperate zone of the Palaearctic, Nearctic and Oriental regions. Among them, 1 species is in the Three Provinces of Northeast China.

Mamestra brassicae **(Linnaeus, 1758) 甘蓝夜蛾**

Phalaena brassicae Linnaeus, 1758, *Systema Naturae* (Edn. 10) 1: 516. Type locality: Europe.

Synonymy: *Phalaena omicron* Geoffrey, 1785; *Noctua albidilinea* Haworth, 1809; *Mamestra brassicae andalusica* Staudinger, 1871 (var.); *Mamestra brassicae decolorata* Staudinger, 1889 (var.); *Hypobarathra unicolor* Marumo, 1917; *Mamestra*? *romieuxi* Culot, 1924.

Distribution. China (Heilongjiang, Jilin, Liaoning, Nei Mongol, Beijing, Hebei, Henan, Shandong, Shanxi, Hubei, Sichuan, Xinjiang, Taiwan); Russia (European part, Ural, Siberia), Mongolia, Korea, Japan, India, Nepal; Central Asia, Middle East, Europe.

寄主：甘蓝及多种蔬菜，甘蔗、高粱、棉、桑、麦类（陈一心，1999）；欧洲耧斗菜（*Aquilegia vulgaris*）、东方铁线莲（*Clematis orientalis*）、（*Ranunculus lanuginosus*）、亚欧唐松草（*Thalictrum minus*）、（*Paeonia corallina*）、罂粟（*Papaver somniferum*）、皱叶剪秋罗（*Lychnis chalcedonica*）、异珠女娄菜（*Melandrium album*）、迪奥卡蝇子草（*Silene dioica*）、反枝苋（*Amaranthus retroflexus*）、（*Atriplex laciniatum*）、亮滨藜（*Atriplex nitens*）、草地滨藜（*Atriplex patula*）、甜菜（*Beta vulgaris*）、藜（*Chenopodium album*）、菠菜（*Spinacia oleracea*）、萹蓄（*Polygonum aviculare*）、酸模（*Rumex acetosa*）、土耳其栎（*Quercus cerris*）、夏栎（*Quercus robur*）、三球悬铃木（*Platanus orientalis*）、灰桤木（*Alnus incana*）、毛桦（*Betula pubescens*）、毛黄连花（*Lysimachia vulgaris*）、银白杨（*Populus alba*）、欧洲山杨（*Populus tremula*）、黄花柳（*Salix caprea*）、爆竹柳（*Salix fragilis*）、白泻根（*Bryonia alba*）、黄瓜（*Cucumis sativus*）、西葫芦（*Cucurbita pepo*）、芸苔（*Brassica campestris*）、欧洲油菜（*Brassica napus*）、黑芥（*Brassica nigra*）、甘蓝（*Brassica oleracea*）、白菜（*Brassica pekinensis*）、芜青（*Brassica rapa*）、疣果匙荠（*Bunias orientalis*）、荠（*Capsella bursa-pastoris*）、圆裂碎米荠（*Cardamine hirsuta*）、辣根（*Armoracia rusticana*）、海甘蓝（*Crambe maritima*）、（*Napus rapa*）、野萝卜（*Raphanus raphanistrum*）、萝卜（*Raphanus sativus*）、野芥（*Sinapis arvensis*）、欧洲小叶椴（*Tilia cordata*）、印度棉（*Gossypium indicum*）、木槿（*Hibiscus syriacus*）、冬葵（*Malva verticillata*）、大麻（*Cannabis sativa*）、啤酒花（*Humulus lupulus*）、朴树（*Celtis sinensis*）、裂叶榆（*Ulmus laciniata*）、桑（*Morus alba*）、蓖麻（*Ricinus communis*）、高山茶藨子（*Ribes alpinum*）、红茶藨子（*Ribes rubrum*）、锐刺山楂（*Crataegus oxyacantha*）、紫萼路边青（*Geum rivale*）、(*Malus domestica*）、苹果（Malus pumila）、稠李（*Padus racemosa*）、白桃（*Persica vulgaris*）、蕨麻（*Potentilla anserina*）、匍匐委陵菜（*Potentilla reptans*）、毛叶插田泡（*Rubus coreanus*）、覆盆子（*Rubus idaeus*）、欧洲花楸（*Sorbus aucuparia*）、绣线菊（*Spiraea salicifolia*）、柳兰（*Chamaenerion angustifolium*）、花生（*Arachis hypogaea*）、树锦鸡儿（*Caragana arborescens*）、蚕豆（*Faba bona*）、大豆（*Glycine max*）、黄香草木樨（*Melilotus officinalis*）、菜豆（*Phaseolus vulgaris*）、豌豆（*Pisum sativum*）、白车轴草（*Trifolium repens*）、广布野豌豆（*Vicia cracca*）、救荒野豌豆（*Vicia sativa*）、欧洲七叶树（*Aesculus hippocastanum*）、梣叶槭（*Acer negundo*）、火炬树（*Rhus typhina*）、亚麻（*Linum usitatissimum*）、金叶山梅花（*Philadelphus coronarius*）、盾叶天竺葵（*Pelargonium peltatum*）、马蹄纹天竺葵（*Pelargonium zonale*）、旱金莲（*Tropaeolum majus*）、朝鲜当归（*Angelica gigas*）、旱芹（*Apium graveolens*）、野胡萝卜（*Daucus carota*）、（*Heracleum sphondylium*）、欧防风（*Pastinaca sativa*）、洋香菜（*Petroselinum sativum*）、西洋接骨木（*Sambucus nigra*）、金叶接骨木（*Sambucus racemosa*）、欧洲山萝卜（*Knautia arvensis*）、（*Succisa pratensis*）、颠茄草（*Atropa belladonna*）、辣椒（*Capsicum annuum*）、毛曼陀罗（*Datura innoxia*）、曼陀罗（*Datura stramonium*）、天仙子（*Hyoscyamus niger*）、番茄（*Lycopersicon esculentum*）、黄花烟草（*Nicotiana rustica*）、烟草（*Nicotiana tabacum*）、茄（*Solanum melongena*）、马铃薯（*Solanum tuberosum*）、田旋花（*Convolvulus arvensis*）、新疆花荵（*Polemonium caeruleum*）、南美天芥菜（*Heliotropium arborescens*）、毛地黄（*Digitalis purpurea*）、阿拉伯婆婆纳（*Veronica persica*）、大车前（*Plantago major*）、神香草（*Hyssopus officinalis*）、野薄荷（*Mentha arvensis*）、皱叶薄荷（*Mentha crispa*）、辣薄荷（*Mentha piperita*）、鼠尾草（*Salvia officinalis*）、牛蒡（*Arctium lappa*）、毛头牛蒡（*Arctium tomentosum*）、中亚苦蒿（*Artemisia absinthium*）、魁蒿（*Artemisia princeps*）、欧洲金盏花（*Calendula arvensis*）、金盏花（*Calendula officinalis*）、杭菊（*Chrysanthemum morifolium*）、刺儿菜

（*Cirsium arvense*）、大丽花（*Dahlia pinnata*）、向日葵（*Helianthus annuus*）、莴苣（*Lactuca sativa*）、野莴苣（*Lactuca scariola*）、（*Mycelis muralis*）、蜂斗菜（*Petasites japonicus*）、窄叶黄菀（*Senecio inaequidens*）、苣荬菜（*Sonchus arvensis*）、麝香百合（*Lilium longiflorum*）、洋葱（*Allium cepa*）、葱（*Allium fistulosum*）、高粱（*Sorghum bicolor*）、小麦（*Triticum aestivum*）、玉米（*Zea mays*）、大黄属（*Rheum*）、荨麻属（*Urtica*）、大戟属（*Euphorbia*）、樱属（*Cerasus*）、草莓属（*Fragaria*）、梨属（*Pyrus*）、蔷薇属（*Rosa*）、忍冬属（*Lonicera*）、川续断属（*Dipsacus*）、蓝盆花属（*Scabiosa*）、野芝麻属（*Lamium*）、金光菊属（*Rudbeckia*）、蒲公英属（*Taraxacum*）、唐菖蒲属（*Gladiolus*）、天门冬属（*Asparagus*）、梯牧草属（*Phleum*）植物（Matov & Kononenko, 2012）。

Genus ***Sideridis*** Hübner, [1821] 1816 **寡夜蛾属**

Sideridis Hübner, [1821] 1816, *Verzeichniss bekannter Schmettlinge* [sic!]: 232. Type species: *Noctua evidens* Hübner, [1808] [Europe], homonym of *Noctua evidens* Thunberg, 1784, repl. name is *Sideridis anapheles* Nye, 1975.

Synonymy: *Heliophobus* Boisduval, 1828 (subgen.); *Neuria* Guenée, 1841; *Aneda* Sukhareva, 1973 (subgen.); *Colonsideridis* Beck, [1992] 1991; *Dianthivora* Varga & Ronkay, 1991 (subgen.); *Mamestra* auct., nec Ochsenheimer, 1816; *Trichoclea* auct., nec Grote, 1883.

The genus contains 32 species, distributed mainly in the Central-Asian subregion of the Palaearctic, partially in the Nearctic and Oriental regions. Among them, 8 species of 3 subgenera are in the Three Provinces of Northeast China.

Subgenus ***Sideridis*** Hübner, [1821] 1816 **寡夜蛾亚属**

Sideridis* (*Sideridis*) *turbida **(Esper, [1790]) 灰褐寡夜蛾**

Phalaena (*Noctua*) *turbida* Esper, [1790], *Die Schmetterlinge in Abbildungen nach der Natur mit Beschreibungen* 4 (2/1): 495, pl. 151, Noct.72, fig.1. Type locality: Europe [Germany].

Synonymy: *Noctua albicolon* Hübner, [1813]; *Trichoclea boursini* Agenjo, 1941.

Distribution. China (Heilongjiang, Jilin, Qinghai, Xinjiang); Russia (European part, Ural, Siberia), Mongolia, Korea, Japan, India, Nepal; Central Asia, Europe.

寄主：芒柄花、旋花属等植物（陈一心，1999）；繁缕（*Stellaria media*）、反枝苋（*Amaranthus retroflexus*）、（*Atrpilex patulum*）、甜菜（*Beta vulgaris*）、藜（*Chenopodium album*）、萹蓄(*Polygonum aviculare*)、甘蓝（*Brassica oleracea*）、海滨芥（*Cakile maritima*）、播娘蒿（*Descurainia sophia*）、匍匐芒柄草（*Ononis repens*）、肾叶打碗花（*Calystegia soldanella*）、葛缕子（*Carum carvi*）、短柄野芝麻（*Lamium album*）、酸模属（*Rumex*）、大蒜芥属（*Sisymbrium*）、大戟属（*Euphorbia*）、烟草属（*Nicotiana*）、车前属（*Plantago*）、蒿属（*Artemisia*）、菊苣属（*Cichorium*）、山柳菊属（*Hieracium*）、蒲公英属（*Taraxacum*）、禾本科（Poaceae）植物（Matov &

Kononenko, 2012）。

注：《中国动物志》（夜蛾科）中引用本种同物异名“*Sideridis albicolon* Hübner, [1813]”，在此予以变更。

Sideridis (_Sideridis_) _incommoda_ (Staudinger, 1888) 烦寡夜蛾

Hadula (*Mamestra*?) *incommoda* Staudinger, 1888, *Entomologische Zeitung. Entomologischen Vereine zu Stettin*, 49: 250. Type locality: Russia, [Primorye terr.] Askold I., Vladivostok, [Bezverkhovo] Sidemi, [Razdolnaya river] Siufun, [Partizansk] Sutschan, Bikin.

Synonymy: *Mamestra lacrimosa* Graeser, 1889 [1890], unavailable name.

Distribution. China (Heilongjiang, Jilin); Russia (Siberia), Korea, Japan.

寄主：蒿属（*Artemisia*）植物（Matov & Kononenko, 2012）。

Sideridis (_Sideridis_) _unica_ (Leech, 1889) 楔斑寡夜蛾

Hadena unica Leech, 1889, *Proceedings of the Zoological Society of London*, 1889: 509, pl. 51: 12. Type locality: Japan, Yokohama.

Synonymy: *Polia suavina* Draudt, 1950.

Distribution. China (Jilin); Russia (Siberia), Korea, Japan.

注：首次给出中文种名。

Subgenus *Aneda* Sukhareva, 1973 唳盗夜蛾亚属

Aneda Sukhareva, 1973, *Entomologicheskoe Obozrenie* [*Entomological Review*], 52: 404, 410. Type species: *Noctua rivularis* Fabncius 1775 [Germany].

注：首次给出中文亚属名。

Sideridis (_Aneda_) _rivularis_ (Fabricius, 1775) 唳盗夜蛾

Noctua rivularis Fabricius, 1775, *Entomologiae Systematica emendata et Aucta. Secundum Classes, Ordines, Genera, Species Adjectis Synonimis, Locis, Observationibus, Descriptionibus*: 613. Type locality: Germany, Sachsen.

Synonymy: *Noctua cucubali* [Denis et Schiffermüller], 1775; *Phalaena* (*Noctua*) *rivosa* Ström, 1783; *Hadena behenis* Freyer, 1845; *Sideridis rivularis pacifica* Hacker, 1996 (ssp.).

Distribution. China (Heilongjiang, Jilin, Hebei, Zhejiang, Hunan, Sichuan); Russia (European part, Ural, Siberia), Mongolia; Central Asia, Middle East, Europe.

寄主：麦瓶草、剪秋罗属植物（陈一心，1999）；麦仙翁（*Agrostemma githago*）、狗筋蔓（*Cucubalus baccifer*）、头石竹（*Dianthus barbatus*）、康乃馨（*Dianthus caryophyllus*）、常夏石竹（*Dianthus plumarius*）、丁香石竹（*Dianthus superbus*）、皱叶剪秋罗（*Lychnis chalcedonica*）、剪秋罗

（*Lychnis flos-cuculi*）、（*Lychnis arkwrightii*）、异珠女娄菜（*Melandrium album*）、肥皂草（*Saponaria officinalis*）、（*Silene dichotoma*）、迪奥卡蝇子草（*Silene dioica*）、（*Silene inflata*）、叉枝蝇子草（*Silene latifolia*）、海滨蝇子草（*Silene maritima*）、夜花蝇子草（*Silene noctiflora*）、欧亚蝇子草（*Silene nutans*）、单花雪轮（*Silene uniflora*）、白玉草（*Silene venosa*）、狗筋麦瓶草（*Silene vulgaris*）、（*Viscaria vulgaris*）、荷包牡丹（*Diclytra spectabilis*）、红缬草（*Centranthus ruber*）、（*Melampyrum nemorosum*）、卷耳属（*Cerastium*）、苦苣菜属（*Sonchus*）植物（Matov & Kononenko, 2012）。

注：《中国动物志》（夜蛾科）中将本种归入"*Hadena* Schrank, 1802 盗夜蛾属"，在此予以变更。

***Sideridis* (*Aneda*) *honeyi* (Yoshimoto, 1989) 拟唳盗夜蛾**

Hadena rivularis honeyi Yoshimoto, 1989, *Japan Heterocerists' Journal*, 155: 74, fig. 1, 3, 6. Type locality: Japan, Yamanashi Pref.; China, Taiwan, Nantou Hsien; North Korea, Gwangweondo [coll. Yoshimoto, Tokyo].

Synonymy: *Hadena rivularis* auct., nec Fabricius, 1775; *Hadena cucubali mandarina*: Draudt, 1950, nec Leech, 1900, misidentification.

Distribution. China (Heilongjiang, Jilin); Russia (Tuva, Transbaikalia, Siberia), Mongolia, Korea, Japan.

寄主：坚硬女娄菜（*Silene firma*）（Matov & Kononenko, 2012）。

注：首次给出中文种名。

***Sideridis* (*Aneda*) *mandarina* (Leech, 1900) 亚唳盗夜蛾**

Dianthoecia mandarina Leech, 1900, *Transactions of the Entomological Society of London*, 1900: 91. Type locality: China, Pu-tsu-fong.

Synonymy: *Harmodia confucii* Draudt, 1950.

Distribution. China (Jilin, Sichuan, Taiwan); Russia (Siberia), Korea, Japan, Vietnam(North).

注：首次给出中文种名。

Subgenus ***Heliophobus*** Boisduval, 1828 网夜蛾亚属

Heliophobus Boisduval, 1828, *Europaeorum Lepidopterorum Index Methodicus*: 69. Type species: *Phalaena saponariae* Borkhausen, 1792 [Europe].

注：《中国动物志》（夜蛾科）中将本亚属引用为属级单元"网夜蛾属"，当前应为"*Sideridis* Hübner, [1821] 1816 寡夜蛾属"的亚属单元，在此予以变更。

***Sideridis* (*Heliophobus*) *kitti* (Schawerda, 1913) 织网夜蛾**

Mamestra kitti Schawerda, 1913, *Verhandlungen des Zoologisch-Botanischen vereins in Wien*, 24: 125. Type locality: Austria, Neubruck.

Synonymy: *Hadena texturata silbernageli* Tykac, 1940 (ab.).

Distribution. China (Heilongjiang, Jilin, Nei Mongol, Qinghai, Sichuan, Yunnan, Xizang); Russia (European part, Ural, Altai, Transbaikalia, Siberia), Mongolia, Kazakhstan.

寄主：黄芪属、巢菜属植物（陈一心，1999）；沙打旺（*Astragalus adsuegens*）、鹰嘴紫云英（*Astragalus cicer*）、（*Astragalus glycyphyllus*）、广布野豌豆（*Vicia cracca*）、苕菜（*Vicia hirsuta*）、（*Vicia silvatica*）、山黧豆属（*Lathyrus*）植物（Matov & Kononenko, 2012）。

***Sideridis* (*Heliophobus*) *unicolor* (Alphéraky, 1889) 网寡夜蛾**

Mamestra unicolor Alphéraky, 1889, *in* N.M. Romanoff (ed.), *Mémoires sur les Lépidoptères*, 5: 147. Type locality: Tadjikistan, Pamir, Sarykol.

Distribution. China (Heilongjiang, Nei Mongol); Russia (European part, Ural, Altai, Transbaikalia), Tajikistan.

Genus ***Conisania*** Hampson, 1905 **侃夜蛾属**

Conisania Hampson, 1905, *Catalogue of the Lepidoptera Phalaenae in the British Museum*, 5: xiv, 472. Type species: *Apamea leineri* Freyer, 1836 [Austria].

Synonymy: *Trichospolas* Draudt, 1936; *Luteohadena* Beck, [1992] 1991 (subgen.); *Renisania* Beck, 1996.

The genus contains 23 species, distributed mainly in the Central Asian subregion of the Palaearctic region. Among them, 1 species is in the Three Provinces of Northeast China.

***Conisania suavis* (Staudinger, 1892) 紫褐侃夜蛾**

Mamestra suavis Staudinger, 1892, *in* N.M. Romanoff (ed.), *Mémoires sur les Lépidoptères*, 6: 428, pl. 7: 4. Type locality: Russia, [Primorye terr.] Vladivostok, [Bezverkhovo] Sidemi.

Distribution. China (Heilongjiang, Jilin, Qinghai, Hebei); Russia (Transbaikalia, Siberia), Korea.

注：《中国动物志》（夜蛾志）中将本种归入“*Polia* Ochsenheimer, 1816 灰夜蛾属”，在此予以变更。

Genus ***Hecatera*** Guenée, 1852 **延夜蛾属**

Hecatera Guenée, 1852, *in* Boisduval & Guenée, *Histoire Naturelle des Insectes. Species Général des Lépidoptères*, 6: 27. Type species: *Noctua dysodea* [Denis & Schiffermüller], 1775 [Austria].

Synonymy: *Epipsammia* Staudinger, 1879; *Aethria* Hübner, [1821] 1816, nec. Hübner, [1819].

The genus contains 24 species, mainly distributed in the Mediterranean and Central-Asian subregions of the Palaearctic region. Among them, 1 species is in the Three Provinces of Northeast China.

Hecatera bicolorata (Hufnagel, 1766) 两色延夜蛾

Phalaena bicolorata Hufnagel, 1766, *Berlinisches Magazine*, 3: 410. Type locality: Germany, Berlin.

Synonymy: *Polia leuconota* Eversmann, 1837; *Phalaena* (*Noctua*) *placida* Esper, [1791]; *Noctua hieracii* Borkhausen, 1793; *Phalaena par* Donovan, 1801; *Noctua monticola* Duponchel, 1826; *Hecatera intermedia* Walker, 1858; *Mamestra serena obscura* Staudinger, 1861 (var.).

Distribution. China (Heilongjiang); Russia (European part, Ural, Siberia), Korea, Mongolia, Kazakhstan; Middle East, Europe.

寄主：(*Malus domestica*)、(*Euphrasia rostkoviana*)、荒野蒿（*Artemisia campestris*）、金盏花（*Calendula officinalis*）、卷心菜蓟（*Cirsium oleraceum*）、粗糙足阳参（*Crepis biennis*）、毛叶还阳参（*Crepis capillaris*）、窄叶还阳参（*Crepis tectorum*）、大麻叶泽兰（*Eupatorium cannabinum*）、（*Hieracium auranthiacum*）、（*Hieracium boreale*）、绿毛山柳菊（*Hieracium pilosella*）、（*Hieracium sylvaticum*）、山柳菊（*Hieracium umbellatum*）、（*Hieracium vulgatum*）、莴苣（*Lactuca sativa*）、野莴苣（*Lactuca serriola*）、秋蒲公英（*Leontodon autumnalis*）、多毛鹰齿菊（*Leontodon hispidus*）、（*Mycelis muralis*）、单毛毛连菜（*Picris hieracioides*）、（*Pilosella prealta*）、（*Prenanthes purpurea*）、新疆千里光（*Senecio jacobaea*）、苣荬菜（*Sonchus arvensis*）、花叶滇苦菜（*Sonchus asper*）、苦苣菜（*Sonchus oleraceus*）、（*Trinia*）、疆菊属（*Tanacetum*）植物（Matov & Kononenko, 2012）。

Genus ***Hadena*** Schrank, 1802 盗夜蛾属

Hadena Schrank, 1802, *Fauna Boica. Durchgedachte Geschichte der in Baiera Einheimischen und Zahmen Thiere.*, 2 (2): 158. Type species: *Noctua capsincola* [Denis & Schinermüller], 1775 [Austria] =(*Phalaena bicruris* Hufnagel, 1766).

Synonymy: *Miselia* Ochsenheimer, 1816; *Harmodia* Hübner, [1820] 1816; *Zeteolyga* Billberg, 1820; *Epia* Hübner, [1821] 1816, preocc. (Hübner, [1820] 1816 [Lepidoptera, Bombycidae]); *Dianthoecia* Boisduval, 1834; *Pronotestra* Hampson, 1905 (subgen.); *Anepia* Hampson, 1918 (subgen.); *Pinkerichola* Hacker, 1987 (subgen.); *Maghadena* Beck, [1992] 1991; *Caeshadena* Beck, [1992] 1991; *Kuruschia* Boursin, 1940; *Perplexhadena* Beck, [1992] 1991; *Paraperplexia* Beck, [1992] 1991; *Maschukia* Hacker, 1996 (subgen.); *Klaperichola* Hacker, 1996 (subgen.); *Sinotibetana* Hacker, 1996 (subgen.); *Albhadena* Beck, 1999; *Adena*: Agassiz, 1846, emend..

The genus includes 134 species, mainly distributed in the Mediterranean and Central Asian subregions of the Palearctic region, a few species known from Nearctic region; 39 species of uncertain status are known from the Neotropical and Ethiopian regions. 125 species are in the Palaearctic region; among them, 4 species of 2 subgenera are in the Three Provinces of Northeast China.

Subgenus *Hadena* Schrank, 1802 盗夜蛾亚属

Hadena Schrank, 1802, *Fauna Boica. Durchgedachte Geschichte der in Baiera Einheimischen und Zahmen Thiere*, 2 (2): 158. Type species: *Noctua capsincola* [Denis & Schiffermüller], 1775 [Austria] =(*Phalaena bicruris* Hufnagel, 1766).

Synonymy: *Miselia* Ochsenheimer, 1816; *Harmodia* Hübner, 1820; *Zeteolyga* Billberg, 1820; *Dianthoecia* Boisduval, 1834, emend.; *Maghadena* Beck, [1992] 1991; *Caeshadena* Beck, [1992] 1991.

***Hadena* (*Hadena*) *compta* ([Denis & Schiffermüller], 1775) 白条盗夜蛾**

Noctua compta [Denis & Schiffermüller], 1775, *Ankündung eines Systematisches Werkes von den Schmetterlinge der Wiener Gegend. Wien*: 70. Type locality: Austria, Vienna.

Synonymy: *Dianthoecia compta viscariae* Guenée, 1852 (var.); *Dianthoecia armeriae* Guenée, 1852; *Dianthoecia compta galactina* Turati, 1907 (var.); *Harmodia compta grisescens* Turati, 1923 (f.); *Harmodia compta persica* Schwingenschuss, 1938 (ssp.); *Harmodia compta kashgaia* Brandt, 1947 (r.); *Hadena compta almoravida* Hacker, 1996 (ssp.).

Distribution. China (Heilongjiang); Russia (European part, Ural, Siberia), Mongolia, Korea, Japan; Central Asia, Middle East, Europe, Africa.

寄主：头石竹（*Dianthus barbatus*）、丹麦石竹（*Dianthus carthusianorum*）、康乃馨（*Dianthus caryophyllus*）、西洋石竹（*Dianthus deltoides*）、（*Dianthus glacialis*）、（*Dianthus gracilis*）、（*Dianthus henteri*）、（*Dianthus pavonius*）、（*Dianthus petraeus*）、常夏石竹（*Dianthus plumarius*）、（*Dianthus prolifer*）、（*Dianthus pungens*）、丁香石竹（*Dianthus superbus*）、迪奥卡蝇草子（*Silene dioica*）、欧亚蝇子草（*Silene nutans*）、狗筋麦瓶草（*Silene vulgaris*）、剪秋罗属（*Lychnis*）植物（Matov & Kononenko, 2012）。

***Hadena* (*Hadena*) *variolata* (Smith, 1888) 齿斑盗夜蛾**

Mamestra variolata Smith, 1888, *Proceedings of the United States National Museum*, 10: 467. Type locality: USA, Washington state.

Synonymy: *Dianthoecia nana dealbata* Staudinger, 1892 (var.); *Polia conspersa chosensis* Bryk, 1948 (ssp.); *Hadena dealbata kogurei* Sugi, 1958 (ssp.).

Distribution. China (Heilongjiang, Jilin, Sichuan, Yunnan); Russia (Southeast European part, Ural, Siberia), Mongolia, Korea, Japan; Central Asia, North America.

寄主：剪秋罗属（*Lychnis*）植物（Matov & Kononenko, 2012）。

注：《中国动物志》（夜蛾科）中引用同物异名"*Hadena dealbata* Staudinger, 1892"，在此予以变更。

Subgenus *Anepia* Hampson, 1918 亮盗夜蛾亚属

Anepia Hampson, 1918, *Novitates Zoologicae*, 25:116. Type species *Phalaena echii* Borkhausen, 1792 [Europe] (=*Phalaena irregularis* Hufnagel, 1766).

Synonymy: *Epia* Hübner, 1818; *Kuruschia* Boursin, 1940; *Perplexhadena* Beck, [1992] 1991; *Paraperplexia* Beck, [1992] 1991.

注：首次给出中文亚属名。

***Hadena* (*Anepia*) *aberrans* (Eversmann, 1856) 梳跗盗夜蛾**

Dianthoecia aberrans Eversmann, 1856, *Bulletin de la Société Impériale des Naturalistes de Moscou*, 29: 104. Type locality: Russia, South Transbaikalia, Kiachta.

Synonymy: *Dianthoecia admiranda* Oberthür, 1880, *Dianthoecia claripennis* Butler, 1886.

Distribution. China (Heilongjiang, Jilin, Liaoning, Shaanxi, Shandong); Russia (Siberia), Mongolia, Korea, Japan.

寄主：坚硬女娄菜（*Silene firma*）（Matov & Kononenko, 2012）。

***Hadena* (*Anepia*) *corrupta* (Herz, 1898) 间盗夜蛾**

Dianthoecia corrupta Herz, 1898, *Deutsche Entomologische Zeitschrift Iris*, 11: 256. Type locality: Russia, Northeast Siberia; South Siberia: Altai; Central Asia, "Ortu Pamir"; Primorye terr., [Bezverkhovo] Sidemi.

Synonymy: *Polia subviolacea* Matsumura, 1925; *Harmodia corrupta siculimima* Bryk, 1948 (ssp.); *Harmodia corrupta splendida* Draudt, 1950 (ssp.).

Distribution. China (Heilongjiang, Jilin, Qinghai, Hebei); Russia (Ural, Siberia).

Tribe **Leucaniini** Guenée, 1837 粘夜蛾族

Leucanidi Guenée, 1837, *Annales de la Société Entomologique de France* 6: 321, 331. Type genus: *Leucania* Ochsenheimer, 1816.

Synonymy: Heliophilinae Grote, 1896; Mythimnini Rungs, 1956; Mythimnini Hacker, Ronkay & Hreblay, 2002, not valid.

The Leucaniini is one of the biggest tribe of Noctuinae comprising more than 400 species, mainly distributed in the subtropical area of the eastern Palaearctic, and the tropical area of the Oriental regions. The members of the subtribe have a surprisingly uniform, concolourous, longitudinal streaking, and similar external and internal appearance. Many Palaearctic and Oriental taxa of Leucaniini were revised in the series of publications by Hreblay (Hreblay *et al.*, 1996, 1998, 1999 and other). The tribe Leucaniini represented in the Three Provinces of Northeast China by 4 genera and 30 species.

References. Hampson, 1905 (Hadeninae, part); Chen, 1982, 1999 (Noctuidae: Hadeninae, part); Sugi, 1982,

1987 (Hadeninae, part); Kononenko, 1990, 2003 (Hadeninae, part); Yoshimatsu, 1987, 1994 (Hadeninae, part); Kitching, Rawlins, 1998 (Hadeninae); Kononenko, 1990 (Hadeninae, part); Hreblay, 1996 (Hadeninae); Hacker *et al.*, 2002 (Hadeninae: Mythimnini); Kononenko, 2003 (Hadeninae, part); Fibiger, Lafontaine, 2005 (Hadeninae: Leucaniini); Fibiger, Hacker, 2005 (Hadeninae: Leucaniini); Lafontaine, Fibiger, 2006 (Hadeninae: Leucaniini); Kononenko, 2005, 2010 (Hadeninae: Leucaniini); Kononenko, Han, 2007 (Hadeninae: Leucaniini); Matov *et al.*, 2008 (Hadeninae: Leucaniini); Lafontaine, Schmidt, 2010 (Noctuinae: Leucaniini); Fibiger *et al.*, 2011 (Hadeninae: Leucaniini).

Genus ***Sarcopolia*** Sugi, 1982 **红棕灰夜蛾属**

Sarcopolia Sugi, 1982, *Moths of Japan*, 1: 704, 2: 355, 462. Type species: *Agrotis illoba* Burler, 1878 [Japan].

The genus includes 1 species is distributed in the Chinese subregion of the Palaearctic and Oriental regions, and in the Three Provinces of Northeast China.

注：首次给出中文属名。

Sarcopolia illoba **(Butler, 1878) 红棕灰夜蛾**

Agrotis illoba Butler, 1878, *Annals and Magazine of Natural History*, (5)1: 162. Type locality: Japan, Hakodate.

Synonymy: *Graphiphor pacifica* Butler, 1878; *Hydraecia khasiana* Moore, 1881; *Mamestra declinans* Staudinger, 1888; *Polia assamica* Warren, 1912; *Aplecta differentiata* Bryk, 1948; *Polia illoba pallida* Draudt, 1950 (f.).

Distribution. China (Heilongjiang, Jilin, Liaoning, Shaanxi, Hebei, Shandong, Jiangsu, Zhejiang, Jiangxi, Fujian, Taiwan); Russia (Altai, Siberia), Korea, Japan, India.

寄主：桑、棉、苜蓿、甜菜、大豆、荞麦、胡萝卜、紫苏等（陈一心，1999）；繁缕（*Stellaria media*）、甜菜（*Beta vulgaris*）、藜（*Chenopodium album*）、荞麦（*Fagopyrum esculentum*）、虎杖（*Polygonum cuspidatum*）、酸模（*Rumex acetosa*）、甘蓝（*Brassica oleracea*）、萝卜（*Raphanus sativus*）、印度棉（*Gossypium indicum*）、木槿（*Hibiscus syriacus*）、桑（*Morus alba*）、毛叶插田泡（*Rubus coreanus*）、蚕豆（*Faba bona*）、大豆（*Glycine max*）、紫苜蓿（*Medicago sativa*）、豌豆（*Pisum sativum*）、野胡萝卜（*Daucus carota*）、毛地黄（*Digitalis purpurea*）、紫苏（*Perilla frutescens*）、牛蒡（*Arctium lappa*）、杭菊（*Chrysanthemum morifolium*）、洋葱（*Allium cepa*）、葱（*Allium fistulosum*）、车轴草属（*Trifolium*）、百合属（*Lilium*）植物（Matov & Kononenko, 2012）。

注：《中国动物志》（夜蛾科）中将本种归入“*Polia* Ochsenheimer, 1816 灰夜蛾属”，在此予以变更。

Genus ***Mythimna*** Ochsenheimer, 1816 **秘夜蛾属**

Mythimna Ochsenheimer, 1816, *Die Schmetterlinge von Europa* 4: 78. Type species: *Phalaena turca*

Linnaeus, 1761[Sweden].

Synonymy: *Heliophila* Hübner, [1806], suppr. (ICZN 1926; 1954 Op. 97); *Heliophilae* Ochsenheimer, 1816, unavail. (Code, Art. 11d); *Philostola* Billberg, 1820; *Aletia* Hübner, [1821] 1816; *Hyphilare* Hübner, [1821] 1816 (subgen.); *Heliophila* Hübner, 1822, preocc. (Klug, 1807 [Hymenoptera]); *Mithimna* Sodoffsky, 1837, emend.; *Borolia* Moore, 1881; *Hyperiodes* Warren, 1910; *Hypopteridia* Warren, 1912 (subgen.); *Pseudaletia* Franclemont, 1951 (subgen.); *Acantholeucania* Rungs, 1953 (subgen.); *Boursinania* Rungs, 1955; *Omphalestra* Fletcher, 1961 (subgen.); *Analetia* Calora, 1966 (subgen.); *Sablia* Sukhareva, 1973 (subgen.); *Xipholeucania* Sugi, 1970 (subgen.); *Anapoma* Berio, 1980 (subgen.); *Apoma* Berio, 1980, preocc. (Beck, 1938 [Mollusca]); *Aletis* Chang, 1991, missp.; *Morphopoliana* Hreblay & Legrain, 1996 (subgen.); *Dysaletia* Sugi, 1982 (subgen.); *Allitoria* Beck, 1996, nom. nud.; *Conthimna* Beck, 1999; *Gruathimna* Beck, 1999; *Foehstia* Beck, 1999; *Pudothimna* Beck, 1999; *Ferrayhimna* Beck, 1999.

The genus includes of over 400 species in the world; according to other sources up to 600 species (subtropical and temperate zones in part of the Palearctic, Nearctic, Oriental, Ethiopian, and Neotropical regions). There are about 120-140 species are in the Palaearctic region which are concentrated mainly in the subtropics of its western and eastern sectors. There are 26 species of 7 sunbgenera occurs in the Three Provinces of Northeast China.

Subgenus *Mythimna* Ochsenheimer, 1816 秘夜蛾亚属

Mythimna Ochsenheimer, 1816, *Die Schmetterlinge von Europa* 4: 78. Type species: *Phalaena turca* Linnaeus, 1761[Sweden].

Synonymy: *Philostola* Billberg, 1820; *Aletia* Hübner, [1821] 1816; *Heliophila* Hübner, 1822; *Leucania* Boisduval, 1828; *Borolia* Moore, 1881; *Hyperiodes* Warren, 1910; *Sideridis* auct., nec Hübner, 1821.

Mythimna (*Mythimna*) *turca* (Linnaeus, 1761) 秘夜蛾

Phalaena turca Linnaeus, 1761, *Fauna Suecica* (Edn. 2): 322. Type locality: Sweden.

Synonymy: *Noctua volupia* Hufnagel, 1766; *Mythimna limbata* Butler, 1881; *Leucania turca turcella* Staudinger, 1897 (var.); *Leucania camuna* Turati, 1915; *Hyperioides turca matsumuriana* Bryk, 1948 (ssp.).

Distribution. China (Heilongjiang, Jilin, Liaoning, Beijing, Shandong, Hubei, Hunan, Jiangxi, Sichuan); Russia (European part, Ural, Siberia), Mongolia, Korea, Japan; Central Asia, Middle East, Europe.

寄主：复活节种草（*Stellaria holostea*）、地杨梅（*Luzula campestris*）、疏毛地杨梅（*Luzula pilosa*）、（*Carex brizoides*）、（*Carex hirta*）、凌风草（*Briza media*）、鸭茅（*Dactylis glomerata*）、荻（*Miscanthus sacchariflorus*）、日本麦氏草（*Molinia japonica*）、早熟禾（*Poa annua*）、林地早熟禾（*Poa nemoralis*）、芦苇（*Phragmites australis*）、报春花属（*Primula*）、黑麦草属（*Lolium*）植物（Matov & Kononenko, 2012）。

Mythimna (*Mythimna*) *monticola* Sugi, 1958 深山秘夜蛾

Mythimna monticola Sugi, 1958, *Tinea*, 4: 189, pl. 25: 3, 4. Type locality: Japan, Gumma Pref., Kita-

Karuisawa.

Distribution. China (Heilongjiang, Jilin, Liaoning, Beijing); Russia (Siberia), Korea, Japan.

Mythimna (*Mythimna*) *grandis* Butler, 1878 宏秘夜蛾

Mythimna grandis Butler, 1878, *Annals and Magazine of Natural History*, (5)1: 79. Type locality: Japan, Hakotate.

Synonymy: *?Segetia biundulata* Motschulsky, [1861] 1860; *Eriopyga fuliginosa* Hampson, 1905; *Hyperiodes sachalinensis* Matsumura, 1925; *Hyperiodes grandis coreana* Matsumura, 1926 (ab.); *Hyperiodes grandis ogasawarae* Matsumura, 1926 (ab.); *Hyperiodes grandis suffusa* Matsumura, 1926 (ab.); *Hyperiodes grandis hirayamae* Matsumura, 1926 (ab.); *Hyperioides sachalinensis kurilesnsis* Bryk, 1942 (ssp.); *Hyperioides grandis coreana* Bryk, 1949 (ssp.); *Hyperioides grandis chidisana* Bryk, 1949 (ssp.).

Distribution. China (Heilongjiang, Jilin, Gansu, Hubei, Sichuan); Russia (Siberia), Korea, Japan.

寄主：地杨梅属植物（陈一心，1999）；禾本科（Poaceae）植物（Matov & Kononenko, 2012）。

Mythimna (*Mythimna*) *divergens* Butler, 1878 曲线秘夜蛾

Mythimna divergens Butler, 1878, *Annals and Magazine of Natural History*, (5)1: 79. Type locality: Japan, Hakodate.

Synonymy: *Hyperiodes divergens sidemiensis* Kardakoff, 1928 (ssp.).

Distribution. China (Heilongjiang, Jilin,Shaanxi, Taiwan); Russia (Siberia), Korea, Japan.

寄主：禾本科（Poaceae）植物（Matov & Kononenko, 2012）。

Mythimna (*Mythimna*) *curvata* Leech, 1900 曲纹秘夜蛾

Mythimna grandis curvata Leech, 1900 (var.), *Transactions of the Entomological Society of London*, 1900: 130. Type locality: China; Korea.

Synonymy: *Mythimna grandis*: Poole, 1989.

Distribution. China (Heilongjiang, Jilin,Beijing, Shaanxi, Zhejiang, Hubei, Sichuan); Russia (Siberia), Korea.

Mythimna (*Mythimna*) *placida* Butler, 1878 柔秘夜蛾

Mythimna placida Butler, 1878, *Annals and Magazine of Natural History*, (5)1: 79. Type locality: Japan.

Synonymy: *Cirphis placida suavis* Draudt, 1950 (f.).

Distribution. China (Jilin, Jiangsu, Zhejiang, Hubei, Hunan, Guangxi, Hainan, Sichuan, Yunnan); Russia (Siberia), Korea, Japan.

注：《中国动物志》（夜蛾科）中将本种归入“*Aletia* Hübner, [1821] 1816 研夜蛾属”在此予以变更。

Mythimna (_Mythimna_) _velutina_ (Eversmann, 1846) 绒秘夜蛾

Leucania velutina Eversmann, 1846, *Bulletin de la Société Impériale des Naturalistes de Moscou*, 19: 87, pl. 2: 3. Type locality: Russia, Ural, Orenburg.

Synonymy: *Hyphilare coreana* Matsumura, 1926; *Sideridis velutina enervata* Warnecke, 1930 (ssp.); *Sideridis velutina kukunorensis* Bryk, 1948 (ssp.).

Distribution. China (Heilongjiang, Jilin, Liaoning, Nei Mongol, Xinjiang, Hebei); Russia (Ural, Siberia), Mongolia, Korea, Japan; Central Asia.

寄主：萹蓄（*Polygonum aviculare*）、大看麦娘（*Alopecurus pratensis*）、拂子茅（*Calamagrostis epigeios*）、发草（*Deschampsia caespitosa*）（Matov & Kononenko, 2012）。

注：《中国动物志》（夜蛾科）中将本种归入"*Leucania* Ochsenheimer, 1816 粘夜蛾属"，在此予以变更。

Mythimna (_Mythimna_) _pudorina_ ([Denis & Schiffermüller], 1775) 苇秘夜蛾

Noctua pudorina [Denis & Schifferm ü ller], 1775, *Ankündung eines Systematisches Werkes von den Schmetterlinge der Wiener Gegend. Wien*: 85 *Noctua*. Type locality: Austria, Vienna.

Synonymy: *Noctua impudens* Hübner, [1803]; *Leucania impudens obscurata* Staudinger, 1901 (var.); *Borolia subrosea* Matsumura, 1926; *Sideridis insecuta bergmani* Bryk, 1948 (ssp.); *Sideridis insecuta tancrei* Bryk, 1948 (ssp.).

Distribution. China (Heilongjiang, Jilin); Russia (European part, Ural, Siberia), Mongolia, Korea, Japan, Turkey; Central Asia, Europe.

寄主：芦苇、杂草（陈一心，1999）；疏毛地杨梅（*Luzula pilosa*）、尖叶苔草（*Carex acutiformis*）、（*Carex pulicaris*）、拂子茅（*Calamagrostis epigeios*）、鸭茅（*Dactylis glomerata*）、酸沼草（*Molinia caerulea*）、丝带草（*Phalaris arundinacea*）、芦苇（*Phragmites australis*）、甘蔗（*Saccharum officinarum*）、短柄草属（*Brachypodium*）、雀麦属（*Bromus*）植物（Matov & Kononenko, 2012）。

注：《中国动物志》（夜蛾科）中将本种归入"*Aletia* Hübner, [1821] 1816 研夜蛾属"，在此予以变更。

Mythimna (_Mythimna_) _rufipennis_ Butler, 1878 红翅秘夜蛾

Mythimna rufipennis Butler, 1878, *Annals and Magazine of Natural History*, (5)1: 79. Type locality: Japan, Yokohama.

Synonymy: *Leucania semicircula* Graeser, 1888 [1889]; *Eriopyga cirphidia* Draudt, 1950; *Leucania cirphidoides* Poole, 1989, unnecessary repl. name for *cirphidia*.

Distribution. China (Liaoning, Beijing, Zhejiang); Russia (Altai, Transbaikalia, Siberia), Korea, Japan.

寄主：日本结缕草（*Zoysia japonica*）（Matov & Kononenko, 2012）。

Mythimna (_Mythimna_) _conigera_ ([Denis & Schiffermüller], 1775) 角线秘夜蛾

Noctua conigera [Denis & Schiffermüller], 1775, *Ankündung eines Systematisches Werkes von den*

Schmetterlinge der Wiener Gegend. Wien: 84. Type locality: Austria, Vienna.

Synonymy: *Noctua floccida* Esper, 1786.

Distribution. China (Heilongjiang, Jilin, Liaoning, Nei Mongol, Hebei, Shanxi); Russia (European part, Ural, Siberia), Korea, Japan, Kazakhstan; Europe.

寄主：茶（*Camellia sinensis*）、甘蓝（*Brassica oleracea*）、异株荨麻（*Urtica dioica*）、烟草（*Nicotiana tabacum*）、翠菊（*Bellis perennis*）、疏毛地杨梅（*Luzula pilosa*）、大看麦娘（*Alopecurus pratensis*）、直立雀麦（*Bromus erectus*）、拂子茅（*Calamagrostis epigeios*）、大叶章（*Calamagrostis purpurea*）、鸭茅（*Dactylis glomerata*）、小穗发草（*Deschampsia caespitosa*）、曲芒发草（*Deschampsia flexuosa*）、葡萄冰草（*Elymus repens*）、绒毛草（*Holcus lanatus*）、黑麦草（*Lolium perenne*）、多花黑麦草（*Lolium multiflorum*）、丝带草（*Phalaris arundinacea*）、芦苇（*Phragmites australis*）、早熟禾（*Poa annua*）、偃麦草（*Triticum repens*）、玉米（*Zea mays*）、草莓属（*Fragaria*）、酸模属（*Rumex*）、蒲公英属（*Taraxacum*）、偃麦草属（*Elytrigia*）、羊茅属（*Festuca*）植物（Matov & Kononenko, 2012）。

注：《中国动物志》（夜蛾科）中将本种归入"*Aletia* Hübner, [1821] 1816 研夜蛾属"，在此予以变更。

***Mythimna* (*Mythimna*) *pallens* (Linnaeus, 1758) 苍秘夜蛾**

Phalaena pallens Linnaeus, 1758, *Systema Naturae* (Edn. 10) 1: 510. Type locality: Europe.

Synonymy: *Noctua ectypa* Hübner, [1803]; *Noctua rufescens* Haworth, 1809; *Leucania arcuata* Stephens, 1829; *Leucania ochracea* Stephens, 1829; *Leucania suffusa* Stephens, 1829; *Leucania pallens infumata* Alphéraky, 1889 (var.); *Sideridis pallescens* [sic!] *orientasiae* Bryk, 1942.

Distribution. China (Heilongjiang, Jilin, Xinjiang, Ningxia, Qinghai, Zhejiang); Russia (European part, Ural, Siberia), Mongolia, Korea, Japan; Central Asia, Middle East, Europe.

寄主：繁缕（*Stellaria media*）、甜菜（*Beta vulgaris*）、疏毛地杨梅（*Luzula pilosa*）、大看麦娘（*Alopecurus pratensis*）、大叶章（*Calamagrostis purpurea*）、鸭茅（*Dactylis glomerata*）、小穗发草（*Deschampsia caespitosa*）、曲芒发草（*Deschampsia flexuosa*）、沙丘野麦（*Elymus arenarius*）、偃麦草（*Elytrigia repens*）、草甸羊茅（*Festuca pratensis*）、浮甜茅（*Glyceria fluitans*）、大麦（*Hordeum vulgare*）、黑麦草（*Lolium perenne*）、虉草（*Phalaris arundinacea*）、早熟禾（*Poa annua*）、草地早熟禾（*Poa pratensis*）、黑麦（*Secale cereale*）、偃麦草（*Triticum repens*）、玉米（*Zea mays*）、翠雀属（*Delphinium*）、大黄属（*Rheum*）、酸模属（*Rumex*）、蔷薇属（*Rosa*）、车轴草属（*Trifolium*）、野芝麻属（*Lamium*）、蒲公英属（*Taraxacum*）、冰草属（*Agropyrum*）、梯牧草属（*Phleum*）、芦苇属（*Phragmites*）植物（Matov & Kononenko, 2012）。

注：《中国动物志》（夜蛾科）中将本种归入"*Aletia* Hübner, [1821] 1816 研夜蛾属"，在此予以变更。

***Mythimna* (*Mythimna*) *atrata* Remm & Viidalep, 1979 黑脉秘夜蛾**

Mythimna atrata Remm & Viidalep 1979, *Uchenye zapiski Tartuskogo University*; 483: 58-59. Type locality: Russia, Tuva.

Synonymy: *Mythimna jutka* Hreblay, 1990.

Distribution. China (Jilin); Russia (Tuva, Altai, Transbaikalia), Mongolia, Korea, Japan, Kazakhstan, Afghanistan; Central Asia.

注：首次给出中文种名。

Mythimna (*Mythimna*) *impura* (Hübner, [1808]) 污秘夜蛾

Noctua impura Hübner, [1808], *Sammlung Europaischer Schmetterlinge. Eulen*, 2, 3: pl. 85: 396. Type locality: Europe.

Synonymy: *Noctua fuliginosa* Haworth, 1809; *Noctua punctina* Haworth, 1809; *Leucania dungana* Alphéraky, 1882; *Leucania impura transbaicalensis* Staudinger, 1892 (var.); *Leucania impura amurensis* Staudinger, 1892 (var.); *Leucania impura scotica* Cockayne, 1944 (ssp.).

Distribution. China (Heilongjiang, Jilin, Nei Mongol, Xinjiang, Qinghai); Russia (European part, Ural, Siberia, Altai), Mongolia, Korea, Japan, Kazakhstan, Ukraine, Belarus; Central Asia, Middle East, the Baltic States, Europe.

寄主：白鳞酢浆草（*Oxalis acetosella*）、（*Juncus acutiflorus*）、地杨梅（*Luzula campestris*）、（*Luzula luzuloides*）、疏毛地杨梅（*Luzula pilosa*）、褐鞘苔草（*Carex acuta*）、尖叶苔草（*Carex acutiformis*）、（*Carex brizoides*）、丛苔草（*Carex cespitosa*）、（*Carex disticha*）、黑苔草（*Carex nigra*）、小糠草（*Agrostis alba*）、燕麦草（*Arrhenatherum elatius*）、（*Bromus hordeacus*）、无芒雀麦（*Bromus inermis*）、野青茅（*Calamagrostis arundinacea*）、拂子茅（*Calamagrostis epigeios*）、鸭茅（*Dactylis glomerata*）、小穗发草（*Deschampsia caespitosa*）、沙丘野麦（*Elymus arenarius*）、偃麦草（*Elytrigia repens*）、紫羊茅（*Festuca rubra*）、绒毛草（*Holcus lanatus*）、酸沼草（*Molinia caerulea*）、虉草（*Phalaris arundinacea*）、芦苇（*Phragmites australis*）、早熟禾（*Poa annua*）、草地早熟禾（*Poa pratensis*）、普通早熟禾（*Poa trivialis*）、芸苔属（*Brassica*）、银须草属（*Aira*）、看麦娘属（*Alopecurus*）、短柄草属（*Brachypodium*）、狗牙根属（*Cynodon*）、雀稗属（*Paspalum*）植物（Matov & Kononenko, 2012）。

注：《中国动物志》（夜蛾科）中将本种归入“*Aletia* Hübner, [1821] 1816 研夜蛾属”，在此予以变更。

Subgenus ***Pseudaletia*** Franclemont, 1951 拟粘夜蛾亚属

Pseudaletia Franclemont, 1951, *Proceedings of the Entomological Society of Washington*, 53: 64. Type species: *Leucania unipuncta* Haworth, 1809 [Great Britain].

注：《中国动物志》（夜蛾科）中将本亚属按属级单元引用为“*Pseudaletia* Franclemont, 1951 拟粘夜蛾属”，在此予以变更。

Mythimna (*Pseudaletia*) *separata* (Walker, 1865) 分秘夜蛾（粘虫）

Leucania separata Walker, 1865, *List of the Specimens of Lepidopterous Insects in the Collection of the*

British Museum, 32: 626. Type locality: China, Shanghai.

Synonymy: ?*Leucania luteomacutala* Bremer et Grey, 1853; ?*Mythimna luteomaculata* Bremer & Grey, 1853; *separans* auct.; *unipuncta* auct., nec Haworth, 1809.

Distribution. China (except Xinjiang); Russia (Transbaikalia, Siberia), Korea, Japan, India, Nepal, Pakistan, Afghanistan, Indonesia, Philippines, Australia, New Zealand.

寄主：稻、麦、高粱、玉米等（陈一心，1999）；笋瓜（*Cucurbita maxima*）、黄瓜（*Cucumis sativus*）、苦瓜（*Momordica charantia*）、甘薯（*Ipomoea batatas*）、番茄（*Lycopersicon esculentum*）、青菜（*Brassica chinensis*）、（*Manihot maritima*）、花生（*Arachis hypogaea*）、菜豆（*Phaseolus vulgaris*）、日本蒲公英（*Taraxacum japonicum*）、葱（*Allium fistulosum*）、莎草（*Cyperus rotundus*）、豇豆（*Vigna sinensis*）、华北剪股颖（*Agrostis clavata*）、（*Anchaus echinatus*）、燕麦（*Avena sativa*）、地毯草（*Axonopus compressus*）、（*Beckmania sizygachne*）、（*Brachiaria distachya*）、巴拉草（*Brachiaria mutica*）、拂子茅（*Calamagrostis epigeios*）、大叶章（*Calamagrostis lansdorfi*）、薏苡（*Coix lachryma-jobi*）、曲序香茅（*Cymbopogon flexuosus*）、枫茅（*Cymbopogon winterianus*）、狗牙根（*Cynodon dactylon*）、（*Dactyloctanium aegyptium*）、龙常草（*Dyarrhena manshurica*）、孔雀稗（*Echinochloa cruspavonis*）、牛筋草（*Eleusine indica*）、偃麦草（*Elytrigia repens*）、大麦（*Hordeum vulgare*）、稻（*Oryza sativa*）、大黍（*Panicum maximum*）、紫黍草（*Panicum purpurascens*）、铺地黍（*Panicum repens*）、（*Paspalidum flavidum*）、两耳草（*Paspalidum flavidum*）、毛花雀稗（*Paspalidum dilatatum*）、杂交狼尾草（*Pennisetum americanum*）、紫御谷（*Pennisetum glaucum*）、象草（*Pennisetum purpureum*）、筒轴茅（*Rottboellia exaltata*）、甘蔗（*Saccharum officinarum*）、黑麦（*Secale cereale*）、高粱（*Sorghum bicolor*）、苏丹草（*Sorghum sudanense*）、番麦（*Sorghum vulgare*）、小麦（*Triticum aestivum*）、玉米（*Zea mays*）、荞麦属（*Fagopyrum*）、萝卜属（*Raphanus*）、棉属（*Gossypium*）、桑属（*Morus*）、烟草属（*Nicotiana*）、番薯属（*Ipomoea*）、冰草属（*Agropyron*）、须芒草属（*Andropogon*）、鸭茅属（*Dactylis*）、白茅属（*Imperata*）、赖草属（*Leymus*）、芦苇属（*Phragmites*）、早熟禾属（*Poa*）、狗尾草属（*Setaria*）、菰属（*Zizania*）植物（Matov & Kononenko, 2012）。

Subgenus ***Sablia*** Sukhareva, 1973 **壮秘夜蛾亚属**

Sablia Sukhareva, 1973, *Entomologicheskoe Obozrenie* [*Entomological Review*], 52: 405, 413. Type species: *Leucania anderreggii* Boisduval, 1840 [Switzerland].

注：首次给出中文亚属名。

Mythimna **(*Sablia*)** ***albiradiosa*** **(Eversmann, 1852) 白辐秘夜蛾**

Leucania albiradiosa Eversmann, 1852, *Bulletin de la Société Impériale des Naturalistes de Moscou*, 25 (1): 158. Type locality: Russia, Volga reg..

Synonymy: *Leucania seifersi* Rangnow, 1930; *alboradiosa* auct., nec Eversmann, 1852.

Distribution. China (Jilin, Nei Mongol, Xinjiang); Russia (Southeast European part, Ural, Siberia), Mongolia,

Kazakhstan.

寄主：小穗发草（*Deschampsia caespitosa*）、偃麦草（*Elytrigia repens*）、梯牧草（*Phleum pratense*）（Matov & Kononenko, 2012）。

注：《中国动物志》（夜蛾科）中将本种归入“*Leucania* Ochsenheimer, 1816 粘夜蛾属”，在此予以变更。

***Mythimna* (*Sablia*) *nigrilinea* (Leech, 1899) 黑纹秘夜蛾**

Leucania nigrilinea Leech, 1899, *Proceedings of the Zoological Society of London*, 1899: 483, pl. 50: 8. Type locality: Japan, Loo-choo Isl., Yokohama.

Synonymy: *Axylia fasciata* Moore 1881.

Distribution. China (Jilin, Hubei, Taiwan); South Korea, Japan, India, Nepal, Pakistan, Thailand, Philippines, Indonesia.

注：《中国动物志》（夜蛾科）中将本种归入“*Aletia* Hübner, [1821] 1816 研夜蛾属”，在此予以变更。

Subgenus ***Hyphilare*** Hübner, [1821] 1816 **研夜蛾亚属**

Hyphilare Hübner, [1821] 1816, *Verzeichniss bekannter Schmettlinge* [sic!]: 239. Type species: *Noctua albipuncta* [Denis & Schiffermüller], 1775 [Austria, Vienna district].

***Mythimna* (*Hyphilare*) *simplex* (Leech, 1889) 单秘夜蛾**

Leucania simplex Leech, 1889, *Transactions of the Entomological Society of London*, 1889: 130. Type locality: China, jiujiang.

Synonymy: *Leucania impuncta* Staudinger, 1892; *Leucania semiusta* Hampson, 1891; *Leucania simplex corrugata* Hampson, 1894 (ssp.); *Sideridis incognita* Draudt, 1934; *Mythimna simplex incognita* (Draudt, 1934) (ssp.); *Mythimna simplex japonica* Yoshimatsu, 1994 (ssp.); *Mythimna simplex curta* Hreblauy, 1998 (ssp.).

Distribution. China (Jilin, Hubei, Hunan, Jiangxi); Russia (Siberia), Korea, Japan, Vietnam, North Thailand.

注：《中国动物志》（夜蛾科）中将本种归入“*Aletia* Hübner, [1821] 1816 研夜蛾属”，在此予以变更。

***Mythimna* (*Hyphilare*) *radiata* (Bremer, 1861) 辐秘夜蛾**

Leucania radiata Bremer, 1861, *Bulletin de la Classe Physico-Mathématique de l'Académie Impériale des Sciences de St.-Pétersbourg*, 3: 484. Type locality: Russia, [Primorye terr.], “am “Ussuri”, zwischen Noor und Ema”.

Synonymy: *Borolia stellata* Hampson, 1905.

Distribution. China (Heilongjiang, Jilin, Hunan); Russia (Siberia), Korea, Japan, India, Nepal, Thailand, Vietnam.

注：《中国动物志》（夜蛾科）中将本种归入“*Aletia* Hübner, [1821] 1816 研夜蛾属”，在此予以变更。

***Mythimna* (*Hyphilare*) *lineatipes* (Moore, 1881) 线秘夜蛾**

Leucania lineatipes Moore, 1881, *Proceedings of the Zoological Society of London*, 1881: 335. Type locality: India [?Bangladesh], East Bengalia, Cherra Punji.

Distribution. China (Heilongjiang, Jilin, Xizang); South Korea, India, Nepal, Pakistan, Thailand.

注：《中国动物志》（夜蛾科）中将本种归入"*Leucania* Ochsenheimer, 1816 粘夜蛾属"，在此予以变更。

***Mythimna* (*Hyphilare*) *proxima* (Leech, 1900) 白钩秘夜蛾**

Leucania proxima Leech, 1900, *Transactions of the Entomological Society of London*, 1900: 124. Type locality: West China, "Ni-tou".

Distribution. China (Jilin, Hebei, Henan, Sichuan, Yunnan, Xizang); Korea.

注：《中国动物志》（夜蛾科）中将本种归入"*Leucania* Ochsenheimer, 1816 粘夜蛾属"，在此予以变更。

***Mythimna* (*Hyphilare*) *flavostigma* (Bremer, 1861) 黄斑秘夜蛾**

Xanthia flavostigma Bremer, 1861, *Bulletin de la Classe Physico-Mathématique de l'Académie Impériale des Sciences de St.-Pétersbourg*, 3: 488. Type locality: Russia, [Primorye terr.] "Ussuri" "Noor".

Synonymy: *Leucania singularis* Butler, 1878; *Leucania macaria* Rebel, 1916.

Distribution. China (Heilongjiang, Jilin, Jiangsu, Zhejiang, Hunan, Fujian, Jiangxi, Yunnan); Russia (Siberia), Korea, Japan, India, Thailand, Vietnam.

寄主：禾本科（Poaceae）植物（Matov & Kononenko, 2012）。

注：《中国动物志》（夜蛾科）中将本种归入"*Aletia* Hübner, [1821] 1816 研夜蛾属"，在此予以变更。

***Mythimna* (*Hyphilare*) *salebrosa* (Butler, 1878) 崎秘夜蛾**

Leucania salebrosa Butler, 1878, *Annals and Magazine of Natural History*, (5)1: 80. Type locality: Japan, Jokohama.

Distribution. China (Heilongjiang, Jilin, Hubei, Zhejiang, Jiangxi, Fujian, Sichuan, Taiwan); Korea, Japan.

注：《中国动物志》（夜蛾科）中将本种归入"*Aletia* Hübner, [1821] 1816 研夜蛾属"，在此予以变更。

***Mythimna* (*Hyphilare*) *chosenicola* (Bryk, 1948) 朝鲜秘夜蛾**

Sideridis chosenicola Bryk, 1948, *Arkiv för Zoologi*, 41 (A) 1: 86. Type locality: North Korea, Shuotsu [near Kyeongseong, Hamgyeong bugdo].

Synonymy: *Aletia insalebrosa* Sugi, 1982.

Distribution. China (Heilongjiang, Jilin, Liaoning); Russia (Siberia), Korea, Japan.

寄主：拂子茅属（*Calamagrostis*）、鸭茅属（*Dactylis*）、发草属（*Deschampsia*）、羊茅属

（*Festuca*）、虉草属（*Phalaris*）植物（Matov & Kononenko, 2012）。

Subgenus ***Dysaletia*** Sugi, 1982 庸秘夜蛾亚属

Dysaletia Sugi, 1982, *Moths of Japan*, 1: 719; 2: 358, 462. Type species: *Leucania inanis* Oberthür, 1880 [Russia, East].

注：新给出中文亚属名。

***Mythimna* (*Dysaletia*) *inanis* (Oberthür, 1880) 庸秘夜蛾**

Leucania inanis Oberthür, 1880, *Études d'Entomologie*, 5: 70, pl. 3: 4. Type locality: Russia, [Primorye terr.] Askold I..

Synonymy: *Leucania gigas* Staudinger, 1901; *Cirphis mesotrostina* Draudt, 1950.

Distribution. China (Heilongjiang, Jilin, Liaoning, Shaanxi, Yunnan); Russia (Siberia), Korea, Japan.

注：《中国动物志》（夜蛾科）中将本种归入"*Leucania* Ochsenheimer, 1816 粘夜蛾属"，在此予以变更。

Genus ***Anapoma*** Berio, 1980 珀夜蛾属

Anapoma Berio, 1980, *Bolletino della Societa Entomologica Italiana*, 112: 40. Type pecies: *Leucania riparia* Rambur, 1829 [France].

Synonymy: *Apoma* Berio, 1980, preocc. Beck, 1938 (Mollusca).

***Anapoma postica* (Hampson, 1905) 后珀夜蛾**

Cirphis postica Hampson, 1905, *Catalogue of the Lepidoptera Phalaenae in the British Museum*, 5: 535, pl. 96: 12. Type locality: Japan, Yokohama.

Synonymy: *Sideridis incognita draudtiana* Bryk, 1942 (ssp.); *Sideridis incognita draudtiphila* Bryk, 1948 (ssp.); *Anapoma postica* Hacker *et al.*, 2002; Zhang *et al.*, 2019.

Distribution. China (Heilongjiang, Jilin, Jiangxi, Xizang); Russia (Siberia), Korea, Japan.

Genus ***Leucania*** Ochsenheimer, 1816 粘夜蛾属

Leucania Ochsenheimer, 1816, *Die Schmetterlinge von Europa*, 4: 81. Type species: *Phalaena comma* Linnaeus, 1761 [Sweden].

Synonymy: *Leucania* Boisduval, 1828, preocc. (Ochsenheimer, 1816 [Lepidoptera, Noctuidae]); *Donachlora* Sodoffsky, 1837, repl. name; *Leucadia* Sodoffsky, 1837, emend.; *Donacochlora* Agassiz, 1846; *Pudorina* Gistl, 1848, repl. name; *Cirphis* Walker, 1865; *Donacochlora* Agassiz, [1874], emend.; *Eurypsyche* Butler, 1886; *Neoborolia* Matsumura, 1926; *Acantholeucania* Rungs, 1953 (subgen.); *Xyphroleucania* Sugi, 1970 (subgen.); *Donochlora* Poole, 1989, missp.; *Broszkusia* Beck, 1999.

The genus contains about 200 species, distributed in all biogeographical regions, except Arctic; the genus is most diverse in the subtropical montane regions. Among them, 2 species are in the Three Provinces of Northeast China.

Leucania comma **(Linnaeus, 1761) 粘夜蛾**

Phalaena comma Linnaeus, 1761, *Fauna Suecica* (Edn. 2) : 316. Type locality: Sweden.

Synonymy: *Noctua turbida* Hübner, [1803]; *Noctua congener* Hübner, [1817]; *Leucania suffusa* Tutt, 1888; *Leucania nigrofasciata* Hampson, 1894; *Leucania comma rhodocomma* Püngeler, 1900 (var.); *Cirphis propensa* Püngeler, 1906; *Leucania comma engadensis* Wagner, 1909 (var.).

Distribution. China (Heilongjiang, Qinghai); Russia (European part, Ural, Siberia), Mongolia; Central Asia, Middle East, Europe.

寄主：鸭茅、酸模（陈一心，1999）；大看麦娘（*Alopecurus pratensis*）、小穗发草（*Deschampsia caespitosa*）、曲芒发草（*Deschampsia flexuosa*）、高羊茅（*Festuca arundinacea*）、羊茅（*Festuca ovina*）、草甸羊茅（*Festuca pratensis*）、早熟禾（*Poa annua*）、酸模属（*Rumex*）、蒲公英属（*Taraxacum*）、薹草属（*Carex*）、拂子茅属（*Calamagrostis*）、鸭茅属（*Dactylis*）、虉草属（*Phalaris*）植物（Matov & Kononenko, 2012）。

Leucania loreyi **(Duponchel, 1827) 白点粘夜蛾**

Noctua loreyi Duponchel, 1827, *in* Godard & Duponchel, *Histoire Naturelle des Lépidoptères ou Papillons de France*, 7 (1): 81, pl. 105: 7. Type locality: France, Dijon.

Synonymy: *Noctua caricis* Treitschke, 1835; *Leucania curvula* Walker, 1856; *Leucania collecta* Walker, 1856; *Leucania exterior* Walker, 1856; *Leucania thoracica* Walker, 1856; *Leucania designata* Walker, 1856; *Leucania denotata* Walker, 1856; *Borolia melanostrotoides* Strand, 1915.

Distribution. China (Heilongjiang, Jilin, Central, South, East); Russia (South Ural), Korea, Japan, India, Nepal, Pakistan; Southeast Asia, Middle East, Europe, Africa.

寄主：稻（陈一心，1999）。

Leucania obsoleta **(Hübner, [1803]) 合粘夜蛾**

Noctua obsoleta Hübner, [1803], *Sammlung Europaischer Schmetterlinge. Eulen*, 2: pl. 48: 233. Type locality: Europe.

Synonymy: *Leucania insecuta* Walker, 1865; *Leucania intermissa* Walker, 1865; *Neoborolia nohirae* Matsumura, 1926; *Borolia griseola* Matsumura, 1926.

Distribution. China (Jilin); Russia (European part, Ural, Siberia), Mongolia; Central Asia, Middle East, Europe.

寄主：萹蓄 (*Polygonum aviculare*)、天仙子（*Hyoscyamus niger*）、烟草（*Nicotiana tabacum*）、野青

茅（*Calamagrostis arundinacea*）、拂子茅（*Calamagrostis epigeios*）、芦苇（*Phragmites australis*）、阿尔泰旱禾（*Poa altaica*）、草地早熟禾（*Poa pratensis*）、玉米（*Zea mays*）、芦竹属（*Arundo*）植物（Matov & Kononenko, 2012）。

Genus ***Senta*** Stephens, 1834 **縻夜蛾属**

Senta Stephens, 1834, *Illustrations of British Entomology*, 4: 297. Type species: *Melia flammea* Curtis, 1828 [England].

Synonymy: *Meliana* Curtis, 1836.

Monotypic genus; 1 species is in the Three Provinces of Northeast China.

Senta flammea **(Curtis, 1828) 縻夜蛾**

Melia flammea Curtis, 1828, *British Entomology; being Illustrations and Descriptions of the Genera of Insects found in Great Britain and Ireland*, 5:201. Type locality: England, London distr..

Synonymy: *Simyra dubiosa* Treitschke, 1835; *Nonagria arundinicola* Doubleday, 1848; *Meliana flammea stenoptera* Staudinger, 1892 (var.).

Distribution. China (Jilin); Russia (European part, Ural, Siberia), Korea, Japan; Europe.

寄主：稻（*Oryza sativa*）、芦苇（*Phragmites australis*）（陈一心，1999；Matov & Kononenko, 2012）。

Tribe **Eriopygini** Fibiger & Lafontaine, 2005 **茸夜蛾族**

Eriopygini Fibiger & Lafontaine, 2005, *Esperiana*, 11: 12. Type genus: *Eriopygodes* Hampson, 1905.

Synonymy: Eriopygini Franclemont & Todd, 1983, nomen nudum.

The tribe includes about 30 genera, mainly distributed in the temperate and partially in the subtemperate zones of the Nearctic region. The genus *Lasionycta* has Holarctic distribution, mainly in the boreal zone. 1 genus and 2 species are in the Three Provinces of Northeast China.

注：首次给出中文族名。

Genus ***Lasionycta*** Aurivillius, 1892 **茸夜蛾属**

Lasionycta Aurivillius, 1892, *Entomologisk Tidskrift*, 13: 285. Type species: *Phlogophora skraelingia* Herrich-Schaffer, 1852 [Scandinavia, Lappland].

Synonymy: *Lasiestra* Hampson, 1905; *Eriopygodes* Hampson, 1905; *Lascionycta* Hill, 1927, missp.; *Pseudanarta* Kozhanchikov, 1947, preocc.; *Anartomima* Boursin, 1952; *Lasionhada* Berio, 1980; *Clemathada* Beck, 1991; *Hada* auct..

Holarctic distributed species includes about 80 species associated mainly with subarctic, boreal and alpine zones. Some species associated with boreal or montane bogs. 3 species are occurring in the Three Provinces of

Northeast China.

Lasionycta skraelingia (Herrich–Schäffer, 1852) 北欧茸夜蛾

Phlogophora skraelingia Herrich–Schäffer, 1852, *Systematische Bearbeitung der Schmetterlinge von Europa* 2: 121, pl. 624: 6 Nachtrag: 57. Type locality: "Lappland".

Synonymy. *Lasionycta scraelingia* Hampson, 1905, emend.; *Lasionycta skroelingia* Warren 1910 (missp.).

Distribution. China (Heilongjiang); Russia (Siberia), Japan (Hokkaido); North Europe (Fennoscandia); North America (Alaska, Yukon).

寄主：幼虫杂食性，危害矮桦（*Betula nana*）、笃斯越橘（*Vaccinium uliginosa*）、萹蓄（*Polygonum aviculare*）和许多苗圃植物。

注：首次给出本种中文种名。

Lasionycta hospita Bang–Haas, 1912 苏茸夜蛾

Lasionycta hospita Bang–Haas, 1912, *Deutsche Entomologische Zeitschrift. Gesellschaft Iris zu Dresden*, 26: 150, pl. 6: 14. Type locality: Russia, [Buryatia], Sayan Mts..

Synonymy. *Lasionycta ardua* Filipjev, 1925.

Distribution. China (Heilongjiang); Russia (Altai, Tuva, Krasnoyarsk terr., Sayan, Baikal area, Transbaikalia, Amur reg., Khabarovsk and Primorye terr., Sakhalin Isl.), Northern Mongolia.

Lasionycta proxima (Hübner, 1808–1809) 近灰茸夜蛾

Noctua proxima Hübner, 1808–1809, *Sammlung Europaischer Schmetterlinge. Eulen*, 2: pl 87: 409. Type locality: Europe.

Synonymy: *Hadena cana* Eversmann, 1841; *Hadena ochrostigma* Eversmann, 1842; *Hoplodrina noguera* deLaever, 1976.

Distribution. China (Heilongjiang, Xinjiang); Russia (Ural, Siberia).

寄主：皱叶酸模（*Rumex crispus*）、西洋石竹（*Dianthus deltoides*）、膀胱麦瓶草（*Silene cucubalus*）、迪奥卡蝇子草（*Silene dioica*）、单花雪轮（*Silene uniflora*）、狗筋麦瓶草（*Silene vulgaris*）、黑果越橘（*Vaccinium myrtillus*）、黄花柳（*Salix caprea*）、羽衣草（*Alchemilla vulgaris*）、广布野豌豆（*Vicia cracca*）、欧洲柳穿鱼（*Linaria vulgaris*）、荒野蒿（*Artemisia campestris*）、车前属（*Plantago*）、风铃草属（*Campanula*）、风毛菊属（*Leontodon*）、蒲公英属（*Taraxacum*）植物（Matov & Kononenko, 2012）。

Tribe **Noctuini** Latreille, 1809 夜蛾族

Noctuidae Latreille, 1809, *Genera Crustaceorum Insectorum*, 4: 191, 224 (as Noctuaelites), incorrect original spelling which had to be emended in Noctuidae; official list of family-group names, opinion 450. Type genus: *Noctua* Linnaeus, 1758.

Synonymy: Noctuida Leach, [1815] 1830 (incorrect subsequent spelling); Noctuides Leach, [1815] 1830 (incorrect subsequent spelling); Noctuelidi Boisduval, [1828] 1829 (incorrect subsequent spelling); Noctuelidae Duponchel, [1845] 1844 (incorrect subsequent spelling); Noctuelites Guenée, 1852 (incorrect subsequent spelling); Noctues Swinhoe, 1890 (incorrect subsequent spelling).

The tribe is probably the most species-rich unit of all Noctuoidea distributed over the World, in all natural zones. Traditionally, it was treated by authors as subfamily, before its status was downgraded to tribe (Lafontaine & Schmidt, 2010). The subtribes Agrotina and Noctuina were treated as a tribes. The main morphological feature of this group is presence of spines on tibia of mid and forelegs or on all legs in combination with typical triphinae hind wing venation (M2 on hindwing always missing). The members of this tribe are most diverse in the cool temperate regions of both hemispheres. Many genera and species have Holarctic distribution. The tribe represented in the Three Provinces of Northeast China by 28 genera and 80 species occurring in various habitats.

References. Hampson, 1903 (Agrotinae); Kozhanchikov, 1937 (Agrotinae); Boursin, 1948, 1954a, 1963 ets. (Agrotinae); Kostrowicki, 1959 (Agrotinae); Zolotarenko, 1970 (Agrotinae); Maschenko, 1980 (Noctuinae); Chen, 1982, 1999 (Noctuidae: Noctuinae); Sugi, 1982, 1995 (Noctuinae); Lafontaine *et al.,* 1983, 1987a, 1998, 2004 (Noctuinae); Lafontaine, Kononenko, 1986, 1988a (Noctuinae); Mikkola *et al.*, 1987, 1991, 1989 (Noctuinae); Fibiger, 1990, 1993, 1997 (Noctuinae); Kononenko, 1990, 2003, 2005, 2010, 2016 (Noctuinae); Kullberg *et al.*, 1995 (Noctuinae); Kononenko *et al.,* 1998 (Noctuinae); Lafontaine *et al.*, 1998 (Noctuinae); Kononenko, Han, 2007 (Noctuinae); Matov *et al.*, 2008 (Noctuinae); Lafontaine, Schmidt, 2010 (Noctuinae: Noctuini); Fibiger *et al.*, 2011 (Noctuinae: Noctuini); Varga *et al.*, 2013 (Noctuinae); Zahiri *et al.*, 2013b (Noctuinae: Noctuini).

Subtribe **Agrotina** Rambur, 1848 地夜蛾亚族

Agrotides Rambur, 1848, *Annales de la Société Entomologique de France*, (2)6: 65. Type genus: *Agrotis* Ochsenheimer, 1816.

Synonymy: Agrotidae Heinemann, 1859; Brotinae Grote, 1882; Euxoinae Warren, 1909, repl. name; Aniclini Franclemont & Todd, 1983, nomen nudum; Dichagyrina Beck, 1996; Euxoina Beck, 1996.

The subtribe represented in the Three Provinces of Northeast China by 6 genera and 25 species occurring in the various habitats.

注：首次给出中文亚族名。

Genus ***Peridroma*** Hübner, [1821] 1816 疆夜蛾属

Peridroma Hübner, [1821] 1816, *Verzeichniss bekannter Schmettlinge* [sic!]: 380. Type species: *Noctua saucia* Hübner, [1808] [Europe].

The genus contains 14 species, mainly distributed in the arid and tropic areas of the Ethiopian and Neotropic regions, with 1 species are in the Palaearctic. 1 species is in the Three Provinces of Northeast China.

Peridroma saucia **(Hübner, 1808) 疆夜蛾**

Noctua saucia Hübner, 1808, *Sammlung Europaischer Schmetterlinge*, [4] : pl. 81, f. 378. Type locality: Europe.

Synonymy: *Noctua majuscula* Haworth, 1809; *Noctua margaritosa* Haworth, 1809; *Noctua aequa* Hübner, 1813; *Brotis orophila* Geyer, 1837; *Agrotis inermis* Harris, 1841; *Spaelotis infuscata* Blanch, 1854; *Spaelotis stictica* Blanchard, 1854; *Agrotis impacta* Walker, 1856; *Agrotis intecta* Walker, 1856; *Agrotis angulifera* Wallengren, 1860; *Agrotis ortonii* Packard, 1869; *Lycophotia ochronata* Hampson, 1903; *Agrotis saucia tenebricosa* Schawerda, 1929 (var.).

Distribution. China (Heilongjiang, Jilin, Gansu, Ningxia, Sichuan, Yunnan, Xizang, Taiwan); South Korea, Japan, India, Nepal, Pakistan, Philippines, Indonesia, Thailand, New Zealand, Australia; Near East, Europe, America.

寄主：玉米、马铃薯、小麦、豆类、瓜类、蔬菜、高粱、牧草等（陈一心，1999）；欧洲蕨（*Pteridium aquilinum*）、香脂冷杉（*Abies balsamea*）、巨冷杉（*Abies grandis*）、美国黑松（*Pinus contorta*）、花旗松（*Pseudotsuga taxifolia*）、加拿大铁杉（*Tsuga canadensis*）、美国尖叶扁柏（*Chamaecyparis thyoides*）、单子麻黄（*Ephedra nebrodensis*）、鳄梨（*Persea americana*）、头石竹（*Dianthus barbatus*）、康乃馨（*Dianthus caryophyllus*）、繁缕（*Stellaria media*）、萹蓄（*Polygonum aviculare*）、（*Rheum rhabarbaram*）、皱叶酸模（*Rumex crispus*）、钝叶酸模（*Rumex obtusifolius*）、甜菜（*Beta vulgaris*）、白栎（*Quercus alba*）、美国赤杨（*Alnus rubra*）、狭叶越橘（*Vaccinium angustifolium*）、（*Vaccinium myrtilloides*）、三色堇（*Viola tricolor*）、斯考勒氏柳（*Salix scouleriana*）、（*Salix hookeriana*）、黄瓜（*Cucumis sativus*）、褐芥末（*Brassica hirta*）、欧洲油菜（*Brassica napus*）、黑芥（*Brassica nigra*）、甘蓝（*Brassica oleracea*）、白菜（*Brassica pekinensis*）、芜青（*Brassica rapa*）、荠菜（*Capsella bursapastoris*）、紫罗兰（*Matthiola incana*）、萝卜（*Raphanus sativus*）、菠菜（*Spinacia oleracea*）、蜀葵（*Althaea rosea*）、草棉（*Gossypium herbaceum*）、美国榆（*UlmuSouth Americana*）、橙桑（*Maclura pomifera*）、啤酒花（*Humulus lupulus*）、桂叶茶藨子（*Ribes lacustre*）、多花醋栗（*Ribes sanguineum*）、红果茶藨子（*Ribes sativum*）、（*Amelanchier florida*）、杏（*Armeniaca vulgaris*）、欧洲酸樱桃（*Cerasus vulgaris*）、草莓（*Fragaria chiloensis*）、（*Malus domestica*）、苹果（*Malus pumila*）、白桃（*Persica vulgaris*）、欧洲李（*Prunus domestica*）、苦樱桃(*Prunus emarginata*)、（*Rubus alleghiensis*）、覆盆子（*Rubus idaeus*）、喜阴悬钩子（*Rubus occidentalis*）、柳兰（*Chamaenerion angustifolium*）、美国皂荚（*Gleditsia triacanthos*）、大豆（*Glycine max*）、香豌豆（*Lathyrus odoratus*）、紫苜蓿（*Medicago sativa*）、白花草木樨（*Melilotus alba*）、印度草木樨（*Melilotus indica*）、（*Ononis hirta*）、豌豆（*Pisum sativum*）、刺槐（*Robinia pseudoacacia*）、红车轴草（*Trifolium pratense*）、梣叶槭（*Acer negundo*）、糖槭（*Acer saccharum*）、柠檬（*Citrus limon*）、甜橙（*Citrus sinensis*）、多花酢浆草（*Oxalis martiana*）、印楝（*Azadirachta indica*）、旱金莲

（*Tropaeolum majus*）、葡萄（*Vitis vinifera*）、旱芹（*Apium graveolens*）、野胡萝卜（*Daucus carota*）、茜草（*Rubia peregrina*）、辣椒（*Capsicum annuum*）、曼陀罗（*Datura stramonium*）、番茄（*Lycopersicon esculentum*）、近缘烟草（*Nicotiana affinis*）、烟草（*Nicotiana tabacum*）、马铃薯（*Solanum tuberosum*）、田旋花（*Convolvulus arvensis*）、长叶车前（*Plantago lanceolata*）、大车前（*Plantago major*）、白网纹草（*Fittonia verschaffeltii*）、辣薄荷（*Mentha piperita*）、鼠尾草（*Salvia officinalis*）、豚草（*Ambrosia artemisiifolia*）、臭春黄菊（*Anthemis cotula*）、翼蓟（*Cirsium vulgare*）、小蓬草（*Conyza canadensis*）、洋蓟（*Cynara scolymus*）、泽兰属（*Eupatorium*）、向日葵（*Helianthus annuus*）、莴苣（*Lactuca sativa*）、（*Onopordum palaestinum*）、灰白银胶菊（*Parthenium argentatum*）、（*Senecio aethnensis*）、雪叶莲（*Senecio cineraria*）、（*Senecio erraticus*）、（*Solidago leavenworthii*）、苍耳（*Xanthium strumarium*）、洋葱（*Allium cepa*）、山韭（*Allium senescens*）、亮绿天门冬（*Asparagus medeoloides*）、石刁柏（*Asparagus officinalis*）、鸡绒芝（*Asparagus plumosa*）、圆叶菝葜（*Smilax rotundifolia*）、菖蒲（*Acorus calamus*）、梯牧草（*Phleum pratense*）、小麦（*Triticum aestivum*）、玉米（*Zea mays*）、胡椒属（*Piper*）、马齿苋属（*Portulaca*）、菠菜属（*Spinacia*）、南瓜属（*Cucurbita*）、桑属（*Morus*）、荨麻属（*Urtica*）、醋栗属（*Grossularia*）、蔷薇属（*Rosa*）、蚕豆属（*Vicia*）、桉属（*Eucalyptus*）、老鹳草属（*Geranium*）、金盏花属（*Calendula*）、菊属（*Chrysanthemum*）、菊苣属（*Cichorium*）、高粱属（*Sorghum*）植物（Matov & Kononenko, 2012）。

Genus ***Actebia*** Stephens, 1829 翠色狼夜蛾属

Actebia Stephens, 1829 [June], *The Nomenclature of British Insectes*: 41. Type species: *Phalaena praecox* Linnaeus, 1758 [Europe].

Synonymy: *Hapalia* Hübner, [1821] 1816, preocc. (Hübner, 1818 [Lepidoptera, Pyralidae]); *Perissandria* Warren, 1909 (subgen.); *Dissmactebia* Beck, [1992] 1991; *Ochropleura* auct., nec Hübner, [1821] 1816; *Protexarnis* McDunnough, 1928 (subgen.); *Hemiexarnis* Boursin (subgen.); *Parexarnis* McDunnough, 1929 (subgen.); *Ledereragrotis* Varga, 1990 (subgen.); *Actobia*: Agassiz, 1846, emend..

The genus contains about 50 species divided for 6 subgenera. Most species are distributed in the arid zone. 4 species of 2 subgenera are in the Three Provinces of Northeast China.

注：《中国动物志》（夜蛾科）中引用本属的同物异名“*Ochropleura* Hübner, [1821] 1816”，在此予以变更。

Subgenus ***Actebia*** Stephens, 1829 翠色狼夜蛾亚属

***Actebia* (*Actebia*) *praecox* (Linnaeus, 1758) 翠色狼夜蛾**

Phalaena praecox Linnaeus, 1758, *Systema Naturae* (Edn. 10) 10: 517. Type locality: Europe.

Synonymy: *Noctua praeceps* [Denis & Schiffermüller], 1775; *Agrotis praecox flavomaculata* Graeser, 1888 [1889] (var.).

Distribution. China (Heilongjiang, Jilin, Liaoning, Hebei, Taiwan); Russia (European part, Ural, Siberia),

Mongolia, Korea, Japan, Kazakhstan; Middle East, Europe.

寄主：桃、梨、柳、蒿属等植物（陈一心，1999）；（*Honkenya peploides*）、仰卧漆姑草（*Sagina procumbens*）、繁缕（*Stellaria media*）、小酸模（*Rumex acetosella*）、核桃楸（*Juglans manshurica*）、笃斯越橘（*Vaccinium uliginosum*）、三色堇（*Viola tricolor*）、白柳（*Salix alba*）、匍匐柳（*Salix repens*）、蒿柳（*Salix viminalis*）、黄瓜（*Cucumis sativus*）、甘蓝（*Brassica oleracea*）、柏大戟（*Euphorbia cyparissias*）、（Malus domestica）、白桃（*Persica vulgaris*）、覆盆子（*Rubus idaeus*）、大豆（*Glycine max*）、海边香豌豆（*Lathyrus maritimus*）、百脉根（*Lotus corniculatus*）、树羽扇豆（*Lupinus arboreus*）、驴食草（*Onobrychis viciifolia*）、匍匐芒柄草（*Ononis repens*）、菜豆（*Phaseolus vulgaris*）、豌豆（*Pisum sativum*）、白车轴草（*Trifolium repens*）、黄檗（*PhellodendroNorth of Amur reg.ense*）、（*Vitis kaemferi*）、葡萄（*Vitis vinifera*）、水芹（*Oenanthe javanica*）、番茄（*Lycopersicon esculentum*）、（*Convolvulus soldanella*）、药用牛舌草（*Anchusa officinalis*）、蓝蓟（*Echium vulgare*）、大车前（*Plantago major*）、红辣椒千叶蓍（*Achillea millefolium*）、荒野蒿（*Artemisia campestris*）、北艾（*Artemisia vulgaris*）、菊蒿（*Tanacetum vulgare*）、石刁柏（*Asparagus officinalis*）、欧洲海滨草（*Ammophila arenaria*）、沙丘野麦（*Elymus arenarius*）、（*Festuca arenaria*）、紫羊茅（*Festuca rubra*）、紫草属（*Lithospermum*）、柳穿鱼属（*Linaria*）、槭树属（*Acer*）、山柳菊属（*Hieracium*）、卷耳属（*Cerastium*）、滨藜属（*Atriplex*）、酸模（*Rumex acetosa*）、山芥属（*Barbarea*）、茶藨子属（*Ribes*）、山楂属（*Crataegus*）、梨属（*Pyrus*）、锦鸡儿属（*Caragana*）、苜蓿属（*Medicago*）、梾木属（*Swida*）、苦苣菜属（*Sonchus*）、蒲公英属（*Taraxacum*）植物（Matov & Kononenko, 2012）。

Actebia (*Actebia*) *praecurrens* (Staudinger, 1888) 黑齿翠色狼夜蛾

Agrotis praecurrens Staudinger, 1888, *Entomologische Zeitung. Entomologischen Vereine zu Stettin*, 49: 248. Type locality: Russia, Primorye terr., Vladivostok, [Ussuriisk] Nikolsk, Baranovsky, [Razdolnaya river] Suifun, [Bezverkhovo] Sidemi.

Synonymy: *Agrotis bisagittata* Graeser, 1888 [1889].

Distribution. China (Heilongjiang, Jilin, Liaoning, Nei Mongol); Russia (Siberia), Korea, Japan.

寄主：甜菜（*Beta vulgaris*）、甘蓝（*Brassica oleracea*）、萝卜（*Raphanus sativus*）、紫苜蓿（*Medicago sativa*）、野胡萝卜（*Daucus carota*）、辣椒（*Capsicum annuum*）、番茄（*Lycopersicon esculentum*）、马铃薯（*Solanum tuberosum*）、车前属（*Plantago*）、蒲公英属（*Taraxacum*）、车轴草属（*Trifolium*）、藜芦属（*Veratrum*）植物（Matov & Kononenko, 2012）。

Actebia (*Actebia*) *fennica* (Tauscher, 1806) 黑地翠色狼夜蛾

Noctua fennica Tauscher, 1806, *Mémoires de la Société Impériale des Naturalistes de Moscou*, 1: 210; [1812], *Mémoires de la Société Impériale des Naturalistes de Moscou*, (2nd edn.) 1: 177, pl. 13: 4. Type locality: Russia, St.-Petersburg reg. ["Petropolis"].

Synonymy: *Hadena intracta* Walker, 1857; *Agrotis eversmanni* Krulikowsky, 1893; *Rhyacia unicolora*

Kozhantshikov, 1925.

Distribution. China (Heilongjiang, Jilin, Nei Mongol, Xinjiang); Russia (Northeast of European part, Ural, Siberia), Mongolia, Korea, Japan; Europe, North America.

寄主：美国黑松（*Pinus contorta*）、（*Corydalis fabacea*）、巨紫堇（*Corydalis gigantea*）、山延胡索（*Corydalis solida*）、甜菜（*Beta vulgaris*）、黄瓜（*Cucumis sativus*）、甘蓝（*Brassica oleracea*）、芜青（*Brassica rapa*）、萝卜（*Raphanus sativus*）、（Malus domestica）、覆盆子（*Rubus idaeus*）、柳兰（*Chamaenerion angustifolium*）、番茄（*Lycopersicon esculentum*）、马铃薯（*Solanum tuberosum*）、玉米（*Zea mays*）、柳叶菜属（*Epilobium*）、绣线菊属（*Spiraea*）、越橘属（*Vaccinium*）、茶藨子属（*Ribes*）、草莓属（*Fragaria*）、亚麻属（*Linum*）、葱属（*Allium*）植物（Matov & Kononenko, 2012）。

Subgenus ***Ledereragrotis*** Varga, 1990 **莱狼夜蛾亚属**

Ledereragrotis Varga, 1990, *in* Varga, Ronkay & Yela, *Acta Zoologica Hungarica*, 36 (3-4): 336. Type species: *Agrotis multifida* Lederer, 1870 [Transcaucasus].

***Actebia* (*Ledereragrotis*) *difficilis* (Erschoff, 1887) 鹜甲卡狼夜蛾**

Agrotis difficilis Erschoff, 1887, *Horae Societatis Entomologicae Rossicae*, 9: 377. Type locality: Russia, Irkutsk.

Synonymy: *Agrotis alpherakii* Christoph, 1893; *Agrotis albirenata* Staudinger, 1893.

Distribution. China (Heilongjiang, Nei Mongol, Qinghai, Xizang); Russia (Altai, Transbaikalia), Mongolia.

寄主：禾本科（Poaceae）植物（Matov & Kononenko, 2012）。

注：《中国动物志》（夜蛾科）中将本种归入"*Chersotis* Boisduval, 1840 卡夜蛾属"，在此予以变更。

Genus ***Dichagyris*** Lederer, 1857 **分狼夜蛾属**

Dichagyris Lederer, 1857, *Die Noctuinen Europa's, mit Zuziehung einger bisher mesit dazu gezahlter Arten des Asiatischen Russland's, Kleinasien's, Syrien's, und Labrador's*:45, 206. Type species: *Agrotis melanura* Kollar, 1846 [Yugoslavia, Dalmatia].

Synonymy: *Ogygia* Hübner, [1821]; *Pseudorthosia* Grote, 1874 (subgen.); *Phleboeis* Christoph, 1887 (subgen.); *Stenosoma* Turati, 1924; *Mesembragrotis* Barnes & Benjamin, 1927 (subgen.); *Loxagrotis* MacDunnough, [1929] (subgen.); *Pseudoseptis* McDunnough, [1929]; *Proragrotis* MacDunnough, [1929]; *Stenosomides* Strand, 1942 (subgen.); *Yigoga* Nye, 1975; *Pseudochropleura* Beck, 1992, nom. nud.; *Grisyigoga* Beck, [1992] 1991; *Trumuspis* Beck, 1996; *Renyigoga* Beck, 1996; *Flavyigoga* Beck, 1996; *Nigryigoga* Beck, 1996; *Vallagyris* Beck, 1996; *Stellagyris* Beck, 1996, nom. nud.; *Celagyris* Beck, 1996, nom. nud.; *Constantargyris* Beck, 1996; *Albocosta* Fibiger & Lafontaine, 1997 (subgen.); *Basistriga* Fibiger & Lafontaine, 1997; *Pseudorichia* Lafontaine, 2004; *Ochropleura* auct..

The genus contains over 200 species, mainly distributed in the arid and temperate areas of the Palaearctic,

Ethiopian and Nearctic regions. Among them, 2 species of 1 subgenus are in the Three Provinces of Northeast China.

注：首次给出中文属名。

Subgenus *Albocosta* Fibiger & Lafontaine, 1997 缪狼夜蛾亚属

Albocosta Fibiger & Lafontaine, 1997, *Noctuidae Europaea*, 3: 126. Type species: *Noctua musiva* Hübner, 1803 [Europe].

Synonymy: *Pseudochropleura* Beck, 1992, nom.nud..

注：首次给出中文亚属名。

Dichagyris (*Albocosta*) *stentzi* (Lederer, 1853) 衍分狼夜蛾

Chersotis stentzi Lederer, 1853, *Verhandlungen des Zoologisch-Botanischen vereins in Wien*, 3: 367, pl. 4: 4. Type locality East Kazakhstan, [West Altai] Ust-Kamengorsk, Ust-Buchtarminsk.

Synonymy: *Agrotis stensi* Hampson, 1903, emend..

Distribution. China (Heilongjiang, Jilin, Liaoning, Nei Mongol, Xinjiang, Qinghai, Hebei, Henan, Yunnan, Xizang); Russia (Altai, Transbaikalia, Siberia), Mongolia, Korea, India(North), Nepal; Central Asia.

寄主：黄瓜（*Cucumis sativus*）、甘蓝（*Brassica oleracea*）、番茄（*Lycopersicon esculentum*）、酸模属（*Rumex*）、车轴草属（*Trifolium*）植物（Matov & Kononenko, 2012）。

注：《中国动物志》（夜蛾科）中将本种归入“*Ochropleura* Hübner, [1821] 1816 狼夜蛾属”，在此予以变更。

Dichagyris (*Albocosta*) *triangularis* (Moore, 1867) 基角分狼夜蛾

Ochropleura triangularis Moore, 1867, *Proceedings of the Zoological Society of London*, 1867: 55. Type locality: North India, Sikkim, Darjiling.

Distribution. China (Heilongjiang, Jilin, Gansu, Sichuan, Yunnan, Xizang, Taiwan); Russia (Siberia), Korea, Japan, India, Nepal, Pakistan, Thailand; Central Asia.

注：《中国动物志》（夜蛾科）中将本种归入“*Ochropleura* Hübner, [1821] 1816 狼夜蛾属”，给出中文种名“基角狼夜蛾”，在此予以变更。

Genus *Euxoa* Hübner, [1821] 1816 切夜蛾属

Euxoa Hübner, [1821] 1816, *Verzeichniss bekannter Schmettlinge* [sic!]: 209. Type species: *Noctua decora* [Denis & Schiffermüller], 1775 [Austria].

Synonymy: *Mimetes* Hübner, [1821] 1816, preocc. (Eschscholtz, 1818 [Coleoptera]); *Metaxyja* Hübner, [1821] 1816; *Exarnis* Hübner, [1821] 1816; *Brotis* Hübner, [1821] 1816; *Telmia* Hübner, [1821] 1816; *Pleonectopoda* Grote, 1873 (subgen.); *Carneades* Grote, 1883, preocc. (Bates, 1869 [Mammalia]); *Chorizagrotis*

Smith, 1890 (subgen.); *Paragrotis* Dyar, [1903], repl. name; *Orosagrotis* Hampson, 1903 (subgen.); *Mesoeuxoa* Corti, 1932; *Menada* Kozhanchikov, 1937; *Longivesica* Hardwick, 1970 (subgen.); *Crassivesica* Hardwick, 1970; *Palaeoeuxoa* Lafontaine, 1987 (subgen.); *Heteroeuxoa* Lafontaine, 1987 (subgen.). *Metaxya*: Walker, 1857, missp.; *Mimetis*: Hampson, 1903, missp.; *Metaxyia*: Hampson, 1903, missp.; *Prosagrotis*: Warren, 1911, missp..

The genus contains totally over 340 species, mainly distributed in the arid and temperate areas of the Palaearctic, Nearctic and mountains of the Ethiopian and Oriental regions. Among them 8 species of 3 subgenera are in the Three Provinces of Northeast China.

Subgenus ***Chorizagrotis*** Smith, 1890 **脉切夜蛾亚属**

Chorizagrotis Smith, 1890, *Bulletin of the United States National Museum*, 38: 10, 98. Type species: *Agrotis auxiliaris* Grote, 1873 [USA, Colorado].

Synonymy: *Mesoeuxoa* Draudt, 1932.

注：首次给出中文亚属名。

***Euxoa* (*Chorizagrotis*) *adumbrata* (Eversmann, 1842) 虚切夜蛾**

Agrotis lidia ab. *adumbrata* Eversmann, 1842, *Bulletin de la Société Impériale des Naturalistes de Moscou*, 3: 543. Type locality: Russia, Kazan, Ural.

Synonymy: *Agrotis drewseni* Staudinger, 1857; *Agrotis norwegica* Staudinger, 1861; *Agrotis polygonides* Staudinger, 1874; *Agrotis lidia inexpectata* Alphéraky, 1897 (var.); *Agrotis dreuseni* Hampson, 1903; *Porosagrotis thanatologia* Rangnow, 1904; *Chorizagrotis sordida* Smith, 1908; *Chorizagrotis boretha* Smith, 1908; *Agrotis variegata* Wagner, 1913; *Euxoa thanatologia perfida* Dod, 1916 (var.); *Euxoa arenacea* Kozhantshikov, 1923; *Euxoa expugnata* Corti, 1932; *Euxoa norwegica obscura* Nordström, 1937 (ssp.); *Euxoa drewseni pseudovitta* Boursin, 1959 (f.); *Euxoa friedeli* Pinker, [1980] 1979; *Euxoa lidia* auct., nec Stoll, 1782.

Distribution. China (Heilongjiang, Jilin, Liaoning, Xizang); Russia (European part, Ural, Siberia), Mongolia, Danmark (Greenland), Korea, Japan; Central Asia, Middle East, Europe, North America.

寄主：繁缕（*Stellaria media*）、藜（*Chenopodium album*）、萹蓄（*Polygonum aviculare*）、大叶芥菜（*Brassica juncea*）、柳兰（*Chamaenerion angustifolium*）、大车前（*Plantago major*）、蒲公英属（*Taraxacum*）、禾本科（Poaceae）植物（Matov & Kononenko, 2012）。

Subgenus ***Euxoa*** Hübner, [1821] 1816 **切夜蛾亚属**

Euxoa Hübner, [1821] 1816, *Verzeichniss bekannter Schmettlinge* [sic!]: 209. Type species: *Noctua decora* [Denis & Schiffermüller], 1775 [Austria].

Synonymy: *Mimetes* Hübner, [1821]; *Metaxyja* Hübner, [1821]; *Exarnis* Hübner, [1821]; *Brotis* Hübner, [1821]; *Telmia* Hübner, [1821]; *Carneades* Grote, 1883; *Paragrotis* Dyar, [1903]; *Agrotis* auct., nec Ochsenheimer, 1816; *Metaxya*: Walker, 1857, missp.; *Mimetis*: Hampson, 1903, missp.; *Metaxyia*: Hampson,

1903, missp..

***Euxoa* (*Euxoa*) *sibirica* (Boisduval, 1834) 寒切夜蛾**

Agrotis sibirica Boisduval, 1834, *Icones Historique des Lépidoptéres nouveaux ou peu connu. Collection, avec Figures coloriées, des Papillons d'Eur*ope, 2: pl. 80: 6. Type locality: Russia, Siberia.

Synonymy: *Agrotis lapidosa* Graeser, 1892, *Agrotis sepulcharis* Alphéraky, 1892; *Euxoa japonica* Strand, 1916; *Rhyacia karafutonis* Matsumura, 1925; *Rhyacia isshikii* Matsumura, 1925; *Euxoa intracta kurilintracta* Bryk, 1948 (ssp.).

Distribution. China (Heilongjiang, Jilin, Xizang); Russia (Ural, Siberia), Mongolia, Korea, Japan.

寄主：甘蓝（*Brassica oleracea*）、滨藜属（*Atriplex*）、报春花属（*Primula*）植物（Matov & Kononenko, 2012）。

***Euxoa* (*Euxoa*) *ochrogaster* (Guenée, 1953) 岛切夜蛾**

Noctua ochrogaster Guenée, 1953, *in* Boisduval & Guenée, *Histoire Naturelle des Insectes. Species Général des Lépidoptères*, 5 (1): 327. Type locality: North America.

Synonymy: *Agrotis insignata* Walker, 1857; *Agrotis islandica* Staudinger, 1857; *Agrotis cinereomaculata* Morrison, 1875; *Agrotis turris* Grote, 1875; *Agrotis gularis* Grote, 1875; *Agrotis islandica rossica* Staudinger, 1881 (var.); *Agrotis karschi* Graeser, 1890; *Euxoa sjoestedti* Corti, 1929; *Euxoa sjöstedti* Corti, 1929; *Euxoa sublata* Corti, 1931; *Euxoa derasa* Corti, 1932; *Euxoa islandica yarkenda* Corti & Draudt, 1932 (f.); *Euxoa deserticola* Kozhantshikov, 1937.

Distribution. China (Heilongjiang, Jilin, Qinghai); Russia (European part, Ural, Siberia), Mongolia, Korea, Japan, India(North), Nepal, Pakistan; Central Asia, Middle East, Europe, North America.

寄主：异珠女娄菜（*Melandrium album*）、繁缕（*Stellaria media*）、甜菜（*Beta vulgaris*）、藜（*Chenopodium album*）、萹蓄 (*Polygonum aviculare*)、卷茎蓼（*Polygonum convolvulus*）、酸模（*Rumex acetosa*）、皱叶酸模（*Rumex crispus*）、西瓜（*Citrullus lanatus*）、黄瓜（*Cucumis sativus*）、齿叶南芥（*Arabis serrata*）、大叶芥菜（*Brassica juncea*）、甘蓝（*Brassica oleracea*）、荠（*Capsella bursa-pastoris*）、蚕豆（*Faba bona*）、岩黄芪（*Hedysarum esculentum*）、紫苜蓿（*Medicago sativa*）、番茄（*Lycopersicon esculentum*）、烟草（*Nicotiana tabacum*）、马铃薯（*Solanum tuberosum*）、短柄野芝麻（*Lamium album*）、中亚苦蒿（*Artemisia absinthium*）、向日葵（*Helianthus annuus*）、莴苣（*Lactuca sativa*）、苣荬菜（*Sonchus arvensis*）、燕麦（*Avena sativa*）、黑麦（*Secale cereale*）、玉米（*Zea mays*）、木贼属（*Equisetum*）、罂粟属（*Papaver*）、石竹属（*Dianthus*）、锦葵属（*Malva*）、荨麻属（*Urtica*）、草莓属（*Fragaria*）、委陵菜属（*Potentilla*）、车轴草属（*Trifolium*）、亚麻属（*Linum*）、旋花属（*Convolvulus*）、勿忘草属（*Myosotis*）、柳穿鱼属（*Linaria*）、婆婆纳属（*Veronica*）、车前属（*Plantago*）、青兰属（*Dracocephalum*）、鼬瓣花属（*Galeopsis*）、牛蒡属（*Arctium*）、蓟属（*Cirsium*）、蒲公英属（*Taraxacum*）、葱属（*Allium*）、小麦属（*Triticum*）植物（Matov & Kononenko,

2012）。

注：《中国动物志》（夜蛾科）中引用本种同物异名“*Euxoa islandica* (Staudinger, 1857)”，在此予以变更。

***Euxoa* (*Euxoa*) *karschi* (Graeser, [1890] 1889) 白边切夜蛾**

Agrotis karschi Graeser, [1890] 1889, *Berliner entomologische Zeitschrift*, 35, II: 253. Type locality: Russia, Amur reg., Radde ["Raddefka"].

Synonymy: *Agrotis oberthueri* Sugi, 1982, nec Leech, 1900, misidentification.

Distribution. China (Heilongjiang, Jilin, Liaoning, Nei Mongol, Hebei, Sichuan, Yunnan, Xizang); Russia (Transbaikalia, Siberia), Korea, Japan.

寄主：粟、高粱、玉米、大豆、甜菜等（陈一心，1999）；甜菜（*Beta vulgaris*）、车前（*Plantago asiatica*）、刺儿菜（*Cirsium setosum*）、抱茎苦荬菜（*Ixeris sonchifolia*）、近无刺苍耳（*Xanthium sibiricum*）、高粱（*Sorghum bicolor*）、玉米（*Zea mays*）（Matov & Kononenko, 2012）。

注：《中国动物志》（夜蛾科）中引用本种同物异名“ *Euxoa oberthueri* Leech, 1900”，在此予以变更。

***Euxoa* (*Euxoa*) *intolerabilis* (Püngeler, 1902) 梦切夜蛾**

Agrotis intolerabilis Püngeler, 1902, *Deutsche Entomologische Zeitschrift Iris*, 15: 147. Type locality: China, Qinghai, Kukunor.

Synonymy: *Agrotis predotae* Schawerda, 1922; *Rhyacia arida* Corti & Draudt, 1933.

Distribution. China (Heilongjiang, Jilin, Qinghai, Xizang); Russia (Transbaikalia, Siberia), Mongolia, Korea.

***Euxoa* (*Euxoa*) *tritici* (Linnaeus, 1761) 黑脉切夜蛾**

Phalaena tritici Linnaeus, 1761, *Fauna Suecica* (Edn. 2): 320. Type locality: Sweden.

Synonymy: *Phalaena brunnea* Hufnagel, 1766; *Phalaena* (*Noctua*) *cruda* Esper, [1790]; *Agrotis crypta* Dadd, 1927; *Agrotis obeliscoides* Dadd, 1927 form, nec Guenée, 1852; *Agrotis rufescens* Dadd, 1927; *Agrotis fasciata* Dadd, 1927; *Agrotis caerulescens* Dadd, 1927; *Agrotis crypta caerulescens* Dadd, 1927 (f.); *Mesoeuxoa biscajana* Corti & Draudt, 1932; *Euxoa crypta*: sensu Fibiger, 1997.

Distribution. China (Heilongjiang, Jilin, Nei Mongol, Xinjiang, Hebei, Xizang); Russia (European part, Ural, Siberia); Europe.

寄主：甜菜（*Beta vulgaris*）、荞麦（*Fagopyrum esculentum*）、萹蓄（*Polygonum aviculare*）、西瓜（*Citrullus edulis*）、黄瓜（*Cucumis sativus*）、西葫芦（*Cucurbita pepo*）、（*Melo sativus*）、豌豆（*Pisum sativum*）、葡萄（*Vitis vinifera*）、番茄（*Lycopersicon esculentum*）、阳芋（*Solanum tuberosum*）、大车前（*Plantago major*）、蝶须（*Antennaria dioica*）、蓝蓟（*Echium vulgare*）、向日葵（*Helianthus annuus*）、绿毛山柳菊（*Hieracium pilosella*）、玉米（*Zea mays*）、罂粟属（*Papaver*）、卷耳属（*Cerastium*）、大爪草属（*Spergula*）、繁缕属（*Stellaria*）、藜属（*Chenopodium*）、芸苔属（*Brassica*）、拉拉藤属（*Galium*）、烟草属（*Nicotiana*）、旋花属（*Convolvulus*）、蒿属（*Artemisia*）、菊苣属（*Cichorium*）、

莴苣属（Lactuca）、蒲公英属（*Taraxacum*）、葱属（*Allium*）、拂子茅属（*Calamagrostis*）、大麦属（*Hordeum*）、小麦属（*Triticum*）植物（Matov & Kononenko, 2012）。

Euxoa (*Euxoa*) *nigricans* (Linnaeus, 1761) 暗切夜蛾

Phalaena nigricans Linnaeus, 1761, *Fauna Suecica* (Edn. 2): 322. Type locality: Sweden.

Synonymy: *Noctua fumosa* [Denis et Schiffermüller], 1775; *Phalaena* (*Noctua*) *rubricans* Esper, [1788]; *Noctua dubia* Haworth, 1809; *Noctua vilis* Hübner, [1813]; *Noctua fuliginea* Hübner, [1813]; *Noctua carbonea* Hübner, [1823]; *Noctua ursina* Godart, 1825; *Agrotis rustica* Eversmann, 1842; *Agrotis oppidicola* Krulikovsky, 1907; *Agrotis nigrata* Matsumura, 1925; *Agrotis ishidae* Matsumura, 1926.

Distribution. China (Heilongjiang, Jilin, Qinghai, Xizang); Russia (European part, Ural, Siberia); Near East, Europe, North Africa.

寄主：低矮草本植物（陈一心，1999）；甜菜（*Beta vulgaris*）、萹蓄（*Polygonum aviculare*）、甜瓜（*Cucumis melo*）、黄瓜（*Cucumis sativus*）、欧洲油菜（*Brassica napus*）、甘蓝（*Brassica oleracea*）、萝卜（*Raphanus sativus*）、马蹄豆（*Hippocrepis comosa*）、豌豆（*Pisum sativum*）、红车轴草（*Trifolium pratense*）、白车轴草（*Trifolium repens*）、葡萄（*Vitis vinifera*）、莳萝（*Anethum graveolens*）、野胡萝卜（*Daucus carota*）、牛防风（*Heracleum sphondylium*）、田旋花（*Convolvulus arvensis*）、长叶车前（*Plantago lanceolata*）、大车前（*Plantago major*）、款冬花（*Tussilago farfara*）、葱（*Allium fistulosum*）、玉米（*Zea mays*）、石刁柏（*Asparagus officinalis*）、酸模属（*Rumex*）、棉属（*Gossypium*）、菜豆属（*Phaseolus*）、亚麻属（*Linum*）、欧防风属（*Pastinaca*）、拉拉藤属（*Galium*）、茄属（*Solanum*）、蒿属（*Artemisia*）、莴苣属（Lactuca）、蒲公英属（*Taraxacum*）、燕麦属（*Avena*）、大麦属（*Hordeum*）、黑麦属（*Secale*）、小麦属（*Triticum*）植物（Matov & Kononenko, 2012）。

Subgenus *Orosagrotis* Hampson, 1903 口切夜蛾亚属

Orosagrostis Hampson, 1903, *Catalogue of the Lepidoptera Phalaenae in the British Museum* 4: xi, 135, 658. Type species: *Agrotis montana* Morrison, 1875 [USA, Colorado Mts.].

注：首次给出中文亚属名。

Euxoa (*Orosagrotis*) *tristis* (Staudinger, 1897) 郁切夜蛾

Agrotis tristis Staudinger, 1897, *Deutsche Entomologische Zeitschrift Iris*, 10: 333. Type locality: Russia, [Yablonovyi Range, Transbaikalia] "Apfel-Gebirge".

Synonymy: *Agrotis tritici varia* Alphéraky, 1889 (var.), nec Walker, 1863; *Agrotis deficiens* Wagner, 1913; *Euxoa* (*Mesoeuxoa*) *distracta* Corti, 1932; *Euxoa* (*Mesoeuxoa*) *opportuna* Corti, 1932.

Distribution. China (Heilongjiang, Nei Mongol, Xizang); Russia (Ural, Siberia), Mongolia.

寄主：甘蓝（*Brassica oleracea*）、玉米（*Zea mays*）、甜菜（*Beta vulgaris*）、栎属（*Quercus*）植物

（Matov & Kononenko, 2012）。

Genus *Feltia* Walker, 1856 菲夜蛾属

Feltia Walker, 1856, *List of the Specimens of Lepidopterous Insects in the Collection of the British Museum*, 9: 165, 202. Type species: *Feltia ducens* Walker, 1856 [Canada, Ontario].

The genus contains 10 species, mainly distributed in the boreal zone of the Holarctic region. Among them 1 species of 1 subgenus is in the Three Provinces of Northeast China.

注：首次给出中文属名。

Subgenus *Trichosilia* Hampson, 1918 发夜蛾亚属

Trichosilia Hampson, 1918, *Novitates Zoologicae*, 25: 112. Type species: *Noctua acarnea* Smith, 1905 [Canada, Alberta].

注：首次给出中文亚属名。

***Feltia* (*Trichosilia*) *nigrita* (Graeser, 1892) 孽菲夜蛾**

Agrotis nigrita Graeser, 1892, *Berliner entomologische Zeitschrift*, 1892: 217. Type locality: Russia, Amur reg. Prokovka.

Synonymy: *Ochropleura maerens* Staudinger, 1896; *Noctua acarnea* Smith, 1905; *Rhyacia kononis* Matsumura, 1925; *Rhyacia tzygankovi* I. Kozhantshikov, 1926; *Agrotis tragica* Corti, 1933; *Agrotis grisea* I. Kozhantshikov, 1935; *moerens* nec Staudinger, 1896.

Distribution. China (Heilongjiang); Russia (Southwest of the European part, Ural, Altai, Transbaikalia, Siberia), Mongolia; North America.

寄主：蒲公英属（*Taraxacum*）植物（Matov & Kononenko, 2012）。

注：《中国动物志》（夜蛾科）中将本种归入“*Ochropleura* Hübner, [1821] 1816 狼夜蛾属”，在此予以变更。

Genus *Agrotis* Ochsenheimer, 1816 地夜蛾属

Agrotis Ochsenheimer, 1816, *Die Schmetterlinge von Europa*, 4: 66. Type species: *Noctua segetum* [Denis & Schiffermüller], 1775 [Austria].

Synonymy: *Agrotis* Hübner, [1806], rejected name suppr. ICZN, 1926 Op. 97; *Agronoma* Hübner, 1816; *Georyx* Hübner, [1821] 1816; *Scotia* Hübner, [1821] 1816; *Noctua* Boisduval, 1828, preocc. Linnaeus, 1758 [Lepidoptera, Noctuidae]; *Psammophila* Stephens, 1850, preocc. Brown, 1827 [Mollusca]; *Tetrapyrgia* Walker, 1865; *Elegarda* Walker, 1865; *Porosagrotis* Smith, 1890; *Mesembreuxoa* Hampson, 1903; *Parosagrotis*: Dod, 1910, missp.; *Militagrotis* Beck, [1992] 1991; *Crassagrotis* Beck, [1992] 1991; *Exagrotis* Beck, 1996; *Ripagrotis* Beck, 1996; *Spinagrotis* Beck, 1996 nom. nud.; *Schawagrotis* Beck, 1996 nom. nud.; *Stritagrotis* Beck, 1996 nom. nud..

The genus contains over 280 species, mainly distributed in the arid, temperate and mountain areas of the Palaearctic, Oriental, Ethiopian and Nearctic regions. Among them, 9 species are in the Three Provinces of Northeast China.

Agrotis characteristica Alphéraky, 1892 汉地老虎

Agrotis characteristica Alphéraky, 1892, *in* N.M. Romanoff (ed.), *Mémoires sur les Lépidoptères*, 6: 31, pl. 1: 9. Type locality: China, Ordos.

Synonymy: *Agrotis trifurca* Herrich-Sch ä ffer, 1845, nec Eversmann, 1837; *Agrotis robusta* Eversmann, 1856; *Agrotis mirifica* Wagner, 1913; *Euxoa coreana* Matsumura, 1926; *Agrotis robustana* Poole, 1989, unnecessary repl. name.

Distribution. China (Heilongjiang, Jilin, Xinjiang, Shanxi); Russia (Ural, Siberia), Mongolia, North Korea, Kazakhstan (North); North America.

注：《中国动物志》（夜蛾科）中引用本种同物异名"*Agrotis robustana* Poole, 1989"，在此予以变更。

Agrotis trifurca Eversmann, 1837 三叉地夜蛾

Agrotis trifurca Eversmann, 1837, *Bulletin de la Société Impériale des Naturalistes de Moscou*, 1837: 31. Type locality: Russia [Bashkiria], Ufa.

Synonymy: *Agrotis atra* Corti & Draudt, 1932.

Distribution. China (Heilongjiang, Jilin, Liaoning, Nei Mongol, Qinghai, Xinjiang); Russia (Southeast European part, Ural, Siberia), Mongolia, North Korea; Europe.

寄主：粟、高粱、玉米、甜菜等（陈一心，1999）；（*Atriplex patulum*）、甜菜（*Beta vulgaris*）、黄瓜（*Cucumis sativus*）、菥蓂（*Thlaspi arvense*）、大豆（*Glycine max*）、车前（*Plantago asiatica*）、刺儿菜（*Cirsium setosum*）、抱茎苦荬菜（*Ixeris sonchifolia*）、长裂苦苣菜（*Sonchus brachyotus*）、近无刺苍耳（*Xanthium sibiricum*）、偃麦草（*Elytrigia repens*）、高粱（*Sorghum bicolor*）、玉米（*Zea mays*）、山芥属（*Barbarea*）、芸苔属（*Brassica*）、车轴草属（*Trifolium*）、老鹳草属（*Geranium*）、蒿属（*Artemisia*）、蒲公英属（*Taraxacum*）植物（Matov & Kononenko, 2012）。

Agrotis segetum ([Denis & Schiffermüller], 1775) 黄地老虎

Noctua segetum [Denis & Schifferm ü ller], 1775, *Ankündung eines Systematisches Werkes von den Schmetterlinge der Wiener Gegend. Wien*: 81, 252, fig. a, b. Type locality: Austria, Vienna.

Synonymy: *Noctua sordida* [Denis et Schiffermüller], 1775, emend.; *Bombyx catenatus* Haworth, 1803; *Bombyx connexus* Haworth, 1803; *Bombyx nigrocornutus* Haworth, 1803; *Bombyx pectinatus* Haworth, 1803; *Bombyx subatratus* Haworth, 1803; *Noctua fervida* Hübner, [1824]; *Agrotis segetum affinis* Stephens, 1829 (var.); *Agrotis sicula* Boisduval, 1840; *Agrotis sicania* Guenée, 1852; *Agrotis aversa* Walker, [1857] 1856; *Agrotis correcta* Walker, [1857] 1856; *Agrotis marginalis* Walker, [1857] 1856; *Agrotis oblivosa* Walker, [1857] 1856; *Agrotis denticulosa* Wallengren, 1860; *Agrotis conspurcata* Walker, 1865; *Agrotis repulsa* Walker, 1865; *Agrotis segetum pallida* Staudinger, 1881 (ssp.); *Agrotis ingrata* Butler, 1881; *Agrotis fucosa* Butler, 1881; *Agrotis lassa* Swinhoe, 1886; *Euxoa segetum albifera* Turati, 1921 (f.); *Euxoa glaucina* W. Kozhantshikov, 1923.

Distribution. China (except Southeast region); Russia (European part, Ural, Siberia), Mongolia, Korea, Japan, Nepal, India; Southeast Asia, Central Asia, Middle East, Europe, Africa.

寄主：棉、玉米、小麦、高粱、烟草、甜菜、马铃薯、瓜类、多种蔬菜、栎、山杨、云杉、松、柏等树苗（陈一心，1999）；雪松（*Cedrus deodara*）、欧洲落叶松（*Larix decidua*）、欧洲云杉（*Picea abies*）、甜菜（*Beta vulgaris*）、藜（*Chenopodium album*）、萹蓄 (*Polygonum aviculare*)、食用大黄（*Rheum rhaponticum*）、酸模（*Rumex acetosa*）、皱叶酸模（*Rumex crispus*）、茶（*Camellia sinensis*）、帚石楠（*Calluna vulgaris*）、西瓜（*Citrullus vulgaris*）、甜瓜（*Cucumis melo*）、黄瓜（*Cucumis sativus*）、西葫芦（*Cucurbita pepo*）、芸苔（*Brassica campestris*）、欧洲油菜（*Brassica napus*）、甘蓝（*Brassica oleracea*）、芜青（*Brassica rapa*）、萝卜（*Raphanus sativus*）、菠菜（*Spinacia oleracea*）、磨盘草（*Abutilon indicum*）、草棉（*Gossypium herbaceum*）、玫瑰茄（*Hibiscus sabdariffa*）、大麻（*Cannabis sativa*）、蓖麻（*Ricinus communis*）、（*Malus hissarica*）、新疆野苹果（*Malus sieversii*）、野生樱桃李（*Prunus divaricata*）、新疆野生樱桃李（*Prunus sogdiana*）、秦氏悬钩子（*Rubus caesius*）、花生（*Arachis hypogaea*）、印度棉（*Gossypium indicum*）、紫苜蓿（*Medicago sativa*）、小苕子（*Ornithopus sativus*）、豌豆（*Pisum sativum*）、黑吉豆（*Vigna mungo*）、葡萄（*Vitis vinifera*）、旱芹（*Apium graveolens*）、芫荽（*Coriandrum sativum*）、野胡萝卜（*Daucus carota*）、茴香（*Foeniculum officinale*）、欧防风（*Pastinaca sativa*）、茴芹（*Pimpinella anisum*）、缬草（*Valeriana officinalis*）、颠茄草（*Atropa belladonna*）、辣椒（*Capsicum annuum*）、番茄（*Lycopersicon esculentum*）、烟草（*Nicotiana tabacum*）、茄（*Solanum melongena*）、马铃薯（*Solanum tuberosum*）、田旋花（*Convolvulus arvensis*）、（*Chaenorrhinum minus*）、芝麻（*Sesamum indicum*）、车前（*Plantago asiatica*）、大车前（*Plantago major*）、柚木（*Tectona grandis*）、牛蒡（*Arctium lappa*）、矢车菊（*Centaurea aspera*）、菊苣（*Cichorium intybus*）、刺儿菜（*Cirsium arvense*）、向日葵（*Helianthus annuus*）、野莴苣（*Lactuca serriola*）、洋葱（*Allium cepa*）、葱（*Allium fistulosum*）、韭葱（*Allium porrum*）、石刁柏（*Asparagus officinalis*）、脚板薯（*Dioscorea batatas*）、燕麦（*Avena sativa*）、无芒雀麦（*Bromus inermis*）、小穗发草（*Deschampsia caespitosa*）、偃麦草（*Elytrigia repens*）、二行大麦（*Hordeum distichum*）、大麦（*Hordeum vulgare*）、黍（*Panicum miliaceum*）、梯牧草（*Phleum pratense*）、早熟禾（*Poa annua*）、甘蔗（*Saccharum officinarum*）、黑麦（*Secale cereale*）、高粱（*Sorghum bicolor*）、小麦（*Triticum aestivum*）、硬粒小麦（*Triticum durum*）、小麦（*Triticum vulgare*）、玉米（*Zea mays*）、松属（*Pinus*）、黑种草属（*Nigella*）、滨藜属（*Atriplex*）、山毛榉属（*Fagus*）、白芥属（*Sinapis*）、柳属（*Salix*）、蜀葵属（*Althaea*）、锦葵属（*Malva*）、草莓属（*Fragaria*）、木蓝属（*Indigofera*）、羽扇豆属（*Lupinus*）车轴草属（*Trifolium*）、牻牛儿苗属（*Erodium*）、老鹳草属（*Geranium*）、羊角芹属（*Aegopodium*）、葛缕子属（*Carum*）、咖啡属（*Coffea*）、柳穿鱼属（*Linaria*）、鼻花属（*Rhinanthus*）、蓍属（*Achillea*）、蒿属（*Artemisia*）、飞廉属（*Carduus*）、菊属（*Chrysanthemum*）、风毛菊属（*Leontodon*）、苦苣菜属（*Sonchus*）、蒲公英属

（*Taraxacum*）植物（Matov & Kononenko, 2012）。

Agrotis exclamationis (Linnaeus, 1758) 警纹地老虎

Phalaena exclamationis Linnaeus, 1758, *Systema Naturae* (Edn. 10) 1: 515. Type locality: Europe.

Synonymy: *Phalaena clavus* Hufnagel, 1766; *Phalaena murina* Goeze, 1781; *Phalaena bimaculata* Geoffrey, 1785; *Noctua picea* Haworth, 1809; *Agrotis plaga* Stephens, 1835; *Agrotis informis* Leech, 1889; *Agrotis confluens* Leech, 1889; *Agrotis informis confluens* Leech, 1889 (var.); *Agrotis exclamationis serena* Alphéraky, 1889 (var.); *Lycophotia haruspex* Le Cerf, 1924; *Agrotis exclamationis corsica* Bang-Haas, 1977 (ssp.); *Agrotis exclamationis nigriorbis* Zerny, 1934 (ssp.).

Distribution. China (Heilongjiang, Jilin, Nei Mongol, Qinghai, Xinjiang, Yunnan, Xizang); Russia (European part, Ural, Siberia), Korea; Europe, Africa.

寄主：繁缕（*Stellaria media*）、甜菜（*Beta vulgaris*）、藜（*Chenopodium album*）、菠菜（*Spinacia oleracea*）、萹蓄（*Polygonum aviculare*）、酸模（*Rumex acetosa*）、甜瓜（*Cucumis melo*）、黄瓜（*Cucumis sativus*）、西葫芦（*Cucurbita pepo*）、大叶芥菜（*Brassica juncea*）、甘蓝（*Brassica oleracea*）、芜青（*Brassica rapa*）、菠菜（*Spinacia oleracea*）、草莓（*Fragaria ananassa*）、欧洲花楸（*Sorbus aucuparia*）、紫苜蓿（*Medicago sativa*）、白车轴草（*Trifolium repens*）、（*Trigonella arabica*）、（*Vicia palaestina*）、葡萄（*Vitis vinifera*）、野胡萝卜（*Daucus carota*）、洋香菜（*Petroselinum sativum*）、番茄（*Lycopersicon esculentum*）、烟草（*Nicotiana tabacum*）、马铃薯（*Solanum tuberosum*）、田旋花（*Convolvulus arvensis*）、大车前（*Plantago major*）、红辣椒千叶蓍（*Achillea millefolium*）、荒野蒿（*Artemisia campestris*）、滨蒿（*Artemisia maritima*）、向日葵（*Helianthus annuus*）、莴苣（*Lactuca sativa*）、苣荬菜（*Sonchus arvensis*）、燕麦（*Avena sativa*）、（*Bromis arvensis*）、偃麦草（*Elytrigia repens*）、梯牧草（*Phleum pratense*）、早熟禾（*Poa annua*）、黑麦（*Secale cereale*）、玉米（*Zea mays*）、棉属（*Gossypium*）、荨麻属（*Urtica*）、蔷薇属（*Rosa*）、亚麻属（*Linum*）、前胡属（*Peucedanum*）、牛蒡属（*Arctium*）、紫菀属（*Aster*）、千里光属（*Senecio*）、蒲公英属（*Taraxacum*）、唐菖蒲属（*Gladiolus*）、小麦属（*Triticum*）植物（Matov & Kononenko, 2012）。

Agrotis scotacra (Filipjev, 1927) 暗地夜蛾

Feltia scotacra Filipjev, 1927, *Annuaire du Musée Zoologique de l'Académie des Sciences d'URSS*, 28: 239. Type locality: Russia, Primorye terr., Tigrovoi.

Synonymy: *Euxoa kaolifeltia* Bryk, 1948.

Distribution. China (Heilongjiang); Russia (Siberia), Mongolia, Korea, Japan.

Agrotis clavis (Hufnagel, 1766) 皱地夜蛾

Phalaena clavis Hufnagel, 1766, *Berlinisches Magazine*, 3: 426. Type locality: Germany, Berlin.

Synonymy: *Noctua corticea* [Denis et Schiffermüller], 1775; *Bombyx clavigerus* Haworth, 1803; *Bombyx subfuscus* Haworth, 1803; *Agrotis sincerii* Freyer, 1849; *Agrotis transversa* Walker, [1857]; *Agrotis corticea amurensis* Staudinger, 1892 (var.); *Agrotis corticea neocomensis* Rougemont, 1902 (var.); *Agrotis praedicta* Corti, 1932; *Agrotis corticea grammomima* Bryk, 1949 (ssp.).

Distribution. China (Heilongjiang, Jilin, Qinghai, Hebei, Sichuan); Russia (European part, Ural, Siberia), Mongolia, Korea, Japan; Near East, Europe, North Africa.

寄主：桑及一些树木幼苗（陈一心，1999）；甜菜（*Beta vulgaris*）、藜（*Chenopodium album*）、菠菜（*Spinacia oleracea*）、春蓼（*Persicaria maculosa*）、萹蓄（*Polygonum aviculare*）、卷茎蓼（*Polygonum convolvulus*）、钝叶酸模（*Rumex obtusifolius*）、黑果越橘（*Vaccinium myrtillus*）、大叶芥菜（*Brassica juncea*）、甘蓝（*Brassica oleracea*）、荠菜（*Capsella bursapastoris*）、光果葶苈（*Draba nemorosa*）、萝卜（*Raphanus sativus*）、菠菜（*Spinacia oleracea*）、（*Malva borealis*）、覆盆子（*Rubus idaeus*）、紫苜蓿（*Medicago sativa*）、豌豆（*Pisum sativum*）、白车轴草（*Trifolium repens*）、黄檗（*PhellodendroNorth of Amur reg.ense*）、野胡萝卜（*Daucus carota*）、欧芹（*Petroselinum crispum*）、番茄（*Lycopersicon esculentum*）、烟草（*Nicotiana tabacum*）、鼬瓣根（*Galeopsis bifida*）、大车前（*Plantago major*）、向日葵（*Helianthus annuus*）、莴苣（*Lactuca sativa*）、鸭蚺草（*Paspalum scrobiculatum*）、梯牧草（*Phleum pratense*）、玉米（*Zea mays*）、繁缕属（*Stellaria*）、核桃属（*Juglans*）、椴树属（*Tilia*）、蜀葵属（*Althaea*）、荨麻属（*Urtica*）、大戟属（*Euphorbia*）、羽衣草属（*Alchemilla*）、苹果属（*Malus*）、委陵菜属（*Potentilla*）、李属（*Prunus*）、新缬草属（*Valerianella*）、梣属（*Fraxinus*）、酸浆属（*Physalis*）、旋花属（*Convolvulus*）、毛蕊花属（*Verbascum*）、牛蒡属（*Arctium*）、蒿属（*Artemisia*）、风毛菊属（*Leontodon*）、苦苣菜属（*Sonchus*）、蒲公英属（*Taraxacum*）植物（Matov & Kononenko, 2012）。

***Agrotis tokionis* Butler, 1881 大地老虎**

Agrotis tokionis Butler, 1881, *Transactions of the Entomological Society of London*, 1881: 178. Type locality: Japan, Tokyo.

Synonymy: *Agrotis nigricostata* Staudinger, 1888; *Agrotis tocionis* Hampson, 1903, emend..

Distribution. China; Russia (Siberia), Korea, Japan.

寄主：棉、玉米、高粱、烟草等（陈一心，1999）；甜菜（*Beta vulgaris*）、黄瓜（*Cucumis sativus*）、芸苔（*Brassica campestris*）、芜青（*Brassica rapa*）、萝卜（*Raphanus sativus*）、印度棉（*Gossypium indicum*）、辣椒（*Capsicum annuum*）、烟草（*Nicotiana tabacum*）、马铃薯（*Solanum tuberosum*）、洋葱（*Allium cepa*）、高粱（*Sorghum bicolor*）、玉米（*Zea mays*）、车轴草属（*Trifolium*）植物（Matov & Kononenko, 2012）。

***Agrotis desertorum* Boisduval, 1840 浦地夜蛾**

Agrotis desertorum Boisduval, 1840, *Genera et Index Methodicus Europaeorum Lepidopterorum*: 111. Type locality: South Russia.

Synonymy: *Agrotis deserticola* Eversmann, 1842; *Agrotis farinosa* Staudinger, 1888; *Euxoa cursoroides* Hampson, 1903; *Agrotis perplexa* Bang-Haas, 1910; *Agrotis ripae albovenosa* Tschetverikov, 1925 (ssp.); *Agrotis ripae wagneri* Corti & Draudt, 1933 (f.); *Rhyacia scruposa* Draudt, 1936; *Agrotis ripae* auct., nec Hübner, [1823].

Distribution. China (Heilongjiang, Nei Mongol, Xinjiang, Xizang); Russia (Ural, Siberia).

寄主：倒提壶、猪毛菜属等植物（陈一心，1999）；（*Atriplex patulum*）、藜（*Chenopodium album*）、猪毛菜属（*Salsola*）、酸模属（*Rumex*）植物（Matov & Kononenko, 2012）。

注：《中国动物志》（夜蛾科）中引用本种同物异名“*Agrotis ripae* Hübner, [1823]”，在此予以变更。

Agrotis ipsilon (Hufnagel, 1766) 小地老虎

Phalaena ipsilon Hufnagel, 1766, *Berlinisches Magazine*, 3: 416, fig. 99. Type locality: Germany, Berlin.

Synonymy: *Noctua suffusa* [Denis et Schiffermüller], 1775; *Phalaena* (*Bombyx*) *idonea* Cramer, 1780; *Bombyx spinula* Esper, 1786; *Phalaena* (*Bombyx*) *spinifera* Villers, 1789; *Phalaena* (*Bombyx*) *spinula* Donovan, 1801; *Agrotis telifera* Harris, 1841; *Agrotis bipars* Walker, [1857]; *Agrotis frivola* Wallengren, 1860; *Noctua aneituma* Walker, 1865; *Agrotis suffusa pepoli* Bertolini, 1874 (var.); *Agrotis aureolum* Schaus, 1898; *Noctua ypsilon*: Rott, 1776, missp..

Distribution. China; Russia (European part, Ural, Siberia, Kamchatka), Mongolia, Korea, Japan, India; Oceania, Europe, North and South America.

寄主：棉、玉米、小麦、高粱、烟草、马铃薯、麻、豆类、蔬菜等植物树苗（陈一心，1999）；雪松（*Cedrus deodara*）、思茅松（*Pinus kesiya*）、红松（*Pinus koraiensis*）、（*Pinus markusii*）、罂粟（*Papaver somniferum*）、马齿苋（*Portulaca oleracea*）、皱果苋（*Amaranthus viridis*）、甜菜（*Beta vulgaris*）、藜（*Chenopodium album*）、俄罗斯蓟（*Salsola iberica*）、菠菜（*Spinacia oleracea*）、钝叶酸模（*Rumex obtusifolius*）、核桃楸（*Juglans manshurica*）、茶（*Camellia sinensis*）、狭叶越橘（*Vaccinium angustifolium*）、蔓越橘（*Vaccinium macrocarpum*）、冬瓜（*Benincasa hispida*）、甜瓜（*Cucumis melo*）、黄瓜（*Cucumis sativus*）、笋瓜（*Cucurbita maxima*）、南瓜（*Cucurbita moschata*）、芥蓝（*Brassica alboglabra*）、芸苔（*Brassica campestris*）、擘蓝（*Brassica caulorapa*）、青菜（*Brassica chinensis*）、褐芥末（*Brassica hirta*）、大叶芥菜（*Brassica juncea*）、黑芥（*Brassica nigra*）、甘蓝（*Brassica oleracea*）、菜薹（*Brassica parachinensis*）、白菜（*Brassica pekinensis*）、芜青（*Brassica rapa*）、臭荠（*Coronopus didymus*）、萝卜（*Raphanus sativus*）、草棉（*Gossypium herbaceum*）、印度棉（*Gossypium indicum*）、圆叶锦葵（*Malva rotundifolia*）、大麻（*Cannabis sativa*）、蓖麻（*Ricinus communis*）、草莓（*Fragaria chiloensis*）、（*Fragaria magna*）、苹果（Malus pumila）、白桃（*Persica vulgaris*）、番石榴（*Psidium guajava*）、花生（*Arachis hypogaea*）、鹰嘴豆（*Cicer arietinum*）、蚕豆（*Faba bona*）、大豆（*Glycine max*）、野大豆（*Glycine soja*）、紫苜蓿（*Medicago sativa*）、菜豆（*Phaseolus vulgaris*）、豌豆（*Pisum sativum*）、埃及车轴草（*Trifolium alexandrinum*）、黑吉豆（*Vigna mungo*）、甜橙（*Citrus sinensis*）、亚麻（*Linum usitatissimum*）、葡萄（*Vitis vinifera*）、旱芹（*Apium graveolens*）、野胡萝卜（*Daucus carota*）、中粒咖啡（*Coffea robusta*）、水曲柳（*Fraxinus *manshurica*）、辣椒（*Capsicum annuum*）、番茄（*Lycopersicon esculentum*）、烟草（*Nicotiana tabacum*）、茄（*Solanum melongena*）、龙葵（*Solanum nigrum*）、马铃薯（*Solanum tuberosum*）、田旋花（*Convolvulus arvensis*）、土丁桂（*Evolvulus alsinoides*）、蕹菜（*Ipomoea aquatica*）、红薯（*Ipomoea batatas*）、圆叶牵牛（*Ipomoea purpurea*）、茑萝（*Ipomoea quamoclit*）、柚木（*Tectona grandis*）、辣薄荷（*Mentha piperita*）、紫苏（*Perilla frutescens*）、田野水苏（*Stachys arvensis*）、红辣椒千叶蓍（*Achillea millefolium*）、牛蒡（*Arctium lappa*）、菊苣（*Cichorium intybus*）、牛膝菊（*Galinsoga parviflora*）、（*Hieracium pilosa*）、莴苣（*Lactuca sativa*）、野莴苣（*Lactuca serriola*）、灰白银胶菊（*Parthenium argentatum*）、苣荬菜（*Sonchus arvensis*）、药用蒲公英（*Taraxacum officinale*）、麝香百合（*Lilium longiflorum*）、洋葱（*Allium cepa*）、葱（*Allium fistulosum*）、蒜（*Allium sativum*）、石刁柏（*Asparagus officinalis*）、山药（*Dioscorea batatus*）、疏毛地杨梅（*Luzula pilosa*）、燕麦（*Avena sativa*）、狗牙根（*Cynodon dactylon*）、鸭茅（*Dactylis glomerata*）、俯仰马唐（*Digitaria decumbens*）、黑麦草（*Lolium perenne*）、稻（*Oryza sativa*）、鸭蠋草（*Paspalum scrobiculatum*）、甘蔗（*Saccharum officinarum*）、高粱（*Sorghum bicolor*）、偏序钝叶草（*Stenotaphrum secundatum*）、小麦（*Triticum aestivum*）、玉米（*Zea mays*）、蓼属（*Polygonum*）、大蒜芥属（*Sisymbrium*）、山楂属（*Crataegus*）、车前属（*Plantago*）、鬼针草属（*Bidens*）、矢车菊属（*Centaurea*）、还阳参属（*Crepis*）、唐菖蒲属（*Gladiolus*）、披碱草属（*Elymus*）、羊茅属（*Festuca*）植物（Matov & Kononenko, 2012）。

Subtribe **Noctuina** Latreille, 1809 夜蛾亚族

Synonymy: Anaplectoidina Beck, 1996; Archanartina Beck, 1996; Axyliina Beck, 1996; Cerastina Beck, 1996; Chersotina Beck, 1996; Coenophilina Beck, 1996; Diarsiina Beck, 1996; Eugnorismina Beck, 1996; Eugraphina Beck, 1996; Euroina Beck, 1996; Hadapameini Beck, 1991; Lycophotiina Beck, 1996; Naeniina Beck, 1996; Netrocerocorina Beck, 1996; Opigenina Beck, 1996; Rhyaciina Beck, 1996; Xestiina Beck, 1996; Axyliina Fibiger & Lafontaine, 2005.

The subtribe represented in the Three Provinces of Northeast China by 21 genera and 54 species occurring in various habitats.

Genus *Axylia* Hübner, [1821] 1816 朽木夜蛾属

Axylia Hübher, [1821] 1816, *Verzeichniss bekannter Schmettlinge* [sic!]: 242. Type species: *Phalaena putris* Linnaeus, 1761 [Sweden].

The genus contains about 30 species, distributed mainly in arid, and mountain areas of the Oriental and Ethiopian regions. Among them, 1 species is in the Three Provinces of Northeast China.

Axylia putris **(Linnaeus, 1761)** 朽木夜蛾

Phalaena putris Linnaeus, 1761, *Fauna Suecica* (Edn. 2): 315. Type locality: Sweden.

Synonymy: *Phalaena subcorticalis* Hufnagel, 1766; *Noctua lignosa* Hübner, [1803]; *Axylia triseriata* Moore,

1888; *Rhyacia mundipennis* Warren, 1912.

Distribution. China (Heilongjiang, Jilin, Liaoning, Beijing, Shandong, Xinjiang, Hebei, Shanxi, Hunan, Yunnan, Xizang, Taiwan); Russia (European part, Ural, Siberia), Mongolia, Korea, Japan, India(North), Nepal, Pakistan, Indonesia; Central Asia, Middle East, Europe, North Africa.

寄主：葡萄叶铁线莲（*Clematis vitalba*）、甜菜（*Beta vulgaris*）、藜（*Chenopodium album*）、萹蓄 (*Polygonum aviculare*)、酸模（*Rumex acetosa*）、小酸模（*Rumex acetosella*）、桦叶鹅耳枥（*Carpinus betulus*）、黑点叶金丝桃（*Hypericum perforatum*）、欧洲山杨（*Populus tremula*）、黄花柳（*Salix caprea*）、芸苔（*Brassica campestris*）、异株荨麻（*Urtica dioica*）、苹果（Malus pumila）、覆盆子（*Rubus idaeus*）、紫苜蓿（*Medicago sativa*）、（*Chamanerion angustifolium*）、西洋接骨木（*Sambucus nigra*）、（*Galium album*）、（*Galium mollugo*）、蓬子菜（*Galium verum*）、红花琉璃草（*Cynoglossum officinale*）、田旋花（*Convolvulus arvensis*）、旋花（*Calystegia sepium*）、车前（*Plantago asiatica*）、长叶车前（*Plantago lanceolata*）、大车前（*Plantago major*）、短柄野芝麻（*Lamium album*）、林石蚕（*Teucrium scorodonia*）、棕矢车菊（*Centaurea jacea*）、鸭茅（*Dactylis glomerata*）、大麦（*Hordeum vulgare*）、黑麦草（*Lolium perenne*）、早熟禾（*Poa annua*）、黑麦（*Secale cereale*）、繁缕属（*Stellaria*）、滨藜属（*Atriplex*）、桦木属（*Betula*）、堇菜属（*Viola*）、百脉根属（*Lotus*）、车轴草属（*Trifolium*）、蚕豆属（*Vicia*）、莴苣属（*Lactuca*）、蒲公英属（*Taraxacum*）、小麦属（*Triticum*）植物（Matov & Kononenko, 2012）。

Genus *Ochropleura* Hübner, [1821] 1816 狼夜蛾属

Ochropleura Hübner, [1821] 1816, *Verzeichniss bekannter Schmettlinge* [sic!]: 223. Type species: *Phalaena plecta* Linnaeus, 1761 [Sweden].

The genus contains more than 30 species, mainly distributed in the Oriental, Ethiopian and Neotropic regions. Among them, 1 species is in the Three Provinces of Northeast China.

Ochropleura plecta (Linnaeus, 1761) 狼夜蛾

Phalaena plecta Linnaeus, 1761, *Fauna Suecica* (Edn. 2): 321. Type locality: Sweden.

Synonymy: *Ochropleura vicaria* Walker, [1857] 1856; *Agrotis plecta unimacula* Staudinger, 1859 (ssp.); *Agrotis plecta anderssoni* Lampa, 1885 (var.); *Agrotis plecta glaucimacula* Graeser, 1888 [1889] (var.); *Ochropleura ignota* Swinhoe, 1889; *Agrotis vibora* Dognin, 1897; *Diarsia plecta urupplecta* Bryk, 1942 (ssp.).

Distribution. China (Heilongjiang, Jilin, Liaoning, Qinghai, Yunnan, Xizang); Russia (European part, Ural, Siberia), Mongolia, Korea, Japan, India, Sri Lanka; Southeast Asia, Central Asia, Middle East, Europe, Africa, North and Central America.

寄主：繁缕（*Stellaria media*）、榆钱菠菜（*Atriplex hortensis*）、（*Atriplex patulum*）、甜菜（*Beta vulgaris*）、藜（*Chenopodium album*）、萹蓄 (*Polygonum aviculare*)、酸模（*Rumex acetosa*）、皱叶酸模（*Rumex crispus*）、钝叶酸模（*Rumex obtusifolius*）、黄花柳（*Salix caprea*）、黄瓜（*Cucumis sativus*）、啤酒花（*Humulus lupulus*）、覆盆子（*Rubus idaeus*）、小冠花（*Coronilla varia*）、大豆（*Glycine max*）、百脉根（*Lotus corniculatus*）、（*Trifolium hybridus*）、白车轴草（*Trifolium repens*）、柳兰（*Chamaenerion angustifolium*）、旱芹（*Apium graveolens*）、四叶葎（*Galium mollugo*）、沼猪殃殃（*Galium uliginosum*）、蓬子菜（*Galium verum*）、番茄（*Lycopersicon esculentum*）、聚合草（*Symphytum officinale*）、大车前（*Plantago major*）、栽培菊苣（*Cichorium endivia*）、菊苣（*Cichorium intybus*）、莴苣（*Lactuca sativa*）、欧洲千里光（*Senecio vulgaris*）、蒜（*Allium sativum*）、早熟禾（*Poa annua*）、蝇子草属（*Silene*）、金丝桃属（*Hypericum*）、芸苔属（*Brassica*）、椴树属（*Tilia*）、车叶草属（*Asperula*）、牛蒡属（*Arctium*）、旋覆花属（*Inula*）、苦苣菜属（*Sonchus*）、蒲公英属（*Taraxacum*）植物（Matov & Kononenko, 2012）。

Genus *Diarsia* Hübner, [1821] 1816 歹夜蛾属

Diarsia Hübner, [1821] 1816, *Verzeichniss bekannter Schmettlinge* [sic!]: 222. Type species: *Noctua dahlii* Hübner, 1813 [Europe].

Synonymy: *Oxyra* Walker, 1865; *Brunnarsia* Beck, 1992 nom. nud.; *Rubarsia* Beck, 1996, nom. nud.; *Menarsia* Beck, 1996, nom. nud..

The genus contains about 100 species, mainly distributed in the Palaearctic, Nearctic, mountains of the Oriental and Australian regions; the center of *Diarsia* specific richness and diversity is mountains of the Southwest China and Himalaya. Among them, 7 species are in the Three Provinces of Northeast China.

Diarsia dahlii **(Hübner, [1813]) 歹夜蛾**

Noctua dahlii Hübner, [1813], *Sammlung Europaischer Schmetterlinge. Eulen*, 2: 99, fig. 465, 466. Type locality: Europe.

Synonymy: *Agrotis dahlii nana* Staudinger, 1892 (ssp.); *Rhyacia provincialis* Corti & Draudt, 1933; *Diarsia dahlii tibetica* Boursin, 1954 (ssp.).

Distribution. China (Heilongjiang, Jilin, Qinghai, Xinjiang, Shandong, Sichuan, Yunnan); Russia (European part, Ural, Siberia), Mongolia, Korea, Japan, Kazakhstan; Europe.

寄主：柳、山楂等（陈一心，1999）；皱叶酸模（*Rumex crispus*）、（*Rumex pulcher*）、帚石楠（*Calluna vulgaris*）、黑果越橘（*Vaccinium myrtillus*）、黄花柳（*Salix caprea*）、东陵山柳（*Salix phylicifolia*）、异株荨麻（*Urtica dioica*）、树莓（*Rubus fruticosus*）、覆盆子（*Rubus idaeus*）、秦氏悬钩子（*Rubus caesius*）、（*Valeriana tripteris*）、大车前（*Plantago major*）、北车前（*Plantago media*）、阿尔泰藜芦（*Veratrum lobelianum*）、桦木属（*Betula*）、报春花属（*Primula*）、山楂属（*Crataegus*）、草莓属（*Fragaria*）、蔷薇属（*Rosa*）、蒲公英属（*Taraxacum*）植物（Matov & Kononenko, 2012）。

Diarsia brunnea **([Denis & Schiffermüller], 1775) 棕色歹夜蛾**

Noctua brunnea [Denis & Schifferm ü ller], 1775, *Ankündung eines Systematisches Werkes von den*

Schmetterlinge der Wiener Gegend. Wien: 83. Type locality: Austria, Vienna.

Synonymy: *Phalaena* (*Noctua*) *lucifera* Esper, 1789; *Phalaena* (*Noctua*) *lucifera* Esper, 1798; *Phalaena* (*Noctua*) *arvensis* Gmelin, [1790]; *Noctua carnea* Thunberg, 1792; *Phalaena* (*Noctua*) *fragariae* Borkhausen, 1792; *Rhyacia collina urupina* Bryk, 1942 (ssp.); *Diarsia brunnea distinctissima* Bryk, 1942 (ssp.).

Distribution. China (Heilongjiang, Jilin, Shanxi, Shaanxi, Sichuan); Russia (European part, Ural, Siberia), Korea, Japan; Europe (North and mountainous areas).

寄主：欧洲鳞毛蕨（*Dryopteris filix-mas*）、欧洲蕨（*Pteridium aquilinum*）、蹄盖蕨（*Athyrium filix-femina*）、欧洲小檗（*Berberis vulgaris*）、白屈菜（*Chelidonium majus*）、（*Stellaria holostea*）、拳参（*Polygonum bistorta*）、酸模（*Rumex acetosa*）、小酸模（*Rumex acetosella*）、钝叶酸模（*Rumex obtusifolius*）、欧洲桤木（*Alnus glutinosa*）、桦叶鹅耳枥（*Carpinus betulus*）、帚石楠（*Calluna vulgaris*）、黑果越橘（*Vaccinium myrtillus*）、笃斯越橘（*Vaccinium uliginosum*）、越橘（*Vaccinium vitis-idaea*）、牛舌樱草（*Primula elatior*）、黄花九轮草（*Primula veris*）、黄华柳（*Salix atrocinerea*）、耳柳（*Salix aurita*）、黄花柳（*Salix caprea*）、灰柳（*Salix cinerea*）、草甸碎米荠（*Cardamine pratensis*）、钝裂叶山楂（*Crataegus laevigata*）、欧亚路边青（*Geum urbanum*）、稠李（*Padus racemosa*）、黑刺李（*Prunus spinosa*）、秦氏悬钩子（*Rubus caesius*）、树莓（*Rubus fruticosus*）、覆盆子（*Rubus idaeus*）、欧洲花楸（*Sorbus aucuparia*）、银叶老鹳草（*Geranium sylvaticum*）、羊角芹（*Aegopodium podagraria*）、林当归（*Angelica sylvestris*）、欧银花（*Lonicera xylosteum*）、（*Succisa pratensis*）、林生玄参（*Scrophularia nodosa*）、春玄参（*Scrophularia vernalis*）、大车前（*Plantago major*）、吊钟千里光（*Senecio fuchsii*）、阿尔泰藜芦（*Veratrum lobelianum*）、丛林地杨梅（*Luzula sylvatica*）、短柄草（*Brachypodium sylvaticum*）、曲芒发草（*Deschampsia flexuosa*）、酸沼草（*Molinia caerulea*）、毛蕊花属（*Ranunculus*）、栎属（*Quercus*）、桦木属（*Betula*）、荨麻属（*Urtica*）、蔷薇属（*Rosa*）、车轴草属（*Trifolium*）、野芝麻属（*Lamium*）、水苏属（*Stachys*）、羽衣草属（*Alchemilla*）、山柳菊属（*Hieracium*）、蒲公英属（*Taraxacum*）植物（Matov & Kononenko, 2012）。

***Diarsia dewitzi* (Graeser, [1889] 1888) 狄歹夜蛾**

Agrotis dewitzi Graeser, [1889]1888, *Berliner entomologische Zeitschrift*, 32, II: 316. Type locality: Russia, Khabarovsk ["Chabarofka"].

Synonymy: *Noctua tarda* Leech, 1889.

Distribution. China (Heilongjiang, Jilin); Russia (Siberia), Korea, Japan.

寄主：冷杉属（*Abies*）、云杉属（*Picea*）植物（Matov & Kononenko, 2012）。

***Diarsia canescens* (Butler, 1878) 灰歹夜蛾**

Graphiphora canescens Butler, 1878, *Annals and Magazine of Natural History*, 1 (5): 165; illustrated in *Illustrations of Typical Specimens of Lepidoptera Heterocera in the Collection of the British Museum* 2: 28, pl. 30: 1. Type locality: Japan, Hakodate, Yokohama.

Synonymy: *Cerastis subdolens* Butler, 1881; *Lycophotia brunnescens* Hampson, 1903; *Rhyacia ishidella* Matsumura, 1926; *Rhyacia subcanescens* Corti & Draudt, 1933.

Distribution. China (Heilongjiang, Liaoning, Nei Mongol, Qinghai, Xinjiang, Hebei, Henan, Hubei, Jiangxi, Sichuan, Taiwan); Russia (Transbaikalia, Siberia), Korea, Japan, India, Pakistan.

寄主：紫云英、多种蔬菜、茶（陈一心，1999）；甜菜（*Beta vulgaris*）、酸模（*Rumex acetosa*）、黄瓜（*Cucumis sativus*）、甘蓝（*Brassica oleracea*）、萝卜（*Raphanus sativus*）、覆盆子（*Rubus idaeus*）、紫云英（*Astragalus sinicus*）、大豆（*Glycine max*）、黄檗（*PhellodendroNorth of Amur reg.ense*）、阳芋（*Solanum tuberosum*）、车前（*Plantago asiatica*）、玉米（*Zea mays*）、接骨木属（*Sambucus*）、核桃属（*Juglans*）、榛属（*Corylus*）、报春花属（*Primula*）、杨属（*Populus*）、椴树属（*Tilia*）、榆属（*Ulmus*）、蚕豆属（*Vicia*）、路边青属（*Geum*）、蒲公英属（*Taraxacum*）、百合属（*Lilium*）、黄芪属（*Astragalus*）、报春花属（*Primula*）、茄属（*Solanum*）植物（Matov & Kononenko, 2012）。

Diarsia pacifica **Boursin, 1943 黑龙江歹夜蛾**

Diarsia pacifica Boursin, 1943, *Zeitschrift der Wiener Entomologischen Gesellschaft*, 28: 344, pl. 43: 1, 2. Type locality: Russia, “Amur”.

Synonymy: *Diarsia ruficauda ottonisbanghaasi* Bryk, 1948 (ssp.).

Distribution. China (Heilongjiang); Russia (Siberia), Korea, Japan.

注：首次给出中文种名。

Diarsia ruficauda **(Warren, 1909) 赭尾歹夜蛾**

Rhyacia ruficauda Warren, 1909, *in* A.Seitz (ed.), *Die Gross-Schmetterlinge der Erde*, 3: 46, pl. 10: f. Type locality: Japan; West China.

Synonymy: *Agrotis rubicilia*: Hampson, 1894; *Diarsia ruficauda steni* Bryk, 1948 (ssp.).

Distribution. China (Heilongjiang, Jiangsu, Zhejiang, Hunan, Jiangxi, Fujian, Yunnan); Russia (Siberia), Korea, Japan.

Diarsia deparca **(Butler, 1879) 分歹夜蛾**

Mythimna deparca Butler, 1879, *Annals and Magazine of Natural History* 5, 4: 358. Type locality: Japan, Yokohama.

Synonymy: *Rhyacia interca* Matsumura, 1926; *Rhyacia takamukui* Matsumura, 1926.

Distribution. China (Heilongjiang, Jilin, Sichuan, Yunnan, Xizang); Russia (Siberia), Korea, Japan, India, Nepal.

Genus *Cerastis* Ochsenheimer, 1816 弓夜蛾属

Cerastis Ochsenheimer, 1816, *Die Schmetterlinge von Europa*, 4: 84. Type species: *Noctua rubricosa* [Denis & Schiffermüller], 1775 [Austria].

Synonymy: *Glaeae* Ochsenheimer, 1816, unavail.; *Matuta* Grote, 1874, preocc. Weber, 1835 [Crustacea]; *Cerastia* Stephens, 1850, emend.; *Sora* Heinemann, 1859, preocc. Walker, 1859 [Coleoptera]; *Gypsitea* Tams, 1939, repl. name.

The genus contains about 12 species, distributed in the Palaearctic, Nearctic, and mountains of the Oriental region, among them 3 species are in the Three Provinces of Northeast China.

Cerastis leucographa ([Denis & Schiffermüller], 1775) 茸弓夜蛾

Noctua leucographa [Denis & Schifferm ü ller], 1775, *Ankündung eines Systematisches Werkes von den Schmetterlinge der Wiener Gegend. Wien*: 83. Type locality: Austria, Vienna.

Synonymy: *Noctua cervina* Hübner, [1821]; *Noctua lepetitii* Boisduval, [1837].

Distribution. China (Heilongjiang, Jilin); Russia (European part, Ural, Siberia), Korea, Japan; Middle East, Europe.

寄主：繁缕（*Stellaria media*）、酸模（*Rumex acetosa*）、胡榛（*Corylus avellana*）、黑果越橘（*Vaccinium myrtillus*）、笃斯越橘（*Vaccinium uliginosum*）、黄花柳（*Salix caprea*）、黑刺李（*Prunus spinosa*）、欧洲花楸（*Sorbus aucuparia*）、四叶葎（*Galium mollugo*）、（*Rhinanthus minor*）、短柄野芝麻（*Lamium album*）、棕矢车菊（*Centaurea jacea*）、多花黄精（*Polygonatum multiflorum*）、（*Alsine*）、栎属（*Quercus*）、帚石楠属（*Calluna*）、悬钩子属（*Rubus*）、车前属（*Plantago*）、蒲公英属（*Taraxacum*）植物（Matov & Kononenko, 2012）。

Cerastis orientalis Boursin, 1948 东弓夜蛾

Cerastis orientalis Boursin, 1948, *Zeitschrift der Wiener Entomologischen Gesellschaft*, 33 (10-12): 128, pl. 3: 11, 12. Type locality: Russia, [Primorye terr., Ussuriisk] Nikolsk-Ussurijsk.

Distribution. China (Heilongjiang); Russia (Siberia), Korea.

Cerastis pallescens (Butler, 1878) 淡弓夜蛾

Semiophora pallescens Butler, 1878, *Annals and Magazine of Natural History*, (5)1: 166. Type locality: Japan, Yokohama.

Synonymy: *Taeniocampa lata* Staudinger, 1892.

Distribution. China (Heilongjiang, Jilin); Russia (Siberia), Korea, Japan.

寄主：延龄草（*Trillium tschonoskii*）（Matov & Kononenko, 2012）。

注：首次给出中文种名。

Genus *Paradiarsia* McDunnough, [1929] 1928 芒夜蛾属

Paradiarsia McDunnough, [1929] 1928, *Bulletin of the Canada Department of Mines*, 55: 48. Type species: *Agrotis littoralis* Packard, 1867 [North America, Canada].

The genus contains 3 species, distributed in the temperate zone of Palaearctic and Nearctic regions. Among them, 1 species is in the Three Provinces of Northeast China.

***Paradiarsia punicea* (Hübner, 1803) 泛紫芒夜蛾**

Noctua punicea Hübner, 1803, *Sammlung Europaischer Schmetterlinge. Eulen*, 2: pl. 25: 115. Type locality: Europe.

Synonymy: *Rhyacia exustiformis* Matsumura, 1925.

Distribution. China (Heilongjiang, Jilin); Russia (European part, Ural, Siberia), Mongolia, Korea, Japan, Kazakhstan; Europe.

寄主：悬钩子、蒲公英等属植物（陈一心，1999）；榕叶毛茛（*Ranunculus ficaria*）、驴蹄草（*Caltha palustris*）、（*Quercus sessiliflora*）、帚石楠（*Calluna vulgaris*）、（*Betua pubescens*）、黑点叶金丝桃（*Hypericum perforatum*）、笃斯越橘（*Vaccinium uliginosum*）、毛黄连花（*Lysimachia vulgaris*）、牛舌樱草（*Primula elatior*）、蒿柳（*Salix viminalis*）、豆瓣菜（*Nasturtium officinale*）、异株荨麻（*Urtica dioica*）、旋果蚊子草（*Filipendula ulmaria*）、野草莓（*Fragaria vesca*）、稠李（*Padus racemosa*）、黑刺李（*Prunus spinosa*）、秦氏悬钩子（*Rubus caesius*）、树莓（*Rubus fruticosus*）、覆盆子（*Rubus idaeus*）、千屈菜（*Lythrum salicaria*）、大豆（*Glycine max*）、百脉根（*Lotus corniculatus*）、广布野豌豆（*Vicia cracca*）、欧鼠李（*Frangula alnus*）、药鼠李（*Rhamnus cathartica*）、林当归（*Angelica silvestris*）、（*Peucedanum palustris*）、缬草（*Valeriana officinalis*）、欧银花（*Lonicera xylosteum*）、沼猪殃殃（*Galium uliginosum*）、（*Euphrasia rostkoviana*）、（*Plantago lanceolatum*）、百里香（*Thymus serpyllum*）、卷心菜蓟（*Cirsium oleraceum*）、蓟（*Cirsium palustre*）、大麻叶泽兰（*Eupatorium cannabinum*）、莴苣（*Lactuca sativa*）、早生一枝黄花（*Solidago gigantea*）、药用蒲公英（*Taraxacum officinale*）、款冬花（*Tussilago farfara*）、白花秋水仙（*Colchicum autumnale*）、草甸羊茅（*Festuca pratensis*）、芦苇（*Phragmites australis*）、酸模属（*Rumex*）、羽衣草属（*Alchemilla*）、路边青属（*Geum*）、委陵菜属（*Potentilla*）、车轴草属（*Trifolium*）、苔属（*Carex*）植物（Matov & Kononenko, 2012）。

Genus *Lycophotia* Hübner, [1821] 1816 烈夜蛾属

Lycophotia Hübner, [1821] 1816, *Verzeichniss bekannter Schmettlinge* [sic!]: 215. Type species: *Noctua porphyrea* [Denis & Schiffermüller], 1775 [Austria].

Synonymy: *Scotophila* Stephens, 1829, preocc. Hübner, [1821] [Lepidoptera, Noctuidae]; *Violaphotia* Beck, [1992] 1991; *Paugraphia* Beck, [1992] 1991; *Ericathia* Beck, 1996; *Licophotia*: Kozh., 1937, missp..

The genus contains 6 species, distributed in the temperate zone of Palaearctic region. Among them, 1 species is in the Three Provinces of Northeast China.

Lycophotia cissigma (Ménétriès, 1859) 顺标烈夜蛾

Noctua cissigma Ménétriès, 1859, *Bulletin de la Classe Physico-Mathématique de l'Académie Impériale des Sciences de St.-Pétersbourg*, 17: 220. Type locality: Russia, East Siberia, [Vilui] "Viloui".

Synonymy: *Agrotis umbra* Staudinger, 1892.

Distribution. China (Heilongjiang, Jilin); Russia (Ural, Siberia), Mongolia, Korea, Japan, Kazakhstan; Europe.

注：首次给出中文种名。

Genus *Pseudohermonassa* Varga, 1990 黑卡夜蛾属

Pseudohermonassa Varga, 1990, *Acta Zoologica Hungarica*, 36: 337. Type species: *Chersotis melancholica* Lederer, 1853 [East Kazakhstan, West Altai].

The genus contains 5 species, distributed in the temperate zone of the Palaearctic and Nearctic region. Among them, 2 species are in the Three Provinces of Northeast China.

注：首次给出中文属名。

Pseudohermonassa melancholica (Lederer, 1853) 黑卡夜蛾

Caradrina melancholica Lederer, 1853, *Verhandlungen des Zoologisch-Botanischen vereins in Wien*, 1853: 367, pl. 4: 3. Type locality: East Kazakhstan, [West Altai] Ust-Kamengorsk, Ust-Buchtarminsk.

Distribution. China (Heilongjiang, Xinjiang, Qinghai); Russia (Ural, Siberia), Mongolia, Korea, Japan, India.

寄主：偃麦草（*Elytrigia repens*）、（*Atriplex patulum*）、山芥属（*Barbarea*）、车轴草属（*Trifolium*）、蒿属（*Artemisia*）、苦苣菜属（*Sonchus*）、蒲公英属（*Taraxacum*）植物（Matov & Kononenko, 2012）。

Pseudohermonassa ononensis (Bremer, 1864) 漫卡夜蛾

Agrotis ononensis Bremer, 1864, *Memoires de l'Académie Imperiale des Sciences de St.-Petersbourg*: 50, pl. 4: 9. Type locality: Russia, Transbaikalia, Onon.

Synonymy: *Agrotis scaramangae* Alphéraky, 1882; *Agrotis praecipua* Staudinger, 1892; *Agrotis cicatricosa* Graeser, 1892; *Agrotis scaramangoides* Barnes & Benjamin, 1926.

Distribution. China (Heilongjiang, Nei Mongol, Xinjiang, Qinghai, Gansu, Xizang); Russia (Siberia, Kamchatka), Mongolia, Kazakhstan; North America.

Genus *Hermonassa* Walker, 1865 狭翅夜蛾属

Hermonassa Walker, 1865, *List of the Specimens of Lepidopterous Insects in the Collection of the British Museum*, 32: 631. Type species: *Hermonassa consignata* Walker, 1865 [India].

The genus contains 85 species, distributed in the mountain areas of the Eastern Palaearctic and Oriental regions; the center of *Hermonassa* species richness and diversity are the Southwest China and Himalayas region. Among them, 2 species are in the Three Provinces of Northeast China.

Hermonassa cecilia Butler, 1878 茶色狭翅夜蛾

Hermonassa cecilia Butler, 1878, *Annals and Magazine of Natural History*, (5)1: 164. Type locality: Japan, Yokohama.

Distribution. China (Jilin, Jiangxi, Sichuan, Xizang); Russia (Siberia), Korea, Japan, India(North).

寄主：直穗酸模（*Rumex longifolius*）（Matov & Kononenko, 2012）。

Hermonassa arenosa (Butler, 1881) 仑狭翅夜蛾

Opigena arenosa Butler, 1881, *Transactions of the Entomological Society of London*, 1881: 179. Type locality: Japan, Tokyo.

Synonymy: *Hermonassa arenosa amurensis* Kozhantschikov, 1935 (ssp.); *Hermonassa yeterofuna* Bryk, 1942.

Distribution. China (Heilongjiang, Jilin); Russia (Siberia), Korea, Japan.

Genus *Rhyacia* Hübner, [1821] 1816 沁夜蛾属

Rhyacia Hübner, [1821] 1816, *Verzeichniss bekannter Schmettlinge* [sic!]: 210. Type species: *Noctua lucipeta* [Denis & Schiffermüller], 1775 [Austria].

Synonymy: *Epipsilia* Hübner, 1821 (subgen.); *Pararhyacia* Kocak, 1980; *Antirhyacia* Beck, 1992; *Stenorhyacia* Varga, 2011 (subgen.).

Predominantly Palaearctic distributed genus containing 45 species. It divided for the subgenera *Rhyacia* (s. str.), *Stenorhyacia* and *Epipsilia*. Most species occurs in the arid, semiarid and mounati areas of the temperate zone. Among them, 1 species of 1 subgenus is in the Three Provinces of Northeast China.

Subgenus *Stenorhyacia* Varga, 2011 窄沁夜蛾亚属

Stenorhyacia Varga, 2011, *Esperiana*, Bd. 16: 257. Type species: *Agrotis electra* Staudinger, 1888. Type locality: Uzbekistan, Margelan.

注：首次给出中文亚属名。

***Rhyacia ledereri* (Erschoff, 1870) 来沁夜蛾**

Agrotis ledereri Erschoff, 1870, *Horae Societatis Entomologicae Rossicae*, 4: 195. Type locality: Russia, Irkutsk.

Synonymy: *Agrotis rava mus* Alphéraky, 1882 (var.); *Rhyacia pallidifrons* Hampson, 1903; *Rhyacia* (*Epipsilia*) *punctinotata* Warren, 1914; *Rhyacia punctinotata* Warren, 1914.

Distribution. China (Heilongjiang, Xinjiang, Xizang); Russia (Ural, Altai, Transbaikalia, Siberia), Mongolia, Korea, Japan, North Kazakhstan.

Genus ***Chersotis*** Boisduval, 1840 卡夜蛾属

Chersotis Boisduval, 1840, *Genera et Index Methodicus Europaeorum Lepidopterorum*: 103. Type species: *Noctua rectangula* [Denis & Schiffermüller], 1775 [Austria].

Synonymy: *Alpsotis* Beck, [1992] 1991; *Margasotis* Beck, [1992] 1991; *Elesotis* Beck, [1992] 1991; *Larixotis* Beck, [1992] 1991; *Fimbriosotis* Beck, [1992] 1991; *Cupreositis* Beck, [1992] 1991.

Predominantly Palaearctic distributed genus containing 75 species. The majority of species occur in the arid or semiarid zones, in habiting dry and rocky habitats. The center of specific diversity is obviously the Mediterranean and Central-Asian subregions of the Palaearctic region. The genus globally has been revised by Varga et al. (2013). 1 species is in the Three Provinces of Northeast China.

***Chersotis deplanata* (Eversmann, 1843) 融卡夜蛾**

Episema deplanata Eversmann, 1843, *Bulletin de la Société Impériale des Naturalistes de Moscou*, 16: 545. Type locality: Russia, Ural.

Synonymy: *Agrotis deplana* Freyer, 1845; *Agrotis autumnalis* Oberthür, 1880; *Manobia sachalinensis* Matsumura, 1925; *Chersotis deplanata columbina* Bryk, 1949 (ssp.).

Distribution. China (Heilongjiang, Xinjiang, Qinghai, Hebei, Zhejiang); Russia (Ural, Siberia), Mongolia, Korea, Japan.

Genus ***Cryptocala*** Benjamin, 1921 模夜蛾属

Cryptocala Benjamin, 1921, *Bulletin of the Southern California Acaddemy of Sciences*, 20: 133. Type species: *Agrotis gilvipennis* Grote, 1874 [North America], a junior synonym of *Anarta acadiensis* Bethune, 1884.

Synonymy: *Noctua* auct., nec Linnaeus, 1758.

The genus contains 2 species, distributed in the temperate zone of the Palaearctic and Nearctic region. Among them, 1 species is in the Three Provinces of Northeast China.

注：《中国动物志》（夜蛾科）中引用本属异名“*Noctua* Linnaeus, 1758”，在此予以变更。

***Cryptocala chardinyi* (Boisduval, 1829) 拱模夜蛾**

Triphaena chardinyi Boisduval, 1829, *Europaeorum Lepidopterorum Index Methodicus*: 94. Type locality: Russia, Moscow reg..

Synonymy: *Noctua hetaera* Eversmann, 1837; *Agrotis chardinyi fuchsii* Wendlandt, 1901 (var.).

Distribution. China (Heilongjiang, Jilin, Nei Mongol, Xinjiang, Gansu, Hebei); Russia (European part, Ural, Siberia), North Korea; Europe.

寄主：萹蓄（*Polygonum aviculare*）、直穗酸模（*Rumex longifolius*）、帚石楠（*Calluna vulgaris*）、欧洲山杨（*Populus tremula*）、异株荨麻（*Urtica dioica*）、覆盆子（*Rubus idaeus*）、柳兰（*Chamaenerion angustifolium*）、（*Epilobium montanum*）、红辣椒千叶蓍（*Achillea millefolium*）、桦木属（*Betula*）、堇菜属（*Viola*）、柳属（*Salix*）、茶藨子属（*Ribes*）、蚕豆属（*Vicia*）、车前草属（*Plantago*）、野芝麻属（*Lamium*）、莴苣属（*Lactuca*）、蒲公英属（*Taraxacum*）植物（Matov & Kononenko, 2012）。

Genus ***Spaelotis*** Boisduval, 1840 **矛夜蛾属**

Spaelotis Boisduval, 1840, *Genera et Index Methodicus Europaeorum Lepidopterorum*: 106. Type species: *Noctua ravida* [Denis & Schiffermüller], 1775 [Austria].

Synonymy: *Amphitrota* Warren, 1909 (subgen.).

The genus divided for the subgenera *Spaelotis* (s. str.) and *Amphitrota.* It contains 24 species, mainly distributed in the Central Asian subregion of the Palaearctic and Nearctic regions. Among them, 3 species are in the Three Provinces of Northeast China.

***Spaelotis ravida* ([Denis & Schiffermüller], 1775) 矛夜蛾**

Noctua ravida [Denis & Schiffermüller], 1775, *Ankündung eines Systematisches Werkes von den Schmetterlinge der Wiener Gegend. Wien*: 80. Type locality: Austria, Vienna.

Synonymy: *Phalaena* (*Noctua*) *obducta* Esper, 1789; *Phalaena* (*Noctua*) *austera* Esper, 1789; *Phalaena* (*Noctua*) *bigramma* Esper, 1790; *Noctua obscura* Brahm, 1791; *Agrotis stabulorum* Bienert, 1869; *Agrotis nipona* Felder et Rogenhofer, 1874; *Agrotis glis* Christoph, 1887; *Rhyacia salva* Corti & Draudt, 1933.

Distribution. China (Heilongjiang, Jilin, Nei Mongol, Xinjiang, Qinghai, Hebei, Jiangsu); Russia (European part, Ural, Siberia), Mongolia, Korea, Japan, India(North), Pakistan; Central Asia, Middle East, Europe.

寄主：反枝苋（*Amaranthus retroflexus*）、藜（*Chenopodium album*）、萹蓄（*Polygonum aviculare*）、小酸模（*Rumex acetosella*）、西瓜（*Citrullus lanatus*）、黄瓜（*Cucumis sativus*）、西葫芦（*Cucurbita pepo*）、（*Melo sativus*）、大叶芥菜（*Brassica juncea*）、荠（*Capsella bursa-pastoris*）、异株荨麻（*Urtica dioica*）、黑茶藨子（*Ribes nigrum*）、树锦鸡儿（*Caragana arborescens*）、番茄（*Lycopersicon esculentum*）、烟草（*Nicotiana tabacum*）、蓝蓟（*Echium vulgare*）、毛头牛蒡（*Arctium tomentosum*）、荒野蒿（*Artemisia *campestris*）、向日葵（*Helianthus annuus*）、秋蒲公英（*Leontodon autumnalis*）、药用蒲公英（*Taraxacum officinale*）、羊茅（*Festuca ovina*）、松属（*Pinus*）、（*Alsine*）、拟漆姑属（*Spergularia*）、繁缕属（*Stellaria*）、山芥属（*Barbarea*）、驴食草属（*Onobrychis*）、亚麻属（*Linum*）、蓝盆花属（*Scabiosa*）、车前属（*Plantago*）、鼬瓣花属（*Galeopsis*）、飞廉属（*Carduus*）、蓟属（*Cirsium*）、苦苣菜属（*Sonchus*）、短柄草属（*Brachypodium*）、黑麦草属（*Lolium*）、早熟禾属（*Poa*）植物（Matov & Kononenko, 2012）。

Spaelotis lucens Butler, 1881 鲁矛夜蛾

Spaelotis lucens Butler, 1881, *Transactions of the Entomological Society of London*, 1881: 179. Type locality: Japan, Tokyo.

Distribution. China (Heilongjiang); Russia (Siberia), Korea, Japan.

注：首次给出中文种名。

Spaelotis valida (Walker, 1865) 褐矛夜蛾

Graphiphora valida Walker, 1865, *List of the Specimens of Lepidopterous Insects in the Collection of the British Museum*, 33: 711. Type locality: China, Shanghai.

Synonymy: *Graphiphora caliginea* Butler, 1878.

Distribution. China (Jilin, Liaoning); South Korea, Japan.

注：首次给出中文种名。

Genus *Graphiphora* Ochsenheimer, 1816 割夜蛾属

Graphiphora Ochsenheimer, 1816, *Die Schmetterlinge von Europa*, 4: 68. Type species: *Noctua augur* Fabricius, 1775 [Germany].

Synonymy: *Graphophora* Agassiz, 1846, emend..

The genus contains 2 species, distributed in the Palaearctic and Nearctic regions. Among them 1 species is in the Three Provinces of Northeast China.

Graphiphora augur (Fabricius, 1775) 割夜蛾

Noctua augur Fabricius, 1775, *Entomologiae Systematica emendata et Aucta. Secundum Classes, Ordines, Genera, Species Adjectis Synonimis, Locis, Observationibus, Descriptionibus*: 604. Type locality: Germany.

Synonymy: *Phalaena* (*Noctua*) *omega* Esper, 1788; *Phalaena* (*Noctua*) *omega* Esper, 1796; *Phalaena* (*Noctua*) *assimulans* Borkhausen, 1792; *Noctua hippophaes* Geyer, [1832]; *Agrotis haruspica* Grote, 1875; *Agrotis unimacula* Morrison, 1875; *Agrotis augur grandis* Speyer, 1875 (var.); *Agrotis sierrae* Harvey, 1876; *Agrotis abdita* Joannis, 1891; *Graphiphora augur helvetina* Tutt, 1892 (var.); *Noctua inopinatus* Smith, 1898; *Agrotis augur nigra* Vorbrodt, 1912 (ab.); *Agrotis augur conjuncta* Schille, 1924 (ab.); *Rhyacia augur tobolskensis* Sheljuzhko, 1929 (ssp.); *Agrotis augur bivirga* Ceton, 1935 (ab.); *Graphiphora augur juncta* Lempke, 1939 (f.); *Rhyacia augur striata* Petersen, 1951 (ab.); *Graphiphora augur basilinea* van Wisselingh, 1959 (f.); *Graphiphora augur rufescens* Lempke, 1962 (f.); *Graphiphora augur pallida* Lempke, 1962 (f.); *Graphiphora augur fuscolimbata* Lempke, 1962 (f.); *Graphiphora augur protensa* Lempke, 1962 (f.); *Graphiphora augur semiconfluens* Lempke, 1962 (f.); *Graphiphora augur cruda* Lempke, 1962 (f.); *Graphiphora augur obsoleta* Lempke, 1962 (f.); *Graphiphora augur marginata* Lempke, 1964 (f.).

Distribution. China (Heilongjiang, Xinjiang); Russia (European part, Ural, Siberia), Mongolia, North Korea, Japan, Pakistan, Kazakhstan; Europe.

寄主：甜菜（*Beta vulgaris*）、藜（*Chenopodium album*）、酸模（*Rumex acetosa*）、夏栎（*Quercus robur*）、北美杨树（*Populus balsamifera*）、欧洲山杨（*Populus tremula*）、黄华柳（*Salix atrocinerea*）、耳柳（*Salix aurita*）、黄花柳（*Salix caprea*）、灰柳（*Salix cinerea*）、东陵山柳（*Salix phylicifolia*）、蒿柳（*Salix viminalis*）、圆叶锦葵（*Malva pusilla*）、钝裂叶山楂（*Crataegus laevigata*）、锐刺山楂（*Crataegus oxyacantha*）、稠李（*Padus racemosa*）、黑刺李（*Prunus spinosa*）、刺蔷薇（*Rosa acicularis*）、秦氏悬钩子（*Rubus caesius*）、覆盆子（*Rubus idaeus*）、北美树莓（*Rubus odoratus*）、绣线菊（*Spiraea salicifolia*）、欧洲花楸（*Sorbus aucuparia*）、欧鼠李（*Frangula alnus*）、欧洲红瑞木（*Cornus sanguinea*）、洋常春藤（*Hedera helix*）、林当归（*Angelica sylvestris*）、蓝果忍冬（*Lonicera coerulea*）、繁果忍冬（*Lonicera tatarica*）、欧丁香（*Syringa vulgaris*）、蓝蓟（*Echium vulgare*）、菊芋（*Helianthus tuberosus*）、山毛榉属（*Fagus*）、桤木属（*Alnus*）、桦木属（*Betula*）、报春花属（*Primula*）、芸苔属（*Brassica*）、椴树属（*Tilia*）、茶藨子属（*Ribes*）、苹果属（*Malus*）、独活属（*Heracleum*）、毛核木属（*Symphoricarpos*）、车前属（*Plantago*）、苦苣菜属（*Sonchus*）、蒲公英属（*Taraxacum*）、葱属（*Allium*）植物（Matov & Kononenko, 2012）。

Genus ***Eurois*** Hübner, [1821] 1816 东风夜蛾属

Eurois Hübner, [1821] 1816, *Verzeichniss bekannter Schmettlinge* [sic!]: 218. Type species: *Phalaena occulta* Linnaeus, 1758 [Europe].

Synonymy: *Eurois* Agassiz, 1846, emend..

Holarctic genus containing 7 species distributed in the temperate zone; among them, 1 species is in the Three Provinces of Northeast China.

Eurois occulta **(Linnaeus, 1758) 东风夜蛾**

Phalaena occulta Linnaeus, 1758, *Systema Naturae* (Edn. 10) 1: 514. Type locality: Europe.

Synonymy: *Hadena implicata* Lefevbre, 1836; *Hadena extricata* Zetterstedt, [1839]; *Polia prolixa* Zetterstedt, [1839]; *Polia puncticosta* Zetterstedt, [1839]; *Agrotis occulta passetii* Thierry-Mieg, 1886 (ab.); *Agrotis occulta pallida* Spuler, 1908 (ab.); *Agrotis occulta coerulescens* Meves, 1914 (ab.); *Eurois occulta grisea* Hannemann, 1915 (f.); *Agrotis* (*Rhyacia*) *occulta rectangularis* Stephan, 1925 (ab.); *Rhyacia occulta nigra* Metz, 1928 (var.);

Agrotis occulta roseovirgata Dannehl, 1929 (var.); *Rhyacia occulta ichinosawana* Matsumura, 1931 (f.); *Eurois occulta fumea* Corti et Draudt, 1933 (f.); *Eurois occulta tibetica* Corti et Draudt, 1933 (f.); *Eurois occulta semiconfluens* Lempke, 1939 (f.); *Eurois occulta paradoxa* Lempke, 1962 (f.); *Eurois occulta brunnescens* Lempke, 1962 (f.); *Eurois occulta juncta* Lempke, 1962 (f.); *Eurois occulta obsoleta* Lempke, 1962 (f.).

Distribution. China (Heilongjiang, Jilin, Liaoning); Russia (European part, Ural, Siberia), Mongolia, Korea, Japan, Kazakhstan; Europe, North America (boreal zone).

寄主：报春、蒲公英等属植物（陈一心，1999）；美洲落叶松（*Larix laricina*）、欧洲赤松（*Pinus sylvestris*）、异珠女娄菜（*Melandrium album*）、萹蓄 (*Polygonum aviculare*)、卷茎蓼（*Polygonum convolvulus*）、甜菜（*Beta vulgaris*）、灰桤木（*Alnus incana*）、帚石楠（*Calluna vulgaris*）、黑果越橘（*Vaccinium myrtillus*）、笃斯越橘（*Vaccinium uliginosum*）、越橘（*Vaccinium vitis-idaea*）、欧洲山杨（*Populus tremula*）、美洲山杨（*Populus tremuloides*）、耳柳（*Salix aurita*）、黄花柳（*Salix caprea*）、五蕊柳（*Salix pentandra*）、东陵山柳（*Salix phylicifolia*）、黄瓜（*Cucumis sativus*）、大叶芥菜（*Brassica juncea*）、甘蓝（*Brassica oleracea*）、芜青（*Brassica rapa*）、萝卜（*Raphanus sativus*）、异株荨麻（*Urtica dioica*）、紫叶景天（*Sedum telephium*）、毛茶藨（*Ribes spicatum*）、稠李（*Padus racemosa*）、欧洲李（*Prunus domestica*）、黑刺李（*Prunus spinosa*）、树莓（*Rubus fruticosus*）、覆盆子（*Rubus idaeus*）、欧洲花楸（*Sorbus aucuparia*）、绣线菊（*Spiraea salicifolia*）、柳兰（*Chamaenerion angustifolium*）、牧地山黧豆（*Lathyrus pratensis*）、（*Lathyrus silvaticum*）、羊角芹（*Aegopodium podagraria*）、蔓长春花（*Vinca major*）、阳芋（*Solanum tuberosum*）、大车前（*Plantago major*）、北车前（*Plantago media*）、短柄野芝麻（*Lamium album*）、百里香（*Thymus serpyllum*）、棕矢车菊（*Centaurea jacea*）、（*Cirsium cetusum*）、莴苣（*Lactuca sativa*）、毛果一枝黄花（*Solidago virgaurea*）、苣荬菜（*Sonchus arvensis*）、普通剪股颖（*Agrostis canina*）、小穗发草（*Deschampsia caespitosa*）、酸沼草（*Molinia caerulea*）、酸模属（*Rumex*）、桦木属（*Betula*）、香杨梅（*Myrica gale*）、报春花属（*Primula*）、樱属（*Cerasus*）、蔷薇属（*Rosa*）、柳叶菜属（*Epilobium*）、毛核木属（*Symphoricarpos*）、风毛菊属（*Leontodon*）、蒲公英属（*Taraxacum*）植物（Matov & Kononenko, 2012）。

Genus *Anaplectoides* McDunnough, [1929] 1928 组夜蛾属

Anaplectoides McDunnough [1929] 1928, *Bulletin of the Canada Department of Mines*, 55. 65. Type species: *Eurois pressus* Grote, 1874, [USA, Canada].

The genus contains 14 species, distributed in the Palaearctic, Oriental and Nearctic regions. Among them, 2 species are in the Three Provinces of Northeast China.

Anaplectoides prasina ([Denis & Schiffermüller], 1775) 绿组夜蛾

Phalaena (*Noctua*) *prasina* [Denis & Schifferm ü ller], 1775, *Ankündung eines Systematisches Werkes von den Schmetterlinge der Wiener Gegend. Wien*: 82. Type locality: Austria, Vienna.

Synonymy: *Noctua herbida* [Denis et Schiffermüller], 1775; *Noctua egregia* Esper, 1787; *Phalaena* (*Noctua*) *jaspidea* Borkhausen, 1792; *Noctua mixta* Haworth, 1809; *Polyphaenis herbacea* Guenée, 1852; *Aplecta prasina pallida* Tutt, 1892 (var.); *Aplecta prasina suffusa* Tutt, 1892 (var.); *Agrotis prasina albimacula* Hormuzaki, 1894 (var.); *Aplecta prasina eboraci* Kane, 1896 (var.); *Agrotis prasina lugubris* Petersen, 1902 (ab.); *Agrotis prasina viridior* Spuler, 1908 (ab.); *Agrotis prasina vittata* Heinrich, 1916 (ab.); *Agrotis prasina medionigra* Lenz, 1927 (var.); *Agrotis prasina olivacea* Lenz, 1927 (var.); *Agrotis prasina obscura* Lenz, 1927 (var.); *Anaplectoides prasina effusa* Cockayne, 1952 (ab.); *Anaplectoides prasina leucozona* Cockayne, 1952 (ab.); *Anaplectoids prasina masseyi* Cockayne, 1952 (ab.); *Anaplectoides prasina demuthi* Richardson, 1958 (ab.); *Anaplectoides prasina protensa* Lempke, 1962 (f.); *Anaplectoides prasina cuneata* Lempke, 1962 (f.); *Anaplectoides prasina juncta* Lempke, 1962 (f.); *Anaplectoides prasina clausa* Lempke, 1962 (f.); *Anaplectoides prasina signata* Lempke, 1962 (f.); *Anaplectoides prasina delineata* Lempke, 1962 (f.).

Distribution. China (Heilongjiang, Jilin, Nei Mongol, Xinjiang, Henan); Russia (European part, Ural, Siberia), Mongolia, Korea, Japan, Kazakhstan; Europe, North America (boreal zone).

寄主：欧洲鳞毛蕨（*Dryopteris filix-mas*）、欧洲蕨（*Pteridium aquilinum*）、蹄盖蕨（*Athyrium filix-femina*）、叶升麻（*Actaea spicata*）、白屈菜（*Chelidonium majus*）、萹蓄 (*Polygonum aviculare*)、桦叶鹅耳枥（*Carpinus betulus*）、胡榛（*Corylus avellana*）、帚石楠（*Calluna vulgaris*）、黑果越橘（*Vaccinium myrtillus*）、珍珠花（*Lyonia ovalifolia*）、笃斯越橘（*Vaccinium uliginosum*）、越橘（*Vaccinium vitis-idaea*）、（*Primula cuneifolia*）、牛舌樱草（*Primula elatior*）、黄花九轮草（*Primula veris*）、北美杨树（*Populus balsamifera*）、欧洲山杨（*Populus tremula*）、耳柳（*Salix aurita*）、黄花柳（*Salix caprea*）、东陵山柳（*Salix phylicifolia*）、绣线菊（*Spiraea salicifolia*）、甘蓝（*Brassica oleracea*）、萝卜（*Raphanus sativus*）、旋果蚊子草（*Filipendula ulmaria*）、（*Malus domestica*）、黑刺李（*Prunus spinosa*）、秦氏悬钩子（*Rubus caesius*）、树莓（*Rubus fruticosus*）、覆盆子（*Rubus idaeus*）、北美树莓（*Rubus odoratus*）、库页悬钩子（*Rubus sachalinensis*）、刺毛白叶莓（*Rubus triphyllus*）、腺地榆（*Sanguisorba officinalis*）、水金凤（*Impatiens noli-tangere*）、野胡萝卜（*Daucus carota*）、繁果忍冬（*Lonicera tatarica*）、毛地黄（*Digitalis purpurea*）、蓝蓟（*Echium vulgare*）、大车前（*Plantago major*）、短柄野芝麻（*Lamium album*）、单毛毛连菜（*Picris hieracioides*）、毛果一枝黄花（*Solidago virgaurea*）、（*Juncus acutiflorus*）、酸沼草（*Molinia caerulea*）、虉草（*Phalaris arundinacea*）、毛蕊花属（*Ranunculus*）、酸模属（*Rumex*）、桦木属（*Betula*）、榉属（*Zelkova*）、羽衣草属（*Alchemilla*）、草莓属（*Fragaria*）、委陵菜属（*Potentilla*）、柳叶菜属（*Epilobium*）、金合欢属（*Acacia*）、羽扇豆属（*Lupinus*）、车轴草属（*Trifolium*）、槭树属（*Acer*）、梣属（*Fraxinus*）、牛蒡属（*Arctium*）、蒲公英属（*Taraxacum*）植物（Matov & Kononenko, 2012）。

Anaplectoides virens **(Butler, 1878) 黄绿组夜蛾**

Eurois virens Butler, 1878, *Annals and Magazine of Natural History*, (5)1: 194. Type locality: Japan, Hakodate.

Distribution. China (Heilongjiang, Jilin, Gansu, Hubei, Yunnan); Russia (Siberia), Korea, Japan.

Genus *Xestia* Hübner, 1818 鲁夜蛾属

Xestia Hübner, 1818, *Zutrage Zuträge zur Sammlung Exotischer Schmetterlinge*, 1: 16. Type species: *Noctua ochreago* Hübner, 1809 [Europe].

Synonymy: *Amathes* Hübner, [1821] 1816; *Megasema* Hübner, [1821] 1816 (subgen.); *Lythaea* Stephens, 1829; *Segetia* Stephens, 1829; *Hiptelia* Guenée, 1852; *Pachnobia* Guenée, 1852 (subgen.); *Anomogyna* Staudinger, 1871; *Pteroscia* Morrison, 1875; *Agrotiphila* Grote, 1876; *Schöyenia* Aurivillius, 1883 (incorr. spell. Code, Art. 32c); *Schoyenia* Aurivillius, 1883; *Platagrotis* Smith, 1890; *Hyptioxesta* Rebel, 1901; *Lena* Herz, 1903, preocc. Casey, 1886 ([Coleoptera]); *Hypoxestia* Hampson, 1904; *Barrovia* Barnes & McDunnough, 1916; *Archanarta* Barnes & Benjamin, 1929; *Agrotimorpha* Barnes & McDunnough, 1929; *Epipsiliamorpha* Barnes & Benjamin, 1929; *Knappia* Nye, 1975, repl. name; *Lankilaia* Beck, 1996, nom. nud.; *Xenopachnobia* Beck 1996, nom. nud.; *Lorezia* Beck, 1996, nom. nud.; *Ashworthia* Beck, 1996, nom. nud.; *Monticollia* Beck, 1996, nom. nud.; *Cenigra* Beck, 1996 nom. nud.; *Megharomba* Beck, 1996, nom. nud.; *Castanasta* Beck, 1996, nom. nud.; *Caloxestia* Beck, 1996, nom. nud.; *Palkermes* Beck, 1996, nom. nud.; *Synanomogyna* Beck, 1996, nom. nud.; *Calanomogyna* Beck, 1996, nom. nud.; *Peranomogyna* Beck, 1996, nom. nud.; *Rhyacia* auct.; *Graphiphora* auct..

The genus contains more than 200 species, distributed in all biogeographic regions. Among them, 16 species of 4 subgenera and 3 species of incertae sedis are in the Three Provinces of Northeast China.

Subgenus *Xestia* Hübner, 1818 鲁夜蛾亚属

Xestia Hübner, 1818, *Zutrage Zuträge zur Sammlung Exotischer Schmetterlinge*, 1: 16. Type species: *Noctua ochreago* Hübner, 1809 [Europe].

Xestia (_Xestia_) _baja_ ([Denis & Schiffermüller], 1775) 劳鲁夜蛾

Noctua baja [Denis & Schiffermüller], 1775, *Ankündung eines Systematisches Werkes von den Schmetterlinge der Wiener Gegend. Wien*: 77. Type locality: Austria, Vienna.

Synonymy: *Agrotis baja bajula* Staudiger, 1881 (var.); *Taeniocampa tabida* Butler, 1878; *Agrotis baja punctata* Arivillius, 1888 (var.); *Hypoxestia ohtanensis* Matsumura, 1925; *Hypoxestia sachalinensis* Matsumura, 1925; *Hypoxestia sachalinensis rikovskensis* Matsumura, 1925 (ssp.); *Agrotis baja cinigera* Filipjev, 1927 (f.); *Diarsia baja chosenbaja* Bryk, 1948 (f.); *Agrotis baja brevipennis* (Lempke, 1962) (f.); *Agrotis baja centrifasciata* (Lempke, 1962) (f.); *Agrotis baja juncta* (Lempke, 1962) (f.); *Agrotis baja nictitans* (Lempke, 1962) (f.); *Xestia tabida* auct..

Distribution. China (Jilin, Nei Mongol, Xinjiang, Shaanxi); Russia (European part, Ural, Siberia), Mongolia, Korea, Japan; Central Asia, Europe, North Africa.

寄主：葡萄叶铁线莲（*Clematis vitalba*）、繁缕（*Stellaria media*）、甜菜（*Beta vulgaris*）、藜（*Chenopodium album*）、拳参（*Polygonum bistorta*）、食用大黄（*Rheum rhaponticum*）、波叶大黄

（*Rheum undulatum*）、垂枝桦（*Betula pendula*）、桦叶鹅耳枥（*Carpinus betulus*）、胡榛（*Corylus avellana*）、香杨梅（*Myrica gale*）、帚石楠（*Calluna vulgaris*）、黑果越橘（*Vaccinium myrtillus*）、笃斯越橘（*Vaccinium uliginosum*）、越橘（*Vaccinium vitis-idaea*）、牛舌樱草（*Primula elatior*）、欧洲报春（*Primula vulgaris*）、黄华柳（*Salix atrocinerea*）、黄花柳（*Salix caprea*）、东陵山柳（*Salix phylicifolia*）、黄瓜（*Cucumis sativus*）、甘蓝（*Brassica oleracea*）、山楂（*Crataegus pinnatifida*）、稠李（*Padus racemosa*）、黑刺李（*Prunus spinosa*）、密刺蔷薇（*Rosa pimpinellifolia*）、秦氏悬钩子（*Rubus caesius*）、树莓（*Rubus fruticosus*）、覆盆子（*Rubus idaeus*）、北美树莓（*Rubus odoratus*）、欧洲花楸（*Sorbus aucuparia*）、绣线菊（*Spiraea salicifolia*）、石榴（*Punica granatum*）、柳兰（*Chamaenerion angustifolium*）、大豆（*Glycine max*）、白车轴草（*Trifolium repens*）、野胡萝卜（*Daucus carota*）、蔓生盘叶忍冬（*Lonicera caprifolium*）、繁果忍冬（*Lonicera tatarica*）、绵毛荚蒾（*Viburnum lantana*）、（*Galium album*）、颠茄草（*Atropa belladonna*）、番茄（*Lycopersicon esculentum*）、旋花（*Calystegia sepium*）、金鱼草（*Antirrhinum majus*）、大车前（*Plantago major*）、草地鼠尾草（*Salvia pratensis*）、向日葵（*Helianthus annuus*）、（*Juncus acutiflorus*）、酸沼草（*Molinia caerulea*）、玉米（*Zea mays*）、酸模属（*Rumex*）、核桃属（*Juglans*）、榉属（*Zelkova*）、荨麻属（*Urtica*）、茶藨子属（*Ribes*）、路边青属（*Geum*）、委陵菜属（*Potentilla*）、车轴草属（*Trifolium*）、梣属（*Fraxinus*）、野芝麻属（*Lamium*）、牛蒡属（*Arctium*）、山柳菊属（*Hieracium*）、蒲公英属（*Taraxacum*）、羊茅属（*Festuca*）植物（Matov & Kononenko, 2012）。

***Xestia* (*Xestia*) *descripta* (Bremer, 1861) 杂绿鲁夜蛾**

Noctua descripta Bremer, 1861, *Bulletin de l'Academie Impériale des Sciences de St.-Petersbroug*, 3: 487. Type locality: Russia, Amur reg., [Bureinsky Range] "Bureja-Gebirge".

Synonymy: *Agrotis pachnobides* Staudinger, 1888; *Rhyacia pelita* Corti et Draudt, 1933.

Distribution. China (Heilongjiang, Jilin); Russia (Siberia), Mongolia, Korea, Japan.

寄主：柳、山楂、桦、报春等及多种草本植物（陈一心，1999）；茶（*Camellia sinensis*）、甘蓝（*Brassica oleracea*）、萝卜（*Raphanus sativus*）、兴安薄荷（*Mentha dahurica*）、芍药属（*Paeonia*）、酸模属（*Rumex*）、报春花属（*Primula*）、路边青属（*Geum*）、车轴草属（*Trifolium*）、车前属（*Plantago*）、风铃草属（*Campanula*）、莴苣属（*Lactuca*）、蒲公英属（*Taraxacum*）、山茶属（*Camellia*）植物（Matov & Kononenko, 2012）。

Subgenus ***Megasema*** Hübner, [1821] 1816 三角鲁夜蛾亚属

Megasema Hübner, [1821] 1816, *Verzeichniss bekannter Schmettlinge* [sic!]: 232. Type species: *Phalaena triangulum* Hufnagel, 1766 [Germany].

Synonymy: *Amathes* Hübner, [1821] 1816; *Lythaea* Stephens, 1829; *Segetia* Stephens, 1829; *Hiptelia* Guen é e, 1852; *Hypoxestia* Hampson, 1904; *Agrotimorpha* Barnes & McDunnough, 1929; *Ashworthia* Beck, 1996, nomen nudum.; *Monticollia* Beck, 1996, nomen nudum.; *Cenigra* Beck, 1996 (subgen.), nomen nudum.;

Megharomba Beck, 1996 (subgen.), nomen nudum.; *Castanasta* Beck, 1996, nomen nudum.; *Caloxestia* Beck, 1996, nomen nudum.; *Palkermes* Beck, 1996 (subgen.), nomen nudum..

注：新给出中文亚属名。

***Xestia* (*Megasema*) *c-nigrum* (Linnaeus, 1758) 八字地老虎**

Phalaena c-nigrum Linnaeus, 1758, *Systema Naturae*, (Edn. 10) 1: 516. Type locality: Europe.

Synonymy: *Bombyx gothica nunatrum* Esper, 1786 (var.); *Bombyx gothica singularis* Esper, 1786 (var.); *Agrotis degenerata* Staudinger, 1889; *Noctua c-nigrum rosea* Tutt, 1892 (var.); *Agrotis c-nigrum umbrata* Schultz, 1908 (ab.); *Agrotis c-nigrum fritschi* Culot, 1910 (ab.); *Agrotis c-nigrum depravata* A. Bang-Haas, 1912 (var.); *Agrotis c-nigrum nigrescens* Buresch, 1914 (f.); *Agrotis c-nigrum maerens* Dannehl, 1925 (f.); *Amathes c-nigrum immaculata* Lempke, 1939 (f.); *Amathes c-nigrum juncta* Lempke, 1939 (f.); *Amathes c-nigrum confluens* Lempke, 1939 (f.); *Noctua c-nigrum grisea* Turner, 1939 (ab.); *Diarsia c-nigrum kurilana* Bryk, 1942 (ssp.); *Amathes c-nigrum albinotica* Cockayne, 1952 (ab.); *Amathes c-nigrum pallida* Lempke, 1962 (f.); *Amathes c-nigrum fuscolimbata* Lempke, 1962 (f.); *Amathes c-nigrum semiconfluens* Lempke, 1962 (f.); *Amathes c-nigrum cruda* Lempke, 1962 (f.); *Amathes c-nigrum signata* Lempke, 1962 (f.); *Amathes c-nigrum brevipennis* Lempke, 1962 (f.); *Amathes c-nigrum ignorata* Eitschberger, 1972 (ssp.); *Xestia c-nigrum dolosa* Franclemont, 1980 (ssp.).

Distribution. China; Russia (European part, Ural, Siberia), Mongolia, Korea, Japan, India, Nepal, Pakistan, Kazakhstan; Europe, North Africa, North America.

寄主：禾谷类、柳、葡萄等（陈一心，1999）；繁缕（*Stellaria media*）、甜菜（*Beta vulgaris*）、藜（*Chenopodium album*）、荞麦（*Fagopyrum esculentum*）、酸模（*Rumex acetosa*）、钝叶酸模（*Rumex obtusifolius*）、三球悬铃木（*Platanus orientalis*）、水晶兰（*Monotropa uniflora*）、红莓苔子（*Oxycoccus palustris*）、黑果越橘（*Vaccinium myrtillus*）、黄花柳（*Salix caprea*）、东陵山柳（*Salix phylicifolia*）、匍匐柳（*Salix repens*）、泻根（*Bryonia dioica*）、黄瓜（*Cucumis sativus*）、芸苔（*Brassica campestris*）、甘蓝（*Brassica oleracea*）、荠（*Capsella bursa-pastoris*）、萝卜（*Raphanus sativus*）、草棉（*Gossypium herbaceum*）、异株荨麻（*Urtica dioica*）、黑茶藨子（*Ribes nigrum*）、红茶藨子（*Ribes rubrum*）、（*Cotoneaster acutiuscula*）、（*Cotoneaster hissarica*）、（*Cotoneaster insignis*）、（*Cotoneaster multiflora*）、准噶尔栒子（*Cotoneaster suavis*）、黑刺李（*Prunus spinosa*）、树莓（*Rubus fruticosus*）、覆盆子（*Rubus idaeus*）、柳兰（*Chamaenerion angustifolium*）、紫苜蓿（*Medicago sativa*）、豌豆（*Pisum sativum*）、苕菜（*Vicia hirsuta*）、亚麻（*Linum usitatissimum*）、欧鼠李（*Frangula alnus*）、葡萄（*Vitis vinifera*）、（*Aegopodium lappa*）、羊角芹（*Aegopodium podagraria*）、峨参（*Anthriscus sylvestris*）、旱芹（*Apium graveolens*）、野胡萝卜（*Daucus carota*）、牛防风（*Heracleum sphondylium*）、欧防风（*Pastinaca sativa*）、洋香菜（*Petroselinum sativum*）、缬草（*Valeriana officinalis*）、猪殃殃（*Galium aparine*）、水曲柳（*Fraxinus manshurica*）、颠茄草（*Atropa belladonna*）、番茄（*Lycopersicon esculentum*）、烟草（*Nicotiana tabacum*）、茄（*Solanum melongena*）、马铃薯（*Solanum tuberosum*）、聚合草（*Symphytum officinale*）、毛地黄（*Digitalis purpurea*）、毛蕊花（*Verbascum thapsus*）、长叶车前（*Plantago lanceolata*）、大车前（*Plantago *major*）、匍筋骨草（*Ajuga reptans*）、短柄野芝麻（*Lamium album*）、皱叶薄荷（*Mentha crispa*）、辣薄荷（*Mentha piperita*）、田旋花（*Convolvulus arvensis*）、红辣椒千叶蓍（*Achillea millefolium*）、牛蒡（*Arctium lappa*）、棕矢车菊（*Centaurea jacea*）、栽培菊苣（*Cichorium endivia*）、刺儿菜（*Cirsium arvense*）、向日葵（*Helianthus annuus*）、欧洲千里光（*Senecio vulgaris*）、毛果一枝黄花（*Solidago virgaurea*）、苣荬菜（*Sonchus arvensis*）、洋葱（*Allium cepa*）、葱（*Allium fistulosum*）、燕麦草（*Arrhenatherum elatius*）、贫育雀麦（*Bromus sterilis*）、曲芒发草（*Deschampsia flexuosa*）、偃麦草（*Elytrigia repens*）、大麦（*Hordeum vulgare*）、酸沼草（*Molinia caerulea*）、梯牧草（*Phleum pratense*）、小麦（*Triticum aestivum*）、偃麦草（*Triticum repens*）、玉米（*Zea mays*）、铁线莲属（*Clematis*）、小檗属（*Berberis*）、米努草属（*Minuartia*）、滨藜属（*Atriplex*）、大黄属（*Rheum*）、桦木属（*Betula*）、报春花属（*Primula*）、大戟属（*Euphorbia*）、醋栗属（*Grossularia*）、羽衣草属（*Alchemilla*）、柳叶菜属（*Epilobium*）、绒毛花属（*Anthyllis*）、车轴草属（*Trifolium*）、当归属（*Angelica*）、忍冬属（*Lonicera*）、勿忘草属（*Myosotis*）、马樱丹属（*Lantana*）、蝶须属（*Antennaria*）、金盏花属（*Calendula*）、大翅蓟属（*Onopordon*）、蒲公英属（*Taraxacum*）植物（Matov & Kononenko, 2012）。

***Xestia* (*Megasema*) *ditrapezium* ([Denis & Schiffermüller], 1775) 兀鲁夜蛾**

Noctua ditrapezium [Denis & Schifferm ü ller, 1775), *Ankündung eines Systematisches Werkes von den Schmetterlinge der Wiener Gegend. Wien*: 312. Type locality: Austria, Vienna.

Synonymy: *Noctua tristigma* Treitschke, 1825; *Agrotis ditrapezium orientalis* Strand, 1916 (ab.); *Amathes ditrapezium orientalis* Boursin, 1963 (ssp.).

Distribution. China (Heilongjiang, Jilin, Nei Mongol, Xinjiang, Hebei, Sichuan); Russia (European part, Ural, Siberia), Mongolia, Kazakhstan, Ukraine; Central Asia, Europe.

寄主：繁缕（*Stellaria media*）、酸模（*Rumex acetosa*）、胡榛（*Corylus avellana*）、黑果越橘（*Vaccinium myrtillus*）、黄花九轮草（*Primula veris*）、欧洲报春（*Primula vulgaris*）、黄花柳（*Salix caprea*）、蒿柳（*Salix viminalis*）、甘蓝（*Brassica oleracea*）、异株荨麻（*Urtica dioica*）、黑刺李（*Prunus spinosa*）、黑草莓（*Rubus armeniacus*）、秦氏悬钩子（*Rubus caesius*）、树莓（*Rubus fruticosus*）、覆盆子（*Rubus idaeus*）、悬钩子（*Rubus palmatus*）、石生悬钩子（*Rubus saxatilis*）、柳兰（*Chamaenerion angustifolium*）、欧洲红瑞木(（*Swida sanguinea*）、烟草（*Nicotiana tabacum*）、匍筋骨草（*Ajuga reptans*）、短柄野芝麻（*Lamium album*）、林石蚕（*Teucrium scorodonia*）、翠菊（*Bellis perennis*）、莴苣（*Lactuca sativa*）、药用蒲公英（*Taraxacum officinale*）、玉米（*Zea mays*）、栎属（*Quercus*）、桤木属（*Alnus*）、桦木属（*Betula*）、鹅耳枥属（*Carpinus*）、茶藨子属（*Ribes*）、山楂属（*Crataegus*）、苹果属（*Malus*）、芒柄花属（*Ononis*）、地锦属（*Parthenocissus*）、拉拉藤属（*Galium*）、车前草属（*Plantago*）、牛蒡属（*Arctium*）、风毛菊属（*Leontodon*）、苦苣菜属（*Sonchus*）植物（Matov & Kononenko, 2012）。

***Xestia* (*Megasema*) *kollari* (Lederer, 1853) 大三角鲁夜蛾**

Graphiphora kollari Lederer, 1853, *Verhandlungen des Zoologisch-Botanischen vereins in Wien*, 3: 366, pl.

4: 1. Type locality: [East Kazakhstan, West Altai] Ust-Kamengorsk, Ust-Buchtarminsk.

Synonymy: *Ochropleura plumbata* Butler, 1881; *Agrotis collari* Hampson, 1903, emend..

Distribution. China (Heilongjiang, Jilin, Liaoning, Nei Mongol, Xinjiang, Hebei, Hunan, Jiangxi, Yunnan); Russia (Ural, Altai, Transbaikalia, Siberia), Mongolia, Korea, Japan.

Xestia (_Megasema_) _stupenda_ (Butler, 1878) 前黄鲁夜蛾

Ochropleura stupenda Butler, 1878, *Annals and Magazine of Natural History*, 5 (1): 166. Type locality: Japan, Yokohama.

Synonymy: *Noctua stupens* Oberthür, 1880.

Distribution. China (Heilongjiang, Jilin, Shaanxi, Hebei, Jiangsu, Zhejiang, Jiangxi, Hunan, Guangdong, Xizang); Russia (Siberia), Mongolia, Korea, Japan, Kazakhstan.

寄主：兰屿木耳菜（*Reynoutria elliptica*）、芸苔（*Brassica campestris*）、萝卜（*Raphanus sativus*）、蜂斗菜（*Petasites japonicus*）（Matov & Kononenko, 2012）。

Xestia (_Megasema_) _fuscostigma_ (Bremer, 1861) 褐纹鲁夜蛾

Agrotis fuscostigma Bremer, 1861, *Bulletin de la Classe Physico-Mathématique de l'Académie Impériale des Sciences de St.-Pétersbourg*, 3: 387. Type locality: Russia, Primorye terr., [Khanka lake] "Kengka-See".

Synonymy: *Graphiphora lubentia* Butler, 1879; *Noctua hysgina* Oberthür, 1880.

Distribution. China (Heilongjiang, Liaoning, Shaanxi, Henan, Hunan); Russia (Siberia), Korea, Japan.

寄主：（*Polygonum album*）、钝叶酸模（*Rumex obtusifolius*）、黄瓜（*Cucumis sativus*）、甘蓝（*Brassica oleracea*）、柔毛路边青（*Geum japonicum*）、覆盆子（*Rubus idaeus*）、月见草（*Oenothera biennis*）、白车轴草（*Trifolium repens*）、毛蕊花（*Verbascum thapsus*）、野芝麻（*Lamium barbatum*）、（*Breea setosa*）、向日葵（*Helianthus annuus*）、蜂斗菜（*Petasites japonicus*）、补血草属（*Limonium*）、白藜芦（*Veratrum album*）、蔷薇属（*Rosa*）、薯蓣属（*Dioscorea*）植物（Matov & Kononenko, 2012）。

Xestia (_Megasema_) _dilatata_ (Butler, 1879) 润鲁夜蛾

Mesogona dilatata Butler, 1879, *Annals and Magazine of Natural History*, 5 (4): 364. Type locality: Japan, Yokohama.

Synonymy: *Megasema cinnamomea* Moore, 1881.

Distribution. China (Heilongjiang, Jilin, Hebei, Jiangsu, Hunan, Taiwan); Russia (Siberia), Korea, Japan, Indonesia.

寄主：长叶酸模（*Rumex longifolius*）（Matov & Kononenko, 2012）。

Xestia (_Megasema_) _vidua_ (Staudinger, 1892) 单鲁夜蛾

Agrotis vidua Staudinger, 1892, *in* N.M. Romanoff (ed.), *Mémoires sur les Lépidoptères*, 6: 409, pl. 7: 3.

Type locality: Russia, [Primorye terr., Partizansk], "Sutschan-Gebiet", Askold Isl..

Synonymy: *Graphiphora obscura* Zolotarenko, 1970.

Distribution. China (Heilongjiang, Jilin, Sichuan); Russia (Siberia), Korea, Japan.

Subgenus *inlertae sedis* 隶属的亚属未明确

Xestia semiherbida **(Walker, 1857) 绿鲁夜蛾**

Triphaena semiherbida Walker, 1857, *List of the Specimens of Lepidopterous Insects in the Collection of the British Museum*, 11: 473. Type locality: North India.

Synonymy: *Epilecta decorata* Butler, 1879.

Distribution. China (Heilongjiang, Jilin, Anhui, Fujian, Sichuan, Xizang); Russia (Siberia), Korea, Japan, Pakistan, India, Nepal.

Xestia efflorescens **(Butler, 1879) 彩鲁夜蛾**

Triphaenopsis efflorescens Butler, 1879, *Annals and Magazine of Natural History*, (5)4: 361. Type locality: Japan, Yokohama.

Synonymy: *Triphaena jankowskii* Oberthür, 1884.

Distribution. China (Heilongjiang, Jilin, Hunan); Russia (Siberia), Korea, Japan.

Xestia undosa **(Leech, 1889) 波鲁夜蛾**

Agrotis undosa Leech, 1889, *Proceedings of the Zoological Society of London* 1889: 501, pl. 50: 3. Type locality: Japan, Nikko.

Distribution. China (Heilongjiang, Jilin, Liaoning); Russia (Siberia), Korea, Japan.

注：《中国动物志》（夜蛾科）中将本种误归入"*Noctua* Linnaeus, 1758 模夜蛾属"，并称为"波模夜蛾"，在此予以变更。

Subgenus ***Pachnobia*** Guenée, 1852 厚鲁夜蛾亚属

Pachnobia Guenée, 1852, *in* Boisduval & Guenée, *Histoire Naturelle des Insectes. Species Général des Lépidoptères*, 5: 341. Type species: *Pachnobia carnea* Thunberg sensu Guenée, 1852, a misidentification of *Noctua tecta* Hübner, 1808 [North Europe].

Synonymy: *Anomogyna* Staudinger, 1871; *Pteroscia* Morrison, 1875; *Schöyenia* Aurivillius, 1883, incorrect original spelling ICZN Art. 32c; *Schoyenia* Aurivillius, 1883; *Platagrotis* Smith, 1890; *Hyptioxesta* Rebel, 1901; *Lena* Herz, 1903; *Barrovia* Barnes & McDunnough, 1916; *Archanarta* Barnes & Benjamin, 1929; *Epipsiliamorpha* Barnes & Benjamin, 1929; *Knappia* Nye, 1975, repl. name for *Lena* Herz, 1903; *Synanomogyna*

Beck, 1996, nomen nudum.; *Peranomogyna* Beck, 1996, nomen nudum.; *Calanomogyna* Beck, 1996, nomen nudum; *Lankilaia* Beck, 1996; *Xenopachnobia* Beck 1996, nomen nudum.; *Lorezia* Beck 1996, nomen nudum.

注：首次给出亚属名。

Xestia (*Pachnobia*) *kolymae* (Herz, 1903) 晕鲁夜蛾

Agrotis kolymae Herz, 1903, *Annuaire du Musée Zoologique de l'Académie Imperiale des Sciences de St.-Petersbourg*, 8: 79. Type locality: Russia, "Janthal"[Yakutia, Verkhoyansk].

Synonymy: *Anomogyna veruta* Corti & Draudt, 1933; *Anomogyna helenae* Corti, 1933.

Distribution. China (Heilongjiang); Russia (Sayan Mt., Baikal area, Transbaikalia, Magadan), Mongolia; North America.

注：首次给出中文种名。

Xestia (*Pachnobia*) *speciosa* (Hübner, [1813]) 暗花鲁夜蛾

Noctua speciosa Hübner, [1813], *Sammlung Europaischer Schmetterlinge. Eulen*, 2: 491, pl. 104. Type locality: Europe.

Synonymy: *Haena arctica* Zetterstedt, [1839]; *Agrotis speciosa obscura* Frey, 1880 (var.); *Agrotis speciosa millierei* Culot, 1909 (ab.); *Platagrotis speciosa aegrota* Alphéraky, 1897 (var.); *Platagrotis speciosa janae* Herz, 1903 (ssp.); *Aplectoides speciosa rybatchiensis* Kotzsch, 1933 (ssp.); *Anomogyna apropita* Benjamin, 1933; *Anomogyna aklavicensis* Benjamin, 1933; *Agrotis speciosa baltica* Valle, 1940 (ssp.); *Aplectoides speciosa modesta* Warnecke, 1962 (ssp.); *Xestia speciosa ussurica* Kononenko, 1984 (ssp.); *Aplecta schoenherri*: Guenée, 1852, nec Herrich-Schäffer, 1845.

Distribution. China (Heilongjiang, Jilin); Russia (North of the European part, Ural, Siberia), Mongolia(North), North Korea; North and Central Europe (in mountains).

寄主：繁缕（*Stellaria media*）、萹蓄（*Polygonum aviculare*）、拳参（*Polygonum bistorta*）、矮桦（*Betula nana*）、熊果（*Arctostaphylos uva-ursi*）、（*Vaccinium myrtilloides*）、黑果越橘（*Vaccinium myrtillus*）、越橘（*Vaccinium vitis-idaea*）、岩高兰（*Empetrum nigrum*）、秦氏悬钩子（*Rubus caesius*）、覆盆子（*Rubus idaeus*）、北美树莓（*Rubus odoratus*）、白车轴草（*Trifolium repens*）、黑果忍冬（*Lonicera nigra*）、繁果忍冬（*Lonicera tatarica*）、大车前（*Plantago major*）、毛果一枝黄花（*Solidago virgaurea*）、鸭茅（*Dactylis glomerata*）、林地早熟禾（*Poa nemoralis*）、山柳菊属（*Hieracium*）植物（Matov & Kononenko, 2012）。

注：首次给出中文种名。

Xestia (*Pachnobia*) *rhaetica* (Staudinger, 1871) 拉鲁夜蛾

Agrotis rhaetica Staudinger, 1871, *Catalog der Lepidopteren des Europaeischen Faunengebiets. Dresden ed.* 2: 82. Type locality: Schwitzerland, Rhaetia.

Synonymy: *Agrotis obscura* Hellbiger, 1932; *Anomogyna rhaetica fennica* (Brandt, 1936) (ssp.); *Anomogyna rhaetica norica* (Lüberbauer, 1952) (spp.).

Distribution. China (Jilin); Russia (Northeast of European part, Ural, Siberia), North Korea, Mongolia (North); North and Central Europe (in mountains).

寄主：幼虫为杂食性，多危害桦木科、杜鹃科、伞形科、蔷薇科、豆科、摩羯科、石竹科、何首科、菊科植物（陈一心，1999）。

Xestia (_Pachnobia_) _penthima_ (Erschoff, 1870) **阴鲁夜蛾**

Erastria penthima Erschoff, 1870, *Horae Societatis Entomologicae Rossicae* 4: 196. Type locality: [Russia, Irkutsk reg.] Balangansk.

Distribution. China (Heilongjiang, Nei Mongol); Russia (Krasnoyarsk terr., Irkutsk reg., Sayan and Baikal area, Yakutia, Transbaikalia, Amur reg., Khabarovsk terr., Magadan reg., Chukotka, Siberia), Mongolia(North).

Xestia (_Pachnobia_) _albonigra_ (Kononenko, 1981) 花鲁夜蛾

Amathes albonigra Kononenko, 1981, *Proceedings of Zoological Institute of the Academy of Sciences of the USSR*, 92: 92, figs. 1, 4. Type locality: Russia, Primorye terr., South Sikhote-Alin, Partizansky Range, Lysaya Benevskaya Mt..

Synonymy: *Amathes albonigra distincta* Kononenko, 1981 (ssp.).

Distribution. China (Jilin); Russia (Ural, Krasnoyarsk terr., Sayan and Baikal area, Siberia).

注：首次给出中文种名。

Xestia (_Pachnobia_) _sincera_ (Herrich–Schäffer, 1851) 真鲁夜蛾

Agrotis sincera Herrich-Schäffer, 1851, *Systematische Bearbeitung der Schmetterlinge von Europa*, 2: 412, pl. 100: 513, 514 [in error for figs. 512, 513]. Type locality: Germany, Dresden.

Synonymy: *Agrotis zelleri* Herrich-Schäffer, 1851.

Distribution. China (Jilin); Russia (Northeast of European part, Ural, Siberia), Mongolia, North Korea, Japan; North and Central Europe, North America.

寄主：萹蓄 (*Polygonum aviculare*)、欧洲落叶松（*Larix decidua*）、欧洲云杉（*Picea abies*）、黑果越橘（*Vaccinium myrtillus*）、笃斯越橘（*Vaccinium uliginosum*）、冷杉属（*Abies*）植物（Matov & Kononenko, 2012）。

Xestia (_Pachnobia_) _albuncula_ (Eversmann, 1851) 漂鲁夜蛾

Cymatophora albuncula Eversmann, 1851, *Bulletin de la Société Impériale des Naturalistes de Moscou*, 24: 627. Type locality: Russia, Siberia, Irkutsk.

Synonymy: *Agrotis vega* Herz, 1903; *Anomogyna tamanukii* Matsumura, 1925; *Anomogyna griseola* Matsumura, 1925; *Anomogyna laetabilis kononis* Matsumura, 1925 (ssp.); *Anomogyna acuminata* Matsumura, 1925; *Euxoa daisetsuzana* Matsumura, 1927; *Anomagyna* [sic!] *mallochi* Benjamin, 1933; *Anomogyna mallochi arufoides* Benjamin, 1933 (var.); *Anomogyna mallochi stejnigeri* Benjamin, 1933 (ssp.); *Anomogyna imperita* auct., nec Hübner.

Distribution. China (Heilongjiang, Nei Mongol); Russia (Ural, Siberia), Mongolia(North), North Korea, Japan, Kazakhstan, USA(Alaska).

寄主：黑果越橘（*Vaccinium myrtillus*）（Matov & Kononenko, 2012）。

Xestia (_Pachnobia_) _subgrisea_ (Staudinger, 1897) 灰鲁夜蛾

Agrotis subgrisea (Staudinger, 1897), *Deutsche Entomologische Zeitschrift. Gesellschaft Iris zu Dresden*, 10: 330, pl. 9: 25. Type locality: Russia, [East Siberia, Yablonovyi Range] "Apfel-Gebirge".

Synonymy: *Platagrotis adducta* Herz, 1898; *Anomogyna excavata* Matsumura, 1925.

Distribution. China (Heilongjiang); Russia (Siberia), Mongolia.

Genus *Eugraphe* Hübner, [1821] 1816 图夜蛾属

Eugraphe Hübner, [1821] 1816, *Verzeichniss bekannter Schmettlinge* [sic!]: 224. Type species: *Noctua sigma* [Denis & Schiffermüller], 1775 [Europe, Austria].

Synonymy: *Hypernaenia* Hampson, 1894.

The genus contains 8 species, distributed mainly in the Central-Asian subregion of the Palaearctic region. Among them, 2 species are in the Three Provinces of Northeast China.

Eugraphe sigma ([Denis & Schiffermüller], 1775) 烙图夜蛾

Noctua sigma [Denis & Schiffermüller], 1775, *Ankündung eines Systematisches Werkes von den Schmetterlinge der Wiener Gegend. Wien*: 78. Type locality: Austria, Vienna.

Synonymy: *Noctua characterea* [Denis & Schiffermüller], 1775; *Phalaena* (*Noctua*) *nubila* Esper, 1786; *Phalaena* (*Noctua*) *characterea* Esper, 1786; *Phalaena* (*Noctua*) *ditrapezium* Esper, 1786; *Noctua signum* Fabricius, 1787; *Phalaena* (*Noctua*) *umbra* Vieweg, 1790; *Diarsia sigma melancholica* Bryk, 1949 (ssp.); *Eugraphe sigma anthracina* Boursin, 1954 (ssp.).

Distribution. China (Heilongjiang, Jilin, Xinjiang); Russia (European part, Ural, Siberia), Mongolia, Korea, Japan, Kazakhstan, Turky; Europe.

寄主：葡萄叶铁线莲（*Clematis vitalba*）、白屈菜（*Chelidonium majus*）、榆钱菠菜（*Atriplex hortensis*）、藜（*Chenopodium album*）、黑果越橘（*Vaccinium myrtillus*）、黄花柳（*Salix caprea*）、啤酒花（*Humulus lupulus*）、异株荨麻（*Urtica dioica*）、黑刺李（*Prunus spinosa*）、秦氏悬钩子（*Rubus caesius*）、树莓（*Rubus fruticosus*）、覆盆子（*Rubus idaeus*）、南欧扫帚豆（*Cytisus austriacus*）、毒豆

（*Laburnum anagyroides*）、胡枝子（*Lespedeza bicolor*）、欧洲红瑞木（*Cornus sanguinea*）、欧银花（*Lonicera xylosteum*）、绵毛荚蒾（*Viburnum lantana*）、普通女贞（*Ligustrum vulgare*）、短柄野芝麻（*Lamium album*）、林地水苏（*Stachys silvatica*）、毛蕊花属（*Ranunculus*）、栎属（*Quercus*）、桤木属（*Alnus*）、榛属（*Corylus*）、报春花属（*Primula*）、茶藨子属（*Ribes*）、羽衣草属（*Alchemilla*）、槭树属（*Acer*）、车前草属（*Plantago*）、羽衣草属（*Alchemilla*）、牛蒡属（*Arctium*）、蒲公英属（*Taraxacum*）、铃兰属（*Convallaria*）、黄精属（*Polygonatum*）植物（Matov & Kononenko, 2012）。

Eugraphe versuta **(Püngeler, 1908) 显图夜蛾**

Agrotis versuta Püngeler, 1908, *Deutsche Entomologische Zeitschrift Iris*, 21: 287, pl. 4: 4. Type locality: China, Qinghai, Kuku-noor.

Synonymy: *Graphiphora invenusta* Kozhantshikov, 1926; *Agrotis kozhantshikovi* Corti et Draudt, 1933; *Agrotis agalmona* Bryk, 1948.

Distribution. China (Heilongjiang, Jilin, Qinghai, Gansu, Shaanxi, Shandong); Russia (Altai, Transbaikalia, Siberia), Mongolia, Korea.

注：《中国动物志》（夜蛾科）中将本种归入"*Xestia* Hübner, 1818 鲁夜蛾属"，在此予以变更。

Genus ***Coenophila*** Stephens, 1850 化图夜蛾属

Coenophila Stephens, 1850, *List of the Specimens of British Animals in the Collection of the British Museum*, 5: 74. Type species: *Graphiphora subrosea* Stephens, 1829 [Great Britain].

The genus contains 2 species, distributed in the temperate zone of the Palaearctic and Nearctic regions. Among them, 1 species is in the Three Provinces of Northeast China.

注：新给出中文属名。

Coenophila subrosea **(Stephens, 1829) 化图夜蛾**

Graphiphora subrosea Stephens, 1829, *The Nomenclature of British Insectes*, 2: 128, 200, pl. 19, fig. 1. Type locality: [Great Britain] England, Wittlesey.

Synonymy: *Graphiphora subrosea rhomboidea* Stephens, 1829 (ssp.); *Agrotis subrosea subcaerulea* Staudinger, 1871 (var.); *Agrotis subrosea latefasciata* Hoyningen-Huene, 1908 (f.); *Eugraphe subrosea kieferi* Rebel, 1912 (f.); *Agrotis subrosea decipiens* Warnecke, 1924 (f.); *Aplectoides furushonis* Matsumura, 1925; *Agrotis subrosea rubrifera* Warnecke, 1931 (ssp.).

Distribution. China (Heilongjiang); Russia (European part, Ural, Siberia), Korea, Japan, Kazakhstan; Europe, North America.

寄主：杨梅、柳、桤木属等植物（陈一心，1999）；矮桦（*Betula nana*）、（*Betula pubescens*）、香杨梅（*Myrica gale*）、安多杜鹃（*Andromeda polifolia*）、帚石楠（*Calluna vulgaris*）、地桂（*Chamaedaphne calyculata*）、格陵兰喇叭茶（*Ledum groenlandicum*）、杜香（*Ledum palustre*）、红莓

苔子（*Oxycoccus palustris*）、笃斯越橘（*Vaccinium uliginosum*）、黑果越橘（*Vaccinium myrtillus*）、欧洲山杨（*Populus tremula*）、白柳（*Salix alba*）、黄花柳（*Salix caprea*）、爆竹柳（*Salix fragilis*）、谷柳（*Salix livida*）、匍匐柳（*Salix repens*）、伪蒿柳（*Salix rossica*）、北美树莓（*Rubus odoratus*）、东方羊胡子草（*Eriophorum angustifolium*）、欧石南属（*Erica*）植物（Matov & Kononenko, 2012）。

注：《中国动物志》（夜蛾科）中将本种归入"图夜蛾属 *Eugraphe* Hübner, [1821] 1816"，在此予以变更。

Genus *Protolampra* McDunnough, [1929] 1928 初夜蛾属

Protolampra McDunnough, [1929] 1928, *Bulletin of the Canada Department of Mines*, 55: 66. Type species: *Agrotis ruflpectus* Morrison, 1874 [USA, New York].

Synonymy: *Paradiarsia* auct., nec Mc Dunnough, [1929].

Holarctic genus containing 4 species, distribited in the temperater zone. Among them, 1 species is in the Three Provinces of Northeast China.

注：新给出中文属名。

***Protolampra sobrina* (Duponchel, 1843) 适初夜蛾**

Noctua sobrina Duponchel, 1843, *in* Godard & Duponchel, *Histoire Naturelle des Lépidoptères ou Papillons de France*, suppl. 4: 224, pl. 69: 5. Type locality: South France, East Pyrenees.

Synonymy: *Noctua lapponica* Freyer, 1845; *Noctua sobrina gruneri* Guenée, 1852 (var.); *Cerastis sobrina confina* W. Kozhantshikov, 1925 (ssp.); *Hypoxestia nyiwonis* Matsumura, 1925; *Cerastis sobrina shoutsensis* Bryk, 1949 (ssp.); *Estimata dailingensis* Chen, 1984.

Distribution. China (Heilongjiang, Qinghai, Xinjiang); Russia (European part, Ural, Siberia), Mongolia, North Korea, Kazkhstan.

寄主：欧洲小檗（*Berberis vulgaris*）、繁缕（*Stellaria media*）、萹蓄（*Polygonum aviculare*）、帚石楠（*Calluna vulgaris*）、枞枝欧石楠（*Erica cinerea*）、轮生叶欧石楠（*Erica tetralix*）、笃斯越橘（*Vaccinium uliginosum*）、黑果越橘（*Vaccinium myrtillus*）、黑茶藨子（*Ribes nigrum*）、刺蔷薇（*Rosa acicularis*）、覆盆子（*Rubus idaeus*）、欧洲花楸（*Sorbus aucuparia*）、酸沼草（*Molinia caerulea*）、桦木属（*Betula*）、杨属（*Populus*）、柳属（*Salix*）、山楂属（*Crataegus*）、车前草属（*Plantago*）、蒲公英属（*Taraxacum*）植物（Matov & Kononenko, 2012）。

注：《中国动物志》（夜蛾科）中将本种归入"*Paradiarsia* McDunnough, [1929] 1928 芒夜蛾属"，在此予以变更。

Genus *Sineugraphe* Boursin, 1954 扇夜蛾属

Sineugraphe Boursin, 1954, *Bonner Zoologische Beiträge*, 5: 266. Type species: *Eugraphe disgnosta* Boursin, 1948, [Japan] (=*Sineugraphe bipartita* Graeser, 1888 [1889]).

Synonymy: *Sineugrapha* Poole, 1989, missp..

The genus contains 7 species, mainly distributed in the Chinese subregion of the Palaearctic region. Among them, 3 species are in the Three Provinces of Northeast China.

Sineugraphe exusta (Butler, 1878) 紫棕扇夜蛾

Graphiphora exusta Butler, 1878, *Annals and Magazine of Natural History*, 5 (1): 164. Type locality: Japan, Yokohama, Hakodate.

Synonymy: *Agrotis exusta nigromaculata* Graeser, 1888 [1889] (ssp.); *Eugraphe exusta sinica* Boursin, 1948 (ssp.).

Distribution. China (Heilongjiang, Jilin, Liaoning, Hubei, Guizhou); Russia (Southeast European part, Ural, Siberia), Mongolia, Korea, Japan.

寄主：（*Betula tauschii*）、硕桦（*Betula ulmifolia*）、（*Salix hultenii*）、龙江柳（*Salix sachalinensis*）、树锦鸡儿（*Caragana arborescens*）、金莲花属（*Trollius*）、勿忘草属（*Myosotis*）、藜芦属（*Veratrum*）、铃兰属（*Convallaria*）、黄精属（*Polygonatum*）、杓兰属（*Cypripedium*）植物（Matov & Kononenko, 2012）。

Sineugraphe bipartita (Graeser, [1889] 1888) 扇夜蛾

Agrotis bipartita Graeser, [1889]1888, *Berliner entomologische Zeitschrift*, 32, II: 318. Type locality: Russia, Vladivostok.

Synonymy: *Eugraphe disgnosta* Boursin, 1948; *Diarsia exusta nolimetangere* Bryk, 1948 (ssp.); *Sineugraphe dysgnosta* Boursin, 1954, emend.; *Sineugraphe disgnosta boursini* Ogata, 1956 (f.).

Distribution. China (Heilongjiang, Jilin, Henan); Russia (Siberia), Korea, Japan.

寄主：滨藜属（*Atriplex*）、蓼属（*Polygonum*）、酸模属（*Rumex*）、报春花属（*Primula*）、荨麻属（*Urtica*）、植物（Matov & Kononenko, 2012）。

注：《中国动物志》（夜蛾科）中引用本种同物异名“*Eugraphe disgnosta* Boursin, 1948”，在此予以变更。

Sineugraphe oceanica (Kardakoff, 1928) 华长扇夜蛾

Rhyacia oceanica Kardakoff, 1928, *Entomologische Mitteilungen*, 17: 418, pl. 9: 1. Type locality: Russia, Vladivostok, Okeanskaja station.

Synonymy: *Eugraphe longipennis* Boursin, 1948; *Sineugraphe longipennis sinensis* Boursin, 1954 (ssp.).

Distribution. China (Heilongjiang, Jilin, Jiangsu, Zhejiang, Jiangxi); Russia (Siberia), Korea, Japan.

寄主：长叶酸模（*Rumex longifolius*）（Matov & Kononenko, 2012）。

Genus *Naenia* Stephens, 1827 宽翅夜蛾属

Naenia Stephens, 1827, *in* Anonymous, *The Retrospective Review and Historical and Antiquarian Magazine*, 2 (1): 243. Type species: *Phalaena typicoides* Donovan sensu Stephens, 1827, a misidentiflcation for *Phalaena typica* Linnaeus, 1758 [Europe].

Synonymy: *Phalaena* Linnaeus, 1758, suppr. ICZN, 1957 Op. 450; *Phalena* Radermacher, 1779, unavail.; *Phalena* Rambur, 1829, missp.; *Naenia* Walker, [1858] 1857, unavail..

The genus contains 2 species, distributed in the temperate zone of the Palaearctic region. Among them, 1 species is in the Three Provinces of Northeast China.

Naenia contaminata **(Walker, 1865) 褐宽翅夜蛾**

Graphiphora contaminata Walker, 1865, *List of the Specimens of Lepidopterous Insects in the Collection of the British Museum*, 33: 710. Type locality: China, Shanghai.

Synonymy: *Atrachea sasayamana* Matsumura, 1931.

Distribution. China (Heilongjiang, Jiangsu, Shanghai, Jiangxi, Sichuan); Russia (Transbaikalia, Siberia), Korea, Japan.

寄主：酸模（*Rumex acetosa*）、直穗酸模（*Rumex longifolius*）、胡桃楸（*Juglans mandshurica*）、甘蓝（*Brassica oleracea*）、荠（*Capsella bursa-pastoris*）、苹果（*Malus pumila*）、大豆（*Glycine max*）、豌豆（*Pisum sativum*）、黄檗（*PhellodendroNorth of Amur reg.ense*）、葡萄（*Vitis vinifera*）、车前（*Plantago asiatica*）、杭菊（*Chrysanthemum morifolium*）、一年蓬（*Erigeron annuus*）、山莴苣（*Lactuca indica*）、台湾蒲公英（*Taraxacum platycarpum*）、补血草属（*Limonium*）、山芥属（*Barbarea*）、卫矛属（*Euonymus*）、接骨木属（*Sambucus*）、梣属（*Fraxinus*）、蒿属（*Artemisia*）、苦苣菜属（*Sonchus*）、薯蓣属（*Dioscorea*）、酸模属（*Rumex*）、苹果属（*Malus*）、豌豆属（*Pisum*）、车前草属（*Plantago*）、菊属（*Chrysanthemum*）、飞蓬属（*Erigeron*）、莴苣属（*Lactuca*）、苦苣菜属（*Sonchus*）植物（Matov & Kononenko, 2012）。

Genus ***Nyssocnemis*** Lederer, 1857 芒胫夜蛾属

Nyssocnemis Lederer, 1857, *Die Noctuinen Europa's, mit Zuziehung einger bisher mesit dazu gezahlter Arten des Asiatischen Russland's, Kleinasien's, Syrien's, und Labrador's*: 35, 117. Type species: *Hadena eversmanni* Lederer, 1853 the objective repl. name for *Noctua obesa* Eversmann, 1846 [East Kazakhstan, West Altai].

Monotypic genus, distributed in the temperate zone of Eastern Palaearctic region. 1 species is in the Three Provinces of Northeast China.

Nyssocnemis eversmanni **(Lederer, 1853) 芒胫夜蛾**

Hadena eversmanni Lederer, 1853, *Verhandlungen des Zoologisch-Botanischen vereins in Wien*, 3: 369, pl. 4: 1. Type locality: [East Kazakhstan, West Altai] Ust-Kamengorsk, Ust-Buchtarminsk, repl. name for *Noctua obesa* Everesmann, 1846, nec Boisduval, 1829.

Synonymy: *Noctua obesa* Eversmann, 1846, nec Boisduval, 1829; *Hadena ledereri* Hampson, 1903, nomen nudum.; *Hadena ledereri*: Herrich-Schäffer, nomen nudum..

Distribution. China (Heilongjiang, Jilin, Liaoning, Hebei, Xinjiang); Russia (Ural, Siberia), Mongolia, Korea, Japan, Kazakhstan.

寄主：覆盆子（*Rubus idaeus*）、林地水苏（*Stachys sylvatica*）、车前草属（*Plantago*）、车前草属（*Plantago*）、蒲公英属（*Taraxacum*）植物（Matov & Kononenko, 2012).

References （参考文献）

[1]Bänziger H. A taxonomic revision of the fruit–piercing and blood–sucking moth genus *Calyptra* Ochsenheimer = *Calpe* Treitschke (Lepidoptera, Noctuidae). Entomologica Scandinavica, 1983, 14: 467–491.

[2]Bänziger H. Skin–piercing blood sucking moths V: Attack on man by five *Calyptra* spp. (Lepidoptera, Noctuidae) in S and SE Asia. Mitteilungen des Schweizerischen Entomologischen Gesellschaft, 1989, 62: 215–233.

[3]Behounek G. Die holarktischen Arten der Gattung *Lacanobia* (Lepidoptera, Noctuidae). Esperiana, 1993, 3: 33–65.

[4]Behounek G, Han H L, Kononenko V S. A review of the genus *Belciades* Kozhanchikov, 1950 (Lepidoptera, Noctuidae, Pantheinae), with descriptions of two new species from China and Thailand. Revision of Pantheinae, contribution II. Zootaxa, 2011. 3070: 40–50.

[5]Behounek G, Kononenko V S, A revision of the "*Belciana*" *kala* Prout, 1924 species group with description of a new genus and two new species (Lepidoptera, Noctuidae: Pantheinae) from East Asia. Revision of Pantheinae, contribution VI. Zootaxa, 2012, 3590: 1–15.

[6]Behounek G, Kononenko V S. Revision of the Old World genera *Panthea* Hübner, [1820]1816 and *Pantheana* Hreblay, 1998 with description two new species from China (Lepidoptera, Noctuidae: Pantheinae). Revision of Pantheinae, contribution IX. Zootaxa, 2013, 3746: 422–438.

[7]Behounek G, Ronkay G, Ronkay L. The Witt Catalogue – Plusiinae II. A taxonomic atlas of the Eurasian and North African Noctuoidea. 1. Budapest: Heterocera Press, 2010, 320.

[8]Berio E, Fletcher D S. Monografia dell'antico genere *Sypna* (Lepidoptera, Noctuidae). Annali del Museo Civico di Storia Naturale Giacomo Doria, 1958, 70: 323–402.

[9]Boursin Ch. Neue palearktische *Agrotis*–Arten aus dem Naturhistorischen Museum in Wien nebst Synonymie–Notizen. Zeitschrift der Wiener Entomologischen Gesellschaft, 1948, 33: 97–136, Taf. 1–14.

[10]Boursin Ch. Die *Cucullia*–Arten aus Dr. Höne's China Ausbeuten, 1931–1938. Deutsche Entomologische Zeitschrift Iris, 1941, 55: 28–84.

[11]Boursin Ch. Eine neue *Cryphia* Hb. (*Bryophila* auct.) aus Ost–Asien. Zeitschrift der Wiener Entomologischen Gesellschaft, 1951, 36: 155–158, Taf. 11.

[12]Boursin Ch. Die "*Agrotis*" Arten aus Dr. h. s. H. Höne's ausbeuten (Beitrage zur Fauna Sinica). Bonner Zoologische Beiträge, 1954a, 5: 213–309, Taf. 14.

[13]Boursin Ch. Eine neue palaarktische (und europaische) *Cryphia* Hb. (*Bryophila*) art. Zeitschrift der Wiener Entomologischen Gesellschaft, 1954b, Bd39: 78–85, Tafs. 5–6.

[14]Boursin Ch. Die *Apopestes* Hb. und *Autophila* Hb. Arten aus Dr. h. c. Höne's China–Ausbeuten. Zeitschrift der Wiener Entomologischen Gesellschaft, 1955, 40: 164–171, Tafs. 13–15.

[15]Boursin Ch. Drei neue palaarctischen *Cryphia* Hb; Arten. Zeitschrift der Wiener Entomologischen Gesellschaf, 1961, 46: 139–144, Taf. 10.

[16]Boursin Ch. Die 'Noctuidae'–Arten aus Dr. h. c. Höne's China–Ausbeuten. V. Der '*Amathes*–Komplex'; VI Die Gattungen: *Raddea* Alph., *Estimata* Kozh., *Erebophasma* n. gen., *Spinipalpa* Alph. und *Perissandra* Warren; VII. Die Gattung *Ochropleura* Hb. Forschungsberichte des Landes Nordrhein-Westfalen, 1963, 1170: 1–107, Tafs. 1–22.

[17]Chen Y X. Noctuidae. Iconographia Insectorum Sinicorum. III. Beijing: Science press, 1982. pp. 237–390, pls 76–118 (in Chinese).

[18]Chen Y X. Lepidoptera Noctuidae (Fauna Sinica. Insecta 16). Beijing: Science press, i–lxxiii, 1999, 1596pp., pls. 1–68 (in Chinese).

[19]Deng G, Qi M J, Han H L. Three Species of the Genus *Meganola* Dyar, 1898 (Lepidoptera, Noctuidae, Nolinae) New to China. Korean Journal of Applied Entomology, 2012, 51(3): 231-234.

[20]Dufay C. Mise au point de la synonymie des *Nycteola* Hb. (Sarrothripus Curt.) europeenes (Lep., Noctuidae, Nycteolinae) (Contributions á l' é tude des Noctuidae Quadrifinae, 8). Bulletin Mensuel de la Société Linneenne de Lyon, 1958, 27: 108–120.

[21]Dufay C. Revision des Plusiinae Paléarctiques, I. Monographie du genre *Euchalcia.* Veröffentlichungen der Zoologischen Staatssammlung München, 1968, 12: 21–154.

[22]Eichlin T D, Cunnigham H B. The Plusiinae (Lepidoptera; Noctuidae) of America North of Mexico, emphasizing genitalia and larval morphology. United States Department of Agriculture Technical Bulle, 1567. U.S. Printing Office, Washington, D.C., 1978, 122pp, figs. 227.

[23]Fibiger M. Noctuinae 2. Noctuidae Europaeae, Vol. 2. Søro: Entomological Press, 1993, 230pp.

[24]Fibiger M. Noctuinae 3. Noctuidae Europaeae, Vol. 3. Søro: Entomological Press, 1997, 418pp.

[25]Fibiger M. Revision of the Micronoctuidae (Lepidoptera: Noctuoidea) Part 1, Taxonomy of the Pollexinae. Zootaxa, 2007, 1567, 1–116.

[26]Fibiger M. Revision of the Micronoctuidae (Lepidoptera: Noctuoidea) Part 2, Taxonomy of the Belluliinae, Magninae, and Parachrostiinae. Zootaxa, 2008, 1867: 1–136.

[27]Fibiger M. Revision of the Micronoctuidae (Lepidoptera: Noctuoidea) Part 3, Taxonomy of the Tactusinae. Zootaxa, 2010, 2583: 1–119.

[28]Fibiger M. Revision of the Micronoctuidae (Lepidoptera: Noctuoidea) Part 4, Taxonomy of the subfamilies Tentaxinae and Micronoctuinae. Zootaxa, 2011, 2842: 1–188.

[29]Fibiger M, Hacker H. Systematic List of the Noctuoidea of Europe (Notodontidae, Nolidae, Arctiidae, Lymantriidae, Erebidae, Micronoctuidae, and Noctuidae). Esperiana, 2005, 11: 83–172.

[30]Fibiger M, Hacker H. Amphipyrinae, Condicinae, Eriopinae, Xyleninae (Part). Noctuidae Europae, Vol. 9. Sorø: Entomological Press, 2007, 410pp.

[31]Fibiger M, Kononenko V S. Revision of the Micronoctuidae occuring in the Russian Far East and neighbouring countries with description of a new species (Lepidoptera, Noctuoidea). Zootaxa, 2008a, 1890: 50–58.

[32]Fibiger M, Kononenko V S. 2008b. A revision of the subfamily Araeopteroninae Fibiger, 2005 in the Russian Far East and neighbouring countries with a description of four new species (Lepidoptera, Noctuidae). Zootaxa, 1891: 39–54.

[33]Fibiger M, Kononenko V S. Nilsson D. 2008. Description of a new species of *Lygephila* Billberg, 1820 (Lepidoptera: Noctuidae, Catocalinae) from Russian Far East and North China. Zootaxa, 1922:62-68.

[34]Fibiger M., Lafontaine J D. A review of the higher classification of the Noctuoidea (Lepidoptera) with special reference to the Holarctic fauna. Esperiana, 2005, 11: 7–92.

[35]Fibiger M, et al. Pantheinae, Dilobinae, Acronicitinae, Nolinae, Eustrotiinae, Acontiinae, Metoponiinae, Heliothinae and Bryophilinae. Noctuidae Europae, Vol. 11. Sorø: Entomological press, 2009, 504pp.

[36]Fibiger M, et al. Rivulinae-Eutelinae and Supplement vols 1-11. Noctuidae Europaea, Vol. 12. Sorø: Entomological press, 2010, 452pp.

[37]Fibiger M, et al. Check list of the quadrifid Noctuoidea of Europe. // Witt T., L. Ronkay (Eds). Lymantriinae and Arctiinae including phylogeny and check list of the quadrifid. Noctuidae Europaea, Vol. 13. Sorø: Entomological press, 2012, 23-44.

[38]Fu Y C, et al. The Study of Lepidoptera of Forestry Botanical Garden in Heilongjiang Province. Forestry Science and Technology Information, 2011, 43(3): 31, 33.

[39]Goater B, et al. Catocalinae and Plusiinae. Noctuidae Europeae, Vol.10. Sorø: Entomological Press, 2003, 452pp.

[40]Gui C M. The Noctuid moths group in Guzhuling. Journal of Jilin Agricultural Science, 1987, 4: 28-37.

[41]Guo M, et al. Two New Record Species of Noctuinae (Lepidoptera, Noctuidae) from Northeast China. Journal of Northeast Forestry University, 2010, 38(7): 129-130.

[42]Guo Z Q, Wang Y. Taxonomy Study on the genus *Zanclognatha* Lederer, 1859 (Lepidoptera, Noctuidae) from Da Hinggan Ling Region, Prov. Heilongjiang. Forestry Science and Technology Information, 2013, 45(1): 38-43.

[43]Hacker H. Revision der Gattung *Hadena* Schrank, 1802 (Lepidoptera). Esperiana, 1990, 3: 243–361.

[44]Hacker H. Revision der Gattung *Hadena* Schrank, 1802 (Lepidoptera: Noctuidae). Esperiana, 1996, 5: 7–775.

[45]Hacker H. Revision der Gattung *Hadula* Staudinger, 1889 (= *Discestra* Hampson, 1905; *Aglossestra*, Hampson, 1905; = *Cardiestra* Boursin, 1963), *Anartomorpha* Alpheraky, 1892, *Trichanarta* Hampson, 1895, *Anarta* Ochsenheimer, 1816 und *Cardepia* Hampson, 1905 mit Beschreibung einer neuen Gattung *Hadumorpha* gen. n. (Lepidoptera, Noctuidae). Esperiana, 1998, 6: 577–843.

[46]Hacker H. Revision of the genus *Caradrina* Ochsenheimer, 18165 with notes on other genera of the tribe Caradrinini (Lepidopetra, Noctuidae). Esperiana, 2004, 10:7- 690.

[47]Hacker H. Biogeography, Boletobiinae (Lepidoptera, Noctuoidea, Erebidae). - Moths of Africa. Systematic and illustrated Catalogue of the Heterocera (Lepidoptera) of Africa. Vol. 1. Bad Staffelstein. 2019: 815pp.

[48]Hacker H, Legrain A, Fibiger M. Revision of the genus *Acontia* Ochsenheimer. 1816 and tribus Acontiini Guenèe, 1841 (Old World) (Lepidoptera: Nictuidae: Acontiinae). Esperiana, 2008, 4: 7–533.

[49]Hacker H, Ronkay L. Hreblay M. et al. Noctuidae Europaeae, Vol. 4. Entomological Press, Søro, 2002,

419pp. + 15 pls.

[50Hacker H, Zilll A. Revisional notes on the genus *Thiacidas* Walker, 1855, with description of Thiacidinae subfam. nov. and eleven new species (Lepidoptera, Noctuidae). Esperiana (Memoirs), 2007, 3:179–264.

[51]Hampson G F. Catalogue of the Lepidoptera Phalaenae in the British Museum. London, 1903. Vol. 4 (Agrotinae). I–XX+689 p.; 1905. Vol. 5 (Hadeninae). I–XVI+633 p.; 1906. Vol. 6 (Cucullianae). I–XIV+532 p.; 1908. Vol. 7 (Acronyctinae). I–XV+709 p.; 1909. Vol. 8 (Acronyctinae). I–XIV+709 p.; 1910a. Vol. 9 (Acronyctinae). I–XV+552 p.; 1910b. Vol. 10 (Erastrianae). I–XIX+829 p.; 1912. Vol. 11 (Eutelianae, Stictopterinae, Sarrothripinae, Acontianae). I–XVII+689 p.; 1913a. Vol. 12 (Catocalinae). I–XIII+858 p.; 1913b. Vol. 13 (Catocalinae, Mominae, Phytometrinae). I–XIV+609. Pls 1–239.

[52]Han H L. Progress of Taxonomy of Erebidae (part), Euteliidae, Nolidae and Noctuidae (Lepidoptera) in the Northeast China. Journal of Environmental Entomology, 2019, 41(6): 1181-1195.

[53]Han H L, Ding Y, Jiang J Q. Taxonomic study of the genus *Conistra* (Lepidoptera, Noctuidae) from Northeast China. Journal of Asia-Pacific Entomology, 2008a, 11(1): 21-24.

[54]Han H L, Duan Z Y, Feng Z W. Three species of the Family Noctuidae (Lepidoptera) New to China. Entmological Research, 2008b, 38(3): 239-242.

[55]Han H L, Gao W T, Meng Q F. Three Newly Recorded Species of the Family Noctuidae (Lepidoptera) from China. Entomotaxonomia, 2010, 32(1): 77-80.

[56]Han H L, Jiang J Q, Ding Y. Three species of the Noctuidae moths new to China, and a newly recorded species from Northeast China (Lepidoptera, Noctuidae). Korean Journal of Applied Entomology, 2008, 47(1): 5-8.

[57]Han H L, et al. Six species of the Subfamily Herminiinae (Lepidoptera, Noctuidae) New to China, with seven new records from Mt. Changbai. Korean Journal of Applied Entomology, 2006, 45(2): 131-137.

[58]Han H L, et al. Plusiinae in Mt. Changbai (Lepidoptera, Noctuidae), with Six New Records from China. Korean Journal of Applied Entomology, 2005, 44(1): 13-20.

[59]Han H L, Kononenko V S. New species of the genera *Acronicta* Ochsenheimer, 1816 and *Craniophora* Snellen, 1867 from China with notes on synonymy and checklist (Lepidoptera, Noctuidae: Acronictinae). Zootaxa, 2010, 2678: 48–68.

[60]Han H L, Kononenko V S. Twelve new species of *Athetis* Hübner, [1821]1816 from China (Lepidoptera, Noctuidae). Zootaxa, 2011, 3068: 49–68.

[61]Han H L, et al. Six species of the Family Noctuidae (Lepidoptera) New to China. Korean Journal of Systematic Zoology, 2006, 22(2): 139-143.

[62]Han H L, Li C D. Taxonomic study of the genus *Meganephria* Hübner (Lepidoptera: Noctuidae) from north-east China. Entomological Research, 2008, 38(2008): 131-134.

[63]Han H L, Li C D. Description of a new species of Nolinae, *Meganola liaoningensis* from China (Lepidoptera, Noctuidae). Tinea, 2008, 20(3): 152-154.

[64]Han H L, Li C D, Kononenko V. Three species of the Family Noctuidae (Lepidoptera) New to China. Journal of Asia-Pacific Entomology, 2007, 10(1): 17-19.

[65]Han H L, Li C D, Ronkay L. Report on two species of noctuid moths (Lepidoptera, Noctuidae s.l.) new to China and one new record species in Continental fauna. Journal of Forestry Research, 2007, 18(2): 144-146.

[66]Han H L, Park K T, Lu L S. *Znaclognatha* species in Mt. Changbai, with Description of a new species and Two unknown species from China (Lepidoptera, Noctuidae). Korean Journal of Systematic Zoology, 2005, 21(1): 1-10.

[67]Han H L, et al. Taxonomic Notes on the genus *Chasminodes* Hampson (Lepidoptera, Noctuidae) of Mt. Changbai, China. Insecta Koreana, 2003, 20(3-4): 301-308.

[68]Hardwick D F. The corn earworn complex. Memoirs of the Entomological Society of Canada, 1965, 40: 1–247.

[69]Hardwick D F. A generic revision of the North American Heliothinae (Lepidoptera, Noctuidae). Memoirs of the Entomological Society of Canada, 1970, 73: 1–59.

[70]Holloway J D. The moths of Borneo Noctuidae: Euteliinae, Stictopterinae, Plusiinae, Pantheinae. Malayan Nature Journal [Separately published as Moths of Borneo 14], 1985, 38: 157–317.

[71]Holloway J D. The moths of Borneo: Family Arctiidae, subfamilies Syntominae, Euchrominae, Arctiinae; Noctuidae misplaced in Arctiidae (Camptoloma, Aganainae). Moths of Borneo, 1988, 6, 101pp, figs. 157, pls. 6.

[72]Holloway J D. The moths of Borneo: family Noctuidae, trifinae subfamilies: Noctuinae, Heliothinae, Hadeninae, Acronictinae, Amphipirynae, Agaristinae. Malayan Nature Journal [Separately published as Moths of Borneo 12], 1989, 42, 57–288 , 40 b/w pls, pls. 8.

[73]Holloway J D. The classification of the Sarrothripinae, Chloephorinae, Camptolominae and Nolinae as Nolidae (Lepidoptera: Noctuoidea). Quadrifina, 1998. 1: 247–276.

[74]Holloway J D. The Moths of Borneo 18: Nolidae, Nolinae, Chloephorinae, Westermanniinae, Collomeninae, Bleninae, Risobinae. Kuala Lumpur, Southdene, 2003, 279pp.

[75]Holloway J D. The moths of Borneo 15 & 16: Noctuidae Catocalinae. Malayan Nature Journal, 2005, 58(1–4), 529pp, figs. 3-983, pls. 28.

[76]Holloway J D, The Moths of Borneo Noctuidae 17: Rivulinae, Phytometrinae, Herminiinae, Hypeninae and Hypenodinae. Malayan Nature Journal, 2008. 60(1–4), 268pp.

[77]Holloway J D. The Moths of Borneo 13: family Noctuidae, subfamilies Pantheinae (part), Bagisarinae, Acontiinae, Aediinae, Eustrotiinae, Bryophilinae, Araeopteroninae, Aventiinae, Eublemminae and further miscellaneous genera. Malayan Nature Journal, 2009, 62(1&2), 240pp.

[78]Holloway J D. The Moths of Borneo. Part 2. Phautidae, Himantopteridae, Zygaenidae. Complete Checklist, Checklist notes, Historical appendix, Index. Malayan Nature Journal, 2011, 63(1–2), 545pp.

[79]Hreblay, M. Neue taxa aus der Gattung Orthosia Ochsenheimer, 1816 (s.l.) (Lepidoptera, Noctuidae). Acta Zoologica Hungarica, 1991, 37: 193–203.

[80]Hreblay M, Legrain A, Yoshimatsu S. Beschreibung von vierzehn neuen Arten mit Ubersicht der Artengruppen aus dem Komplex *Mythimna* Ochsenheimer, 1816. Festlegungen vierzehn Lectotypen, neue Synonymen und Kombinationen aus der Himalaya Region. Annales Historico–Naturales Musei Nationalis

Hungarici, 1996, 88: 89–126.
[81]Hreblay, M., Legrain, A. Yoshimatsu, S.I. Beschreibung von zwölf neuen Arten und zwei Unterarten mit einer Übersicht über die Artengruppen des *Mythimna* Ochsenheimer, 1816 – Komplexes der südostasiatischen Rgion. (II.) (Lepidoptera: Noctuidae). Esperiana, 1999, 7: 377–398.
[82]Hreblay M, Ronkay L, Plante J. Contribution to the Noctuidae (Lepidoptera) fauna of Tibet and the adjacent regions (II). A systematic survey of the Tibetan Noctuidae fauna based on the material of the Schäfer–expedition (1938–1939) and recent expeditions (1993–1997). Esperiana, 1998, 6: 69–184.
[83]Huang B J, Han H L. One *Hepatica* species of subfamily Hypeninae (Lepidoptera, Erebidae) new record from China. Journal of Agricultural Science Yanbian University, 2014, 36(2): 111-113.
[84]Ichinose T. Studies on the noctuid subfamily Plusiinae of Japan. Bull. Fac. Agric. Tokyo Univ. Agric. Techn., [1963]1962, 6: 1–127.
[85]Ichinose T. A revision of some genera of the Japanese Plusiinae with descriptions of a new genus and two new subgenera (Lepidoptera, Noctuidae). Kontyu, 1973, 4: 135–140.
[86]Inoue H. Three new species of the genus *Araeopteron* (Lepidoptera, Noctuidae) from Japan. Tinea, 1958, 4(1): 229–233.
[87]Inoue H. Two new species of Araeopteron (Lepidoptera, Noctuidae) from Japan with notes on the known species. Tinea, 1965, 7(1), 81–83.
[88]Inoue H. Two new genera of the subfamily Nycteolinae, Noctuidae from East Asia (Lepidoptera). Bulletin of the Japan Entomological Academy, 1970, 37–42.
[89]Inoue H. Genera *Schrankia* and *Luceria* (Lepidoptera, Noctuidae) from Japan. Tinea, 1979, 10: 300–308.
[90]Inoue H. Nolidae. // Inoue et al., Moths of Japan, Tokyo: Kodansha, 1982, 1: 660–668; 2: 342–343, 460–461, pls. 154, 229, 349–354.
[91]Inoue H. Notes on the Nolinae (Lepidoptera, Noctuidae): Four new species from Japan and two species described by Staudinger from southeast Siberia. Tyo to Ga, 1991, 42(2): 63–73.
[92]Inoue, H. On the systematic status of *Nola minutalis* Leech (Noctuidae, Nolinae). Yugato, 2001, 164: 53–60 (in Japanese).
[93]Jin G Q, et al. Investigation of moth species in Suifenhe port of Heilongjiang Province. Journal of Northeast Agricultural University, 2009, 40(2): 19-23.
[94]Jin Z M, Liu T, Li D W. Study on Phalaenae Resource in Eastern Forest of Heilongjiang. Journal of Anhui Agriculture Science, 2011, 39(6): 3330-3331.
[95]Jin Z Y, et al. Study on the Diversity of Big Moths from Jingpohu. Journal of Anhui Agricultural Science, 2014, 42(4): 1063-1065.
[96]Ju C, Gao W, Wang K Y. Biologeical species and distribution in Jilin Province, China. Changchun: Northeast Normal University Press, 1997, 649pp.
[97]Kiriakoff S G. Lepidoptera Noctuiformes Agaristidae I (Palaearctic and Oriental genera). Das Terreich, [1976]1977, 97, 180pp, figs. 39.
[98]Kitching I J. A historical review of the higher classification of the Noctuidae. Bulletin of the British Nuseum

Nature History (Entomology), 1984, 49: 153-234.

[99]Kitching I J. Spectacles and Silver Ys: a synthesis of the systematics, cladistic and biology of the Plusiinae (Lepidoptera: Noctuidae). Bulletin of the British Nuseum Nature History (Entomology), 1987, 54: 75–261.

[100]Kitching I J, Rawlins J E. The Noctuoidea. // N. P. Kristinsen (Ed.) Handbook of zoology, IV Arthropoda: Insecta. Pt. 35. Lepidoptera, Moths and Butterflies. Vol. 1. Evolution, systematics, and biogeography. Berlin, New York, 1999, I–X, 491pp.

[101]Kljuchko Z F, Kononenko V S. Plusiine moths (Lepidoptera, Noctuidae, Plusiinae) of the Far East of the USSR. Syatematics and ecology of the Lepidoptera of the Far East of the USSR. Vladivostok, 1986, 95-112.

[102]Klyuchko Z F. Semeistva Nycteolinae, Hypeninae, Catocalinae, Othreinae, Euteliinae, Pantheinae, Plusiinae, Jaspidiinae. Sovki kvadrifinoidnogo kompleksa. Fauna Ukrainy. Tom 19 N 6. Kiev, Naukova dumka, 1978, 414 pp (in Ukrainen).

[103]Klyuchko Z F. Podsem. Plusiinae. // P. A. Lehr (ed). Key to insects of Russian Far East. V. Trichoptera and Lepidoptera. Pt. 4. Vladivostok, Dal'nauka, 2003, 187–215(in Russian).

[104]Kononenko V S. The noctuid moths of the subfamily Heliothinae (Lepidoptera, Noctuidae) of the fauna of the Far East. Fauna i ekologia nasekomyh Primorya i Kamchatki. Vladivostok, 1985, 34–54 (in Russian).

[105]Kononenko V S. The noctuids of the genus *Atrachea* Warren (Lepidoptera, Noctuidae) of the fauna of Soviet Far East. Sistematika i ecologia nasekomyh Da'lnego Vostoka. Vladivostok, 1986, 36–43 (in Russian).

[106]Kononenko V S. Sovki podsemeistva Agaristinae (Lepidoptera, Noctuidae) fauny SSSR. Cheshuekrylye Dal'nego Vostoka SSSR. Vladivostok, 1987, 94–101 (in Russian).

[107]Kononenko V S. Synonimic check list of the Noctuidae (Lepidoptera) of the Primorye Territory, the Far East of U.S.S.R. Tinea, 1990, 13(Suppl. 1): 1– 40.

[108]Kononenko V S. A review of Far Eastern species of the *Euplexia licipara* complex with description of a new species from China (Lepidoptera, Noctuidae, Amphipyrinae). Japan Heterocerist's Journal, 1996, 191: 262–266.

[109]Kononenko V S. A revision of the genus *Belciades* Kozhanchikov, 1950 (Lepidoptera, Noctuidae, Acronictinae), with description of three new species from China, Laos and Thailand. Tinea, 1997a, 15(2): 106–118.

[110]Kononenko V S. A revision of the *Hoplodrina implacata* species complex (Lepidoptera, Noctuidae, Amphipyrinae), with description of three new species from China. Japan Heterocerist's Journal, 1997b, 193: 291–297.

[111]Kononenko V S. Revisional notes on the Palaearctic species of the genus *Atrachea* Warren (Lepidoptera, Noctuidae, Amphipyrinae), with description of two new species from China. Insecta Koreana, 2001, 18(1): 51–68.

[112]Kononenko V S. Fam. Noctuidae – Owlet moths. Introduction; 6. Subfam. Euteliinae; 9. Subfam. Acontiinae; 10. Subfam. Pantheinae; 11. Subfam. Acronictinae; 12. Subfam. Bryophilinae; 13. Subfam. Agaristinae; 14. Subfam. Amphypirinae; 15. Subfam. Cuculliinae; 16. Subfam. Hadeninae; 17 Subfam. Noctuinae; 6 Subfam. Heliothinae. *In*: P. A. Lehr. (ed). Key to insects of Russian Far East. V. Trichoptera

and Lepidoptera. Pt.4. Vladivostok, Dal'nauka, 2003, 11–33, 215–217, 237–602 (in Russian).

[113]Kononenko V S. Noctuidae Sibiricae 1. An annotated check list of the Noctuidae (s.l.) (Lepidoptera, Noctuoidae: Nolidae, Erebidae, Micronoctuidae, Noctuidae) of the Asian part of Russia and the Ural region. Sorø: Entomological Press, 2005, 243pp.

[114]Kononenko V S. Micronoctuidae, Noctuidae: Rivulinae – Agaristinae (Lepidoptera); Noctuidae Sibiricae, Vol. 2. Sorø: Entomological Press, 2010, 475pp.

[115]Kononenko V S. Noctuidae: Cuculliinae – Noctuinae, part (Lepidoptera); Noctuoidea Sibiricae, Part 3. Proceedings of the Museum Witt Munich (Munich, Vilnius), 2016, 5, 500pp.

[116]Kononenko V S. Noctuoidea (exept Lymantriinae and Arctiinae). Annotated catalogue of the insects of Russian Far East. Vladivostok, Dalnauka, 2016, 340-341, 369-509.

[117]Kononenko V S, Ahn S B, Ronkay L. Illustrated catalog of Noctuidae in Korea (Lepidoptera). // Park K. T. (Ed.). Insects of Korea, 1998. 3. Jung haeng–Sa,Seoul, 509pp.

[118]Kononenko V S, Behounek G. A revision of the genus *Lophomilia* Warren, 1913 with description of four new species from East Asia (Lepidoptera: Noctuidae, Hypeninae). Zootaxa, 2009, 1989:1-22.

[119]Kononenko V S, Han H L. Atlas genitalia of the Noctuidae in Korea (Lepidoptera). // Park, K. T. (Ed.): Insects of Korea, 2007, 11. Seoul: Jung haeng–Sa, 461pp.

[120]Kononenko V S, Han H L. *Strotihypera* gen. n; new Old World genus of the tribe Elaphriini (Lepidoptera, Noctuidae, Noctuinae). Zootaxa, 2011, 3116: 59–68.

[121]Kononenko V S, Matov A Yu. A review of Palaearctic *Metachrostis* Hübner, [1820] 1816 with description of three new species (Lepidoptera: Noctuidae, Eublemminae). Zootaxa, 2009, 2026: 1–17.

[122]Kononenko V S, Mikkola K. *Panolis flammea* auct; two species (Lepidoptera, Noctuidae, Hadeninae). Nota lepidopterologica, 1989, 12: 29–35.

[123]Kononenko V S, Pinratana A. Moths of Thailand, Vol. 3, Part 1. An illustrated catalogue of the Noctuidae in Thailand (Insecta, Lepidoptera). Subfamilies Herminiinae, Rivulinae, Hypeninae, Catocalinae, Aganainae, Euteliinae, Stictopterinae, Plusiinae, Pantheinae, Acronictinae and Agaristinae. Brother of St Gabriel in Thailand, Bangkok, 2005, 261pp.

[124]Kononenko V S, Pinratana A. Moth of Thailand Vol. 3, Part 2. Noctuoidea. An illustrated Catalogue of Erebidae, Nolidae, Euteliidae and Noctuidae (Insecta, Lepidoptera) in Thailand. Brothers of St. Gabriel in Thailand, Bangkok, 2013, 625pp.

[125]Kononenko V S, Ronkay L. On the *Xylopolia* species of the Manchurian region (Lepidoptera, Noctuidae). Acta Zoologica Academiae Scientiarum Hungaricae, 1995, 41: 119–129.

[126]Kononenko V S, Ronkay L. A revision of the genus *Stenoloba* Staudinger (Lepidoptera, Noctuidae, Bryophilinae) with description of 25 new species and 3 subspecies from East Asia. Part 1. Insecta Koreana, 2000, 17: 137–166.

[127]Kononenko V S, Ronkay L. A revision of the genus *Stenoloba* Staudinger (Lepidoptera, Noctuidae, Bryophilinae) with description of 15 new species and 3 subspecies from East Asia. Part 2. Insecta Koreana, 2001, 18: 95–121.

[128]Kononenko V S, Spitzer K. *Polia malchani* (Draudt) (Lepidoptera. Noctuidae) as a representative of boreal faunal complex, new to fauna of Korea. Tinea, 1993, 13: 289–295.

[129]Kostrowicki A S. Sówki – Noctuidae. Podrodzini: Agrotinae, Melicleptriinae / Klucze do oznaczania owadow Polski. Cz. 27 (Motyle–Lepidoptera), 1959, zesz. 53b, Warszawa, 146 ss.

[130]Kostrowicki A S. Studies on the palearctic species of the subfamily Plusiinae (Lepidoptera, Phalaenidae). Acta Zoologica Cracoviensia., 1961, 6: 367–472, figs. 154.

[131]Kozhantschikov I V. Moths (Subfam. Agrotinae). Nasekomye Cheshuekrylye [Insecta, Lepidoptera], Fauna SSSR, Vol. 13(3). Moskow - Leningrad, Academy of Sciencs of the USSR, 1937, 674pp (in Russian).

[132]Kozhantschikov I V. Tussock Moths (Orgiidae). Nasekomye Cheshuekrylye [Insecta, Lepidoptera], Fauna SSSR, Vol. 12. Moskow - Leningrad, Academy of Sciencs of the USSR, 1950, 581pp, (in Russian).

[133]Kullberg J, et al. Checklist of Finnish Lepidoptera. Sahlbergia, 2001, 6: 45–190.

[134]Lafontaine J D. Noctuoidea: Noctuidae (Part), Noctuinae (Part – *Euxoa*). // Dominik R.B. et al. (Eds). The Moths of America North of Mexico. Wedge Entomological Research Foundation (Washington), 1987a, 27(2), 237pp.

[135]Lafontaine J D. Noctuoidea: Noctuidae (Part), Noctuinae, part. // Dominik R.B. et al. (Eds). The Moths of America North of Mexico. Wedge Entomological Research Foundation (Washington), 1998, 27(3), 348pp.

[136]Lafontaine J D. Noctuoidea: Noctuidae (Part), Noctuinae, part: Noctuinae, Agrotini. // Dominik R.B. et al. (Eds). The Moths of America North of Mexico. Wedge Entomological Research Foundation (Washington), 2004, 27(1), 358pp.

[137]Lafontaine J D, Fibiger M. Revised higher classification of the Noctuoidea (Lepidoptera). Cannadian Entomologist, 2006, 138(5): 610–635.

[138]Lafontaine J D, Kononenko V S. A revision of the genus *Trichosilia* (Hampson) (Lepidoptera, Noctuidae) with description of four new species. Canadian Entomologist, 1986, 118: 1072–1013.

[139]Lafontaine J D, Kononenko V S. A review of the genus *Parabarrovia* Gibson (Lepidoptera: Noctuidae) with description of immature stages and a new species. Canadian Entomologist, 1988a, 120: 507–523.

[140]Lafontaine J D, Kononenko V S. A revision of the *Lasionycta skraelingia* (Herrich–Schaffer) species complex (Lepidoptera, Noctuidae). Canadian Entomologist, 1988b, 120: 903–916.

[141]Lafontaine J D, Mikkola K, Kononenko V S. A revision of the genus *Xestia* subg. *Schoyenia* Auriv. (Lepidoptera, Noctuidae) with descriptions of four new species and a new subspecies. Entomologica Scandinavica, 1983, 14: 337–369.

[142]Lafontaine J D, Mikkola K, Kononenko V S. A revision of the genus *Xestia* subg. *Pachnobia* (Lepi., Lepidoptera, Noctuidae), with description of two new subspecies. Entomologica Scandinavica, 1987a, 18: 305–331.

[143]Lafontaine J D, Mikkola K, Kononenko V S. *Anarta cordigera* (Thunberg) (Lepidoptera: Noctuidae: Hadeninae) a species complex. Canadian Entomologist, 1987b, 119: 931–940.

[144]Lafontaine J D, et al. Subgenus *Pachnobia* Guenee, Lafontaine J.D. Noctuoidea: Noctuidae (Part), Noctuinae, part. // Dominik R.B. et al. (Eds). The Moths of America North of Mexico. Wedge Entomological

Research Foundation (Washington), 1998, 27(3), 139–165.

[145]Lafontaine J D, Poole R. Noctuoidea: Noctuidae (Part), Plusiinae. // Dominik R.B. et al. (Eds). The Moths of America North of Mexico. Wedge Entomological Research Foundation (Washington), 1991, 25(1), 182pp.

[146]Lafontaine J D, Schmidt B C. Annotated check list of the Noctuoidea (Insecta, Lepidoptera) of North America north of Mexico. ZooKeys, 2010, 40: 1–239.

[147]Lafontaine J D, Schmidt B C. Additions and corrections to the check list of the Noctuoidea (Insecta, Lepidoptera) of North America north of Mexico. ZooKeys, 2013, 264: 227–236.

[148]Leech J H. On the Lepidoptera of Japan and Korea. Part III. Heterocera, sect. II. Proceedings of the Zoological Society of London, 1889a: 474–571.

[149]Leech J H. New species of Deltoids and Pyrales from Korea, North China and Japan [J]. Entomologist, 1889b, 22: 62–71, pls. 2–4.

[150]Leech J H. Lepidoptera Heterocera from northern China, Japan and Corea. Part. III . Transactions of the Entomological Society of London, 1900a: 9–161.

[151]Leech J H. Lepidoptera Heterocera from northern China, Japan and Corea. Part. IV. Transactions of the Entomological Society of London, 1900b: 511–663.

[152]Li B S, Han H L. A New Record Species of Genus *Mosopia* (Lepidoptera, Noctuidae) from China. Journal of Northeast Forestry University, 2011, 39(11): 139-140.

[153]Li J H, et al. Preliminary Survey on Noctuidae Insects in Changchun suburb, Jilin Province. Journal of Jilin Agricultural University, 1989, 11(3): 13-18.

[154]Li M W, Han H L. New Record Species of Subfamily Herminiinae (Lepidoptera, Noctuidae), One in China and Five in Northeast China. Journal of Northeast Forestry University, 2013, 41(7): 163-167.

[155]Liu S D, et al. Diversity, of Noctuidae in Changbai Mountain National Nature Reserve. Journal of Beihua University (Natural Science), 2006, 7(4): 355-350.

[156]Lou W, et al. A Preliminary Report of Mothes`s Inverstiga in Hunchun Area on the Frontier of China, Russia and Korea. Journal of Jilin Agricultural University, 1997, 19(2): 35-38.

[157]Lödl M. Remarks on the classification of the genera *Hypena* Schrank, 1802, *Dichromia* Guenée, 1854 and *Harita* Moore, 1882 (Lepidoptera: Noctuidae). Nota Lepidopterologica, 1994, 16: 241–250.

[158]Lödl M, Mayerl B. Revision der Gattung *Zekelita* Walker, 1863 (=*Rhynchodontodes* Warren, 1913, (Lepidoptera: Noctuidae: Hypeninae). Quadrifina, 1998, 1:115-230.

[159]Maschenko N B. Ecological-fauna review of the cutworms (Lepidoptera, Noctuidae) of the Middle Amur basin. Fauna and ecology of the insects of Siberia. Novosibirsk, 1980, 189-217.

[160]Matov A Yu, Kononenko V S. Trophic connections of larvave of the Noctuoidea Lepidoptera of the fauna of Russia (Lepidoptera, Noctuoidea: Nolidae, Erebidae, Euteliidae, Noctuidae). Vladivostok, Dal'nauka, 2012, 346pp.

[161]Matov A Yu, Kononenko V S, Sviridov A V. Family Noctuidae. The Cataloge of Lepidoptera of Russia. St;Peterburgh – Moskow. KMK: 2008, 239– 296 (in Russian).

[162]Matthews M. Classification of the Heliothinae. Bulletin Natural Resources Institute, 1991, 44, i–v, 195pp.

[163]Mayerl B, Lödl M. Revision der Gattung *Rhynchina* Guenée, 1852, (Lepidoptera, Noctuidae, Hypeninae). Quadrifina, 1999, 2:1-124.

[164]McDunnough J. Revision of the North American genera and species of the Phalaenid subfamily Plusiinae (Lepidoptera). Memoirs of the Southern California Academy of Sciences, 1944, 2: 175–232.

[165]McCabe T L. A reclassification of the *Polia* complex for North America (Lepidoptera: Noctuidae). Bulletin University of the state of New York, 1980, 42: 1–141.

[166]Mell R. Beitrage zur Fauna Sinica. XI. Zur Biologie und Systematik der chinesischen *Catocala*. Deutsche Entomologische Zeitschrift, Entomologischen Vereins Iris Zu Dresden, 1936, 50: 49–90.

[167]Mell R. Beitrage zur Fauna Sinica. XXIV. Über Phlogophorinae, Odontodinae, Sarrothripinae, 'Westermanniinae' und Camptolominae (Noctuidae, Lepid.) von Kuantung. Zoologische Jahrbücher, Abteilung für Systematik, Oekologie, und Geographie der Tiere, 1943, 76: 171–266, pl. 1.

[168]Mikkola K, Lafontaine J D, Gill J D. Noctuoidea, Noctuidae (part), Xyleninae (part): Apameini (part – Apamea group of genera). // Hodges R.W. et al. (Eds). The Moths of North America. Wedge Entomological Research Foundation (Washington), 2009, 26(9), 192pp.

[169]Mikkola K, Lafontaine J D, Grotenfeld P. A revision of the Holarctic Chersotis andereggii complex (Lepidoptera, Noctuidae). Nota lepidopterologica, 1987, 10: 140–157.

[170]Mikkola K, Lafontaine J D, Kononenko V S. Zoogeography of the Holarctic species of the Noctuidae (Lepidoptera): importance of the Beringian refuge. Entomologica Fennica, 1991, 2: 157–173.

[171]Mikkola K, Sinevirta M, Vaalamo K. *Xestia brunneopicta* (Matsumura) new to Europe (Lepidoptera, Noctuidae). Notulae Entomology, 1989, 69: 33–37.

[172]Nieukerken E J, et al. Order Lepidoptera Linnaeus, 1758. // Zhang, Z.–Q. (Ed.), Animal Biodiversity: An Outline of Higher–Level Classification and Survey of Taxonomic Richness. Zootaxa, 2011, 3148: 212–221.

[173]Nye I W B. Noctuoidea (part): Noctuidae, Agaristidae, and Nolidae. The Generic Names of Moths of the World. Vol. 1. London, British Museum (Nat. Hist.), 1975, vi+568pp.

[174]Obraztsov N. Neues über Gelastocera–Arten (Lepidoptera, Agrotidae, Hylophilinae). Zeitschrift der Wiener Entomologische Gesellschaft, 1943, 28: 78–81.

[175]Oh S H. A Review of the Subfamily Nolinae (Lepidoptera, Noctuidae) in Korea (1): Genus *Nola* Leech. Insecta Koreana, 2001, 18(2): 123–137.

[176]Owada M. Noctuidae (Herminiinae). // Inoue H., S. Sugi, H. Kuroko, S. Moriuti, and A. Kawabe. Moths of Japan. Tokyo, 1982, Vol. 1: 913–935; Vol. 2: 405–408, pls. 224–226, 356, 381–392.

[177]Owada M. A taxonomic study of the subfamily Herminiinae of Japan (Lepidoptera, Noctuidaea). National Science Museum, Tokyo, 1987, 208pp.

[178]Owada M. On the herminiine moths *Zanclognatha perfractalis* (Noctuidae) discovered in Aomori prefecture, northern Honshu, Japan. Memoirs of the National Science Museum, 1988, 21: 137–143.

[179]Owada M. Synonymic notes on the herminiine moths (Noctuidae) of Japan, with description of three new species. Tinea, 1992a, 13: 183–203.

[180]Owada M. Noctuidae (Herminiinae). Checklist of the Lepidoptera of Taiwan. *In* Heppner J. B, Inoue H.

(eds). Lepidoptera of Taiwan, Taipei, 1992b, 1: 172–173.

[181]Pei H Y, et al. Investigation of moth species under ultraviolet light- trap in Harbin of Heilongjiang Province. Journal of Northeast Agricultural University, 2000, 31(4): 346-351.

[182]Pogue M. A World Revision of the genus *Spodoptera* Guenée (Lepidoptera, Noctuidae). Memoirs of the American Entomological Society, 2002, 43, 202pp.

[183]Pogue M G, Sullivan J B. Re–evaluation of the *Elaphria festivoides* (Guenée) species complex (Lepidoptera: Noctuidae). Proceedings of Entomological Society Washington, 2011, 105: 331–347.

[184]Poole R W. Noctuidae. // Heppner J.B. (Ed.). Lepidopterorum Catalogus (New Series). Fasc. 118. Leiden, 1989, Pt 1: v–xii+1–500pp.; Pt. 2. P. 501–1013; Pt. 3. P. 1014–1314.

[185]Poole R W. Noctuoidea: Noctuidae (part), Cuculliinae (part). // Dominik R.B. et al. (Eds). The Moths of America North of Mexico. Wedge Entomological Research Foundation (Washington), 1995, 26(1), 249pp.

[186]Ronkay G, Ronkay L. Cuculliinae I, Noctuidae Europapeae, Vol. 6. Søro: Entomological Press, 1994, 282pp, 10pls.

[187]Ronkay G, Ronkay L. Cuculliinae II, Noctuidae Europapeae, Vol. 7. Søro: Entomological Press, 1995, 224pp, 4pls.

[188]Ronkay G, Ronkay L, Gyulai P. The Witt Catalogue – Cuculliinae 2 and Psaphidinae. A taxonomic atlas of the Eurasian and North African Noctuoidea, 5. Budapest: Heterocera Press, 2011, 380pp.

[189]Ronkay G, et al. New Orthosiini (Lepidoptera, Noctuidae, Hadeninae) species and genera from the wide sense Himalayan region. Esperiana, 2010, Bd. 15: 127-222.

[190]Ronkay L. Plusiinae (Lepidoptera, Noctuidae) from Korea. Folia entomologica Hungarica, 1982, 43: 137–145.

[191]Ronkay L. On the taxonomy and zoogeography of some Palaearctic and Indo–Australian Plusiinae (Lepidoptera, Noctuidae). Annales Historico-Naturales Musei Nationalis Hungarici, 1986, 33: 157–175.

[192]Ronkay L. A new species of *Antoculeora* Ichinose (Lepidoptera, Noctuidae, Plusiinae) from Taiwan. Bulletin of the National Science Museum, 1997, 23: 35–49.

[193]Ronkay L, et al. Hadeninae II. Noctuidae Europaeae, Vol. 5. Entomological Press, Søro, 2001, 452pp, 5pls.

[194]Ronkay L, Ronkay G. The Witt catalogue – Cuculliinae I. A taxonomic atlas of the Eurasian and North African Noctuoidea, 2. Budapest: Heterocera Press, 2009, 365pp.

[195]Ronkay L, Ronkay G, Behounek G. The Witt Catalogue – Plusiinae I. A taxonomic atlas of the Eurasian and North African Noctuoidea, 1. Budapest: Heterocera Press, 2008, 342pp.

[196]Ronkay L, et al. The Witt Catalogue – Erebidae I. Autophila and Apopestes. A taxonomic atlas of the Eurasian and North African Noctuoidea, 7. Budapest: Heterocera Press, 2014. 313pp.

[197]Ronkay L, et al. The Witt Catalogue – Xyleninae I. The Agrochola generic complex. A taxonomic atlas of the Eurasian and North African Noctuoidea, 9. Budapest: Heterocera Press, 2017, 342pp.

[198]Seitz A. (Ed.). Die Gross–Schmetterlinge der Erde. // Macrolepidoptera of the world. Die Gross–Schmetterlinge des Palaearctischen Faunengebietes, 3. Eulenartige Nachtfalter. Stuttgart: Alfred Kernen, [1907]–1914, i–iv, 511pp, pls1–75.

[199]Seitz, A. (Ed.). Die Gross–Schmetterlinge der Erde. // Die Gross–Schmetterlinge des Palaearctischen Faunengebietes, 3. Supplement. Die Eulenartige Nachtfalter. Stuttgart: Alfred Kernen, 1931–1938, 333pp, pls1–26.

[200]Speidel W, Naunmann C M. A survey of family–group names in noctuoid moths (Insecta: Lepidoptera). Systematics & Biodiversity, 2004, 2(2): 191–221.

[201]Speidel W, Fänger G, Naumann C M. The phylogeny of the Noctuidae (Lepidoptera). Systematic Entomology, 1996a, 21(3): 219–252.

[202]Speidel W, Fänger G, Naumann C M. The surface structure of the noctuid proboscis (Insecta, Lepidoptera). Zoologischer Anzeiger, 1996b, 234: 307–315.

[203]Spuler A. Die Schmetterlinge Europas, Band I. Rhopalocera–Brephidae. XVI. Fam. Noctuidae. Eulen: 133–332; Bd. III, Taf. 28–55. Stuttgart. E. Schweizerbartsche Verlagsbuchhandlung, 1908, 1–424.

[204]Staudinger O, Rebel H. Catalog der Lepidopteren des Palaearctischen Faunengebietes. Famil. Papilionidae – Hepialidae. i. Theil: i–[xxxi]. Berlin. R. Friedlander and Sohn, 1901, 411pp.

[205]Strand E. Noctuidae, Agaristinae. // Wagner, H. (ed.). Lepidopterorum Catalogus, 1912, 3: 1–82.

[206]Sugi S. A revision of the japanese *Orthosia* (Lepidoptera, Noctuidae, Hadeninae). Tinea, 1955, 2: 90–104.

[207]Sugi S. Notes on the genus *Stenoloba* Staudinger, with description of a new genus (Lepidoptera, Noctuidae, Cryphiinae). Kontyû, 1970a, 38: 130–135.

[208]Sugi S. The Japanese species of the genus *Bena* (Lepidoptera, Noctuidae, Chloephorinae), with new synonymies and notes on their seasonal variation. Tyô to Ga, 1970b, 21: 112–117.

[209]Sugi S. New and little known species of *Cryphia* (s. l.) with illustration of male genitalia of the known Northeast Asian species (Lepidoptera: Noctuidae). Tyô to Ga, 1980, 30: 188–195.

[210]Sugi S. Noctuidae (except Herminiinae). // Inoue H., Sugi, S., Kuroko, H., Moriuti, S. & Kawabe, A. Moths of Japan, 1982, 1: 669–913, 2: 334–405, pls. 37, 164–223, 229, 278, 355–280. Kodansha, Tokyo (in Japanese with English synopsis).

[211]Sugi S. Summary: General views on immature stages and life histories of larger moths in Japan. // Sugi S. (Ed.) Larvae of Larger Moths in Japan. Tokyo: Kodansha Co. Ltd., 1987, 265–301.

[212]Sugi S. *Earias erubescens* Staudinger, the valid name replacing *E. jezoensis* Sugi (Noctuidae, Chloephorinae). Japan Heterocerist's Journal, 1990, 159: 154.

[213]Sugi S. Noctuidae (except Aganainae, Herminiinae and Nolinae). Checklist of the Lepidoptera of Taiwan. // Heppner, J. B. and H. Inoue. (Eds), Lepidoptera of Taiwan, 1992, 1(2), 171–202.

[214]Sugi S. *Paraphyllophila confusa* Kononenko (Noctuidae) from Japan. Japan Heterocerists'Journal, 1993, 175: 444– 445.

[215]Sugi S. *Protexarnis balanitis* (Lepidoptera, Noctuidae) found from the Hidaka Range, Hokkaido. Japan Heterocerist's Journal, 1995, 185: 163–164

[216]Sugi S, Jinbo U. 'Post–MJ'. Additions of species and changes in names of Japanese moths. Supplement 1. Edn. 2. Japan Heterocerist's Society, Tokyo, 2004, i–ii +1–60 (in Japanese).

[217]Sugi S, Nakamura M. *Trichosea champa* (Moore) and its allied species (Lepidoptera, Noctuidae). Gekkan–

Mushi, 1989, 1989: 4–12 (in Japanese with English summary).

[218]Sui H, Han H L. Two New Record Species of the Genus *Hypena* (Lepidoptera, Noctuidae, Hypeninae) from Northeast China. Journal of Northeast Forestry University, 2010, 38(5): 138-139.

[219]Sviridov A V. Fam. Noctuidae. 1. Subfam. Herminiinae; 2. Subfam. Rivulinae; 3. Subfam. Hypeninae; 4. Subfam. Catocalinae; 7. Subfam. Sarrothripinae; 8. Subfam. Chloephorinae. // P.A. Lehr (ed). Key to insects of Russian Far East. V. Trichoptera and Lepidoptera. Pt. 4. Vladivostok, Dal'nauka, 2003, 35–187, 217–236 (in Russian).

[220]Tang W Q, Han H L. A New Record Genus from the Mainland of China and Three New Species (Lepidoptera, Noctuidae, Eustrotiinae) in China. Journal of Northeast Forestry University, 2013, 41(7): 160-162.

[221]Troubridge J, Crabo L. New *Oncocnemis* (Lepidoptera: Noctuidae) from the Pacific Northwest. Journal of the Entomological Society of British Columbia, 1998, 95: 33-51.

[222]Troubridge J A. A generic realignment of the Oncocnemidini sensu Hodges (1983) (Lepidoptera: Noctuidae: Oncocnemidinae), with description of a new genus and 50 new species. Zootaxa, 2008, 1903: 1–95.

[223]Ttshistjakov Y A. Fam. Lymantriidae – tussock moths. // Keys for identification of the insects of the Far East of Russia. 5. Trichoptera and Lepidoptera. Part 4. Vladivostok, Dal'nauka, 2003, 603–636(in Russian).

[224]Ttshistjakov Y A. Fam. Nolidae. // Keys for identification of the insects of the East Siberia of Russia. 5. Trichoptera and Lepidoptera. Part 4. Vladivostok, Dal'nauka, 2003, 637–652(in Russian).

[225]Ueda K. A revision of the genus *Deltote* R.L. and its allied genera from Japan and Taiwan (Lepidoptera: Noctuidae, Acontiinae). Pt 1. A generic classification of the genus *Deltote* R.L. and its allied genera. Bulletin of Kitakyushu Nature History Museum, 1984, 5: 91–133.

[226]Ueda K. A revision of the genus *Deltote* R. L. and its allied genera from Japan and Taiwan (Lepidoptera: Noctuidae, Acontiinae). Pt 2. Systematics of the genus *Deltote* R.L. and its allied genera. Bulletin of Kitakyushu Nature History Museum, 1987, 6: 1–117.

[227]Varga Z, et al. The Witt Catalogue – Noctuinae I. Chersotis. A taxonomic atlas of the Eurasian and North African Noctuoidea. Vol. 6. Heterocera Press, Budapest, 2013, 313pp.

[228]Varga Z, Ronkay L. Taxonomic studies on the genera *Sideridis* Hübner, *Saragossa* Staudinger and *Conisania* Hampson (Lepidoptera, Noctuidae: Hadeninae). Acta Zoologica Academiae Scientiarum Hungaricae, 1991, 37: 145–172.

[229]Varga Z, et al. The Witt Catalogue – Noctuinae II. The Eugnorisma – Eugraphe generic complex. A taxonomic atlas of the Eurasian and North African Noctuoidea. Vol. 8. Heterocera Press, Budapest, 2015, 362pp.

[230]Volynkin A N, Matov A Y, Han H L. On the taxonomy of the genus *Acronicta* (Lepidoptera, Noctuidae). IV. *Acronicta digna* species-complex. Zootaxa, 2017, 4344(1): 189-194.

[231]Volynkin A N, et al. A review of the Palaearctic *Mniotype* adusta (Esper, 1790) species-group with description of a new species and six new subspecies (Lepidoptera: Noctuidae). Zootaxa, 2014, 3796(1): 1-32.

[232]Wang J H, Han H L. One species of the genus *Xestia* (Lepidoptera, Noctuidae) new record from China.

Agriculture Science Journal of Yanbian University, 2017, 39(1): 76-78.

[233]Wu K Y, et al. A list of the insects in the forest areas of Changbai mountains II, Noctuidae. Forest Pest and Disease, 1999a, 3: 34-36, 40.

[234]Wu K Y, et al. A list of the insects in the forest areas of Changbai mountains II, Noctuidae (Continue). Forest Pest and Disease, 1999b, 4: 32-36.

[235]Xia C, Zhang H H, Jiang J. The checklist of Family Noctuidae in Meihekou Area of Jilin Province. Forestry Science and Technology Information, 2012, 44(2): 24-26.

[236]Yang Q Y, Li C D, Han H L. Taxonomic revision of the genus *Brachionycha* from China (Lepidoptera: Noctuidae). Entomological Research, 2007, 37(4): 287-289.

[237]Yen S H, Wu S. Noctuoidea, Noctuidae, Eriopinae. Biota Taiwanica, Hexapoda: Lepidoptera. Kaohsiung: National Sun Yat–Sen University, 2009, 88pp.

[238]Yoshimatsu S I. Taxonomic study on *Leucania striata* Leech and its close relatives, with a description of a new species (Lepidoptera, Noctuidae). Tyo Ga, 1987, 38: 56–67.

[239]Yoshimatsu S I. A revision of the genus *Mythimna* (Lepidoptera: Noctuidae) from Japan and Taiwan. Bulletin of the National Institute of Agro-Environmental Sciences, 1994, 11: 81–323.

[240]Zahiri R, et al. Molecular phylogenetics of Erebidae (Lepidoptera, Noctuoidea). Systematic Entomology, 2012, 37(1): 102–124.

[241]Zahiri R, et al. A new molecular phylogeny offers hope for a stable family level classification of the Noctuoidea (Lepidoptera). Zoologica Scripta, 2011, 40(2): 158–173.

[242]Zahiri R, et al. Major lineages of Nolidae (Lepidoptera, Noctuoidea) elucidated by molecular phylogenetics. Cladistics, 2013a, 29: 337–359.

[243]Zahiri R, et al. Relationships among the basal lineages of Noctuidae (Lepidoptera, Noctuoidea) based on eight gene regions. Zoologica Scripta, 2013b, 42: 488–507.

[244]Zaspel J M, Branham M A. World Checklist of Tribe Calpini (Lepidoptera: Noctuidae: Calpinae). Insecta Mundi, 2008, Vol. 47. P. 1-15.

[245]Zaspel J M, Kononenko V S, Goldstein P Z. Another blood feeder? Experimental feeding of a Fruit–Piercing Moth species on human blood in Primorye territory of Far Eastern Russia (Lepidoptera: Noctuidae: Calpinae). Journal of Insect Behavior, 2007, 5: 437–451.

[246]Zaspel J M, et al. A molecular phylogenetic analysis of the vampire moths and their fruit-piercing relatives (Lepidoptera: Erebidae: Calpinae). Molecular Phylogenetics and Evolution, 2012, 65: 786–791.

[247]Zhang C, Han H L. Two New Record Species of Genus *Mythimna* (Lepidoptera, Noctuidae, Hadeninae) from China. Journal of Northeast Forestry University, 2015, 43(9): 125-127.

[248]Zhang C, Han H L, Kononenko V S. The tribe Leucaniini in Southwest China (Lepidoptera, Noctuidae), The Lepidoptera in the collection of Northeast Forestry University II. Harbin: Heilongjiang Science and Technology Press, 2019, 178pp.

[249]Zhang B, Hu C X, Han H L. Two species of the genus *Hypena* Schrank, 1802 (Lepidoptera, Noctuidae) new to China. Korean Journal of Applied Entomology, 2011, 50(4): 301-306.

[250]Zhang S M, Sun Y M. Comparison of Faunal Structure of Noctuids Between Tieling and Nanchang, China. Journal of Shenyang Agricultural University, 1987, 18(3): 39-45.

[251]Zhao Z Y, et al. A New Record Species of *Catocala* (Lepidoptera， Noctuidae) from China. Journal of Northeast Forestry University, 2012, 40(8): 142-143.

[252]Zilli A, Hogenes W. An annotated list of the fruit-percing moth genus *Eudocima* Billberg, 1820 (sensu Poole, 1989) with description of four new species (Lepidoptera: Noctuidae, Catocalinae). Quadrifina, 2002, 5: 153-207.

[253]Zilli A, Ronkay L, Fibiger M. Apameini. Noctuidae Europae Vol. 8. Sorø: Entomological Press, 2005, 323pp.

[254]Zilli A, et al. The Witt catalogue – Apameini 1. A taxonomic atlas of the Eurasian and North African Noctuoidea. 3. Budapest: Heterocera Press, 2009, 393pp.

[255]Zolotarenko G S. Cutworms of the Western Siberia (Lepidoptera, Agrotinae). Novosibirsk, Nauka, 1970, 436pp (in Russian).

索引

中文索引

A

B

C

D

E

F

G

H

J

K

L

M

N

O

p

Q

R

S

T

V

W

X

Y

Z

Index

A

B

C

D

E

F

G

H

I

J

K

L

M

N

O

P

Q

R

S

T

U

V

W

X

Y

Z